Energy Density Functional Methods for Atomic Nuclei

Energy Density Functional Methods for Atomic Nuclei

Edited by Nicolas Schunck
Nuclear and Chemical Science Division, Lawrence Livermore National Laboratory, Livermore, CA, USA

IOP Publishing, Bristol, UK

ISBN 978-0-7503-1422-0 (ebook)
ISBN 978-0-7503-1423-7 (print)
ISBN 978-0-7503-1424-4 (mobi)

DOI 10.1088/2053-2563/aae0ed

Version: 20190101

IOP Expanding Physics
ISSN 2053-2563 (online)
ISSN 2054-7315 (print)

British Library Cataloguing-in-Publication Data: A catalogue record for this book is available from the British Library.

Published by IOP Publishing, wholly owned by The Institute of Physics, London

IOP Publishing, Temple Circus, Temple Way, Bristol, BS1 6HG, UK

US Office: IOP Publishing, Inc., 190 North Independence Mall West, Suite 601, Philadelphia, PA 19106, USA

Contents

Preface

The original purpose of this book was to give an updated presentation of energy density functional methods as they pertain to the description of atomic nuclei. In particular our ambition was to discuss some of the limitations of existing nuclear phenomenology based on Skyrme or Gogny forces, as well as the recent progress in establishing connections with *ab initio* theory and the density functional theory of electronic systems. We also wanted to put more emphasis on several sub-fields such as covariant functionals and time-dependent methods for nuclear dynamics, which have reached maturity and can be applied on a wider scale. Among the recent developments in the field is the increased focus on uncertainty quantification and propagation, and the spectacular progress in computing capabilities: these topics have often been overlooked in the literature and we felt they deserved more exposure.

Another important motivation to write this book was to highlight the profound coherence of the various methods developed over the years. Typical examples include the role of the QRPA matrix both in the QRPA and linear response theory themselves, but also as a probe of the stability of the Hartree–Fock–Bogoliubov (HFB) equation and as the starting point for many calculations of large-amplitude collective motion, or the essential role of the time-dependent HFB equation, which is used to derive both linear response theory and the collective Hamiltonian within the adiabatic time-dependent HFB approach. While there exist excellent review articles on each of these topics, the connection between them may not always be apparent. In this sense, we felt that a textbook would better convey the interconnections between these methods.

So, what should the reader expect to find in this book? Below is a condensed, itemized list of some of its unique features:

- Comprehensive presentation of the covariant energy density functional formalism, from its roots in quantum chromodynamics and effective field theory to its practical formulations in terms of phenomenological Lagrangians, and the construction of relativistic energy functionals.
- Discussion of formal aspects of spontaneous symmetry-breaking in terms of symmetry groups.
- Detailed presentation of multi-reference energy density functional techniques, including the discrete generator coordinate method (GCM), projection on particle number and angular momentum and their combinations.
- Updated description of QRPA and linear response theory including a presentation of the finite-amplitude method, which is becoming a standard to compute the linear response in deformed, heavy nuclei.
- Updated presentation of large-amplitude collective motion including the adiabatic self-consistent generator coordinate method and detailed derivations of the Gaussian overlap approximation of the GCM with HFB generator states.
- Finite-temperature Hartree–Fock, HFB and RPA with derivations.

- Detailed presentation of numerical implementations of the HFB equation, including special techniques related to the harmonic oscillator basis, lattice techniques and gradient methods for the self-consistent loop.
- A full section on statistical aspects including Bayesian methods.

We cannot pretend to have reached excellence. As often the case in endeavors of this kind, we feel that we fell short of our goals in several respects. The reader might thus be disappointed that:

- There is no presentation of the time-dependent Hartree–Fock theory, and the description of the time-dependent HFB approach is less detailed than for other topics. This is partly justified by the relatively recent publication of a complete book on this subject by Lacroix, Avez and Simenel [1].
- We include only a very short discussion of some of the most recent attempts to design non-relativistic energy functionals beyond the traditional Skyrme and Gogny, and focus exclusively on those that have already been applied in large-scale calculations.
- Contrary to our proclaimed goals, we barely cover the connection between nuclear phenomenology and either *ab initio* theory or atomic density functional theory. The community working on these problems is very small and has not reached a consensus on several of these questions.
- Overall, there are relatively few examples of each of the topics covered since we wanted to focus this book on methods rather than applications. While the few we provide, together with the references, may satisfy the curiosity of the reader, this could be deemed a weakness.

Finally, one should emphasize that this book is a collective work. Each chapter was originally written independently by its authors and reflects the opinions of its authors and its authors only. We find that this diversity of opinions is enriching—nuclear theory cannot be solved exactly, and there will always be room for different schools of thought.

Reference

[1] Lacroix D, Avez B and Simenel C 2010 *Quantum Many-Body Dynamics: Applications to Nuclear Reactions* (Saarbrücken: VDM Verlag Dr Müller)

Editor biography

Nicolas Schunck

Nicolas Schunck received his PhD from the University of Strasbourg, France, in 2001. He was a postdoctoral research associate at the University of Surrey, UK (2002–04), the Universidad Autónoma de Madrid, Spain (2005–07) and the University of Tennessee, USA (2007–10). He was also a visiting scientist at the Niels Bohr Institute in Denmark (2004), the Institute of Nuclear Physics in Kraków, Poland, and the Institute of Theoretical Physics in Warsaw, Poland (2005). Schunck's interests are centered on the development and implementation of theoretical methods to describe the structure and decay of heavy atomic nuclei. Most of his work involves research in nuclear density functional theory and its applications, in particular as they pertain to a fundamental description of nuclear fission. He is particularly active in the development of computational nuclear theory and has more than a decade of experience in high-performance computing applications on leadership class computers. In particular, he is leading the development of the density functional theory for nuclei at extreme scale (DFTNESS) computational framework based on the HFBTHO and HFODD computer programs. Schunck has a large network of collaborations, both within the US, notably with Michigan State University (MSU), the University of Washington, Seattle, and Los Alamos National Laboratory (LANL), and throughout the world, in particular the CEA at Bruyères-le-Châtel (France), the University Autónoma de Madrid (Spain) and VECC Kolkata (India).

Contributors

Michael Bender
Institut de Physique Nucléaire de Lyon, Villeurbanne, France

Aurel Bulgac
Department of Physics, University of Washington, Seattle, WA, USA

Thomas Duguet
CEA, DRF, IRFU, SPhN, Gif-sur-Yvette, France

Jean-Paul Ebran
CEA, DAM, DIF, Bruyères-le-Châtel, France

Jon Engel
Department of Physics and Astronomy, University of North Carolina Chapel Hill, NC, USA

Michael McNeil Forbes
Department of Physics and Astronomy, Washington State University, Pullman, WA, USA
and
University of Washington, Seattle, WA, USA

Markus Kortelainen
Department of Physics, University of Jyväskylä, Jyväskylä, Finland

Takashi Nakatsukasa
Center for Computational Sciences University of Tsukuba, Tsukuba, Japan

Nicolas Schunck
Nuclear and Chemical Science Division, Lawrence Livermore National Laboratory, Livermore, CA, USA

Glossary

BCS theory	The BCS theory is an approximation of the full HFB theory where the last transformation of the Bloch–Messiah–Zumino theorem is neglected.
HFB matrix ($\mathcal{H}$)	The HFB matrix contains the variation of the total energy with respect to the generalized density and is discussed extensively in section 3.1. In the SR-EDF approach, its eigenvalues are the quasiparticle energies, which represent an approximation to the excitations of the system.
Adiabatic time-dependent Hartree–Fock theory	The ATDHF theory is a small-velocity limit of the TDHFB theory which aims to build a classical Hamiltonian for collective motion with coordinates $\boldsymbol{q}$ and momenta $\boldsymbol{p}$ related to the one-body density matrix. It yields a Hamiltonian that is quadratic in collective velocities and depends on a collective inertia tensor B. Its extension to superfluid systems is ATDHFB theory. While the ATDHF approach determines, in principle, a collective subspace (or collective path), most applications of the formalism are made with a preset collective space obtained by constrained SR-EDF calculations.
Angular-momentum projection	When the rotational invariance of nuclear forces is spontaneously broken at the EDF level, the HFB vacuum is no longer an eigenstate of the angular-momentum operator. Projection techniques are thus needed to restore these quantum numbers and compute, for example, electromagnetic transitions. The standard approach relies on projectors that involve the Wigner matrices and rotation operators.
Anomalous density (κ)	The anomalous density describes the overlap between an A-body wave function and an $A \pm 2$-body wave function. It is non-zero only when pairing correlations are active, that is, at the BCS or HFB level. κ and κ^* are two of the three degrees of freedom of the EDF theory. The anomalous density is also often called the pairing tensor.
Blocking approximation	The solution of the HFB equation always yields solutions that are fully paired, for which the HFB vacuum is a superposition of states with an even number of particles. The blocking approximation consists in acting on such a fully paired vacuum to generate a HFB vacuum with the proper number parity.
Bogoliubov transformation (W)	The Bogoliubov transformation defines quasiparticle ladder operators from arbitrary particle ladder operators, and is therefore used to define a product state, the HFB vacuum, that can break particle-number symmetry. The matrix W of this transformation is unitary and has a specific block structure. There is a one-to-one correspondence between the Bogoliubov matrix, the generalized density and quasiparticle operators. Bogoliubov transformations can also be introduced to relate two different Bogoliubov vacua, that is, two different sets of quasiparticle operators.

Bohr Hamiltonian The Bohr collective Hamiltonian is a particular case of a (semi-classical) collective Hamiltonian based on two collective variables associated with the (β, γ) quadrupole deformation parameters and another three (of Anderson–Nambu–Goldstone type) associated with rotational motion.

Boundary condition When integrating the HFB equation directly in coordinate space, one needs to specify boundary conditions at the edges of the spatial domain. Very often, Dirichlet boundary conditions are used (the value of the function vanishes), which implies discretizing the continuum.

Broyden method The Broyden method is a numerical technique used to accelerate the convergence of non-linear problems such as the HFB equation or the finite-amplitude method of QRPA. It relies on formulating the self-consistent loop as a fixed-point problem and using a Newton-like method to estimate the next point.

Chemical potential (λ_F) In the context of the EDF approach, the chemical potential, or Fermi energy, is the Lagrange parameter introduced in the HFB theory to constrain the number of particles to a given value. Since in the HFB (and BCS) theory, $\lambda = \partial E/\partial N$, the chemical potential becomes ill-defined when pairing correlations vanish and particle number is actually conserved, i.e., at the Hartree-Fock limit.

Collective inertia tensor In a collective model of the atomic nucleus characterized by N collective variables, the inertia tensor is an $N \times N$ rank-2 tensor that extends the traditional concept of mass to motion in collective space. There are different prescriptions (ATDHF, ASCC, GCM+GOA, etc.; see chapter 6) to compute the inertia tensor that can be benchmarked to the analytical case of translational collective motion.

Collective Schrödinger equation The Hill–Wheeler–Griffin equation of the GCM can be simplified by assuming that the overlap between two generator states $|\Phi(\boldsymbol{q})\rangle$ and $|\Phi(\boldsymbol{q}')\rangle$ has a Gaussian form in terms of $\boldsymbol{q}-\boldsymbol{q}$, and that the energy kernel is proportional to the norm kernel. These approximations then lead to a Schrödinger-like equation for a function related to the weight function of the GCM.

Collective space The collective space is a subspace of the full many-body Hilbert space spanned by a few variables typically given by the expectation value of some operator, e.g. multipole moments, radius, etc, on the HFB product state—hence the adjective collective since all nucleons are involved. In theories of large-amplitude collective motion, one tries to identify such a collective subspace that is decoupled from the rest of Hilbert space.

Contraction ($\overline{AB}$) A contraction between two operators $\hat{A}$ and $\hat{B}$ is given by the difference between the expectation value of $\hat{A}\hat{B}$ on a many-body state $|\Phi\rangle$ and that of the normal-ordered operator $:\hat{A}\hat{B}:$ (recall that normal-ordering means all creation operators are moved to the left). For product states $|\Phi\rangle$, the contraction is a complex number associated with both the operators themselves and $|\Phi\rangle$. The Wick theorem then gives expectation values in terms of all possible contractions of pairwise operators.

Coulomb potential
: Protons in an atomic nucleus are subject to the Coulomb force, which is almost always treated in a non-relativistic framework. The Coulomb potential is discussed in section 1.3.2, where the Coulomb energy functional is also presented.

Coupling constant
: Energy functionals are typically given by the sum of several terms involving the one-body density matrix and its various derivatives (up to second order for Skyrme functionals). Each term is multiplied by a given low-energy coupling constant that needs to be adjusted on experimental data, as discussed in chapter 9. When the energy functional is built out of an effective pseudopotential, these coupling constants can be related to the parameters of the pseudopotential.

Covariance matrix
: Since the parameters of the energy functional must be adjusted on experimental data, there is an intrinsic uncertainty associated with them. The coupling constants can thus be treated as random variables and the covariance matrix captures their correlations and uncertainties. Knowledge of the covariance matrix also allows propagating the uncertainties to model predictions.

Covariant derivative
: In tensor analysis, covariant derivatives generalize the notion of derivatives to curved spaces—spaces with a non-flat metric given by some metric tensor g_{ij}. Covariant derivatives are particularly important in theories of large-amplitude collective motion, either based on the GCM+GOA (where the metric is related to the norm overlap between generator states) or the ASCC theory (where the metric is the inertia tensor). Covariant derivatives also naturally occur in relativistic quantum field theory, where they are associated with the concept of local gauge invariance.

Creation operator (c_i, $c_i^\dagger$)
: Ladder operators create or annihilate particles (or quasiparticles) in configuration space: $c_i^\dagger$ creates a particle in the quantum state i, which is characterized by some quantum numbers. Most of the theoretical methods used in quantum many-body systems are formulated in terms of these operators.

Density operator ($\hat{D}$)
: A quantum system at temperature T is entirely defined by its statistical density operator (not to be confused with the one-body density matrix ρ). For systems described by the statistical grand potential (for which particle number is not conserved), the density operator is formally given by $\hat{D} = \exp[-\beta(\hat{H} - \mu\hat{N})]$ with $\beta = 1/kT$. For a realistic nuclear Hamiltonian, it is extremely difficult to compute the density operator, hence various approximations such as the Hartree–Fock or Hartree–Fock–Bogoliubov approximation are commonly invoked. At finite temperature, the expectation

value of any operator $\hat{O}$ is given by $\mathrm{Tr}\hat{D}\hat{O}$ where the trace refers to the statistical trace—taken over the entire many-particle Fock space.

Dirac equation — The Dirac equation is the equation of motion for spin-1/2 particles in a relativistic framework. In the context of the nuclear covariant EDF approach, it governs the motion of nucleons. Its non-relativistic limit is called the Pauli-Schrodinger equation. These notions are discussed in chapter 2.

Dirac matrix (γ^{μ}) — The four Dirac matrices generalize the concept of Pauli matrices to Minkowski space–time and the Lorentz symmetry group.

Dirac spinor (ψ) — Dirac spinors are the solutions to the Dirac equation. They are composed of two spinors, the large and small components. To describe nuclei, the concept of Dirac spinors can be extended to isospinors, which concatenate the neutron and proton regular Dirac spinor into a single object.

Energy density functional — The total energy of the nucleus can always be written as an integral over space of some energy density $\mathcal{H}(\boldsymbol{r})$. In the most general case of the multi-reference EDF approach, this energy density is a functional of the transition densities $\rho^{(g'g)}$, $\kappa^{(g'g)}$ and $\kappa^{(g'g)*}$. In the single-reference EDF approach, it is a functional of the symmetry-breaking one-body density matrix and anomalous density. In the particular case where the energy density is derived from a zero-range pseudopotential, it only depends on local densities and their derivatives.

Energy kernel ($\mathcal{E}$) — The energy of the nucleus is expressed as a functional of the bra and ket associated with a many-body product state. The resulting functional form is called energy kernel. In the single-reference EDF approach, the energy kernel depends only on the normal and anomalous densities; in a multi-reference EDF approach, it depends on the normal and anomalous transition densities.

Field operator ($\psi(\boldsymbol{r}\sigma)$, $\psi^{\dagger}(\boldsymbol{r}\sigma)$) — Field operators create or annihilate particles (or quasiparticles) with spin σ (and isospin τ) at position $\boldsymbol{r}$. Similar operators can be introduced to create particles with momentum $\boldsymbol{p}$ and, in general, any continuous representation of the single-particle Hilbert space.

Finite-amplitude method (FAM) — The finite-amplitude method is a numerical technique to solve the QRPA equation by linearizing it and turning it into a self-consistent equation. Its main advantage is that its computational cost scales more or less like the cost of the underlying HFB calculation. Discrete QRPA eigenstates can be computed from the FAM solution by contour integration.

Galilean transformation — In non-relativistic mechanics, Galilean transformations relate the coordinates in two reference frames moving at constant velocity with respect to one another. The concept of Lorentz boosts, which are part of Lorentz or Poincaré transformations, generalizes Galilean transformations to arbitrary velocities. Galilean transformations are sometimes invoked to constrain the form of energy functionals and are important in time-dependent applications.

Gaussian overlap approximation (GOA) In the context of the generator coordinate method, the Gaussian overlap approximation is the combination of two assumptions: that the norm kernel between generator states, $\langle \Phi(\boldsymbol{q})|\Phi(\boldsymbol{q}')\rangle$, has a Gaussian form of the type $\sum_{ij}(q_i - q'_i)G_{ij}(q_j - q'j)$, and that the energy kernel is proportional to the norm kernel by a polynomial of order 2. Under these conditions, the Hill–Wheeler–Griffin equation can be simplified into a Schrödinger-like equation for a collective wave-packet.

Gaussian process Gaussian processes generalize the concept of probability distribution function from variables to functions of variables. They are ubiquitous tools in applications of statistics to build emulators of complex models that can be trained on data.

Generalized density ($\mathcal{R}$) The generalized density is a Hermitian matrix that is defined in terms of the one-body density matrix and the anomalous density. In the HFB theory, it contains all degrees of freedom of the system.

Gogny force The Gogny force presented in section 1.2.2 is a model of a finite-range, effective two-body pseudopotential. Its central term involves the sum of two Gaussians of different ranges and is complemented by a zero-range spin–orbit force and a density-dependent force. Because of the finite range of the pseudopotential, The Gogny functional is a functional of the non-local density.

Goodman basis Starting from an arbitrary basis of the single-particle Hilbert space such as, e.g. the harmonic oscillator basis, one can define linear combinations of basis functions that are eigenstates of discrete symmetry operators such as the signature of simplex. Any operator that commutes with these symmetry operators will thus acquire a block-diagonal matrix structure in that new basis, which is called the Goodman basis. Embedding such discrete symmetries in the basis can considerably accelerate calculations without losing too much physics.

Hartree–Fock The Hartree–Fock theory can be viewed as a limiting case of the HFB theory when pairing correlations vanish. The many-body product state is a single Slater determinant and the system is thus entirely characterized by the one-body density matrix only. The single-particle states composing the Slater determinant are determined by requiring that the energy be minimal in the HF state. The HF theory can also be viewed mathematically as a specific change of single-particle basis.

Hessian matrix The Hessian matrix of dimension $N \times N$ collects all second-order partial derivatives of a scalar function of N variables. In the EDF theory and its applications, it is used in models of collective motion where it is related to the collective

	inertia tensor (section 6.3.1); in algorithms to determine the constraints on HFB solutions (section 8.3.1); and in the determination of the covariance matrix of EDF parameters (section 9.3.1).
Hill–Wheeler–Griffin equation	The integral Hill–Wheeler–Griffin equation is obtained by substituting the GCM ansatz in the many-body Schrödinger equation. It is solved either numerically by discretization, or semi-analytically by introducing additional hypotheses about the generator states such as the Gaussian overlap approximation.
Hohenberg–Kohn theorem	In many-electrons systems, the Hohenberg–Kohn existence theorem states that there exists a (universal) functional $F[n]$ of the local density of electrons $n(\boldsymbol{r})$ that gives the ground-state energy when the density of electrons is that of the ground-state. In other words, it establishes a one-to-one mapping between the description of a quantum many-body system in terms of a many-body wave function and in terms of a local density.
Kinetic density (τ)	Like the spin density, the kinetic density also originates from the expansion of $\hat{\rho}$. It is used to define the kinetic energy. The Thomas–Fermi approximation, the ancestor of modern density functional theory, was applied to the kinetic density.
Klein–Gordon equation	The Klein–Gordon equation is the equation of motion for spin-0 particles in a relativistic framework. In the covariant energy density functional theory as described in section 2.5, the Klein–Gordon equation applies to the bosonic (meson) fields.
Kohn–Sham potential	The Kohn–Sham procedure complements the Hohenberg–Kohn existence theorem in that it provides a recipe to compute the actual density of the system—if one knows the exact functional $F[n]$ or an approximation of it. In nuclear physics, the mean field potential h plays the role of an effective Kohn–Sham potential.
Lagrange multiplier	The method of Lagrange multipliers is a standard method of numerical optimization to minimize or maximize a function under constraints. In the context of the HFB equation, it is widely used to fix the average value of particle number—the chemical potential is nothing other than a Lagrange multiplier. When needed, it is also used to constrain the expectation value of any operator of interest. Knowledge of the QRPA matrix (that is, the stability matrix) provides additional control to optimize the choice of the Lagrange multiplier at each iteration of the self-consistent loop.
Lagrangian ($\mathcal{L}$)	The Lagrangian density is the fundamental building block of field theory (relativistic or not, quantum or not). It depends on fields such as nucleon and mesons in low-energy nuclear physics, or quarks and gluons in QCD. It can be used to define equations of motion by variation of the

action, which is constructed by integrating $\mathcal{L}$ over space–time. The QCD Lagrangian is discussed in section 2.2, effective chiral Lagrangian in section 2.3 and phenomenological Lagrangians in section 2.4.

Likelihood function ($\mathcal{L}$) In statistics, the likelihood function gives the probability that a model produces some data for a fixed set of model parameters. It is typically estimated by $\exp(-\frac{1}{2}\chi^2)$ where the χ^2 collects the average deviation between the data and the model prediction; see chapter 9 for more information.

Liouville equation In statistical mechanics, the Liouville equation is the evolution equation for the statistical density operator characterizing the system. It is the equivalent of the time-dependent Schrödinger equation for density operators.

Lipkin–Nogami The Lipkin–Nogami method is an approximate way to restore particle-number symmetry in superfluid systems. It consists in adding a term to the energy functional that cancels the fluctuations $\langle \Delta\hat{N}^2 \rangle$ of particle number.

Lorentz group The Lorentz group is the isometry (i.e. that which preserves the distance) group of the Minkowski space of special relativity. It comprises ordinary rotations and Lorentz boosts. In this book, the Lorentz group refers to what is also called the homogeneous Lorentz group; see Poincaré group.

Mean field (h) The mean field (or Hartree–Fock) potential h is a component of the HFB matrix that represents the variation of the energy with respect to the one-body density matrix. When the coordinate space representation of the mean field is local, it gives an approximation of the average nuclear potential.

Moshinsky transformation The Moshinsky transformation for a product of two harmonic oscillator wave functions of variables x and x' is a change of variables $U = (x + x')/\sqrt{2}$ and $u = (x - x')\sqrt{2}$. It is widely used in calculations of matrix elements of two-body potentials in the harmonic oscillator basis to separate the motion of the center of mass from the relative motion. It also accelerates the calculation of matrix elements of Gaussian-like potentials by reducing the number of active variables.

Norm kernel ($\mathcal{N}$) The norm kernel, or norm overlap, is the overlap between two many-body product states $|\Phi^{(g)}\rangle$ and $|\Phi^{(g')}\rangle$, $N(\boldsymbol{g}, \boldsymbol{g}') = \langle \Phi^{(g)} | \Phi^{(g')} \rangle$. When the two product states are different HFB vacua, the Onishi theorem provides an analytical expression for the norm kernel.

Nuclear matter Nuclear matter is an idealized system of nucleons that is both infinite and homogeneous. The relative proportion of neutrons and protons is controlled by the parameter $\beta = (\rho_n - \rho_p)/(\rho_n + \rho_p)$. If $\beta = 0$, one talks of symmetric nuclear matter; if $\beta = 1$, one talks of neutron matter.

One-body density matrix (ρ)	In configuration space, the one-body density matrix for an arbitrary A-body state is given by $\rho_{ij} = \langle\Psi|c_j^\dagger c_i|\Psi\rangle/\langle\Psi|\Psi\rangle$, with c_i, $c_j^\dagger$ arbitrary single-particle operators. It can be thought of as the matrix elements of a one-body operator $\hat{\rho}$. The one-body density matrix (or density matrix, in short) is one of the cornerstones of the EDF approach. When pairing correlations vanish and the A-body state is a product state $|\Phi\rangle$, it allows computing the expectation value of any observable thanks to the Wick theorem.
Onishi formula	The Onishi formula is an analytical formula giving the overlap between two different HFB vacua $|\Phi_0\rangle$ and $|\Phi_1\rangle$.
Pairing density ($\tilde{\rho}$)	The pairing density $\tilde{\rho}$ is derived from the anomalous density (i.e. pairing tensor). Its mathematical properties are identical to the one-body density matrix, which sometimes facilitates practical calculations.
Pairing field (Δ)	The pairing field Δ is a component of the HFB matrix that represents the variation of the energy with respect to the anomalous density. It is identically zero when pairing correlations are not included (and the HFB theory reduces to the HF approach). The pairing field cannot be interpreted as a one-body operator in contrast to the mean field.
Pairing regularization	In the HFB approach with local pairing forces, the pairing gaps Δ are independent of the momentum $\boldsymbol{k}$. In this case, the gap equation in nuclear matter diverges linearly. The pairing regularization is a method to remove this divergence.
Particle-number projection	In the EDF approach, the nucleus is described by a HFB vacuum, a product state of quasiparticle operators that is not an eigenstate of the particle-number operator. Particle-number projection refers to the techniques used to project such a state onto the subspace of A-body states.
Partition function (Z)	The partition function is the trace of the statistical density operator. In statistical mechanics, the partition function can be used to compute many state variables of the system. In nuclear physics, it is also used to compute level densities.
Poincaré group	The Poincaré group, also called the inhomogeneous Lorentz group, is the (homogeneous) Lorentz group to which is added space translations. In relativistic quantum field theory, one requires that laws of physics be invariant under transformations of the Poincaré group.
Posterior distribution	In a Bayesian formulation, the coupling constants of the energy functional (and more generally of any model) are genuine random variables that are characterized by a certain probability distribution function. The Bayes' theorem allows estimating this probability distribution given a set of experimental data.
Product state ($|\Phi\rangle$)	Product states describe systems of independent particles or quasiparticles. In second quantization, they can be expressed as the direct product of creation or annihilation operators acting on the vacuum. In the Hartree–Fock

theory, the many-body wave function is a product state of particles; in the HFB theory, of quasiparticles.

Pseudospin symmetry — The pseudospin symmetry is an approximate degeneracy of spherical single-particle levels with quantum numbers $(n, l, j = l + 1/2)$ and $(n - 1, l + 2, j = l + 3/2)$. Its origin comes from the 2-spinor structure of Dirac spinors.

QRPA matrix (QRPA) — The QRPA matrix plays a central role in many aspects of the EDF approach. In linear response theory, its eigenvectors give the transition densities induced by external operators; in the HFB theory, it is equal to the stability matrix and determines if the self-consistent solution is a minimum; in theories of large-amplitude collective motion, it is a major ingredient of computing the collective inertia tensor; it is also used in numerical algorithms to determine Lagrange multipliers for constrained calculations and in the gradient method. In the cranking approximation, the QRPA matrix is reduced to its diagonal elements.

Quantization — Nuclear collective models are often defined in terms of classical Hamiltonians involving collective variables (i.e. generalized coordinates) and collective momenta. Examples of such models are the ATDHF, ASCC and Bohr Hamiltonians discussed in chapter 6. To obtain a collective spectrum, one needs to quantize these classical objects, which is usually done with the Pauli prescription. When quantized, the ATDHF Hamiltonian formally resembles the collective Hamiltonian from the GCM+GOA approach.

Redundant state — In multi-reference EDF, the norm kernel matrix may have zero eigenvalues. This implies that the basis of generator states is over-complete and contains redundant states (this is somewhat analogous to the problem of spurious states in the RPA and QRPA theories). When solving the MR-EDF equation numerically, these redundant states must be eliminated by diagonalizing the norm kernel matrix and filtering out the states with zero eigenvalues.

Response function ($R_{ab,\,cd}$) — In the linear response theory, the response function quantifies how the system reacts when a (time-dependent) perturbation operator is applied to it. Knowledge of the response function allows computing transition rates between excited states induced by an operator (for instance, β- or γ-decay). The RPA and QRPA provide good approximations to the exact response function.

Riemannian connection ($\Gamma^{\alpha}_{\beta\gamma}$) — In tensor analysis, the affine Riemannian connection, or Christoffel symbol, characterizes the metric of a curved space and is used to define covariant derivatives in that space. In theories of nuclear collective motion, it appears both in the GCM+GOA theory (when the norm overlap is a generalized Gaussian form) and in the ASCC theory.

Self-consistent symmetry — While the EDF theory is designed to allow for spontaneous symmetry-breaking, this can only happen if the self-consistent loop of the HF or HFB equation is initialized with symmetry-breaking solutions. If the initial condition conserves a given symmetry, then this symmetry will be preserved throughout the iterations: this is called a self-consistent symmetry.

Skyrme functional — The Skyrme force discussed in section 1.2.1 is an effective, zero-range, two-body pseudopotential. Its central term includes gradient terms that simulate the finite range of nuclear forces. The Skyrme force also includes a zero-range spin–orbit force, a density-dependent force and often a tensor force. The Skyrme functional comes about from computing the expectation value of the Skyrme pseudopotential on a Slater determinant, and can be expressed as a functional of the local density only.

Spin density ($\boldsymbol{s}$) — The spin density originates from the expansion of the coordinate space representation of the one-body density matrix operator $\hat{\rho}$ in terms of Pauli matrices in both the spin and isospin channels. The spin density vanishes in time-even systems.

Spin–isospin expansion — Operators relevant for atomic nuclei act in the tensor product of space, spin space and isospin space (or more precisely, in the direct sum of all such tensor products). Since operators in both the spin and isospin space belong to the $SU(2)$ group, they can be expanded as a linear combination of Pauli matrices and the identity matrix. Any N-body operator can thus be expanded as a sum of terms composed of a spatial form factor, a tensor product of N Pauli matrices (for spin) and another tensor product of N Pauli matrices (for isospin). This is the spin–isospin expansion.

Spin–orbit potential — The spin–orbit potential is a component of two-body nuclear potentials. Its couples nucleons with the same orbital angular momentum but different spin projections. A discussed in section 2.6.2, its origin can be explained within a relativistic approach as a relativistic correction induced by the difference between the scalar and vector self-energy.

Spurious excitation — In the QRPA and linear response theory, spurious excitations, also known as Nambu–Goldstone modes, correspond to eigenvectors of the QRPA matrix with eigenvalue zero. These modes correspond to broken symmetries rather than genuine excitations.

Stability condition — The HFB equation obtained after applying the variational principle only guarantees that the energy is an extremum. The expansion of the energy up to second order around the extremum introduces the QRPA matrix, which plays the role of stability matrix and determines whether the energy is a minimum or not.

Sum rule — For an arbitrary operator $\hat{G}$, the transition strength for an excited state $|\nu\rangle$ is simply $|\langle\nu|\hat{G}|0\rangle|^2$, where $|0\rangle$ is the ground-state. Sum rules are sums of all transition strengths weighted by powers of the excitation energy. Linear

	response theory provides good approximations of exact sum rules.
Tensor force	The tensor force (rather, potential) is an important component of two-body nuclear potentials and involves couplings between different spin and isospin projections between two nucleons. As discussed in section 9.1.1, it should not be confused with the tensor EDF, which results from the contribution of both the spin–orbit and tensor potentials.
Thermofield dynamics	Thermofield dynamics (TFD) is an extension of quantum field theory at finite temperature. The objective of TFD is to define the state of a system at finite temperature by a regular state vector rather than a density operator. This is achieved by doubling the size of the Hilbert space.
Thouless matrix (Z)	The Thouless matrix is a complex, skew-symmetric matrix used to relate two different product states, either two different Slater determinants (as in the original work by Thouless) or two HFB vacua. Discussed in more detail in section 8.3.2, it provides an alternative mathematical description of a Bogoliubov transformation.
Time-dependent Hartree–Fock–Bogoliubov	Starting from the time-dependent, many-body Schrödinger equation, one obtains the time-dependent Hartree–Fock–Bogoliubov (TDHFB) equation by assuming that the time-dependent many-body state remains a HFB vacuum at all times. In TDHFB, the total energy is a constant of motion. The small-amplitude limit of TDHFB gives either the QRPA theory or the ATDHFB theory of collective motion.
Two-body density matrix ($\gamma(\boldsymbol{r}_1, \boldsymbol{r}_2)$)	In configuration space, the two-body density matrix for an arbitrary A-body state is given by $\gamma_{ijkl} = \langle\Psi|c_l^\dagger c_k^\dagger c_j c_i|\Psi\rangle / \langle\Psi|\Psi\rangle$. When $|\Psi\rangle$ is a HFB vacuum, the two-body density matrix can be expressed as a function of the one-body density matrix and the anomalous density thanks to the Wick theorem.
Unitary Fermi gas	The unitary Fermi gas is an idealized many-body system composed of spin-1/2 fermions interacting via a zero-range, infinite scattering-length contact interaction. It can be considered as an extremely simplified version of the nucleus.
Variation after projection (VAP)	When restoring spontaneously broken symmetries after variation, that is, after the symmetry-breaking solution has been determined from the variational principle, there is no guarantee that the projected energy is the lowest. In contrast, one can formulate a variational principle directly on symmetry-restored solutions, in which case one will automatically ensure that the energy is a minimum. Such a technique is called variation after projection.
Vector density (ρ_v)	In covariant EDF, the vector density, thus called because it is built of our the four-current j^μ (i.e. a vector), is the analog of the one-body density matrix of non-relativistic EDF.

Acronyms

χEFT	chiral effective field theory
χPT	chiral perturbation theory
χPT$_\sigma$	scale-chiral perturbation theory
3N	three-body
G-matrix	G-matrix
ATDHF	adiabatic time-dependent Hartree–Fock
AD	automatic differentiation
ALM	augmented Lagrangian method
ANM	asymmetric nuclear matter
BCS	Bardeen–Cooper–Schrieffer
BEC	Bose–Einstein condensate
BHF	Brueckner–Hartree–Fock
CEDF	covariant energy density functional
CI	configuration interaction
DBHF	Dirac–Brueckner–Hartree–Fock
DBWA	distorted-wave Born approximation
DFT	density functional theory
EDF	energy density functional
EFA	equal-filling approximation
EFT	effective field theory
EoS	equation of state
FAM	finite-amplitude method
FE	finite elements
FFT	fast Fourier transform
FT-HFB	finite-temperature Hartree–Fock–Bogoliubov
FT-HF	finite-temperature Hartree–Fock
GCM	generator coordinate method
GDR	giant dipole resonance
GOA	Gaussian overlap approximation
GP	Gaussian process
HFB	Hartree–Fock–Bogoliubov
HF	Hartree–Fock
HK	Hohenberg–Kohn
HO	harmonic oscillator
IM-SRG	in-medium similarity renormalization group
INM	infinite nuclear matter
IR	infrared
KS-EDF	Kohn–Sham energy density functional
KS	Kohn–Sham
LDA	local density approximation
LN	Lipkin–Nogami
LQCD	lattice quantum chromodynamics
LSDA	local spin density approximation
MAP	minimization after projection
MBPT	many-body perturbation theory
MCMC	Markov-chain Monte Carlo
MR-EDF	multi-reference energy density functional

MRA	multi-resolution analysis
NEDF	nuclear energy density functional
NN	nucleon–nucleon
PAV	projection after variation
PDE	partial differential equation
PDF	probability distribution function
PSS	pseudospin symmetry
QCD	quantum chromodynamics
QFT	quantum field theory
QMC	quantum Monte Carlo
QRPA	quasiparticle random-phase approximation
RG	Runge–Gross
RHF	relativistic Hartree–Fock
RMF	relativistic mean field
RPA	random-phase approximation
RVAP	restricted variation after projection
SLDA	superfluid local density approximation
SNM	symmetric nuclear matter
SR-EDF	single-reference energy density functional
SRG	similarity renormalization group
SSB	spontaneous symmetry-breaking
TDDFT	time-dependent density functional theory
TDEDF	time-dependent energy density functional
TDGCM	time-dependent generator coordinate method
TDHFB	time-dependent Hartree–Fock–Bogoliubov
TDSLDA	time-dependent superfluid local density approximation
TKE	total kinetic energy
UFG	unitary Fermi gas
UV	ultraviolet
VAP	variation after projection
ANG	Anderson–Nambu–Goldstone
ASCC	adiabatic self-consistent collective coordinate
ATDHFB	adiabatic time-dependent Hartree–Fock–Bogoliubov
TDHF	time-dependent Hartree–Fock
TDKS	time-dependent Kohn–Sham
c.o.m.	center of mass
cDFT	covariant density functional theory
d.o.f.	degree of freedom
irrep	irreducible representation
NG	Nambu–Goldstone
p.h.	particle–hole
p.p.	particle–particle
q.p.	quasiparticle
rms	root-mean-square
s.p.	single-particle
WW	Wigner–Weyl

IOP Publishing

Energy Density Functional Methods for Atomic Nuclei

Nicolas Schunck

Chapter 1

Non-relativistic energy density functionals

Jean-Paul Ebran, Michael Bender, Nicolas Schunck and Thomas Duguet

Nuclear systems exhibit a rich phenomenology that is testimony to the complexity of their structure. There are several reasons that can qualitatively explain this complexity. First, we note that nuclei are four-component, self-bound, Fermi systems: nucleons have both a spin and isospin quantum number (proton spin-up, proton spin-down, neutron spin-up, neutron spin-down). The proton and the neutron also share approximately the same mass. Forces between them are not elementary forces but instead are a consequence of the composite nature of nucleons as many-body, interacting quark systems—which are themselves described within quantum chromodynamics (QCD). As a result, simple symmetry requirements allow nucleons to be coupled in multiple ways: central, spin–orbit, tensor, etc, couplings [1]. The subtle interplay between the repulsive short-range part of their strong interaction and Pauli blocking effects, the attractive intermediate-range interaction and the long-range Coulomb repulsion between protons leaves room for the manifestation of numerous properties, e.g. individual (particle–hole) and collective (rotation, vibration, etc) excitations occurring at the same energy scale, superfluidity, clustering, etc.

The goal of low-energy nuclear theory is to provide a consistent description of the rich phenomenology characterizing nuclear systems. Tensions between what we might call the reductionist viewpoint (i.e. nuclear phenomena can be comprehended on the sole basis of the properties of their constituents considered as elementary particles) and the emergent viewpoint (i.e the behavior of a complex ensemble of elementary particles cannot be comprehended unless new, different degrees of freedom are introduced) have resulted in the elaboration of a plethora of complementary approaches. To name but a few, we could mention the macroscopic–microscopic models [2, 3], the collective [4] and algebraic models [5], and the various microscopic methods [6–13]. Among the efforts aimed at reconciling these different viewpoints, a notable one consists in reformulating what used to be phenomenological models in the language of effective field theory (EFT) (see [14], for example). The resulting approaches, based on the relevant degree of freedom (d.o.f.) associated

doi:10.1088/2053-2563/aae0edch1

with their energetic domain of validity, are connected with one another as members of a tower of nuclear EFTs.

The nuclear energy density functional (EDF) approach is at the intersection between phenomenology and fundamental theory. It contains many elements of a rigorous many-body theory of quantum systems as we will see throughout this book, but at the same time the very concept of an energy functional provides enough flexibility to match the accuracy of phenomenological models. In this chapter, we will give a high-level overview of the EDF approach, in particular in the context of more microscopic theories of nuclear structure. We will also discuss some of the most popular non-relativistic energy functionals (see chapter 2 for covariant functionals). Our presentation of the latter does not aim to be comprehensive, since there are excellent review articles on the topic, e.g. [13, 15, 16].

1.1 Introduction

1.1.1 Definition of a configuration space

In this book, the term configuration space refers to a discrete set of quantum numbers, or configurations, characterizing a (countable) basis of the single-particle (s.p.) Hilbert space; by extension, it denotes the s.p. basis itself. In this section, we recall a number of important properties pertaining to the various representations of the quantum mechanical state of a particle. Some of the material presented in the next two sections is relatively basic quantum mechanics inspired by [17–20].

Hilbert space of single particles

We consider the Hilbert space $\mathcal{L}_2$ of square integrable, complex-value, wave functions. The elements of $\mathcal{L}_2$ will be called one-body, or s.p. states and will be denoted generically by $|\phi\rangle$. The ket notation captures the fact that these elements are vectors of a Hilbert space. At the same time, these vectors represent actual three-dimensional scalar functions $|\phi\rangle\colon \mathbb{R}^3 \to \mathbb{C}$.

If one can find a countable (discrete) basis $|n\rangle$, then any state vector $|\phi\rangle$ of $\mathcal{L}_2$ can be expanded into

$$|\phi\rangle = \sum_n \langle n|\phi\rangle |n\rangle,$$

where the complex number $\langle n|\phi\rangle$ represents the scalar product of the vectors $|n\rangle$ and $|\phi\rangle$. Recall that a basis is said to be countable if its elements can be indexed by integer numbers (i.e. 'counted'). In turn, any given set S is countable if its number of elements is the same as some subset of the set of natural numbers $\mathbb{N}$, i.e. there exists an injective mapping $f\colon S \to \mathbb{N}$. It is also possible to expand the state vector onto uncountable (continuous) bases, the elements of which do not belong to $\mathcal{L}_2$. The best examples are the continuous vectors $|\boldsymbol{r}\rangle$ and $|\boldsymbol{p}\rangle$ which are eigenfunctions of the position, $\hat{\boldsymbol{r}}$, and momentum, $\hat{\boldsymbol{p}}$ operators. The expansion then reads

$$|\phi\rangle = \int d^3\boldsymbol{r}\ \langle \boldsymbol{r}|\phi\rangle |\boldsymbol{r}\rangle,$$

where $\langle \boldsymbol{r}|\phi\rangle = \phi(\boldsymbol{r})$ actually represents the value of the function ϕ at point $\boldsymbol{r}$. Mathematically, one can reconcile the notions of state vectors represented by square integrable wave functions and continuous representations corresponding to unbounded operators by introducing rigged Hilbert spaces.

> Consider a Hilbert space H and an algebra $\mathfrak{a}$ of operators with a continuous spectrum, e.g. $\left[\hat{r}_i, \hat{p}_j\right] = i\hbar\delta_{ij}$. These elements cannot have eigenvectors in H. In general, one can still find a maximal subspace $S \subset H$ such that for any vector $v \in H$ and operator $\hat{a} \in \mathfrak{a}$, $\hat{a}v$ is defined and has a finite semi-norm $||\hat{a}v||$: S is called the space of smooth vectors for $\mathfrak{a}$. From S, one can construct the dual space $S^* \supset H$ of continuous complex-linear functionals on S, which is nothing else than the space of bras (in the Dirac bra-ket sense). The bra $\langle x|$ is the linear functional which maps a state $\phi \in S$ to $\phi(x) = \langle x|\phi\rangle$. Now, the elements of the algebra $\mathfrak{a}$ do have eigenvectors in the space of bras S^* : $v \in S^*$ is an eigenvector of $\hat{a} \in \mathfrak{a}$ with eigenvalue λ if $(\hat{a}v)(\phi) = \lambda v(\hat{a}^*\phi)$ for all $\phi \in S$. Together with the original Hilbert space H, the smooth vector space S of an algebra of operators with a continuous spectrum and its dual vector space S^* form the triplet (S, H, S^*), called a rigged Hilbert space.

We then introduce the spin and isospin d.o.f. and restrict ourselves to spin-1/2 particles. The spin space $\mathcal{E}^\sigma$ is a two-dimensional space made of the spinors $|\chi_\sigma\rangle$. Basis spinors are eigenstates of the spin operator $\hat{s}_z$ with eigenvalues ± 1 and will be denoted by $|\sigma\rangle$. The isospin space $\mathcal{E}^\tau$ has the same algebraic structure, and we will denote its spinors as $|\chi_\tau\rangle$ and its basis spinors as $|\tau\rangle$. The full Hilbert space of s.p. states is defined as the tensor product $\mathcal{H}_1 \equiv \mathcal{H} = \mathcal{L}_2 \otimes \mathcal{E}^\sigma \otimes \mathcal{E}^\tau$. Its elements will be denoted generically as $|\psi\rangle$. We have, most generally,

$$|\psi\rangle = |\phi\rangle \otimes |\chi_\sigma\rangle \otimes |\chi_\tau\rangle.$$

As before, we can expand any state vector $|\psi\rangle$ into either a discrete or continuous basis. For example,

$$|\psi\rangle = \int d^3\boldsymbol{r} \sum_\sigma \sum_\tau \langle \boldsymbol{r}\sigma\tau|\psi\rangle |\boldsymbol{r}\sigma\tau\rangle$$

with $|\boldsymbol{r}\sigma\tau\rangle \equiv |\boldsymbol{r}\rangle \otimes |\sigma\rangle \otimes |\tau\rangle$ a short-hand notation for the full basis spinor. Note that the quantity $\psi(\boldsymbol{r}\sigma\tau) = \langle \boldsymbol{r}\sigma\tau|\psi\rangle$ is a $\mathbb{C}$-number, which is a function of $\boldsymbol{r}$ and of the spin and isospin projections σ and τ. The resolution of the identity for the representation $|\boldsymbol{r}\sigma\tau\rangle$ is

$$\int d^3\boldsymbol{r}\ |\boldsymbol{r}\rangle\langle \boldsymbol{r}| \otimes \sum_\sigma |\sigma\rangle\langle\sigma| \otimes \sum_\tau |\tau\rangle\langle\tau| = 1.$$

From the s.p. Hilbert space, one may construct two-particle vector states by considering tensor products of the kind $|n\rangle^{(1)} \otimes |m\rangle^{(2)}$, where the subscript refers to which particle the state corresponds to. These states may be additionally symmetrized (for identical bosons) or anti-symmetrized (for identical fermions). Two-body vector states live in a new Hilbert space, $\mathcal{H}_2$. One may similarly construct N-particle vector states by considering the (symmetrized/anti-symmetrized) tensor product of N s.p. state vectors. The direct sum of all the possible many-body Hilbert spaces defines the Fock space,

$$\mathcal{F} = \bigoplus_{n=1}^{+\infty} \mathcal{H}_n.$$

We will use the generic notation $|\Psi\rangle$ for an element of $\mathcal{F}$. Among these, Slater determinants play a special role for multi-fermion systems. The N-particle Slater determinant is simply an element of the vector space $\mathcal{H}_N$.

In the following, we will focus on s.p. states $|a\rangle$ that can be identified with the eigenvectors of some Hermitian operator $\hat{A}$, $\hat{A}|a\rangle = a|a\rangle$. For the moment, we do not concern ourselves whether the eigenspectrum is discrete, continuous, or contains both a discrete and continuous part. In the language of second quantization, we introduce the creation operator $c_a^\dagger$ of a particle in the s.p. state $|a\rangle$. The corresponding annihilation operator is denoted by c_a. For fermions, these operators obey the following (anti-)commutation relations,

$$\{c_a, c_b\} = \{c_a^\dagger, c_b^\dagger\} = 0 \qquad \{c_a, c_b^\dagger\} = \{c_a^\dagger, c_b\} = \delta_{ab}.$$

Creation and annihilation operators act in Fock space.

Field quantization consists in substituting creation (annihilation) operators $c_a^\dagger$ (c_b) of a particle in state a (b) by creation (annihilation) operators $c_{r\sigma\tau}^\dagger$ ($c_{r\sigma\tau}$) of a particle of type τ at point $\boldsymbol{r}$ and with spin projection σ. Adopting the compact notation $x \equiv (\boldsymbol{r}, \sigma, \tau)$, we define the field operator through the set of relations

$$c_x^\dagger = \sum_a \psi_a^*(x)\; c_a^\dagger \qquad c_x = \sum_b \psi_b(x)\; c_b \tag{1.1}$$

and the inverse transformation is simply

$$c_a^\dagger = \int dx\; \psi_a(x) c_x^\dagger \qquad c_b = \int dx\; \psi_b^*(x) c_x, \tag{1.2}$$

where the integral over $x \equiv (\boldsymbol{r}, \sigma, \tau)$ is a short-hand notation for

$$\int dx = \int d^3\boldsymbol{r} \sum_\sigma \sum_\tau.$$

Single-particle representations

From a practical point of view, it is very important to realize that such an s.p. state can be represented equivalently by either

- An s.p. 'ket', either in configuration space, $|a\rangle$, where a is a generic notation for the eigenvector of some Hermitian operator $\hat{A}$; or in coordinate space, $|r\sigma\tau\rangle$, where $\boldsymbol{r}$ refers to the position of the particle (i.e.

the eigenvector of the position operator $\hat{\boldsymbol{r}}$), σ to its spin projection and τ to its isospin projection; or in momentum space, $|\boldsymbol{p}\sigma\tau\rangle$, where $\boldsymbol{p}$ refers to the momentum of the particle (i.e. the eigenvector of the momentum operator $\hat{\boldsymbol{p}}$).

- An s.p. 'wave function' $\psi_a(x)$ with $x \equiv (\boldsymbol{r}, \sigma, \tau)$, and a the same generic label referring to an eigenvector of some operator $\hat{A}$. We could equivalently consider $x \equiv (\boldsymbol{p}, \sigma, \tau)$, and relate the two wave functions by a Fourier transform.
- A pair of creation/annihilation operators $c_a^\dagger$, c_a acting in Fock space; or a pair of field operators $c_{\boldsymbol{r}\sigma\tau}^\dagger$, $c_{\boldsymbol{r}\sigma\tau}$ or $c_{\boldsymbol{p}\sigma\tau}^\dagger$, $c_{\boldsymbol{p}\sigma\tau}$.

Relations (1.1) and (1.2) show how wave functions are nothing but the coefficients of the transformations between the representations of the creation/annihilation operators.

Representation of one-body operators

By definition, one-body operators are mathematical operators $\hat{F}$ that only act on s.p. states. If one represents the latter by a generic ket $|\psi\rangle$ in configuration space, then

$$\hat{F}: \begin{cases} L_2 \to L_2 \\ |\psi\rangle \mapsto |\varphi\rangle = \hat{F}|\psi\rangle \end{cases}. \tag{1.3}$$

If we note that $|a\rangle$ is a basis of $\mathcal{L}_2$, then the matrix elements of a one-body operator are simply $F_{ab} = \langle a|\hat{F}|b\rangle$. These matrix elements can be used to define how $\hat{F}$ will act in Fock space,

$$\hat{F} = \sum_{ab} F_{ab} c_a^\dagger c_b, \tag{1.4}$$

where $c_a^\dagger$, c_a are the creation and annihilation operators associated with the (basis) s. p. states. In practice, an operator is most often known through its action on wave functions $\psi(\boldsymbol{r})$. For example, the action of a local, scalar potential $\hat{V}(\boldsymbol{r})$ is simply the multiplication of a wave function by said potential,

$$\hat{V}: \psi(\boldsymbol{r}) \mapsto V(\boldsymbol{r})\psi(\boldsymbol{r}).$$

Similarly, the action of a non-local, scalar potential $\hat{V}(\boldsymbol{r}, \boldsymbol{r}')$ involves the integral

$$\hat{V}: \psi(\boldsymbol{r}) \mapsto \varphi(\boldsymbol{r}) = \int d^3\boldsymbol{r}'\; V(\boldsymbol{r}, \boldsymbol{r}')\psi(\boldsymbol{r}').$$

By applying the previous expressions on a basis $\psi_b(\boldsymbol{r})$ of $\mathcal{L}_2$, using the scalar product associated with the Hilbert space, and the resolution of the identity for kets $|\boldsymbol{r}\rangle$, the matrix elements in configuration space for the scalar operator are given by

$$v_{ab} = \langle a|\hat{V}|b\rangle = \int d^3\boldsymbol{r}\, \psi_a^*(\boldsymbol{r}) V(\boldsymbol{r}) \psi_b(\boldsymbol{r}).$$

To obtain expressions for the matrix elements of the operators in coordinate space, we start with the definition of the operator $\hat{V}$ on an arbitrary ket $|\psi\rangle$, insert the resolution of the identity and project on $|\boldsymbol{r}\rangle$

$$|\varphi\rangle = \hat{V}|\psi\rangle \Rightarrow \langle \boldsymbol{r}|\varphi\rangle = \langle \boldsymbol{r}|\hat{V} \int d^3\boldsymbol{r}' |\boldsymbol{r}'\rangle\langle \boldsymbol{r}'|\varphi\rangle = \int d^3\boldsymbol{r}' \langle \boldsymbol{r}|\hat{V}|\boldsymbol{r}'\rangle \varphi(\boldsymbol{r}')$$

which naturally leads to $\langle \boldsymbol{r}|\hat{V}|\boldsymbol{r}'\rangle = V(\boldsymbol{r}, \boldsymbol{r}')$. Note that while $V(\boldsymbol{r}, \boldsymbol{r}')$ is, quite properly, a $\mathbb{C}$-number (as expected since it is a scalar product), it is at the same time the value of the function $V\colon \mathbb{R}^3 \times \mathbb{R}^3 \to \mathbb{C}$.

Matrix elements of differential operators

Let us now consider the one-body operator $\hat{F} \equiv \hat{D}$, where by convention $\hat{D}$ is the differential operator in the space of square integrable wave functions of one variable: $\hat{D}\colon \psi \in \mathcal{L}_2(\mathbb{R}) \mapsto \psi' \in \mathcal{L}_2(\mathbb{R})$. The matrix elements of this operator in configuration space are straightforward,

$$\langle a|\hat{D}|b\rangle = \int dx\, \psi_a^*(x)\hat{D}\psi_b(x) = \int dx\, \psi_a^*(x)\psi'_b(x).$$

These matrix elements are $\mathbb{C}$-numbers, as expected. The matrix elements in coordinate space are, however, a little less straightforward. Starting from the definition of $\hat{D}$ and inserting the resolution of the identity, we find

$$\begin{aligned}\langle x|\hat{D}|\psi\rangle = \psi'(x) &\Rightarrow \int dx' \langle x|\hat{D}|x'\rangle\langle x'|\psi\rangle \\ &= \int dx' \langle x|\hat{D}|x'\rangle \psi(x') = \psi'(x).\end{aligned}$$

This leads to $\langle x|\hat{D}|x'\rangle = \delta'(x - x')$, that is, it is the distributional derivative of the Dirac function. As a reminder, $\delta'(x)$ is itself a distribution, which associates to any function the value of its derivative at 0. As a consequence, we can write formally

$$\langle x|\hat{D}|x'\rangle \equiv \delta(x - x')\frac{\partial}{\partial x'} \quad \text{or} \quad \langle x|\hat{D}|x'\rangle \equiv \left.\frac{\partial}{\partial x'}\right|_{x'=x}$$

Therefore, in contrast to the configuration space representation, matrix elements in the coordinate-space representation are in fact 'linear functionals', i.e. mappings between functions and numbers instead of plain $\mathbb{C}$-numbers. When contracted with quantities such as $\rho(x, x')$ or $\kappa(x, x')$, these matrix elements will thus behave in practice like operators.

Representation of N-body operators

The previous considerations can be generalized to N-body operators. Such operators act in the space $\bigoplus_{n=1}^{N}\mathcal{H}_n$ which includes 1-, 2-, ..., N-body vectors. The simplest basis of the N-body vector space is made of a tensor product of s.p. basis states of the kind

$$|a_1 \dots a_N\rangle = |a_1\rangle \otimes \cdots \otimes |a_n\rangle$$

and matrix elements in configuration space of a N-body operator $\mathcal{O}$ would thus be written

$$O_{\boldsymbol{ab}} = \langle a_1 \dots a_N|\hat{O}|b_1 \dots b_N\rangle \equiv \langle \boldsymbol{a}|\hat{O}|\boldsymbol{b}\rangle,$$

where we use the short-hand notation $|\boldsymbol{a}\rangle = |a_1 \dots a_N\rangle$. If one deals with fermions, the representation of the operator $\hat{O}$ in Fock space must involve anti-symmetrized matrix elements, which we note as $\bar{O}_{\boldsymbol{ab}}$. In the familiar (and much simpler) case of a two-body operator, we have

$$\overline{O}_{abcd} = \langle ab|\hat{O}|cd\rangle - \langle ab|\hat{O}|dc\rangle = \langle ab|\hat{O}\hat{P}|cd\rangle,$$

where the anti-symmetrizer is defined as

$$\hat{\mathcal{P}} = 1 - \hat{\mathcal{P}}_x\hat{\mathcal{P}}_\sigma,$$

with $\hat{\mathcal{P}}_x$ the space-exchange operator and $\hat{\mathcal{P}}_\sigma$ the spin-exchange operator. Note that for atomic nuclei, an additional isospin exchange operator must be explicitly considered. Given anti-symmetrized matrix elements, the Fock space representation of the operator $\hat{O}$ is then

$$\hat{O} = \sum_{\boldsymbol{ab}} \bar{O}_{\boldsymbol{ab}} c_{a_1}^{\dagger} \cdots c_{a_N}^{\dagger} c_{b_N} \cdots c_{b_1}. \tag{1.5}$$

The practical calculation of matrix elements involves computing terms such as

$$O_{\boldsymbol{ab}} = \int d^3\boldsymbol{r}_1 d^3\boldsymbol{r}'_1 \cdots \int d^3\boldsymbol{r}_N d^3\boldsymbol{r}'_N \psi_{a_1}^{*}(\boldsymbol{r}_1) \cdots \psi_{a_N}^{*}(\boldsymbol{r}_N) \hat{O} \psi_{b_1}(\boldsymbol{r}'_1) \cdots \psi_{b_N}(\boldsymbol{r}'_N).$$

1.1.2 Microscopic approaches of nuclear systems

Among the various methods involved in nuclear structure physics, microscopic approaches aim to account in a unified manner for the various properties of all atomic nuclei in the nuclear chart. They depict nuclear systems as collections of interacting, point-like nucleons whose dynamics is described by the nuclear Hamiltonian

$$\hat{H} = \sum_{i=1}^{A} \frac{\hat{\boldsymbol{p}}_i^2}{2m} + \frac{1}{2!} \sum_{i \neq j=1}^{A} \hat{V}_{ij} + \frac{1}{3!} \sum_{i \neq j \neq k=1}^{A} \hat{V}_{ijk} + \dots. \tag{1.6}$$

The first contribution represents the non-relativistic kinetic energy operator with m the nucleon mass, while the others stand for the 2-, 3-, ..., A-body interaction terms.

Note that in contrast to electronic systems, the nuclear Hamiltonian (1.6) does not contain only one- and two-body potentials. For an atomic nucleus with mass number A, up to A-body terms should in principle be present. The goal of a microscopic approach is to extract from the Hamiltonian (1.6) the relevant information needed to compute the nuclear observables of interest. This leads to two distinct but strongly inter-dependent issues: (i) describing complex inter-nucleon interactions and (ii) handling the nuclear many-body problem.

> The scope of microscopic approaches actually extends beyond finite nuclei *per se*. By providing a consistent framework to describe the structure and reactions of atomic nuclei, they can be applied to the study of fundamental symmetries [21–24] nucleosynthesis mechanisms [25–28], nuclear astrophysics such as the structure of neutron stars [29–35], etc.

The first challenge stems from the very origin of nuclear forces: they are derived from the strong force acting between their quarks and gluons whose theoretical framework is provided by quantum chromodynamics (QCD). Therefore we can expect inter-nucleon forces to be described by the low-energy sector of QCD. However, a number of features of low-energy QCD makes the direct derivation of inter-nucleon interactions extremely difficult. Not only does the typical nuclear structure energy scale correspond to the QCD low-energy non-perturbative regime (asymptotic freedom [36]), but due to the color confinement property [37], the QCD relevant low-energy degrees of freedom (hadrons) differ from the QCD natural degrees of freedom (quarks and gluons). Several alternatives have been explored to circumvent this issue, from lattice QCD [38] to the definition of highly accurate phenomenological interactions (Argonne [39], CD-Bonn [40], etc), including holographic QCD [41] or chiral effective field theory (χEFT) [42]. The latter has become the currently accepted paradigm to rigorously describe nuclear forces.

The second challenge is the many-body problem itself (given a nuclear Hamiltonian). Beyond the complex operator structure displayed by the nucleon–nucleon (NN) force, the latter produces a weakly-bound neutron–proton state, i.e. the deuteron, in the coupled $^3S_1$–$^3D_1$ channels and a virtual di-neutron state in the $^1S_0$ channel.

> We briefly recall that scattering phase shifts are typically noted as $^{2S+1}L_J$, with $L = 0, 1, 2, \ldots$ the relative orbital angular momentum of the two nucleons, J their relative angular momentum (conserved because of the rotational invariance of nuclear forces) and S the relative intrinsic spin of the two nucleons (coupling of two spins 1/2). The spectroscopic notation for L is S, P, D, F, etc for $L = 0, 1, 2, 3, \ldots$

The associated large scattering lengths of such states, together with the short-range repulsion between nucleons, make the nuclear many-body problem highly non-perturbative. In addition to such difficulties, it has become clear over the last twenty years that three-body (3N) interactions are indispensable in a theory of point-like nucleons [43]. Finally, atomic nuclei are mesoscopic systems: the typical number of nucleons in a nucleus is large enough that a quasi-exact description is computationally prohibitive, yet too small for finite-size effects to be negligible and to justify a statistical treatment.

Different strategies can be adopted to tackle the nuclear many-body problem, that is, to fully account for nucleon correlations inside the nucleus, which leads to a rich variety of many-body approaches. A first common feature of these approaches is to turn the correlated many-nucleon system into an auxiliary one-body problem. Such a procedure is made more difficult by the ultraviolet (UV) divergences mainly induced by the hard core of the NN potential. To bypass this issue, one can make the nuclear many-body problem more perturbative, e.g. evolving the nuclear Hamiltonian with the similarity renormalization group (SRG) [44] to decouple the nucleon low- and high-momentum modes. The validity of turning the original, correlated many-body problem into an auxiliary, independent one-body problem can be interpreted as a consequence of the quantum liquid nature of the nuclear ground state. Recall that the NN interaction is strong and short-ranged, with a repulsive part of the order of 1 GeV, an attractive part $V_0 \sim 100$ MeV and a typical range $a \sim 1$ fm. Yet, the gain in energy induced by localizing each nucleon such that the resulting potential energy is minimum is much smaller than the zero-point kinetic energy $\sim\hbar^2/ma^2$ involved in the localization of a nucleon within a volume $\propto a^3$: the ratio of the kinetic energy over the potential energy [45] $\Lambda \equiv \hbar^2/(ma^2V_0)$, is of the order of 0.5 for nuclei. Since the transition between a localized crystalline-like phase and a delocalized quantum liquid phase corresponds to $\Lambda \sim 0.1$ [45], nuclear systems are much like quantum liquids. In other words the nuclear force, as strong as it is believed to be, is still too weak, relative to its range and to the nucleon mass, to localize nucleons in a crystalline-like phase. Consequently, the ground state of nuclear systems involves delocalized individual orbitals that reflect the shape and radial dependence of the potential over the whole nuclear volume. This translates into a large mean free path for nucleons in the nucleus which, in turn, suggests that it should be possible to map the nuclear many-body problem onto an effective one-body one. Put another way, the short-range NN correlations responsible for the UV divergences can be transferred from the many-body state to the NN interaction, either explicitly [46–48] or implicitly (phenomenologically effective in-medium NN interactions [13]).

In any case, the resulting auxiliary one-body system is governed by a Hamiltonian of the form

$$\hat{H}_0 = \sum_{i=1}^{A}\left(\frac{\hat{\boldsymbol{p}}_i^2}{2m} + \hat{U}_i\right) \tag{1.7}$$

with $\hat{U} \equiv \sum_i \hat{U}_i$ a one-body operator that can be chosen in such a way as to maximally absorb the physics of the system. The original Hamiltonian (1.6) can then be recast into

$$\hat{H} = \sum_{i=1}^{A}\left(\frac{\hat{\boldsymbol{p}}_i^2}{2m} + \hat{U}_i\right) + \left(\frac{1}{2!}\sum_{i\neq j=1}^{A}\hat{V}_{ij} + \ldots - \sum_{i=1}^{A}\hat{U}_i\right) \equiv \hat{H}_0 + \hat{V}_{\text{res}}. \tag{1.8}$$

Provided that the one-body term $\hat{U}$ is well chosen, the residual interaction approximately verifies $||\hat{V}_{\text{res}}|| \ll ||\hat{H}_0||$.

The total wave function of the auxiliary one-body system governed by the unperturbed Hamiltonian (1.7) takes the form of a Slater determinant. Such a product state already encompasses one of the most important classes of nucleon correlations, namely the Pauli exclusion principle stemming from their fermionic behavior. However, due to the residual interactions $\hat{V}_{\text{res}}$, nuclei involve additional correlations that a product state cannot account for because of its lack of flexibility. Hence, the underlying strategy of a many-body approach can often be summarized by the question 'How to incorporate such additional correlations on top of a product state?'. As we will argue hereafter, all the answers to this question that theorists proposed are related to one or a combination of the three following developments: vertical expansion, horizontal expansion and considerations pertaining to the inter-nucleon interaction.

1.1.3 Vertical and horizontal philosophies

The Hartree–Fock (HF) method [49] provides a procedure to construct the best one-body potential $\hat{U}$, that is, the one that encodes the maximum amount of correlations and thus leads to the least-energy independent-particle description. In such a framework, the symmetry-restricted HF product state is the eigenstate of the unperturbed Hamiltonian (1.7) with the lowest eigenvalue. It reads

$$|\Phi\rangle_{N=Z,\,N,\,J,\,M,\,\pi\ldots} = \prod_i c_i^\dagger | - \rangle. \tag{1.9}$$

The labels Z, N, J, M, π... refer to good quantum numbers (particle number, total angular momentum, projection of the total angular momentum on a given axis, parity, etc). Particle operators $\{c_i^\dagger,\, c_i\}$ with i a generic index standing for a collection of spatial, spin and isospin quantum numbers, are associated with a basis of the one-body Hilbert space $\mathcal{H}_1$. The state $|-\rangle$ stands for the vacuum of particles.

The vertical philosophy is the dominant view of *ab initio* [7, 9] and configuration-interaction (CI) [12] approaches. It amounts to building on top of the HF reference vacuum (1.9) a set of particle–hole (p.h.) excited states

$$|\Phi_{ijk\ldots}^{mnp\ldots}\rangle_N = c_m^\dagger c_n^\dagger c_p^\dagger \ldots c_k c_j c_i |\Phi\rangle_N. \tag{1.10}$$

Following rather standard practice, we will use the convention that indices m, n, p,... refer to *particle* states with s.p. energies higher than the Fermi energy, and indices i, j, k,... refer to *hole* states with s.p. energies below the Fermi energy. The same convention will be used in chapter 5.

Each of these excited states is itself a Slater determinant corresponding to the same number of particles as the reference vacuum (1.9). One may then expand any many-body state $|\Psi\rangle_N$ of interest, for instance an eigenstate of the original Hamiltonian (1.8), in the orthonormal basis thus formed

$$|\Psi\rangle_N = C^0|\Phi\rangle_N + \sum_{m,i} C_{mi}^{1p1h}|\Phi_i^m\rangle_N + \sum_{mn,ij} C_{mnij}^{2p2h}|\Phi_{ij}^{mn}\rangle_N + \dots. \tag{1.11}$$

In the above vertical expansion, all the Slater determinants preserve the symmetries of the original Hamiltonian (1.8) and differ from one another by essentially non-collective excitations. Considerations relative to the inter-nucleon interaction aim at transferring the maximal weight on the first terms of the vertical expansion (1.11) such that it can be truncated at the lowest possible order. The most relevant criterion to organize the vertical expansion may not be the p.h. order but the difference between the energies of the p.h. configuration and the vacuum.

In doubly closed-shell systems, the HF representation based on a single Slater determinant already yields a faithful qualitative description of their structural properties, at least if the UV divergence associated with the NN potential has been tamed down, that is, if the interaction is 'soft' enough so that it does not diverge for $r \to 0$. In this sense, the orbitals defining the reference symmetry-restricted vacuum encapsulate the essential physics of such systems. Attempting to describe additional dominant nucleon correlations (the so-called dynamical correlations [50]) through numerous rearrangements of the nucleons within the fixed set symmetry-restricted HF orbitals—what a vertical expansion actually does—thus turns out to be very effective. Indeed, such dynamical correlations translate into a large number of p.h. states contributing to the expansion (1.11). However each of them comes with a small weight, rendering their account manageable. If treating dynamical correlations does not radically change the properties of the systems concerned, it is still absolutely necessary in order to reach a highly accurate description.

The situation is very different in open-shell nuclei. There, the occurrence of (quasi-)degeneracies translates into several product states contributing with a comparable weight in the vertical expansion (1.11): this is the hallmark of what is called non-dynamical correlations [50]. The latter induces non-trivial rearrangements in the system, lifting the degeneracies and yielding collective features such as deformation or superfluidity. In such a case, a single-determinant state that preserves the symmetries of the system fails to provide even a qualitative description of its structural properties. In other words, the orbitals defining the symmetry-restricted HF vacuum do not capture the dominant physics of the system and thus should not be employed in a tractable p.h. expansion.

To illustrate our discussion with a concrete example, spherical HF theory is notoriously insufficient to describe heavy open-shell nuclei. Features of such nuclei such as the high energy of the first 2^+ state in even–even systems or the regular pattern of energy levels in a sequence 0^+, 2^+, 4^+,... are qualitatively explained as a consequence of pairing correlations and intrinsic deformation, respectively—both effects induced by symmetry-breaking as we shall see later.

Extensions of theories retaining the vertical philosophy but with a broader scope of application are currently being investigated in order to account for non-dynamical correlations. Some of them give up on employing a simple product state at the zeroth-order, but rather involve from the very beginning a multi-determinant reference state [51]. Others inform the reference product state of additional correlations in a doubly self-consistent procedure (multi-configuration, self-consistent field approach [52]). In any case, these approaches do not scale well with the size of the system, which limits their domain of applicability.

The horizontal philosophy underpinning the EDF approach is an alternative viewpoint that enables a very effective treatment of non-dynamical correlations. Instead of introducing a set of p.h. excitations on top of a symmetry-restricted vacuum, a horizontal expansion introduces a series of non-orthogonal symmetry-unrestricted vacua

$$|\Phi(\boldsymbol{g})\rangle \equiv \prod_{\mu} \beta_{\mu}^{(g)} | - \rangle. \tag{1.12}$$

The collective label $\boldsymbol{g} = ||\boldsymbol{g}||e^{i\varphi_g}$ stands for the order parameters associated with the spontaneously broken symmetries of the original Hamiltonian (1.6) of the system. In nuclear physics, we know that the latter preserves translational, rotational, Galilean, time-reversal and parity invariance, and is also invariant under isospin rotations [1]. The modulus $g = ||\boldsymbol{g}||$ plays the role of a collective coordinate whose value measures the extent to which the symmetries are broken, in other words, the 'deformation' of the system; see figure 1.1 for a visual representation of the mechanism. Its phase $\alpha = \varphi_g$ determines the favored orientation of the system induced by this spontaneous symmetry-breaking (SSB).

Although (1.12) is written in a very general way, the most successful example of such a symmetry-unrestricted vacuum is the quasiparticle (q.p.), or Bogoliubov, vacuum of the Hartree–Fock–Bogoliubov (HFB) theory. The theory is based on introducing q.p. creation and annihilation operators that satisfy the anti-commutation relations $\{\beta_{\mu}^{(g)}, \beta_{\nu}^{(g)\dagger}\} = \delta_{\mu\nu}$ and are related to the particle ladder operators through the so-called Bogoliubov transformation

$$\beta_{\mu}^{(g)} = \sum_{i} U_{i\mu}^{(g)*} c_i + \sum_{i} V_{i\mu}^{(g)*} c_i^{\dagger} \tag{1.13a}$$

$$\beta_{\mu}^{(g)\dagger} = \sum_{i} V_{i\mu}^{(g)} c_i + \sum_{i} U_{i\mu}^{(g)} c_i^{\dagger}. \tag{1.13b}$$

The state (1.12) is a generalized vacuum in the sense that, by construction, $\beta_{\mu}^{(g)}|\Phi(\boldsymbol{g})\rangle = 0$ for all μ. Such a state is allowed to span several irreducible representations (irreps) of the Lie algebra associated with the symmetry group of the Hamiltonian; for a reminder about Lie groups, Lie algebras and their representations, we refer to the discussion in section 2.1.2. Therefore, rather than approximating an eigenstate of the total Hamiltonian (1.8), it is associated with the recombination of a manifold of quasi-degenerate states yielding a very stable wave packet with essentially the nature of a symmetry-breaking state. In that sense, a generalized vacuum of the horizontal philosophy encodes the physics contained in a resummation to all orders of a specific symmetry-conserving Slater determinant

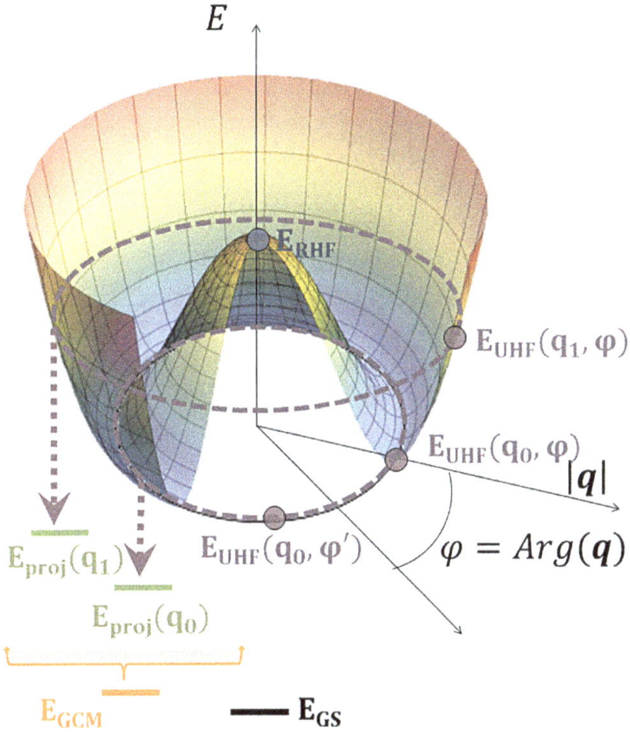

Figure 1.1. Schematic illustration of the concept of SSB in the context of atomic nuclei. On a global energy scale, the true energy of the ground state is marked by the black line E_{GS}. A symmetry-conserving, restricted HF calculation gives the point E_{RH}. Breaking symmetries, for example rotational invariance, lower the energy and lead to the points marked $E_{\text{UHF}}(q_0, \varphi)$ and $E_{\text{UHF}}(q_1, \varphi)$. These solutions are invariant under a gauge transformation, which is represented by the 'Mexican hat' shape; see section 3.1.1 for a list of gauge transformations. The projection on good quantum numbers further lowers the energy of the symmetry-unrestricted solutions and yields $E_{\text{proj}}(q_0)$ and $E_{\text{proj}}(q_1)$. Configuration mixing through the generator coordinate method (GCM) gives an additional correlation energy to the point E_{GCM} that best approximates the true ground-state energy of the system, marked E_{gs}. (Figure produced by J-P Ebran and T Duguet.)

involved in the vertical viewpoint—in a very inexpensive way since the many-body wave function is now a product state.

In the horizontal philosophy, the many-body state associated with the total Hamiltonian (1.8) is then expressed as a mixture of symmetry-unrestricted vacua

$$|\Psi\rangle_N = \int d\boldsymbol{g}\, f(\boldsymbol{g})|\Phi(\boldsymbol{g})\rangle. \tag{1.14}$$

In the above horizontal expansion, the non-orthogonal symmetry-unrestricted vacua differ one from another by collective excitations. For a fixed modulus g of the order parameter, they can also differ by the value of the phase α. Mixing such degenerate modes (projection on good quantum numbers) encodes a rotational-like physics.

Conversely, for a fixed value of the phase, the generalized vacua can also be distinguished by the value of the order parameter g. Mixing such states as, e.g. in the GCM, gives access to a vibrational-like physics. Techniques to restore broken symmetries are discussed in section 3.2.3; configuration mixing techniques such as the GCM are presented in sections 3.2.1 and 3.2.2.

To summarize, by introducing the symmetry-restricted HF one-body potential $\hat{U}$ to partition the total Hamiltonian $\hat{H} = \hat{T} + \hat{V}$, with $\hat{T}$ ($\hat{V}$) the kinetic (potential) energy operator, according to an unperturbed part $\hat{H}_0 = \hat{T} + \hat{U}$ and a residual interaction part $\hat{V}_{\text{res}} = \hat{V} - \hat{U}$, we hope that $||\hat{V}_{\text{res}}|| \ll ||\hat{H}_0||$. Because of non-dynamical correlations contained in $|\Phi(\boldsymbol{g}')\rangle$, the latter relation is not verified for the majority of nuclei (open-shell systems). Allowing the HF one-body potential to spontaneously break the symmetries of the original Hamiltonian $\hat{H}$, $\hat{U} \rightarrow \hat{U}(\boldsymbol{g})$ formally, enables an effective account of non-dynamical correlations, the latter being transferred in the unperturbed part

$$\hat{H} = \hat{H}_0(\boldsymbol{g}) + \hat{V}_{\text{res}}(\boldsymbol{g}) \quad \text{with} \quad \begin{cases} \hat{H}_0(\boldsymbol{g}) = \hat{T} + \hat{U}(\boldsymbol{g}) \\ \hat{V}_{\text{res}}(\boldsymbol{g}) = \hat{V} - \hat{U}(\boldsymbol{g}) \end{cases} \tag{1.15}$$

and in a symbolic notation, $||\hat{V}_{\text{res}}(\boldsymbol{g})|| \leqslant ||\hat{V}_{\text{res}}||$.

1.2 Energy density functional kernels

At the core of the nuclear EDF approach are what are called the off-diagonal energy and norm kernel,

$$\mathcal{E}(\boldsymbol{g}', \boldsymbol{g}) = \mathcal{E}\left[\langle\Phi(\boldsymbol{g}')|, |\Phi(\boldsymbol{g})\rangle\right] \tag{1.16a}$$

$$\mathcal{N}(\boldsymbol{g}', \boldsymbol{g}) = \langle\Phi(\boldsymbol{g}')|\Phi(\boldsymbol{g})\rangle \tag{1.16b}$$

associated with two symmetry-unrestricted vacua $|\Phi(\boldsymbol{g})\rangle$ and $|\Phi(\boldsymbol{g}')\rangle$ of the type (1.12), possibly differing by their corresponding value for the order parameter $\boldsymbol{g}$. Such a form for the energy kernel results from various requirements [53], such as, e.g. being a scalar under any transformation of the symmetry groups commuting with the nuclear Hamiltonian. A sufficient condition to fulfill such requirements is to postulate that the energy kernel is a functional of the normal and anomalous transition density matrices

$$\rho_{ij}^{g'g} = \frac{\left\langle\Phi(\boldsymbol{g}')\middle|c_j^\dagger c_i\middle|\Phi(\boldsymbol{g})\right\rangle}{\langle\Phi(\boldsymbol{g}')|\Phi(\boldsymbol{g})\rangle} \tag{1.17a}$$

$$\kappa_{ij}^{g'g} = \frac{\langle\Phi(\boldsymbol{g}')|c_j c_i|\Phi(\boldsymbol{g})\rangle}{\langle\Phi(\boldsymbol{g}')|\Phi(\boldsymbol{g})\rangle} \tag{1.17b}$$

$$\kappa_{ij}^{gg'*} = \frac{\langle \Phi(g')|c_i^\dagger c_j^\dagger|\Phi(g)\rangle}{\langle \Phi(g')|\Phi(g)\rangle}, \tag{1.17c}$$

with the $\{c, c^\dagger\}$ ladder operators associated with an arbitrary one-body basis of $\mathcal{H}_1$. In equation (1.17), note the order, both of the creation and annihilation operators and of the subscripts $\boldsymbol{g}$ and $\boldsymbol{g}'$ for the normal and anomalous transition densities. Therefore,

$$\mathcal{E}[\boldsymbol{g}', \boldsymbol{g}] = \mathcal{E}\left[\rho_{ij}^{g'g}, \kappa_{ij}^{g'g}, \kappa_{ij}^{gg'*}\right]. \tag{1.18}$$

A possible method to derive the EDF kernel relies on the concept of the pseudo-Hamiltonian, that is, a Hamiltonian whose NN and 3N interaction terms already implicitly resum many-body effects, according to

$$\mathcal{E}\left[\rho_{ij}^{g'g}, \kappa_{ij}^{g'g}, \kappa_{ij}^{gg'*}\right] = \frac{\langle \Phi(g')|\hat{H}_{\text{eff}}|\Phi(g)\rangle}{\langle \Phi(g')|\Phi(g)\rangle}.$$

In the literature, one often uses the term of effective Hamiltonian or effective force for the potential part of said Hamiltonian. Here, we make an additional distinction. While an effective Hamiltonian *stricto sensu* is still a proper operator that can be represented unambiguously in second quantization, we will use the term of pseudo-Hamiltonian to refer to an object that may also contain density dependencies. In this case there can be no such unambiguous second quantized form [54]; see also below.

Indeed, because of their quantum liquid nature, many-nucleon systems can be roughly represented by an effective system of weakly interacting particles. Dressing the initial strongly interacting nucleons by the p.h. excitation cloud they induce in the medium leads to the definition of approximately independent Landau q.p. d.o.f.s [55]. Instead of explicitly performing such a transformation, it has become traditional to define phenomenological effective NN interactions whose form is inspired by the Brueckner G-matrix [48]. The latter is indeed known to be compatible with an independent-particle picture by reintegrating the influence the hard core part of the NN potential induces on the wave functions of two scattering nucleons. In particular, the strong dependence of the G-matrix with respect to the nuclear density is retained in the formulation of effective interaction (while the energy dependence is traditionally disregarded). Yet, enabling a density dependence for an effective in-medium interaction captures N-body features well beyond those encompassed in a conventional G-matrix, e.g. the effects of the internal polarization of nucleons yielding the contribution of more than pairwise interaction in a point-like description.

Such an explicit density dependence of the effective inter-nucleon interaction has several consequences:

- Part of the Hamiltonian is no longer an operator, but rather a term that depends on the solution of the mean-field equation [54]. Such a configuration is beyond the scope of the Ritz variational principle, with the result that the minimization of the corresponding energy with respect to a product state can lead to energies lower than the exact one.

- On the other hand, the presence of a density dependence modifies the mean-field equations by introducing an additional contribution: the re-arrangement term. The latter gives more flexibility to the relation between the energy of the system computed at the single-reference energy density functional (SR-EDF) level and the s.p. contributions [56].
- Together with approximations or modifications made to handle exchange terms and the choice of different interactions in the p.h. and particle–particle (p.p.) channels, an explicit density dependence of the effective NN interaction is the source of spurious self-interaction contributions to the EDF [57–60]. It has been pointed out that when calculating low-lying states in odd nuclei, the contamination of the EDF energy kernel by such spurious terms is in general of the same order of magnitude as the energy from the polarization effects induced by time-odd terms [61]. Furthermore, different procedures to calculate the polarization energy that are equivalent for true Hamilton operators yield different results when using EDF contaminated by self-interactions [61]. Even more importantly, the presence of self-interactions induces many inconsistencies at the level of multi-reference energy density functional (MR-EDF) calculations, which are related to the corresponding EDF being ill-defined when extended to the complex plane [58–60, 62, 63].

These difficulties have triggered recent proposals for novel generations of EDF [64–69].

In any event, the energy functionals are characterized by a small number (5–15) of parameters adjusted at the SR-EDF or the MR-EDF level in order to reproduce various features, e.g. some properties of infinite nuclear matter (saturation point, compressibility, asymmetry parameter), semi-infinite nuclear matter (surface coefficient), or finite nuclei (binding energies, radii, spin–orbit energy gaps). Once determined, those parameters are kept as they are and the resulting energy functional is meant to be employed in any region of the nuclear landscape, irrespective of the zones used for the adjustment of the parameters. We will discuss the choice of experimental observables for the calibration of the energy functional in more detail in section 9.2. In section 9.3, we will also give an overview of the methods used to quantify and propagate, respectively, the uncertainties induced by the calibration process itself.

In general, the nuclear EDF can be split into a kinetic, potential, pairing, Coulomb and center-of-mass (c.o.m.) correction contribution

$$\mathcal{E}\,[\rho, \kappa^*, \kappa] = \mathcal{E}_{\text{kin}}[\rho] + \mathcal{E}_{\text{pot}}[\rho] + \mathcal{E}_{\text{pair}}[\rho, \kappa, \kappa^*] + \mathcal{E}_{\text{Cou}}[\rho] + \mathcal{E}_{\text{c.o.m.}}[\rho], \tag{1.19}$$

where the normal and anomalous density matrices ρ and κ correspond to the diagonal limit of equation (1.17), i.e. to the case where $|\Phi(\boldsymbol{g}')\rangle = |\Phi(\boldsymbol{g})\rangle$. Note that the decomposition (1.19) is arbitrary. In particular, the potential term refers to the part of the energy functional that originates from nuclear forces only, i.e. without Coulomb. In the rare cases where the Coulomb contribution to the p.p. channel is explicitly taken into account [62, 70, 71], this term could be bundled either in $\mathcal{E}_{\text{pair}}$ or $\mathcal{E}_{\text{Cou}}$. In the following subsections, we will give a brief summary of the two most commonly employed parametrizations of $\mathcal{E}_{\text{pot}}$, namely the Skyrme and Gogny EDF. We will devote chapter 2 to covariant density functionals, which have several distinct features worth mentioning.

1.2.1 Skyrme functionals

The Skyrme functional is quasi-local, namely it only depends on local densities such as the matter density $\rho(\boldsymbol{r})$, spin density $\boldsymbol{s}(\boldsymbol{r})$, kinetic density $\tau(\boldsymbol{r})$, current $\boldsymbol{j}(\boldsymbol{r})$ and derivatives of these densities. Through the derivatives contained in their definition, however, local densities such as $\tau(\boldsymbol{r})$ and $\boldsymbol{j}(\boldsymbol{r})$ entering the Skyrme EDF still probe the non-locality of the full one-body density matrix $\rho(\boldsymbol{r}, \sigma; \boldsymbol{r}', \sigma')$. This form of EDF was motivated by the short range of the inter-nucleon interaction as compared to the typical spatial scale where the density matrix displays meaningful variations. From a given finite-range interaction, one can build such a quasi-local functional in the framework of the density matrix expansion [72–74]. There are two possible strategies, one leading to intricate density dependences of the coupling constants [74–78], the other requiring going to higher-order in gradient terms [79]. In one way or the other, any form of effective interaction can be mapped onto a Skyrme-like EDF [80], including covariant EDFs [81, 82], and realistic potentials computed from χEFT either in their local formulation (in momentum space) [83, 84] or non-local one [85].

General form of the Skyrme energy functional

In more details, the potential term corresponding to the Skyrme EDF involves the most general bilinear combinations of the various local densities built from the one-body density matrix and its derivatives, consistent with the invariance under Galilean transformation of the system. These densities are obtained from the isoscalar and isovector non-local densities defined formally according to

$$\rho_0(\boldsymbol{r}, \boldsymbol{r}') = \sum_{ij} \rho_{ji} \sum_{\sigma,\tau} \varphi_i^*(\boldsymbol{r}'\sigma\tau)\, \varphi_j(\boldsymbol{r}\sigma\tau) \tag{1.20a}$$

$$\rho_{1,k}(\boldsymbol{r}, \boldsymbol{r}') = \sum_{ij} \rho_{ji} \sum_{\sigma,\tau,\tau'} \varphi_i^*(\boldsymbol{r}'\sigma\tau')\, \hat{\tau}_k\, \varphi_j(\boldsymbol{r}\sigma\tau) \tag{1.20b}$$

$$\boldsymbol{s}_0(\boldsymbol{r}, \boldsymbol{r}') = \sum_{ij} \rho_{ji} \sum_{\sigma,\sigma',\tau} \varphi_i^*(\boldsymbol{r}'\sigma'\tau)\, \hat{\boldsymbol{\sigma}}\, \varphi_j(\boldsymbol{r}\sigma\tau) \tag{1.20c}$$

$$\boldsymbol{s}_{1,k}(\boldsymbol{r}, \boldsymbol{r}') = \sum_{ij} \rho_{ji} \sum_{\sigma,\sigma',\tau,\tau'} \varphi_i^*(\boldsymbol{r}\sigma'\tau')\, \hat{\boldsymbol{\sigma}}\, \hat{\tau}_k\, \varphi_j(\boldsymbol{r}\sigma\tau), \tag{1.20d}$$

where $\varphi_i(\boldsymbol{r}, \sigma\tau) \equiv \langle \boldsymbol{r}\sigma\tau | i \rangle$ are the spin and isospin components of the spinors representing the single-particle states, whereas ρ_{ij} is the one-body density matrix. In the absence of proton–neutron mixing, only the third component of the isospin densities $\rho_{1,k}(\boldsymbol{r}, \boldsymbol{r}')$ and $s_{1,k}(\boldsymbol{r}, \boldsymbol{r}')$ is non-zero. In this case, which covers the vast majority of applications, it is customary to drop the second index and to define the isovector densities as

$$\rho_{1,1}(\boldsymbol{r}, \boldsymbol{r}') = \rho_{1,2}(\boldsymbol{r}, \boldsymbol{r}') = 0 \quad \rho_{1,3}(\boldsymbol{r}, \boldsymbol{r}') \to \rho_1(\boldsymbol{r}, \boldsymbol{r}')$$
$$s_{1,1}(\boldsymbol{r}, \boldsymbol{r}') = s_{1,2}(\boldsymbol{r}, \boldsymbol{r}') = 0 \quad s_{1,3}(\boldsymbol{r}, \boldsymbol{r}') \to s_1(\boldsymbol{r}, \boldsymbol{r}').$$

Operator structure of the density matrix

We have introduced in (1.17) the most general form for the one-body density matrix—which is often called the transition density matrix or mixed density matrix since it involves two different reference many-body states $|\Phi^{(g)}\rangle$ and $|\Phi^{(g')}\rangle$. Equation (1.17) reduces to (3.1) when the two states are equal, and the resulting one-body density matrix, together with the anomalous density, form then the cornerstone of the SR-EDF theory. The form (3.1) is written in configuration space: it can be transformed to a coordinate-space representation according to equation (1.2); see the discussion in section 1.1.1. The one-body density matrix then reads

$$\rho(x, x') = \frac{\langle \Phi(\boldsymbol{g})|c^\dagger(x')c(x)|\Phi(\boldsymbol{g})\rangle}{\langle \Phi(\boldsymbol{g})|\Phi(\boldsymbol{g})\rangle} \equiv \rho(\boldsymbol{r}\sigma\tau, \boldsymbol{r}'\sigma'\tau'),$$

with $x \equiv (\boldsymbol{r}, \sigma, \tau)$, $\sigma = \pm 1/2$ is the spin projection and $\tau = \pm 1/2$ is the isospin projection. We can interpret the functions $\rho(\boldsymbol{r}\sigma\tau, \boldsymbol{r}'\sigma'\tau')$ as matrix elements of an operator in the space $\mathcal{H}_1 = \mathcal{L}_2 \otimes \mathcal{E}_\sigma \otimes \mathcal{E}_\tau$; see section 1.1.1 for notations. Recall that the space $\mathcal{E}_\sigma$ of the intrinsic spin is a vector space of dimension 2. Any Hermitian operator $\hat{O}$: $\mathcal{E}_\sigma \to \mathcal{E}_\sigma$ can thus be expanded as [17, 19]

$$\hat{O} = \sum_{\mu=0,x,y,z} O_\mu \hat{\sigma}_\mu,$$

where σ_μ are the Pauli spin matrices and $\sigma_0 = I_\sigma$ is the 2×2 identity matrix. A similar expansion holds for operators acting within the isospin space. As a result, the one-body density matrix 'operator' can be written

$$\hat{\rho}(\boldsymbol{r}, \boldsymbol{r}') = \sum_{\mu k} \rho_{\mu k}(\boldsymbol{r}, \boldsymbol{r}')\hat{\sigma}_\mu \hat{\tau}_k. \tag{1.21}$$

By convention, we introduce the isoscalar and isovector non-local densities as follows

$$\hat{\rho}(\boldsymbol{r}, \boldsymbol{r}') = \frac{1}{4}\rho_0(\boldsymbol{r}, \boldsymbol{r}')\hat{I}_\sigma\hat{I}_\tau + \frac{1}{4}\rho_1(\boldsymbol{r}, \boldsymbol{r}')\hat{I}_\sigma\hat{\boldsymbol{\tau}} + \frac{1}{4}s_0(\boldsymbol{r}, \boldsymbol{r}')\hat{\boldsymbol{\sigma}}\hat{I}_\tau + \frac{1}{4}s_1(\boldsymbol{r}, \boldsymbol{r}')\hat{\boldsymbol{\sigma}}\hat{\boldsymbol{\tau}},$$

where products such as $\hat{I}_\sigma\hat{I}_\tau$ should be understood as tensor products, $\hat{I}_\sigma \otimes \hat{I}_\tau$. We refer the reader to [13, 86, 87], among others, for more details. We also discuss in section 8.1.2 how such expansions can be used to compute matrix elements.

It is customary to use the subscript t to refer to the isoscalar ($t = 0$) or isovector ($t = 1$) character of densities and the subscript q to refer to the neutron ($q \equiv n$) or proton ($q \equiv p$) densities. Recall that

$$\rho_0(\boldsymbol{r}, \boldsymbol{r}') = \rho_n(\boldsymbol{r}, \boldsymbol{r}') + \rho_p(\boldsymbol{r}, \boldsymbol{r}')$$
$$\rho_1(\boldsymbol{r}, \boldsymbol{r}') = \rho_n(\boldsymbol{r}, \boldsymbol{r}') - \rho_p(\boldsymbol{r}, \boldsymbol{r}').$$

Using $\rho_t(\boldsymbol{r}, \boldsymbol{r}')$ and $\boldsymbol{s}_t(\boldsymbol{r}, \boldsymbol{r}')$ as elementary building blocks, we can then construct all the local densities up to second order in derivatives. In practice, this means acting on both $\rho_t(\boldsymbol{r}, \boldsymbol{r}')$ and $\boldsymbol{s}_t(\boldsymbol{r}, \boldsymbol{r}')$ with the operators $(\nabla - \nabla')/2i$ and setting $\boldsymbol{r} = \boldsymbol{r}'$ after this operation. Going through this process up to second order in derivatives yields the following set of densities [86–88]

$$\text{local density:}\quad \rho_t(\boldsymbol{r}) \equiv \rho_t(\boldsymbol{r}, \boldsymbol{r}')|_{r=r'} \tag{1.22a}$$

$$\text{spin density:}\quad \boldsymbol{s}_t(\boldsymbol{r}) \equiv \boldsymbol{s}_t(\boldsymbol{r}, \boldsymbol{r}')|_{r=r'} \tag{1.22b}$$

$$\text{kinetic density:}\quad \tau_t(\boldsymbol{r}) \equiv \nabla \cdot \nabla' \rho_t(\boldsymbol{r}, \boldsymbol{r}')|_{r=r'} \tag{1.22c}$$

$$\text{spin kinetic density:}\quad T_{t,\mu}(\boldsymbol{r}) \equiv \nabla \cdot \nabla' s_{t,\mu}(\boldsymbol{r}, \boldsymbol{r}')|_{r=r'} \tag{1.22d}$$

$$\text{current density:}\quad \boldsymbol{j}_t(\boldsymbol{r}) \equiv -\frac{\mathrm{i}}{2}(\nabla - \nabla')\rho_t(\boldsymbol{r}, \boldsymbol{r}')|_{r=r'} \tag{1.22e}$$

$$\text{spin current density:}\quad J_{t,\mu\nu}(\boldsymbol{r}) \equiv -\frac{\mathrm{i}}{2}\left(\nabla_\mu - \nabla'_\mu\right) s_{t,\nu}(\boldsymbol{r}, \boldsymbol{r}')|_{r=r'} \tag{1.22f}$$

$$\text{tensor - kinetic spin density:}\quad F_{t,\mu}(\boldsymbol{r}) \equiv \frac{1}{2}\sum_\nu \left(\nabla_\mu \nabla'_\nu + \nabla'_\mu \nabla_\nu\right) s_{t,\nu}(\boldsymbol{r}, \boldsymbol{r}')|_{r=r'} \tag{1.22g}$$

for $t = 0, 1$. While the non-local density matrices $\rho_t(\boldsymbol{r}, \boldsymbol{r}')$ and $s_{t,\nu}(\boldsymbol{r}, \boldsymbol{r}')$ transform in a complicated manner under time reversal, the local densities $\rho_t(\boldsymbol{r})$, $\tau_t(\boldsymbol{r})$ and $J_{t,\mu\nu}(\boldsymbol{r})$ are invariant under such a transformation, whereas $\boldsymbol{s}_t(\boldsymbol{r})$, $T_{t,\mu}(\boldsymbol{r})$, $jt(\boldsymbol{r})$ and $F_{t,\mu}(\boldsymbol{r})$ change their sign. This allows for an easy identification of the parts of the functional which are always present and those which are only present when the reference state is not time-reversal invariant. We will discuss in chapter 3 the case of nuclei characterized by an odd number of particles (protons, neutrons or both), for which the reference state is not invariant under time reversal.

Given the set of densities (1.22), we can build an energy density by forming couplings between them in such a way that we find all possible bilinear combinations of local densities and currents with up to two gradients that are invariant under rotations in space and isospin, parity, time reversal, translations and Galilean invariance. The result of this procedure yields what might be called the 'canonical' form of the Skyrme EDF, which reads [87]

$$\begin{aligned}\mathcal{E}_{\text{pot}}[\rho] = \int \mathrm{d}^3 r \sum_{t=0,1} \Bigg\{ & C_t^{\rho\rho}\, \rho_t \rho_t + C_t^{\rho\Delta\rho}\, \rho_t \Delta \rho_t + C_t^{\rho\tau}(\rho_t \tau_t - \boldsymbol{J}_t \cdot \boldsymbol{J}_t) \\ & + C_t^{ss}\, \boldsymbol{s}_t \cdot \boldsymbol{s}_t + C_t^{s\Delta s}\, \boldsymbol{s}_t \cdot \Delta \boldsymbol{s}_t + C_t^{\rho\nabla J}\left(\rho_t \nabla \cdot \boldsymbol{J}_t + \boldsymbol{j}_t \cdot \nabla \times \boldsymbol{s}_t\right) \\ & + C_t^{s\nabla s}(\nabla \cdot \boldsymbol{s}_t)^2 + C_t^{JJ}\left(\sum_{\mu\nu} J^t_{\mu\nu} J_{t,\mu\nu} - \boldsymbol{s}_t \cdot \boldsymbol{T}_t\right) \\ & + C_t^{J\bar{J}}\left[\left(\sum_{\mu} J_{t,\mu\mu}\right)\left(\sum_{\mu} J_{t,\mu\mu}\right) + \sum_{\mu\nu} J_{t,\mu\nu} J_{t,\nu\mu} - 2\boldsymbol{s}_t \cdot \boldsymbol{F}_t\right]\Bigg\}. \end{aligned} \tag{1.23}$$

In this expression, the coefficients $C_t^{uu'}$ are the coupling constant for terms containing either isoscalar ($t = 0$) or isovector ($t = 1$) local densities and currents. The vector density $\boldsymbol{J}_t(\boldsymbol{r}) = \sum_{\mu\nu\kappa} \epsilon_{\mu\nu\kappa} J_{t,\mu\nu}(\boldsymbol{r})\, \boldsymbol{e}_\kappa$ entering the spin–orbit term in equation (1.23) is the rank-one contraction of the spin-current pseudotensor density. The requirement that the EDF satisfies Galilean invariance imposes that some specific combinations of terms in equation (1.23) share the same coupling constant [86, 88, 89]. All of the functional's coupling constants might be chosen to be density-dependent, but for the vast majority of parameterizations, only the $C_t^{\rho\rho}$ are chosen to be. The most widely used choice can be expressed in the generic form [75, 90]

$$C_t^{\rho\rho}[\rho_0] = C_t^{\rho\rho}[0] + C_{tD}^{\rho\rho}\left(\frac{\rho_0(\boldsymbol{r})}{\rho_{\text{sat}}}\right)^{\alpha} \tag{1.24a}$$

$$C_t^{ss}[\rho_0] = C_t^{ss}[0] + C_{tD}^{ss}\left(\frac{\rho_0(\boldsymbol{r})}{\rho_{\text{sat}}}\right)^{\alpha}, \tag{1.24b}$$

where ρ_{sat} is the saturation density of homogeneous infinite nuclear matter. The value of $C_{tD}^{\rho\rho}$ can be related to the full, density-dependent coupling constant $C_t^{\rho\rho}[\rho_0]$ at special values for the density

$$C_{tD}^{\rho\rho} = C_t^{\rho\rho}[\rho_{\text{sat}}] - C_t^{\rho\rho}[0].$$

The same remark applies to C_{tD}^{ss}. In principle, coupling constants could also depend on other densities than the isoscalar local density $\rho_0(\boldsymbol{r})$, but the functional form has to be such that the coupling constant is a scalar in position, spin and isospin space.

Skyrme pseudo-potential

The form (1.23) of the Skyrme functional is often motivated as the HF expectation value of a pseudo-potential

$$\hat{v}_{\text{Skyrme}}(\boldsymbol{R}, \boldsymbol{r}, \overrightarrow{k}, \overleftarrow{k}') = [\hat{v}_{\text{centr}} + \hat{v}_{\text{LS}} + \hat{v}_{\text{tens}}](\boldsymbol{R}, \boldsymbol{r}, \overrightarrow{k}, \overleftarrow{k}'), \tag{1.25}$$

where the central contribution reads

$$\hat{v}_{\text{centr}}\left(\boldsymbol{R},\,\boldsymbol{r},\,\overrightarrow{k},\,\overleftarrow{k}'\right) = t_0\left(1 + x_0\hat{P}_\sigma\right)\delta(\boldsymbol{r}) \\ + \frac{1}{2}t_1\left(1 + x_1\hat{P}_\sigma\right)\left[\overleftarrow{k}'^2\delta(\boldsymbol{r}) + \delta(\boldsymbol{r})\overrightarrow{k}^2\right] \\ + t_2\left(1 + x_2\hat{P}_\sigma\right)\overleftarrow{k}' \cdot \delta(\boldsymbol{r})\overrightarrow{k} \\ + \frac{1}{6}t_3\left(1 + x_3\hat{P}_\sigma\right)\rho^\alpha(\boldsymbol{R})\,\delta(\boldsymbol{r}), \tag{1.26}$$

the spin–orbit

$$\hat{v}_{\text{LS}}(\boldsymbol{r},\,\overrightarrow{k},\,\overleftarrow{k}') = \mathrm{i}W_0[\boldsymbol{\sigma}_1 + \boldsymbol{\sigma}_2] \cdot \overleftarrow{k}' \times \delta(\boldsymbol{r})\overrightarrow{k} \tag{1.27}$$

and the tensor

$$\hat{v}_{\text{tens}}(\boldsymbol{r},\,\overrightarrow{k},\,\overleftarrow{k}') = \frac{1}{2}t_e\left\{\left[3\left(\boldsymbol{\sigma}_1 \cdot \overleftarrow{k}'\right)\left(\boldsymbol{\sigma}_2 \cdot \overleftarrow{k}'\right) - (\boldsymbol{\sigma}_1 \cdot \boldsymbol{\sigma}_2)\overleftarrow{k}'^2\right]\delta(\boldsymbol{r}) \right. \\ \left. + \delta(\boldsymbol{r})\left[3\left(\boldsymbol{\sigma}_1 \cdot \overrightarrow{k}\right)\left(\boldsymbol{\sigma}_2 \cdot \overrightarrow{k}\right) - (\boldsymbol{\sigma}_1 \cdot \boldsymbol{\sigma}_2)\overrightarrow{k}^2\right]\right\} \\ + t_o\left[\frac{3}{2}\left(\boldsymbol{\sigma}_1 \cdot \overleftarrow{k}'\right)\delta(\boldsymbol{r})\left(\boldsymbol{\sigma}_2 \cdot \overrightarrow{k}\right) + \frac{3}{2}\left(\boldsymbol{\sigma}_2 \cdot \overleftarrow{k}'\right)\delta(\boldsymbol{r})\left(\boldsymbol{\sigma}_1 \cdot \overrightarrow{k}\right) \right. \\ \left. - (\boldsymbol{\sigma}_1 \cdot \boldsymbol{\sigma}_2)\overleftarrow{k}' \cdot \delta(\boldsymbol{r})\overrightarrow{k}\right], \tag{1.28}$$

where $\overrightarrow{k} = -\frac{\mathrm{i}}{2}(\nabla_1 - \nabla_2)$ stands for the relative momentum operator acting to the right while $\overleftarrow{k}'$ is the Hermitian conjugate of this operator acting to the left. $\hat{P}_\sigma = \frac{1}{2}(1 + \boldsymbol{\sigma}_1 \cdot \boldsymbol{\sigma}_2)$ is the spin-exchange operator, $\boldsymbol{r} = \boldsymbol{r}_1 - \boldsymbol{r}_2$ the relative distance between the two particles and $\boldsymbol{R} = \frac{1}{2}(\boldsymbol{r}_1 + \boldsymbol{r}_2)$ the position of their center-of-mass. The coefficients $(t_i,\,x_i)$, W_0, t_e, t_o and α are the free parameters of the pseudo-potential to be fitted to relevant data; see chapter 9.

As will be discussed in more detail in section 9.1.1, there is in principle a one-to-one correspondence between the (t,x) set of parameters and the coupling constants $C_t^{uu'}$. However, there are only relatively few parameterizations of the Skyrme functional for which this one-to-one relation is enforced. Among the reasons to break their direct link are computational efficiency and larger flexibility to adjust the functional to data [91]. In particular, for many interactions specific time-odd terms have to be dropped [90, 92, 93] in order to avoid spurious finite-size spin instabilities [94].

Because of its simplicity, the Skyrme functional has been very popular for many types of applications. Over 200 parametrizations have been constructed over the years. Only a few of these are frequently used, however. In fact, many were adjusted with particular emphasis on a specific observable and are not meant to be universal.

There are also many 'families' of fits that explore the influence of systematic variations of specific features of the parametrizations on their predictive power. Some noteworthy milestones are:

1. The early and still sometimes used parameterization SIII [95] was adjusted with a linear density dependence ($\alpha = 1$) that mimics a gradient-less 3N force—at least as far as the time-even part of the functional is concerned. This linear density dependence, however, yields too large an incompressibility K^{NM} of nuclear matter.
2. Motivated by the Fermi-momentum dependence of the G-matrix of Brueckner–Hartree–Fock (BHF) theory, the parameterization SKa [96] was constructed with a fractional density dependence $\alpha = 1/3$, which provides the flexibility to obtain a more realistic value for K_∞.
3. A still widely used early fit with $\alpha = 1/6$ is SkM* [97], which was adjusted with particular emphasis on its deformation properties and is very often employed in fission applications [98].
4. The SLy family of fits [99] with a density dependence of the form $\alpha = 1/6$. The parameterizations SLy4, SLy5, SLy6 and SLy7 differ in their choice for the center-of-mass correction and whether the coupling constants C_t^{JJ} are set to zero or not.
5. For the parameterizations SkI3 and SkI4 [100], the usual interdependence of the isoscalar and isovector coupling constants $C_0^{\rho\nabla J}$ and $C_1^{\rho\nabla J}$ of the spin–orbit term is relaxed, meaning that the correspondence between the spin–orbit terms in the functional and the pseudo-potential $\hat{v}_{\mathrm{LS}}$ is cut.
6. The ongoing construction of fits in the BSk series [101–106] aims at a global description of binding energies, radii, fission barrier heights and an ever increasing number of observables of astrophysical interest. Several extensions of the Skyrme EDF have been introduced over the years, as have a number of correction terms that model the contribution of various correlation effects in the nuclear wave function to the total binding energy.
7. The Tij family of fits [107] include the contribution from the tensor force $\hat{v}_{\mathrm{tens}}$, with the corresponding coupling constants t_e and t_o being systematically varied over large intervals. For all other parameterizations mentioned here, t_e and t_o or the possibly corresponding coupling constants $C_t^{\nabla s \nabla s}$ and $C_t^{J\bar{J}}$ of the EDF are zero; see also the discussion in section 9.1.1.
8. The recent family of UNEDF parameterizations [108–110] at the deformed HFB level. The three parametrizations differ from one another by the selection of data considered in the fit protocol and were adjusted using the larger freedom of functionals without keeping a link to an underlying pseudo-potential.

In spite of their differences, all of these parameterizations correspond to a functional of the form (1.23). The only difference in the functional form of the Skyrme EDF is mostly whether specific coupling constants are set to 0 or not. Many extensions of this standard form have been proposed over the years, but none has become frequently used so far. Examples are fits with a different isospin dependence of the density-dependent term [111]; two density dependencies with different exponents [71]; and additional density dependencies of the coupling constants of the gradient terms [103, 112, 113]. All of these lead to a functional with additional terms that are build out of the same local densities as the standard form. When adding also higher-order gradient terms of fourth order (N2LO) and sixth order (N3LO), however, one has to consider additional local densities with more complicated tensor structure and containing additional derivatives [114–117].

While equations (1.23) and (1.25) are the standard form of the Skyrme pseudo-potential as used in nuclear EDF methods, Skyrme's original proposal for an effective interaction [118, 119] used genuine three- and four-body forces instead of density-dependent two-body terms. The earliest fits made, including SIII, could still be interpreted as either a combination of density-independent two- and three-body potentials or as a density-dependent two-body pseudo-potential. Both types differ in the isospin dependence of the time-odd terms, which is the reason why the interpretation of SIII as density-dependent two-body interaction prevails. When the density-dependent term is treated as a three-body force instead, it leads to a spin instability [120]. Still, some authors have combined density-dependent two-body terms and three-body terms with gradients, see for example [121]. Motivated by problems faced when using non-pseudo-potential-based functionals with density dependencies in multi-reference applications, it was only recently that new efforts to construct density-independent Skyrme interactions have been made again that led to the systematic construction of three-body and four-body terms with and without gradients [65, 66]. For these, also the pairing energy is calculated from the same underlying pseudo-potential, something rarely done for standard parameterizations. The SkP parametrization of the Skyrme EDF is the only one for which the same Skyrme pseudo-potential was used both in the p.h. and in the p.p. channels, and adjusted accordingly [122]. In almost all applications of the Skyrme functional, the pairing functional is constructed and parameterized independently; see section 1.3 below.

Conventions for coupling constants

In the literature one can find many different notations for the Skyrme EDF and, consequently, its coupling constants. Many authors choose a representation that expresses the energy in terms of proton and neutron densities instead of the isoscalar and isovector densities used in equation (1.23), but without a consensus on a universal form. Examples of different possible choices can be found [66, 90, 93]. The arguably largest variety of notations has been introduced for extensions of the spin–orbit part of the functional. When calculated from the two-body spin–orbit pseudo-potential $\hat{v}_{\mathrm{LS}}$ of equation (1.25), the isoscalar and isovector coupling constants of the Skyrme EDF have a fixed ratio of three

$$C_0^{\rho \nabla J} = -\frac{3}{4} W_0 \qquad C_1^{\rho \nabla J} = -\frac{1}{4} W_0. \tag{1.29}$$

It has been argued that relaxing this constraint and treating $C_0^{\rho\nabla J}$ and $C_1^{\rho\nabla J}$ as independent parameters can be used to improve the phenomenological description of various observables. Many modern Skyrme parameterizations use such an extension of the spin–orbit interaction that cannot be mapped on the expectation value of the pseudo-potential (1.25), but without consensus on a common notation. The SkI3 and SkI4 parameterizations are defined with parameters b_4 and b_4',

$$C_0^{\rho\nabla J} = -b_4 - \frac{1}{2}\,b_4' \qquad C_0^{\rho\nabla J} = -\frac{1}{2}\,b_4',$$

while SLy10 of [99] is defined through parameters W_1 and W_2,

$$C_0^{\rho\nabla J} = -\frac{1}{2}\,W_1 - \frac{1}{4}\,W_2 \qquad C_0^{\rho\nabla J} = -\frac{1}{4}\,W_2.$$

Some numerical codes [123] use a convention with $b_9 = -W_1$ and $b_9' = -W_2$ instead. The family of SV parameterizations of [124], uses instead parameters t_4 and b_4' [125]

$$C_0^{\rho\nabla J} = -\frac{1}{2}\,t_4 - \frac{1}{2}\,b_4' \qquad C_0^{\rho\nabla J} = -\frac{1}{2}\,b_4'.$$

Only for a few parameterizations is the extended spin–orbit interaction defined directly through the values of $C_0^{\rho\nabla J}$ and $C_1^{\rho\nabla J}$, examples being the UNEDF parameterizations [108–110].

1.2.2 Gogny functional

The Gogny functional [126, 127] was motivated by the will to treat in a unified manner p.h. and p.p. correlations at the HFB level. In this regard, its underlying strategy involves the parametrization of the p.h. and p.p. effective vertexes from the same finite-range two-body phenomenological pseudo-potential $\hat{v}_{\text{Gogny}}(\boldsymbol{R}, \boldsymbol{r}, \overrightarrow{k}, \overleftarrow{k}')$:

$$\begin{aligned} v_{D1X}(\boldsymbol{R}, \boldsymbol{r}, \overrightarrow{k}, \overleftarrow{k}') = & \sum_{i=1}^{2}[W_i + B_i\hat{P}_\sigma - H_i\hat{P}_\tau - M_i\hat{P}_\sigma\hat{P}_\tau]e^{-r^2/\mu_i^2} \\ & + t_0(1 + x_0\hat{P}_\sigma)\rho^\alpha(\boldsymbol{R})\delta(\boldsymbol{r}) \\ & + \mathrm{i}W_{LS}(\boldsymbol{\sigma}_1 + \boldsymbol{\sigma}_2)\overleftarrow{k}' \times \delta(\boldsymbol{r})\overrightarrow{k}. \end{aligned} \tag{1.30}$$

The Gogny pseudo-potential thus has a central term with a finite-range contribution involving the sum of two Gaussians with different ranges (inspired by the pioneering work of Brink and Boeker [128]) and the same zero-range spin–orbit and density-dependent terms as the Skyrme interaction. The finite-range feature of the central term introduces a natural cut-off preventing UV divergences when describing pairing effects [129]; see also the discussion in section 4.3. Setting the parameter $x_0 = 1$ removes the contribution of the density-dependent term in the singlet-even

$(S, T) = (0, 1)$ channel. The spin–orbit contribution (often omitted in the p.p. channel in SR-EDF calculations) and the Coulomb one are patterned after the Skyrme vertex. The finite-range property of the Gogny vertex leads to a non-local EDF. Adapting expressions given in [69] for finite-range terms and in [86, 87] for the density dependence and spin–orbit parts, the p.h. part of the Gogny EDF corresponding to equation (1.30) reads

$$
\begin{aligned}
&\mathcal{E}_{\text{pot}}[\rho] \\
&= \int d^3\boldsymbol{r} \int d^3\boldsymbol{r}' \sum_{t=0,1} \sum_{i=1}^{2} e^{-(\boldsymbol{r}-\boldsymbol{r}')^2/\mu_i^2} \big[A_{it}^{\rho\rho}\, \rho_t(\boldsymbol{r})\, \rho_t(\boldsymbol{r}') + A_{it}^{ss}\, \boldsymbol{s}_t(\boldsymbol{r}) \cdot \boldsymbol{s}_t(\boldsymbol{r}') \\
&+ B_{it}^{\rho\rho}\, \rho_t(\boldsymbol{r}, \boldsymbol{r}')\, \rho_t(\boldsymbol{r}', \boldsymbol{r}) + B_{it}^{ss}\, \boldsymbol{s}_t(\boldsymbol{r}, \boldsymbol{r}') \cdot \boldsymbol{s}_t(\boldsymbol{r}', \boldsymbol{r}) \big] \\
&+ \int d^3\boldsymbol{r} \sum_{t=0,1} [C_t^{\rho\rho}[\rho_0]\rho_t\rho_t + C_t^{ss}[\rho_0]\boldsymbol{s}_t \cdot \boldsymbol{s}_t + C_t^{\rho\nabla J}(\rho_t \nabla \cdot \boldsymbol{J}_t + \boldsymbol{J}_t \cdot \nabla \times \boldsymbol{s}_t)],
\end{aligned}
\tag{1.31}
$$

where the coupling constants of the finite-range part of the two-body interaction combine to

$$A_{i0}^{\rho\rho} = \frac{1}{2} W_i + \frac{1}{4} B_i - \frac{1}{4} H_i - \frac{1}{8} M_i \tag{1.32a}$$

$$A_{i1}^{\rho\rho} = -\frac{1}{4} H_i - \frac{1}{8} M_i \tag{1.32b}$$

$$A_{i0}^{ss} = \frac{1}{4} B_i - \frac{1}{8} M_i \tag{1.32c}$$

$$A_{i1}^{ss} = -\frac{1}{8} M_i \tag{1.32d}$$

$$B_{i0}^{\rho\rho} = -\frac{1}{8} W_i - \frac{1}{4} B_i + \frac{1}{4} H_i + \frac{1}{2} M_i \tag{1.32e}$$

$$B_{i1}^{\rho\rho} = -\frac{1}{8} W_i - \frac{1}{4} B_i \tag{1.32f}$$

$$B_{i0}^{ss} = -\frac{1}{8} W_i + \frac{1}{4} H_i \tag{1.32g}$$

$$B_{i1}^{ss} = -\frac{1}{8} W_i. \tag{1.32h}$$

The local densities $\rho_t(\boldsymbol{r})$, $\boldsymbol{s}_t(\boldsymbol{r})$, $\boldsymbol{j}t(\boldsymbol{r})$ and $\boldsymbol{J}_t(\boldsymbol{r})$ are the same as those entering the Skyrme EDF (1.23). The exchange term of the finite-range central interaction, however, contains the non-local densities $\rho_t(\boldsymbol{r}, \boldsymbol{r}')$ and $\boldsymbol{s}_t(\boldsymbol{r}, \boldsymbol{r}')$. Finite range, local potentials always lead to functionals of the non-local density when exchange terms are properly included. Only for zero-range potentials can the functional be expressed exclusively in terms of local densities; see also the example of the Coulomb potential discussed later in section 1.3.2.

As for the Skyrme EDF, the terms containing local densities can be divided into time-even and time-odd ones. For the terms containing non-local densities, however, this is not possible as the full density matrices do not have a definite behavior under time reversal. Indeed, under the action of the time-reversal operator $\hat{T}$, the non-local densities $\rho(\boldsymbol{r}, \boldsymbol{r}')$ and $\boldsymbol{s}(\boldsymbol{r}, \boldsymbol{r}')$ transform as [86]

$$\begin{aligned} \rho^T(\boldsymbol{r}, \boldsymbol{r}') &= \rho^*(\boldsymbol{r}, \boldsymbol{r}') = \rho(\boldsymbol{r}', \boldsymbol{r}) \\ \boldsymbol{s}^T(\boldsymbol{r}, \boldsymbol{r}') &= -\boldsymbol{s}^*(\boldsymbol{r}, \boldsymbol{r}') = -\boldsymbol{s}(\boldsymbol{r}', \boldsymbol{r}). \end{aligned}$$

Contrary to their local version, they are not invariant or simply change sign. An alternative and more compact notation of the finite-range terms in the EDF will be introduced in section 8.1.2.

In presenting equation (1.31), we have emphasized that only the density-independent, central part of the Gogny EDF is different from the Skyrme one. In its standard formulation, both the density-dependent and spin–orbit contributions are formally identical. However, as can already be seen in equation (1.30), it is customary to use different notations: the parameters of the density-dependent term are noted (t_0, x_0) instead of (t_3, x_3) for the Skyrme pseudo-potential, and the definitions are such that $t_0 \equiv 6t_3$ and $x_0 \equiv x_3 = 1$. Since the spin–orbit contribution to the functional originates from an actual pseudo-potential, the coupling constants $C_t^{\rho\nabla J}$ are related to a single spin–orbit strength noted W_{LS} (instead of W_0 for Skyrme) according to equation (1.29).

The minimization of this EDF involves solving integro-differential equations that are far more complex than in the Skyrme case. In practice, the Gogny EDF is almost always implemented in HFB solvers that solve the HFB equation by expanding the solution on the harmonic oscillator (HO) basis. This basis is well-adapted to computations with Gaussian potentials thanks to various analytical properties, such as the Talmi–Moshinsky transformation; see section 8.1.1 for a quick reminder. We also show in detail in section 8.1.2, how to compute matrix elements for a sum of Gaussians for the most general case of a 3D Cartesian HO basis. As a result of interesting progress in algorithms and computing capabilities, the computing time of, say, full HFB calculations with finite-range pseudo-potentials—or equivalently EDF of the non-local density—has dramatically reduced, making it competitive with quasi-local EDF such as Skyrme [130–132].

There are considerably fewer parametrizations of the Gogny EDF, partly because of the larger computational needs. The most commonly used are listed below:

1. D1S [127] is a parametrization that reproduces globally a variety of nuclear properties well, with the exception of nuclear binding energies for which the root-mean-square (rms) deviation with experimental data is sometimes large. It has also been calibrated on the fission barrier of ^{240}Pu and is often employed in fission calculations.

2. D1N [133] is an adjustment of D1S for which the equation of state (EoS) of neutron matter is used as a constraint. This allows in particular for correcting the strong drift of mass residuals along isotopic chains found for D1S.
3. D1M [134] is a fit on nuclear binding energies of 2149 known nuclei which also attempts to conserve the global predictive power of D1S and D1N. The optimization strategy of D1M is different from the previous ones and allows the incorporation of dynamical collective correlations.

Work is currently under way to generalize the Gogny pseudo-potential. One direction is to introduce a finite-range in the density-dependent term,

$$\hat{v}(\boldsymbol{R}, \boldsymbol{r}, \overrightarrow{k}, \overleftarrow{k}') = \sum_{i=1}^{2}[W_i + B_i\hat{P}_\sigma - H_i\hat{P}_\tau - M_i\hat{P}_\sigma\hat{P}_\tau]e^{-r^2/\mu_i^2}$$
$$+ [W_3 + B_3\hat{P}_\sigma - H_3\hat{P}_\tau - M_3\hat{P}_\sigma\hat{P}_\tau]e^{-r^2/\mu_3^2}\left(\frac{\rho^\alpha(\boldsymbol{r}_1) + \rho^\alpha(\boldsymbol{r}_2)}{2}\right)$$
$$+ \mathrm{i}W_{LS}(\boldsymbol{\sigma}_1 + \boldsymbol{\sigma}_2)\overleftarrow{k}' \times \delta(\boldsymbol{r})\overrightarrow{k}.$$

This leads to the D2 parametrization of [135]. The impact of adding a finite-range tensor force to the Gogny interactions has been explored in [136, 137]. Another direction consists in merging the elementary building blocks of Skyrme and Gogny pseudo-potentials by building a pseudo-potential as an expansion in terms of relative momenta ('à la Skyrme') while taking Gaussian form factors instead of Dirac distributions ('à la Gogny') [64, 67, 69].

1.2.3 Phenomenological functionals

Both the Skyrme and Gogny energy functionals have their historical root in some specific form of an effective NN pseudo-potential. Even though in recent years work has often focused on breaking this connection between effective pseudo-potential and functional, the mathematical structure of the functional, equation (1.23) for Skyrme and equation (1.31) for Gogny, still largely betrays their origin. In this section, we will briefly review energy functionals that are built from the ground up without any reference to some NN potential. By construction, these functionals are 'phenomenological', in the sense that they model, or parametrize, directly the effect of in-medium nuclear forces as a functional form, rather than mathematically derive the form of the functional from a potential.

The notion of a phenomenological model is often contentious and rarely objective. Here, we may draw an analogy with independent-particle models based on the parametrization of the nuclear mean field as a Nilsson, Woods–Saxon or folded-Yukawa potential [138]. These models are more phenomenological than the EDF approach, where the same mean-field potential is mathematically derived from an underlying energy functional. Importantly, in addition to not being derived from NN potentials, most of the energy functionals discussed in this section have non-trivial density dependencies. For these reasons, it is not at all clear how one could apply them to MR-EDF calculations, which are rigorously defined only when the functional derives *strictly* from a true potential; see section 3.2.

There are essentially three main families of phenomenological energy functionals: the functionals built and used by Fayans and collaborators [139–142]; the Barcelona–Catana–Paris–Madrid (BCPM) functional [143, 144] which originates from an early work by Baldo and collaborators [145]; and the recent SeaLL functionals from a Seattle–Livermore collaboration [146]. All three families share a very similar philosophy, which we may summarize as follows: (i) directly parametrize the nuclear EoS as a series of powers of the density, (ii) add corrective terms to account for finite-size and purely quantum effects such as the spin–orbit and (iii) add terms accounting for the Coulomb potential and pairing corrections. The energy functional will thus take the generic form

$$\mathcal{E} = \mathcal{E}_\tau + \mathcal{E}_{\text{hom}} + \mathcal{E}_{\text{surf}} + \mathcal{E}_{\text{Coul}} + \mathcal{E}_{\text{pair}},$$

with $\mathcal{E}_\tau$ the kinetic energy term, $\mathcal{E}_{\text{hom}}$ the part characterizing homogeneous infinite nuclear matter (INM), $\mathcal{E}_{\text{surf}}$ the surface contribution (spin–orbit effects are included in it), and $\mathcal{E}_{\text{Coul}}$ and $\mathcal{E}_{\text{pair}}$ the Coulomb and pairing terms, respectively.

Both the Fayans and SeaLL functionals use only the independent-particle kinetic energy [139, 146]

$$\mathcal{E}_\tau = \sum_{q=n,\,p} \frac{\hbar^2}{2m}\tau_q,$$

with τ_q the kinetic energy density (1.22) for particle q. The BCPM also introduces an effective mass m_q^* fitted to a BHF calculation of INM [143],

$$\frac{m_q}{m_q^*} = 1 + \frac{m_q}{\hbar^2 k_{\text{F},q}} \frac{dU_q(k)}{dk}\bigg|_{k=k_{\text{F},q}},$$

where $U_q(k)$ is the s.p. potential in momentum space. In practice, the effective masses are also parametrized in terms of the asymmetry β

$$\beta = \frac{\rho_n - \rho_p}{\rho_n + \rho_p} \qquad \rho = \rho_n + \rho_p \tag{1.33}$$

as

$$\frac{m_q}{m_q^*} = a_0(\rho) \pm a_1(\rho)\beta,$$

where the superscript is + for neutrons and − for protons. This effective mass enters the kinetic energy which becomes

$$\mathcal{E}_\tau = \sum_{q=n,p} \frac{\hbar^2}{2m_q^*}\tau_q.$$

The homogeneous part $\mathcal{E}_{\text{hom}}$ reflects bulk, or volume, nuclear properties. In practice, this term is a parametrization of the nuclear EoS of symmetric nuclear matter (SNM) or asymmetric nuclear matter (ANM). In the case of the Fayans functional, this term is written as [139]

$$\mathcal{E}_{\text{hom}} = \frac{2}{3}\varepsilon_F \rho_0 \left(a_+ x_+^2 f_+ + a_- x_-^2 f_- \right)$$

with

$$x_\pm = \frac{1}{2} \frac{\rho_n \pm \rho_p}{\rho_c} \qquad f_\pm = \frac{1 - h_2 x_+}{1 + h_2 x_+}.$$

For both the BCPM and the SeaLL functionals, the nuclear EoS is expanded in terms of powers of the asymmetry parameter (1.33) with coefficients that are dependent on the isoscalar density ρ. We thus write [143–146]

$$\mathcal{E}_{\text{hom}} = \sum_{j=0}^{n_{\max}} \mathcal{E}_j(\rho) \beta^{2j}.$$

The number of coefficients and their particular dependence on the density is what distinguishes these functionals. For the BCPM functional, the expansion stops at $n = 1$ and the coefficients read

$$\mathcal{E}_0(\rho) = \sum_{n=1}^{5} a_n \left(\frac{\rho}{\rho_0} \right)^n \qquad \mathcal{E}_1(\rho) = \sum_{n=1}^{5} b_n \left(\frac{\rho}{\rho_0 n} \right)^n - \mathcal{E}_0(\rho).$$

Note that the term

$$\sum_q \frac{\hbar^2}{2m_q} \left(\frac{m_q}{m_q^*} - 1 \right) \tau_q^\infty$$

with

$$\tau_q^\infty = \frac{3}{5} (3\pi^2)^{2/3} \rho_q^{5/3}(\boldsymbol{r})$$

should be subtracted to the homogeneous term. This is done to ensure that the introduction of the effective mass does not break the EoS of the functional [143]. In the case of the SeaLL functional, the expansion goes up to $n = 2$ with the following generic form for the coefficients,

$$\mathcal{E}_j(\rho) = a_j \rho^{5/3} + b_j \rho^2 + c_j \rho^{7/3}.$$

Finally, the surface term is designed to represent finite-size effects in nuclei. As mentioned above, this includes both 'geometrical' effects which, for example, the $C_t^{\rho\Delta\rho}$ and $C_t^{\rho\tau}$ terms of the Skyrme functionals would simulate, and the spin–orbit term. The latter is identical to one used with both Skyrme and Gogny. In the BCPM functional, geometrical effects are modeled through the use of a finite-range potential [143–145],

$$\mathcal{E}_{\text{surf}} = \sum_{qq'}\Big(\int d^3r' \rho_q(\boldsymbol{r}) V_{qq'}(\boldsymbol{r} - \boldsymbol{r}')\rho_{q'}(\boldsymbol{r}') - \gamma_{qq'}\rho_q(\boldsymbol{r})\rho_{q'}(\boldsymbol{r}') + \mathcal{E}_{\text{SO}}\Big),$$

with $V_{qq'}$ taken in the form of a simple Gaussian. In the case of the SeaLL functional, surface terms are expressed instead as a sum of gradient corrections in the spirit of atomic density functional theory (DFT) [146]:

$$\mathcal{E}_{\text{surf}} = \eta_s \sum_{q=n,p} \frac{\hbar^2}{2m}|\nabla\rho_q|^2 + \mathcal{E}_{SO}.$$

The paper on SeaLL [146] also contains suggestions to encode an isovector dependence in the gradient term in the form

$$\mathcal{E}_{\text{grad.}} = \eta_0\frac{\hbar^2}{2m}|\nabla\rho_n + \nabla\rho_p|^2 + \eta_1\frac{\hbar^2}{2m}|\nabla\rho_n - \nabla\rho_p|^2.$$

One may argue that the modeling of surface effects for the BCPM functional somewhat resembles the Gogny pseudo-potential, while it is closer to what a Skyrme pseudo-potential would give in the case of the SeaLL functional.

1.3 Pairing and Coulomb functionals

The non-relativistic Skyrme and Gogny functionals (as well of the different versions of covariant functionals discussed in chapter 2) describe the term $\mathcal{E}_{\text{pot}}[\rho]$ in equation (1.19)—although the Gogny pseudo-potential is also used to compute the pairing contribution, see section 1.3.1 below. This term is the contribution of the p.h. channel (i.e. the mean field) to the total energy and gives by far the largest contribution to the latter. Since it should be designed to encode as much as possible the many-body correlations induced by in-medium nuclear forces, it is also a term with a large uncertainty and, for this reason, most of the effort in designing new EDFs is spent toward improving $\mathcal{E}_{\text{pot}}[\rho]$; see among others [108–110, 135, 139–143, 145, 146] for various recent efforts.

However, the other contributions to the total energy, in particular the pairing $\mathcal{E}_{\text{pair}}$ and Coulomb $\mathcal{E}_{\text{Cou}}$ contributions, play also an important role in determining properties of atomic nuclei. In a strict Hamiltonian picture of the EDF theory, the pairing functional (p.p. channel) should be derived from the same pseudo-Hamiltonian as the nuclear potential term $\mathcal{E}_{\text{pot}}[\rho]$ (p.h. channel). There are pragmatic arguments why this condition may be relaxed and a different pseudo-Hamiltonian could be chosen in the p.p. channel at least at the SR-EDF level [147]—although this choice has dire consequences when going to the MR-EDF level [62, 148]. In particular, as we will discuss in section 9.2 when addressing the problem of calibration of EDF, it is not always straightforward to identify physical observables that could constrain every term in the pairing functional. If the latter does not have a simple structure, this task becomes particularly difficult. In contrast, the Coulomb potential is well-known, but its contributions to many-body effects and/or the pairing channel are often neglected.

1.3.1 Pairing functionals

The p.p. channel of the nuclear EDF is only relevant in full HFB or HF+Bardeen–Cooper–Schrieffer (BCS) calculations, that is, where the order parameter $g = ||\kappa||$ is non-zero, see discussion in section 3.1.1, and the ground state has the form (3.3)—which makes this spontaneous symmetry -breaking possible in the first place. In that case, one must specify a functional form for $\mathcal{E}_{\text{pair}}[\rho, \kappa, \kappa^*]$. There are essentially two strategies:

- If the EDF is derived strictly from an effective pseudo-potential, one option is to simply take the pairing EDF from the direct application of the Wick theorem. In other words, we compute the expectation value of $\hat{V}$ on the HFB ground state with the Wick theorem, collect all the terms that are proportional to $\kappa^*\kappa$: this defines the form of $\mathcal{E}_{\text{pair}}$. Note that if the pseudo-potential has a finite range, then the resulting EDF will depend on the non-local pairing tensor $\kappa(\boldsymbol{r}, \sigma; \boldsymbol{r}'\sigma')$. This is the default choice for practitioners of the Gogny EDF. The SkP parametrization [122] of the Skyrme EDF was also obtained in this framework. The main advantage of this approach is that by ensuring that the same pseudo-Hamiltonian is used in both channels, it strongly mitigates some of the formal problems when going to the MR-EDF level [58–60, 62, 148, 149]. The SLyMR0 parametrization [65] designed for MR-EDF has been constructed in this spirit.
- If the EDF is not strictly derived from an effective pseudo-potential, or one wishes explicitly to severe the connection between the p.h. and p.p. channels, one can simply choose a convenient two-body potential only for the p.p. channel and compute the EDF from it. This is the default choice for practitioners of covariant density functional theory (cDFT) and the Skyrme EDF, with the exceptions of SkP already mentioned above. This choice is mostly motivated by pragmatism: it provides more flexibility to tune pairing properties to data. As discussed in section 9.2, it is difficult to actually isolate experimental data that are only sensitive to the pairing channel. In this respect, using a simpler pairing functional such as the one of equation (1.34) below makes it easier to calibrate it on data.

The most widely used form for the pairing functional arguably reads [139, 150]

$$\mathcal{E}_{\text{pair}}[\rho, \kappa, \kappa^*] = \int d^3\boldsymbol{r}\; A^{\kappa\kappa}[\rho] \sum_q \tilde{\rho}_q^*(\boldsymbol{r})\, \tilde{\rho}_q(\boldsymbol{r}), \tag{1.34}$$

where $\tilde{\rho}(\boldsymbol{r}) = \sum_\sigma (-2\sigma)\kappa_q(\boldsymbol{r}, \sigma; \boldsymbol{r}, -\sigma)$ is a local pairing density [151]. The pairing density $\tilde{\rho}(\boldsymbol{r}\sigma q, \boldsymbol{r}'\sigma' q')$ is obtained from the anomalous density $\kappa(\boldsymbol{r}\sigma q, \boldsymbol{r}'\sigma' q')$ according to

$$\tilde{\rho}(\boldsymbol{r}\sigma q, \boldsymbol{r}'\sigma' q') = (-2\sigma')\kappa(\boldsymbol{r}\sigma q, \boldsymbol{r}' - \sigma' q').$$

Its advantage is that it behaves mathematically exactly like the one-body density. In particular, it is Hermitian and time-even in time-even systems, and transforms like operators under a unitary change of basis; see also the discussion in section 3.1.5. When time-reversal symmetry is broken, the pairing density is no longer Hermitian and the local pair density $\tilde{\rho}_q(\boldsymbol{r})$ thus becomes complex. The coupling function $A^{\kappa\kappa}$ is in general chosen to be density-dependent,

$$A^{\kappa\kappa}[\rho] = \frac{\tilde{t}_0}{4}\left[1 - \eta\left(\frac{\rho_0(\boldsymbol{r})}{\rho_{\rm sat}}\right)^{\alpha}\right].$$

The above expression can be derived from the two-body, zero-range, p.p. vertex

$$\hat{v}_{\rm pp}(\boldsymbol{R}, \boldsymbol{r}) = \tilde{t}_0\frac{1}{2}\left(1 - \hat{P}_\sigma\right)\left[1 - \eta\left(\frac{\rho_0(\boldsymbol{R})}{\rho_{\rm sat}}\right)^{\alpha}\right]\delta(\boldsymbol{r}), \tag{1.35}$$

where $\hat{P}_\sigma$ is the spin-exchange operator defined above. In this case the pairing strength $\tilde{t}_0$ is identical for both the proton–proton and neutron–neutron pairings. In many applications, however, this relation is relaxed for practical purposes, and a different pairing strength is used for protons and neutrons [108–110, 152–155]. This is sometimes justified by the need to account effectively for the contribution of the Coulomb potential to pairing, which is often neglected (see next section) [156]. By construction, the vertex (1.35) and the functional (1.34) only give pair correlations in the $S = 0$, $T = 1$ channels. This is sufficient for almost all applications, as it is common practice to omit the possibility of proton–neutron pairing in the $S = 0$, $T = 1$ and $S = 1$, $T = 0$ channels.

A zero value for the η parameter corresponds to a uniformly distributed pairing force in the nuclear volume (volume pairing). On the other hand, the case $\eta = 1$ amounts to a pairing force whose contribution is much more pronounced near the nuclear surface (surface pairing). It is standard practice to take $\eta = 1/2$, although large-scale studies appear only weakly sensitive to the various prescriptions [156–158]. It is also standard practice to take $\alpha = 1$. Values of α lower than 1 induce strong pairing correlations at sub-saturation density, and it has been suggested that this could lead to non-physical predictions of giant halos caused by artificially enhanced pairing correlations [159].

Because of its zero-range, a cut-off $E_{\rm cut}$ must be introduced *ad hoc* in order to define a pairing window, i.e. a region in the individual q.p. spectrum affected by the pairing correlations. In the context of the HFB theory, this means that the summation (3.14) over the q.p. defining the one-body density matrix becomes

$$\rho_{ij} = \sum_{\mu=1}^{M} V_{i\mu}^* V_{j\mu} \to \rho_{ij} = \sum_{\epsilon_\mu \leqslant E_{\rm cut}} V_{i\mu}^* V_{j\mu},$$

where ϵ_μ is the equivalent s.p. energy that one can extract from the q.p. spectrum according to $\epsilon_\mu = (1 - 2|V_\mu|^2)E_\mu + \lambda_F$ [122]. As a result of this procedure, HFB calculations, including the calibration of the parameters of the pairing functional, become cut-off-dependent [151]. A more advanced technique to remove the dependence on the cut-off energy and at the same time regularize the divergence of the anomalous density was proposed in [129, 160] and validated in [161]. It is implemented in some HFB solvers [132, 162]. It is particularly important for time-dependent applications; see the discussion in section 4.3.

1.3.2 Treatment of the Coulomb potential

The potential energy given by equation (1.19) also contains a contribution from the Coulomb repulsion between the protons,

$$\hat{V}_{\text{Cou}}(\boldsymbol{r}, \boldsymbol{r}') = \frac{e^2}{|\boldsymbol{r} - \boldsymbol{r}'|}. \tag{1.36}$$

We note here that the treatment of the Coulomb potential is notably different from that of nuclear forces. As discussed in section 1.1.3, one of the main goals of the EDF approach is to combine computational efficiency and predictive power by encoding into the energy functional complex many-body correlations induced by the nuclear Hamiltonian ('horizontal philosophy'). Although the local, two-body Coulomb potential is known exactly, there have only been very few exploratory studies aiming at mapping out the form of the many-body correlations induced by this potential in the context of nuclear physics [163–166]. Instead, the Coulomb contribution to the energy is treated only either in the HF or HFB approximation.

Since the Coulomb potential has a finite range, like the Gogny force, it gives rise to an energy functional of the non-local densities. More precisely, computing the expectation value of the Coulomb potential on an HFB ground state of the form (3.3) gives the following terms: a direct, or Hartree, contribution to the energy that only depends on the local charge density

$$\mathcal{E}_{\text{Cou}}^{\text{dir}} = \frac{1}{2}\int d^3\boldsymbol{r} \int d^3\boldsymbol{r}' \frac{\rho_{\text{ch}}(\boldsymbol{r})\rho_{\text{ch}}(\boldsymbol{r}')}{|\boldsymbol{r} - \boldsymbol{r}'|}$$

and an exchange, or Fock, contribution that depends on the non-local charge density

$$\mathcal{E}_{\text{Cou}}^{\text{exc}} = -\frac{1}{4}\int d^3\boldsymbol{r} \int d^3\boldsymbol{r}' \left[\frac{\rho_{\text{ch}}(\boldsymbol{r}, \boldsymbol{r}')\rho_{\text{ch}}(\boldsymbol{r}', \boldsymbol{r})}{|\boldsymbol{r} - \boldsymbol{r}'|} + \frac{\boldsymbol{s}_{\text{ch}}(\boldsymbol{r}, \boldsymbol{r}') \cdot \boldsymbol{s}_{\text{ch}}(\boldsymbol{r}', \boldsymbol{r})}{|\boldsymbol{r} - \boldsymbol{r}'|}\right]. \tag{1.37}$$

The various terms entering the non-local density are given by equation (1.20) in section 1.2.1. Section 8.1.2 shows how to compute these terms in practice. In principle, the charge density matrices $\rho_{\text{ch}}(\boldsymbol{r}, \boldsymbol{r}')$ and $\boldsymbol{s}_{\text{ch}}(\boldsymbol{r}, \boldsymbol{r}')$ should be constructed by folding the density matrices of point protons and neutrons with their respective charge form factor. However, in most applications the charge density matrices are approximated by multiplying the point-proton density matrices with the electromagnetic coupling constant e. The BSk series of Skyrme interactions [101–106] is one of the rare cases where the charge density is computed by folding the proton density with a Gaussian form factor.

There is also a contribution of the Coulomb potential to the pairing energy. One can show that pseudo-potentials that are local, such as the Coulomb force (1.36), lead to pairing fields that are non-local, and that the contribution of these fields to the energy is thus given by

$$\mathcal{E}_{\mathrm{Cou}}^{\mathrm{pair}} = \frac{e^2}{2} \int d^3\boldsymbol{r} \int d^3\boldsymbol{r}' \sum_{\sigma\sigma'} \frac{\kappa_p^*(\boldsymbol{r}\sigma, \boldsymbol{r}'\sigma')\kappa_p(\boldsymbol{r}\sigma, \boldsymbol{r}'\sigma')}{|\boldsymbol{r} - \boldsymbol{r}'|}, \tag{1.38}$$

where $\kappa_p(\boldsymbol{r}\sigma, \boldsymbol{r}'\sigma')$ is the anomalous density matrix of protons. Note that the expression of the Coulomb pairing energy is obtained by using the proton anomalous density matrix, since we have no estimate of how the latter would differ from a possible 'charge anomalous density matrix'.

In practice, the direct contribution to the Coulomb energy is often computed 'exactly' by various numerical techniques such as direct integration (in particular in spherical symmetry where this is computationally inexpensive [162]) or Fourier transform [167, 168]. Users of Skyrme EDF often treat the exchange term approximately by resorting to the Slater approximation, which allows simulating the non-local exchange term by a local one,

$$\mathcal{E}_{\mathrm{Cou}}^{\mathrm{Slater}} = -\frac{3e^2}{4}\left(\frac{3}{\pi}\right)^{1/3} \int d^3\boldsymbol{r}\, \rho_{\mathrm{ch}}^{4/3}(\boldsymbol{r}).$$

Practitioners of the Gogny force often use the Slater approximation, but also use separation methods introduced in [169] to compute the exchange term exactly, for instance at convergence. Finally, we note that to date there are only very few SR-EDF studies that take the full Coulomb exchange and pairing term into account in the self-consistent fields, an example being [70]. Many recent MR-EDF calculations, however, include the Coulomb exchange and pairing terms exactly in order to avoid non-physical contributions to the energy [62, 170–172].

References

[1] Ring P and Schuck P 2004 *The Nuclear Many-Body Problem Texts and Monographs in Physics* (Berlin: Springer)

[2] Brack M *et al* 1972 Funny hills: the shell-correction approach to nuclear shell effects and its applications to the fission process *Rev. Mod. Phys.* **44** 320–405

[3] Möller P *et al* 1995 Nuclear ground-state masses and deformations *Atom. Data Nuc. Data Tab.* **59** 185

[4] Villars F 1957 The collective model of nuclei *Ann. Rev. Nucl. Sc.* **7** 185–230

[5] Rowe D J 1994 Algebraic models of nuclear collective motion *Nucl. Phys.* A **574** 253

[6] Pieper S C, Wiringa R B and Carlson J 2004 Quantum Monte Carlo calculations of excited states in $A = 6$–8 nuclei *Phys. Rev.* C **70** 054325

[7] Navrátil P *et al* 2009 Recent developments in no-core shell-model calculations *J. Phys. G: Nucl. Part. Phys.* **36** 083101

[8] Epelbaum E and Meißner U G 2012 Chiral dynamics of few- and many-nucleon systems *Ann. Rev. Nucl. Part. Sci.* **62** 159
[9] Hagen G *et al* 2010 *Ab initio* coupled-cluster approach to nuclear structure with modern nucleon–nucleon interactions *Phys. Rev.* C **82** 034330
[10] Dickhoff W H and Barbieri C 2004 Self-consistent Green's function method for nuclei and nuclear matter *Prog. Part. Nucl. Phys.* **52** 377–496
[11] Tsukiyama K, Bogner S K and Schwenk A 2011 In-medium similarity renormalization group for nuclei *Phys. Rev. Lett.* **106** 222502
[12] Caurier E *et al* 2005 The shell model as a unified view of nuclear structure *Rev. Mod. Phys.* **77** 427–88
[13] Bender M, Heenen P-H and Reinhard P-G 2003 Self-consistent mean-field models for nuclear structure *Rev. Mod. Phys.* **75** 121–80
[14] Papenbrock T and Weidenmüller H A 2014 Effective field theory for finite systems with spontaneously broken symmetry *Phys. Rev.* C **89** 014334
[15] Stone J R and Reinhard. P-G 2007 The Skyrme interaction in finite nuclei and nuclear matter *Prog. Part. Nucl. Phys.* **58** 587
[16] Nikšić T, Vretenar D and Ring P 2011 Relativistic nuclear energy density functionals: mean-field and beyond *Prog. Part. Nucl. Phys.* **66** 519
[17] Messiah A 1961 *Quantum Mechanics* vol 1 (Amsterdam: North-Holland)
[18] Messiah A 1962 *Quantum Mechanics* vol 2 (Amsterdam: North-Holland)
[19] Cohen-Tannoudji C, Diu B and Laloe F 1991 *Quantum Mechanics* vol 1 (New York: Wiley)
[20] Cohen-Tannoudji C, Diu B and Laloe F 1992 *Quantum Mechanics* vol 2 (New York: Wiley)
[21] Engel J *et al* 2003 Time-reversal violating Schiff moment of 225Ra *Phys. Rev.* C **68** 025501
[22] Dobaczewski J and Olbratowski P 2005 Solution of the Skyrme–Hartree–Fock–Bogolyubov equations in the Cartesian deformed harmonic-oscillator basis.: (V) HFODD (v2.08k) *Comput. Phys. Commun.* **167** 214
[23] Ban S *et al* 2010 Fully self-consistent calculations of nuclear Schiff moments *Phys. Rev.* C **82** 015501
[24] Avignone F T, Elliott S R and Engel. J 2008 Double beta decay, Majorana neutrinos, and neutrino mass *Rev. Mod. Phys.* **80** 481
[25] Langanke K and Martínez-Pinedo G 2003 Nuclear weak-interaction processes in stars *Rev. Mod. Phys.* **75** 819
[26] Arnould M, Goriely S and Takahashi K 2007 The *r*-process of stellar nucleosynthesis: astrophysics and nuclear physics achievements and mysteries *Phys. Rep.* **450** 97
[27] Grawe H, Langanke K and Martínez-Pinedo G 2007 Nuclear structure and astrophysics *Rep. Prog. Phys.* **70** 1525
[28] Thielemann F-K *et al* 2011 What are the astrophysical sites for the *r*-process and the production of heavy elements? *Prog. Part. Nucl. Phys.* **66** 346
[29] Magierski P and Heenen P-H 2002 Structure of the inner crust of neutron stars: crystal lattice or disordered phase? *Phys. Rev.* C **65** 045804
[30] Magierski P, Bulgac A and Heenen P-H 2002 Neutron stars and the fermionic Casimir effect *Int.J. Mod. Phys.* A **17** 1059–64
[31] Magierski P, Bulgac A and Heenen P-H 2003 Exotic nuclear phases in the inner crust of neutron stars in the light of Skyrme–Hartree–Fock theory *Nucl. Phys.* A **719** C217–20
[32] Magierski P and Bulgac A 2004 Nuclear structure and dynamics in the inner crust of neutron stars *Nucl. Phys.* A **738** 143–9

[33] Paar N *et al* 2014 Neutron star structure and collective excitations of finite nuclei *Phys. Rev.* C **90** 011304

[34] Chen W-C and Piekarewicz J 2014 Building relativistic mean field models for finite nuclei and neutron stars *Phys. Rev.* C **90** 044305

[35] Sagert I *et al* 2016 Quantum simulations of nuclei and nuclear pasta with the multi-resolution adaptive numerical environment for scientific simulations *Phys. Rev.* C **93** 055801

[36] Gross D J and Wilczek F 1973 Ultraviolet behavior of non-Abelian gauge theories *Phys. Rev. Lett.* **30** 1343–6

[37] Gross D J 1996 Asymptotic freedom, confinement and QCD *History of Original Ideas and Basic Discoveries in Particle Physics* (Boston, MA: Springer) pp 75–99

[38] Ishii N, Aoki S and Hatsuda T 2007 Nuclear force from lattice QCD *Phys. Rev. Lett.* **99** 022001

[39] Wiringa R B, Stoks V G J and Schiavilla R 1995 Accurate nucleon–nucleon potential with charge-independence breaking *Phys. Rev.* C **51** 38–51

[40] Machleidt R 2001 High-precision, charge-dependent Bonn nucleon–nucleon potential *Phys. Rev.* C **63** 024001

[41] Kim K-Y and Zahed I 2009 Nucleon–nucleon potential from holography *J. High Energy Phys.* **2009** 131

[42] Epelbaum E 2006 Few-nucleon forces and systems in chiral effective field theory *Prog. Part. Nucl. Phys.* **57** 654–741

[43] Kalantar-Nayestanaki N *et al* 2012 Signatures of three-nucleon interactions in few-nucleon systems *Rep. Prog. Phys.* **75** 016301

[44] Jurgenson E D, Navrátil P and Furnstahl R J 2009 Evolution of nuclear many-body forces with the similarity renormalization group *Phys. Rev. Lett.* **103** 082501

[45] Nifenecker H *et al* 1998, *Trends in Nuclear Physics, 100 Years Later Proc. NATO Advanced Study Institute* vol 66, *Les Houches, France 30 July–30 August 1996* (Amsterdam: North Holland)

[46] Brueckner K A 1955 Two-body forces and nuclear saturation. III. Details of the structure of the nucleus *Phys. Rev.* **97** 1353–66

[47] Day B D 1967 Elements of the Brueckner–Goldstone theory of nuclear matter *Rev. Mod. Phys.* **39** 719

[48] Day B D 1978 Current state of nuclear matter calculations *Rev. Mod. Phys.* **50** 495–521

[49] Slater J C 1951 A simplification of the Hartree–Fock method *Phys. Rev.* **81** 385–90

[50] Mok D K W, Neumann R and Handy N C 1996 Dynamical and nondynamical correlation *J. Phys. Chem.* **100** 6225–30

[51] Tichai A, Gebrerufael E and Roth R 2017 Open-shell nuclei from no-core shell model with perturbative improvement, arXiv: 1703.05664

[52] Szalay P G *et al* 2012 Multiconfiguration self-consistent field and multireference configuration interaction methods and applications *Chemical Reviews* **112** 108–81

[53] Duguet T 2014 The nuclear energy density functional formalism *The Euroschool on Exotic BeamsLecture Notes in Physics* vol 879 ed C Scheidenberger and M Pfütznervol 4 (Berlin: Springer) p 293

[54] Erler J, Klüpfel P and Reinhard P-G 2010 Misfits in Skyrme–Hartree–Fock *J. Phys. G: Nucl. Part. Phys.* **37** 064001

[55] Neilson D 1996 Landau Fermi liquid theory *Aust. J. Phys.* **49** 79

[56] Negele J W 1982 The mean-field theory of nuclear structure and dynamics *Rev. Mod. Phys.* **54** 913–1015

[57] Stringari S and Brink D M 1978 Constraints on effective interactions imposed by antisymmetry and charge independence *Nucl. Phys.* A **304** 307

[58] Lacroix D, Duguet T and Bender M 2009 Configuration mixing within the energy density functional formalism: removing spurious contributions from nondiagonal energy kernels *Phys. Rev.* C **79** 044318

[59] Bender M, Duguet T and Lacroix D 2009 Particle-number restoration within the energy density functional formalism *Phys. Rev.* C **79** 044319

[60] Duguet T *et al* 2009 Particle-number restoration within the energy density functional formalism: nonviability of terms depending on noninteger powers of the density matrices *Phys. Rev.* C **79** 044320

[61] Tarpanov D *et al* 2014 Polarization corrections to single-particle energies studied within the energy-density-functional and quasiparticle random-phase approximation approaches *Phys. Rev.* C **89** 014307

[62] Anguiano M, Egido J L and Robledo L M 2001 Particle number projection with effective forces *Nucl. Phys.* A **696** 467

[63] Dobaczewski J *et al* 2007 Particle-number projection and the density functional theory *Phys. Rev.* C **76** 054315

[64] Dobaczewski J, Bennaceur K and Raimondi F 2012 Effective theory for low-energy nuclear energy density functionals *J. Phys. G: Nucl. Part. Phys.* **39** 125103

[65] Sadoudi J *et al* 2013 Skyrme pseudo-potential-based EDF parametrization for spuriousity-free MR EDF calculations *Phys. Scr.* **T154** 014013

[66] Sadoudi J *et al* 2013 Skyrme functional from a three-body pseudopotential of second order in gradients: formalism for central terms *Phys. Rev.* C **88** 064326

[67] Raimondi F, Bennaceur K and Dobaczewski J 2014 Nonlocal energy density functionals for low-energy nuclear structure *J. Phys. G: Nucl. Part. Phys.* **41** 055112

[68] Duguet T 2015 Symmetry broken and restored coupled-cluster theory: I. Rotational symmetry and angular momentum *J. Phys. G: Nucl. Part. Phys.* **42** 025107

[69] Bennaceur K *et al* 2017 Nonlocal energy density functionals for pairing and beyond-mean-field calculations *J. Phys.* G **44** 045106

[70] Anguiano M, Egido J L and Robledo L M 2001 Coulomb exchange and pairing contributions in nuclear Hartree–Fock–Bogoliubov calculations with the Gogny force *Nucl. Phys.* A **683** 227

[71] Lesinski T *et al* 2009 Non-empirical pairing energy density functional *Eur. Phys. J.* A **40** 121–6

[72] Negele J W and Vautherin D 1972 Density-matrix expansion for an effective nuclear Hamiltonian *Phys. Rev.* C **5** 1472–93

[73] Negele J . W and Vautherin D 1975 Density-matrix expansion for an effective nuclear Hamiltonian. II *Phys. Rev.* C **11** 1031–41

[74] Gebremariam B, Duguet T and Bogner S K 2010 Improved density matrix expansion for spin-unsaturated nuclei *Phys. Rev.* C **82** 014305

[75] Stoitsov M *et al* 2010 Microscopically based energy density functionals for nuclei using the density matrix expansion: implementation and pre-optimization *Phys. Rev.* C **82** 054307

[76] Dobaczewski J, Carlsson B G and Kortelainen M 2010 The Negele–Vautherin density-matrix expansion applied to the Gogny force *J. Phys. G: Nucl. Part. Phys.* **37** 075106

[77] Dyhdalo A, Bogner S K and Furnstahl R J 2017 Applying the density matrix expansion with coordinate-space chiral interactions *Phys. Rev.* C **95** 054314
[78] Navarro Pérez R *et al* 2018 Microscopically based energy density functionals for nuclei using the density matrix expansion. II. Full optimization and validation *Phys. Rev.* C **97** 054304
[79] Carlsson B G and Dobaczewski J 2010 Convergence of density-matrix expansions for nuclear interactions *Phys. Rev. Lett.* **105** 122501
[80] Davesne D *et al* 2016 Infinite matter properties and zero-range limit of non-relativistic finite-range interactions *Ann. Phys. (NY)* **375** 288
[81] Sulaksono A *et al* 2003 The nonrelativistic limit of the relativistic point coupling model *Ann. Phys.* **308** 354–70
[82] Sulaksono A *et al* 2007 From self-consistent covariant effective field theories to their Galilean-invariant counterparts *Phys. Rev. Lett.* **98** 262501
[83] Kaiser N and Weise W 2010 Nuclear energy density functional from chiral pion–nucleon dynamics revisited *Nucl. Phys.* A **836** 256–74
[84] Kaiser N 2010 Nuclear energy density functional from chiral pion–nucleon dynamics: isovector terms *Eur. Phys. J.* **A45** 61–8
[85] Gebremariam B, Bogner S K and Duguet T 2011 Microscopically-constrained Fock energy density functionals from chiral effective field theory. I. Two-nucleon interactions *Nucl. Phys.* A **851** 17–43
[86] Engel Y M *et al* 1975 Time-dependent Hartree–Fock theory with Skyrme's interaction *Nucl. Phys.* A **249** 215–38
[87] Perlińska E *et al* 2004 Local density approximation for proton–neutron pairing correlations: formalism *Phys. Rev.* C **69** 014316
[88] Dobaczewski J and Dudek J 1996 Time-odd components in the rotating mean field and identical bands *Acta Phys. Pol.* B **27** 45
[89] Raimondi F *et al* 2011 Continuity equation and local gauge invariance for the N3LO nuclear energy density functionals *Phys. Rev.* C **84** 064303
[90] Hellemans V, Heenen P-H and Bender M 2012 Tensor part of the Skyrme energy density functional. III. Time-odd terms at high spin *Phys. Rev.* C **85** 014326
[91] Bender M *et al* 2002 Gamow–Teller strength and the spin–isospin coupling constants of the Skyrme energy functional *Phys. Rev.* C **65** 064308
[92] Schunck N *et al* 2010 One-quasiparticle states in the nuclear energy density functional theory *Phys. Rev.* C **81** 024316
[93] Pototzky K J *et al* 2010 Properties of odd nuclei and the impact of time-odd mean fields: a systematic Skyrme–Hartree–Fock analysis *Eur. Phys. J.* A **46** 299
[94] Pastore A, Davesne D and Navarro J 2015 Linear response of homogeneous nuclear matter with energy density functionals *Phys. Rep.* **563** 1
[95] Beiner M *et al* 1975 Nuclear ground-state properties and self-consistent calculations with the Skyrme interaction: (I). Spherical description *Nucl. Phys.* A **238** 29
[96] Köhler H S 1976 Skyrme force and the mass formula *Nucl. Phys.* A **258** 301
[97] Bartel J *et al* 1982 Towards a better parametrisation of Skyrme-like effective forces: a critical study of the SkM force *Nucl. Phys.* A **386** 79–100
[98] Schunck N and Robledo L M 2016 Microscopic theory of nuclear fission: a review *Rep. Prog. Phys.* **79** 116301

[99] Chabanat E *et al* 1998 A Skyrme parametrization from subnuclear to neutron star densities Part II. Nuclei far from stabilities *Nucl. Phys.* A **635** 231–56

[100] Reinhard P-G and Flocard H 1995 Nuclear effective forces and isotope shifts *Nucl. Phys.* A **584** 467

[101] Goriely S *et al* 2003 Further explorations of Skyrme–Hartree–Fock–Bogoliubov mass formulas. II. Role of the effective mass *Phys. Rev.* C **68** 054325

[102] Samyn M *et al* 2004 Further explorations of Skyrme–Hartree–Fock–Bogoliubov mass formulas. III. Role of particle-number projection *Phys. Rev.* C **70** 044309

[103] Chamel N, Goriely S and Pearson J M 2009 Further explorations of Skyrme–Hartree–Fock–Bogoliubov mass formulas. XI. Stabilizing neutron stars against a ferromagnetic collapse *Phys. Rev.* C **80** 065804

[104] Goriely S, Chamel N and Pearson J M 2013 Further explorations of Skyrme–Hartree–Fock–Bogoliubov mass formulas. XIII. The 2012 atomic mass evaluation and the symmetry coefficient *Phys. Rev.* C **88** 024308

[105] Goriely S, Chamel N and Pearson J M 2013 Hartree–Fock–Bogoliubov nuclear mass model with 0.50 MeV accuracy based on standard forms of Skyrme and pairing functionals *Phys. Rev.* C **88** 061302

[106] Goriely S 2015 Further explorations of Skyrme–Hartree–Fock–Bogoliubov mass formulas. XV: The spin–orbit coupling *Nucl. Phys.* A **933**(Suppl. C) 68–81

[107] Lesinski T *et al* 2007 Tensor part of the Skyrme energy density functional: spherical nuclei *Phys. Rev.* C **76** 014312

[108] Kortelainen M *et al* 2010 Nuclear energy density optimization *Phys. Rev.* C **82** 024313

[109] Kortelainen M *et al* 2012 Nuclear energy density optimization: large deformations *Phys. Rev.* C **85** 024304

[110] Kortelainen M *et al* 2014 Nuclear energy density optimization: shell structure *Phys. Rev.* C **89** 054314

[111] Farine M, Pearson J M and Tondeur F 1997 Nuclear-matter incompressibility from fits of generalized Skyrme force to breathing-mode energies *Nucl. Phys.* A **615** 135

[112] Krewald S *et al* 1977 On the use of Skyrme forces in self-consistent RPA calculations *Nucl. Phys.* A **281** 166

[113] Farine M, Pearson J M and Tondeur F 2001 Skyrme force with surface-peaked effective mass *Nucl. Phys.* A **696** 396

[114] Carlsson B G, Dobaczewski J and Kortelainen M 2008 Local nuclear energy density functional at next-to-next-to-next-to-leading order *Phys. Rev.* C **78** 044326

[115] Raimondi F, Carlsson B G and Dobaczewski J 2011 Effective pseudopotential for energy density functionals with higher-order derivatives *Phys. Rev.* C **83** 054311

[116] Becker P *et al* 2015 Tools for incorporating a D-wave contribution in Skyrme energy density functionals *J. Phys. G: Nucl. Part. Phys.* **42** 034001

[117] Becker P *et al* 2017 Solution of Hartree–Fock–Bogoliubov equations and fitting procedure using the N2LO Skyrme pseudopotential in spherical symmetry *Phys. Rev.* C **96** 044330

[118] Skyrme T H R 1956 CVII. The nuclear surface *Phil. Mag.* **1** 1043–54

[119] Skyrme T H R 1959 The effective nuclear potential *Nucl. Phys.* **9** 615

[120] Chang B D 1975 Spin saturation and the Skyrme interaction *Phys. Lett.* B **56** 205

[121] Waroquier M *et al* 1979 Extended-Skyrme-force calculation: theoretical description and application to Ca and Sn isotopes *Phys. Rev.* C **19** 1983

[122] Dobaczewski J, Flocard H and Treiner J 1984 Hartree–Fock–Bogolyubov description of nuclei near the neutron-drip line *Nucl. Phys.* A **422** 103–39
[123] Ryssens W *et al* 2015 Solution of the Skyrme-HF+BCS equation on a 3D mesh, II: a new version of the Ev8 code *Comput. Phys. Commun.* **187** 175
[124] Klüpfel P *et al* 2009 Variations on a theme by Skyrme: a systematic study of adjustments of model parameters *Phys. Rev.* C **79** 034310
[125] Reinhard P-G 2009 private communication
[126] Dechargé J and Gogny D 1980 Hartree–Fock–Bogolyubov calculations with the D_1 effective interaction on spherical nuclei *Phys. Rev.* C **21** 1568–93
[127] Berger J F, Girod M and Gogny D 1991 Time-dependent quantum collective dynamics applied to nuclear fission *Comput. Phys. Commun.* **63** 365–74
[128] Brink D M and Boeker E 1967 Effective interactions for Hartree–Fock calculations *Nucl. Phys.* A **91** 1
[129] Bulgac A and Yu Y 2002 Renormalization of the Hartree–Fock–Bogoliubov equations in the case of a zero range pairing interaction *Phys. Rev. Lett.* **88** 042504
[130] Robledo L M and Bertsch G F 2011 Application of the gradient method to Hartree–Fock–Bogoliubov theory *Phys. Rev.* C **84** 014312
[131] Robert M and Parrish *et al* 2013 Exact tensor hypercontraction: a universal technique for the resolution of matrix elements of local finite-range *N*-body potentials in many-body quantum problems *Phys. Rev. Lett.* **111** 132505
[132] Navarro Perez R *et al* 2017 Axially deformed solution of the Skyrme–Hartree–Fock–Bogolyubov equations using the transformed harmonic oscillator basis (III) HFBTHO (v3.00): a new version of the program *Comput. Phys. Commun.* **220**(Suppl. C) 363–75
[133] Chappert F, Girod M and Hilaire S 2008 Towards a new Gogny force parameterization: impact of the neutron matter equation of state *Phys. Lett.* B **668** 420
[134] Goriely S *et al* 2009 First Gogny–Hartree–Fock–Bogoliubov nuclear mass model *Phys. Rev. Lett.* **102** 242501
[135] Chappert F *et al* 2015 Gogny force with a finite-range density dependence *Phys. Rev.* C **91** 034312
[136] Anguiano M *et al* 2012 Tensor and tensor–isospin terms in the effective Gogny interaction *Phys. Rev.* C **86** 054302
[137] Anguiano M *et al* 2016 Gogny interactions with tensor terms *Eur. Phys. J.* A **52** 183
[138] Nilsson S G and Ragnarsson I 1995 *Shapes and Shells in Nuclear Structure* (Cambridge: Cambridge University Press)
[139] Fayans S A *et al* 1994 Isotope shifts within the energy-density functional approach with density dependent pairing *Phys. Lett.* B **338** 1
[140] Fayans S A, Trykov E L and Zawischa D 1994 Influence of effective spin–orbit interaction on the collective states of nuclei *Nucl. Phys.* A **568** 523
[141] Krömer E *et al* 1995 Energy-density functional approach for non-spherical nuclei *Phys. Lett.* B **363** 12
[142] Fayans S A *et al* 2000 Nuclear isotope shifts within the local energy-density functional approach *Nucl Phys.* A **676** 49
[143] Baldo M *et al* 2013 New Kohn–Sham density functional based on microscopic nuclear and neutron matter equations of state *Phys. Rev.* C **87** 064305
[144] Baldo M *et al* 2017 Barcelona–Catania–Paris–Madrid functional with a realistic effective mass *Phys. Rev.* C **95** 014318

[145] Baldo M, Schuck P and Viñas X 2008 Kohn–Sham density functional inspired approach to nuclear binding *Phys. Lett.* B **663** 390–94

[146] Bulgac A *et al* 2018 Minimal nuclear energy density functional *Phys. Rev.* C **97** 044313

[147] Duguet T and Lesinski T 2008 Non-empirical pairing functional *Eur. Phys. J.* A **156** 207

[148] Stoitsov M V *et al* 2007 Variation after particle-number projection for the Hartree–Fock–Bogoliubov method with the Skyrme energy density functional *Phys. Rev.* C **76** 014308

[149] Egido J L 2016 State-of-the-art of beyond mean field theories with nuclear density functionals *Phys. Scr.* **91** 073003

[150] Chasman R R 1976 Density-dependent delta interactions and actinide pairing matrix elements *Phys. Rev.* C **14** 1935

[151] Dobaczewski J *et al* 1996 Mean-field description of ground-state properties of drip-line nuclei: pairing and continuum effects *Phys. Rev.* C **53** 2809–40

[152] Bender M *et al* 2000 Pairing gaps from nuclear mean-field models *Eur. Phys. J.* A **8** 59

[153] Bertulani C A, Lü H F and Sagawa H 2009 Odd–even mass difference and isospin dependent pairing interaction *Phys. Rev.* C **80** 027303

[154] Bertulani C A, Liu H and Sagawa H 2012 Global investigation of odd–even mass differences and radii with isospin-dependent pairing interactions *Phys. Rev.* C **85** 014321

[155] Yamagami M *et al* 2012 Isoscalar and isovector density dependence of the pairing functional determined from global fitting *Phys. Rev.* C **86** 034333

[156] Bertsch G F *et al* 2009 Odd–even mass differences from self-consistent mean field theory *Phys. Rev.* C **79** 034306

[157] Dobaczewski J, Nazarewicz W and Stoitsov M V 2002 Nuclear ground-state properties from mean-field calculations *Eur. Phys. J.* A **15** 21–6

[158] Dobaczewski J, Nazarewicz W and Werner T R 1995 Closed shells at drip-line nuclei *Phys. Scr.* **T56** 15

[159] Dobaczewski J, Nazarewicz W and Reinhard P-G 2001 Pairing interaction and self-consistent densities in neutron-rich nuclei *Nucl. Phys.* A **693** 361

[160] Bulgac A 2002 Local density approximation for systems with pairing correlations *Phys. Rev.* C **65** 051305(R)

[161] Borycki P J *et al* 2006 Pairing renormalization and regularization within the local density approximation *Phys. Rev.* C **73** 044319

[162] Bennaceur K and Dobaczewski J 2005 Coordinate-space solution of the Skyrme–Hartree–Fock–Bogolyubov equations within spherical symmetry. The program HFBRAD (v1.00) *Comput. Phys. Commun.* **168** 96

[163] Bulgac A and Shaginyan V R 1996 A systematic surface contribution to the ground-state binding energies *Nucl. Phys.* A **601** 103–16

[164] Bulgac A and Shaginyan V R 1999 Proton single-particle energy shifts due to Coulomb correlations *Phys. Lett.* B **469** 1

[165] Bulgac A and Shaginyan V R 1999 Influence of Coulomb correlations on the location of drip line, single particle spectra and effective mass *Eur. Phys. J.* A **5** 247

[166] Shaginyan V R 2001 Coulomb energy of nuclei *Phys. Atom. Nuclei* **64** 471

[167] Dobaczewski J and Dudek J 1997 Solution of the Skyrme–Hartree–Fock equations in the Cartesian deformed harmonic oscillator basis II. The program HFODD *Comput. Phys. Commun.* **102** 183

[168] Maruhn J A *et al* 2014 The TDHF code Sky3D *Comput. Phys. Commun.* **185** 2195–6

[169] Girod M and Grammaticos B 1983 Triaxial Hartree–Fock–Bogolyubov calculations with D1 effective interaction *Phys. Rev.* C **27** 2317

[170] Rodríguez T R and Luis Egido J 2010 Triaxial angular momentum projection and configuration mixing calculations with the Gogny force *Phys. Rev.* C **81** 064323

[171] Bally B *et al* 2014 Beyond mean-field calculations for odd-mass nuclei *Phys. Rev. Lett.* **113** 162501

[172] Borrajo M, Rodríguez T R and Egido J L 2015 Symmetry conserving configuration mixing method with cranked states *Phys. Lett.* B **746** 341

IOP Publishing

Energy Density Functional Methods for Atomic Nuclei

Nicolas Schunck

Chapter 2

Covariant energy density functionals

Jean-Paul Ebran

Non-relativistic energy density functionals (EDFs) of the Skyrme or Gogny type have been very successful in accounting for nuclear phenomenology for more than 40 years. Therefore, one may question the need to build a relativistic version of such approaches. When considering nuclei under extreme conditions of density, isospin asymmetry, temperature or momentum transfer, e.g. in various astrophysical scenarios [1] or in the context of heavy ion collisions [2], a Lorentz-covariant formulation can be motivated by relativistic kinematic arguments. On the other hand, under ordinary conditions, nucleon Fermi velocities are much smaller than the speed of light. A simple estimate of the mean nucleon velocity in nuclear matter gives $\langle v/c \rangle = \langle \partial E(p)/\partial p \rangle = \frac{3}{4} p_{\mathrm{F}}/M^{*} \sim 0.21\text{–}0.28$. The Fermi momentum of nucleons p_{F} is of the order of 262 MeV c^{-1} (corresponding to a saturation density of 0.16 fm^{-3}) and M^{*} stands for the effective nucleon mass, typically $\frac{M^{*}}{M} \sim 0.75$. Therefore, the assumption of non-relativistic EDF with non-relativistic nucleon kinematics and dominant relativistic corrections—such as spin–orbit effects—incorporated perturbatively is fully justified. Yet, such kinematical considerations do not tell the full story. In the effective field theory (EFT) perspective there are infinitely many representations of low-energy quantum chromodynamics (QCD) physics. However not all are equally efficient or physically compelling. In the end, what matters is the ability to account for low-energy QCD phenomenology in the most efficient way. In this respect, the choice between the Lorentz-covariant and non-relativistic formulations should not only be motivated by kinematics issues, but also by an argument of effectiveness. These different formulations are related by the heavy-baryon expansion in chiral effective field theory (χEFT) [3].

doi:10.1088/2053-2563/aae0edch2

Effective field theory

The intuitive idea behind the concept of effective theories is simple: calculate without knowing the exact theory. However, implementing this idea in a mathematically consistent way in the case of an interacting quantum field theory is far from being obvious. The EFT follows from the so-called 'folk theorem' enunciated by Weinberg [4, 5].

> Weinberg's folk theorem reads 'If one writes down the most general possible Lagrangian, including all terms consistent with assumed symmetry principles, and then calculates matrix elements with this Lagrangian to any given order of perturbation theory, the result will simply be the most general possible S −matrix consistent with perturbative unitarity, analyticity, cluster decomposition and the assumed symmetry properties'.

It defines procedures to construct in a controlled manner, below a characteristic energy scale, the most general Lagrangian consistent with relevant degrees of freedom (d.o.f.s) and symmetry patterns of an underlying theory [6, 7]. The general idea behind EFT is to identify a cutoff Λ representing the breakdown scale of the theory, which enables one to define a small parameter governing a perturbative expansion as well as relevant high- and low-energy d.o.f.s that are consistent with the symmetry pattern of the underlying theory,

$$\phi \to \phi_H + \phi_L,$$

with ϕ the general d.o.f. of the underlying theory.

The high-energy modes can be integrated out in the functional integral sense

$$\int \mathcal{D}\phi_L \mathcal{D}\phi_H e^{iS(\phi_L,\, \phi_H)} = \int \mathcal{D}\phi_L e^{iS_{\text{eff}}^{\Lambda}(\phi_L)},$$

where $S(\phi_L, \phi_H)$ is the original action of the underlying theory and $S_{\text{eff}}^{\Lambda}(\phi_L) \equiv \int \mathcal{D}\phi_H e^{iS(\phi_L,\, \phi_H)}$ is the low-energy effective one. Obviously, it is not possible to perform such an integration in general. However, the power of EFT comes from the fact that the effective action can be expanded in terms of local operators $\mathcal{O}_i$

$$S_{\text{eff}}^{\Lambda} = \int d^D x \sum_i g_i \mathcal{O}_i,$$

where the sum runs over all local operators allowed by the symmetries of the problem and the g_i are the low-energy coupling constants. Such an expression would not really be of any practical use without a scheme to rank the contributions by order of importance. This is achieved by a dimensional analysis that yields a so-called power counting and allows one to compute low-energy observables at a desired accuracy.

In some cases, the low-energy d.o.f.s can easily be identified using the high-energy d.o.f.s, e.g. electronic cooper pairs in the Ginzburg–Landau theory of superconductivity or quark bound states such as pions for QCD. In more typical situations, however, the low-energy d.o.f.s are totally different from the high-energy d.o.f.s. For example, in water molecules are the high-energy d.o.f. and the local velocity field of the fluid represents the low-energy d.o.f. The two are related in a very complicated way but in practice we do not need to determine this relation.

So what does a relativistic formulation of the nuclear EDF approach bring compared to a non-relativistic one? Imposing Lorentz covariance reveals the large relevant scales of the order of the QCD mass scales (hidden in non-relativistic treatments) that allow simple, efficient and compelling explanations of nucleon–nucleon scattering data and nuclear properties. Famous examples of physics insight provided by a relativistic treatment are the explanation of the spin–orbit force [8] and the energy dependence of the optical potential for nucleon–nucleus scattering up to 100 MeV, which emerges from the Lorentz structure of the interaction [8]; see section 2.6.

Indeed, when the Poincaré symmetry group governs nuclear systems, the relevant d.o.f.s of the problem behave in a well-defined manner (such as Lorentz scalars, Dirac spinors, four-vectors, etc) under the transformations of the group; see section 2.1 for more details. In the nuclear medium, this translates into the existence of two distinct types of large nucleon self-energies, namely a Lorentz scalar one ($S \sim -400$ MeV) and a four-vector one (V^{μ} with $V \equiv V^{0} \sim +350$ MeV). These two types of nuclear self-energy largely cancel in the central potential channel ($V + S \sim -50$ MeV), hence justifying a non-relativistic kinematics.

In the Dirac equation, equivalent to the Schrödinger equation in a relativistic framework, only the large component of the nucleon spinors is sensitive to the cancellation of the two large fields ($V + S \sim -50$ MeV). The small component of the nucleon spinors, however, feels the strong field $V - S \sim 750$ MeV leading to a large spin–orbit splitting in nuclei [9]. The near cancellation of the scalar and vector self-energies also implies the quasi-realization of pseudospin symmetry (PSS) in nuclei [10]; see section 2.6 for a discussion of some of these effects.

Non-relativistic approaches incorporate these large cancellations from the get-go and hide the underlying QCD mass scales by setting as relevant scales the binding energy of nuclear systems and the central potential (tens of MeV). In contrast, relativistic approaches conserve the large QCD mass scales as relevant scales. They account for the cancellation between the scalar and vector self-energies in some channels, but also for their constructive combination in some others. One should emphasize that there is no approximate symmetry that enforces this near cancellation between scalar and vector contributions. Chiral symmetry alone does not lead to a scalar–vector fine tuning. Rather, the cancellation is accidental, and hiding these underlying scales is not required by first principles.

Therefore, the specific advantage of a covariant approach is the description of nuclear properties in terms of large scalar and vector fields, for which there can be no direct experimental evidence or refutation (as for any nuclear potentials since these are not observables). Note that the relevance of a representation in terms of large fields is supported by QCD phenomenological models such as the Nambu–Jona-Lasinio model [11] (see, e.g. figure 3 of [12]) or finite-density QCD sum rules [13] relating the occurrence of the scalar and vector self-energies to the modifications, with respect to the QCD vacuum case, in the chiral condensate and the quark density induced by the presence of baryonic matter [14]. In practice, one should weigh these advantages against the technical challenges that come with a relativistic extension of the nuclear EDF formalism.

Before addressing these issues and in order to make this chapter self-contained, section 2.1 introduces the Lorentz symmetry group and its basic properties. A modern discussion of the various phenomenological Lagrangian densities underpinning covariant energy density functionals (CEDFS) begins by reinterpreting them as downgraded non-renormalizable low-energy QCD effective Lagrangians [15]. For this reason, we then discuss in section 2.2 the symmetry properties of QCD, which play a prominent role in constraining the form of the EFT. An overview of the construction of both a Lorentz-covariant EFT of QCD at low energy and its various phenomenological counterparts is then included in sections 2.3 and 2.4, respectively. Section 2.5 elaborates on the derivation of a relativistic energy functional from the phenomenological Lagrangians. Finally, typical advantages provided by a relativistic formulation of the nuclear EDF framework are addressed in section 2.6.

> One should be clear about the fact that CEDFS are not genuine EFT. Even if they explicitly retain the symmetry pattern of QCD, they do not include all the interaction terms compatible with the symmetries of the underlying theory (what actually makes an EFT model-independent) and do not come with a power-counting scheme (which enables a controlled and systematic account of the radiative corrections beyond the tree level). Hence, they cannot provide error estimates contrary to EFT.

2.1 Relativistic description of quantum systems

Non-relativistic physical systems are characterized by how they behave under Galilean transformations (space and time translations, rotations, and space and time inversions), e.g. such as a scalar, a vector, etc. Requiring Galilean invariance for the Hamiltonian of a non-relativistic interacting system puts tight constraints on it. It also constrains the acceptable terms entering a systematic expansion of a non-relativistic EDF in powers of various densities and currents and of their derivatives; see, e.g. [16–20]. Likewise, relativistic physical systems are characterized by their transformations under the Poincaré group, that is, the Lorentz group complemented by the group of space–time translations. In this first section, we recall some properties of the Minkowski space–time, of the Poincaré and Lorentz groups and of their representation, as well as of relativistic wave equations. We refer the reader to excellent textbooks such as [21–24] for more in-depth discussions.

2.1.1 Minkowski space–time

The relativity principle, first formulated by Galileo [25], states that the laws of nature retain precisely the same form when expressed in any uniformly moving frame: it is impossible to distinguish the physics in a state at rest from that of a state in uniform motion by means of local experiments. Such a relativity principle, when coupled to the causality one, dictates the very structure of space–time. This is manifested by the general transformation laws that relate the coordinates $(t,\, \boldsymbol{x})$ in one reference frame to the coordinates $(t',\, \boldsymbol{x}')$ in another frame boosted at velocity $\boldsymbol{v}$ with respect to the former. Recall that in special relativity, the term 'boost' is introduced to refer to the transformation that relates two reference frames that move at a constant velocity $\boldsymbol{v}$ relative to each other. Equivalently, this transformation law can be applied to the time and space intervals $(\Delta t,\, \Delta x)$ and $(\Delta t',\, \Delta x')$ in two such reference frames. Consistency with respect to the relativity principle reduces the possible forms of such general transformations to [26, 27]

$$\begin{cases} \Delta t' = \gamma(\nu)[\Delta t - \alpha\nu\Delta x] \\ \Delta x' = \gamma(\nu)[\Delta x - \nu\Delta t] \end{cases}, \tag{2.1}$$

with the function of $\nu = ||\boldsymbol{v}||$,

$$\gamma(\nu) = \frac{1}{\sqrt{1 - \alpha\nu^2}}.$$

There are three possible cases: $\alpha < 0$, $\alpha = 0$ and $\alpha > 0$. The first one is ruled out by the principle of causality; the second one corresponds to Galilean boost transformations and implies absolute simultaneity, which means $\Delta t = \Delta t'$ whatever the boost velocity ν, or equivalently that there are no specific constraints on the velocity at which information can propagate ($\alpha = 1/\nu_{\text{lim}}^2 = 1/\infty = 0$); the third one defines Lorentz boost transformations and implies that information cannot propagate faster than a *finite* limit.

Such a finite maximum velocity is called for historical reasons the speed of light, c, and is manifested by the constant value of c in all inertial frames. Special relativity may thus be summarized by the fact that there is a maximum velocity for all physical systems. Likewise, quantum mechanics can be interpreted as the discovery of the existence of a finite limit of information one can obtain to characterize a physical system.

Therefore, only two of the three cases are consistent with the principles of relativity and causality, and the only type of transformation also consistent with the existence of a 'finite' limit to the velocity at which information can propagate is the case of Lorentz transformations with $\alpha \equiv 1/c^2 > 0$ [28].

The transformations (2.1) represent an isometry group of space–time. We may think separately of time as a one-dimensional manifold $\mathcal{T}$ (the time line) and of space as a three-dimensional manifold $\mathcal{S}$. A manifold is a topological space (i.e. a set of points with some axioms relating them) which is locally Euclidean but globally might have a more involved structure, such as a torus or a sphere. These manifolds are considered smooth and endowed with some notion of distance, possibly with additional structure. One might take for $\mathcal{S}$ the space $\mathbb{R}^3$: it is a vector space endowed with the Euclidean metric, or standard inner product (which makes it a Hilbert space). However, a vector space has the physically undesirable feature to have a special element: the origin. In nature, there is no center of the Universe, or put better, everywhere can be a possible center of the Universe. Therefore the vector space is replaced by its affine space (it amounts to forgetting the special role played by the zero vector), which defines what is called the Euclidean space, usually noted as $\mathbb{E}^3$. It leads us to identify $\mathcal{T} \equiv \mathbb{E}$, the real Euclidean line endowed with the positive-definite distance $|dt|$, and $\mathcal{S} \equiv \mathbb{E}^3$, the three-dimensional real Euclidean space endowed with a scalar product. Lorentz kinematics then suggests that space–time can be constructed as the product manifold

$$\mathbb{M} = \mathbb{E} \times \mathbb{E}^3 \equiv \mathbb{E}^{1,3} \tag{2.2}$$

endowed with a (pseudo)metric. We call such a space–time the Minkowski space–time.

Some words about the notation: in the following, we use the Einstein convention, i.e. implicit summation over repeated indices: $x^\mu e_\mu \equiv \sum_{\mu=0}^{3} x^\mu e_\mu$. Greek letter indices will span the values $\{0, 1, 2, 3\}$ while Latin letter indices will be restricted to $\{1, 2, 3\}$. $x^0 \equiv ct$, with c the speed of light and t the time, represents the time-like coordinate. x^i stands for the space-like coordinate. From now on we set $c = 1$.

In order to be consistent with Lorentz transformations, special relativity requires a metric structure that can lead both to positive and to negative squared distances between events, according to whether or not they are reachable by a ray of light. In other words, the metric structure of Minkowski is not positive definite. Denoting

$$e_0 \equiv (1, 0, 0, 0) \quad e_1 \equiv (0, 1, 0, 0) \quad e_3 \equiv (0, 0, 1, 0) \quad e_3 \equiv (0, 0, 0, 1)$$

the canonical basis of $\mathbb{M}$, a generic element $x = (ct, \boldsymbol{x}) = (x^0, x^1, x^2, x^3)$ of $\mathbb{M}$ can be expanded:

$$x = x^\mu e_\mu \qquad \mu = 0, 1, 2, 3.$$

It is called a four-vector. Calling η: $\mathbb{M} \times \mathbb{M} \rightarrow \mathbb{E}$ the pseudometric (a bi-linear form) of the Minkowski manifold, its matrix $I_{1,3} \equiv (\eta_{\mu\nu})_{0\leqslant\mu,\ \nu\leqslant 3}$ in the canonical basis reads

$$I_{1,3} = \begin{pmatrix} -1 & 0 & 0 & 0 \\ 0 & 1 & 0 & 0 \\ 0 & 0 & 1 & 0 \\ 0 & 0 & 0 & 1 \end{pmatrix} \equiv \begin{pmatrix} -1 & 0 \\ 0 & I_3 \end{pmatrix}, \tag{2.3}$$

where in the second equality we note by I_3 the 3×3 identity matrix. From equation (2.3), we deduce that η is symmetric (its matrix $I_{1,3}$ is symmetric) and non-degenerate ($I_{1,3}$ is invertible), with the signature $(-, +, +, +)$: the eigenvalues of $I_{1,3}$, which are real and non-zero because the matrix is diagonalizable on $\mathbb{R}$ and invertible, have the signs $-, +, +, +$ to within a permutation.

For any two events, i.e. four-vectors, $(x, y) \in \mathbb{M}$ in the Minkowski space, we note the inner product

$$x \cdot y = \eta(x, y) = x^\mu \eta_{\mu\nu} y^\nu = -x^0 y^0 + x^1 y^1 + x^2 y^2 + x^3 y^3 \equiv x^\mu y_\mu .$$

In particular, the Minkowski norm of a four-vector x is

$$x \cdot x = -(x^0)^2 + (x^1)^2 + (x^2)^2 + (x^3)^2 .$$

If $x \cdot x < 0$, the four-vector x is said to be time-like. If $x \cdot x > 0$, x is called space-like while $x \cdot x = 0$ corresponds to a light-like four-vector. A non-zero light-like or time-like four-vector is future-(past-)oriented if its time-like component x^0 is strictly positive (negative).

2.1.2 Lorentz and Poincaré groups

From the previous section, it follows that the laws of physics are governed by the Lorentz symmetry group or, equivalently, that the dynamical content of the world evolves in Minkowski space–time (the effects of gravitational interaction are disregarded here). As a consequence, physical systems should be compatible with the symmetries of the Lorentz group. In other words, they should behave in a well-defined manner under Lorentz transformations, which in turn implies certain relations among observable quantities that should be obeyed with great precision. For that purpose, we discuss in more detail the properties of Lorentz and Poincaré groups. Note that invariance under Galilean transformations, discussed in sections 1.2.1 and 4.6, can be viewed as the low-velocity, non-relativistic limit of invariance under Lorentz transformations.

The Lorentz group

The Lorentz group is the isometric group of the Minkowski space–time. In other words, it is the orthogonal group $O(1, 3)$ of transformations $\mathbf{\Lambda}$ that preserve the pseudometric η of the Minkowski manifold

$$\eta(\mathbf{\Lambda} x, \mathbf{\Lambda} y) = \eta(x, y) \quad \forall x, y \in \mathbb{M} .$$

Under a Lorentz transformation $\mathbf{\Lambda}$, a generic element of Minkowski space–time $x \in \mathbb{M}$ behaves according to

$$x^{\mu} \overset{\Lambda}{\mapsto} x'^{\mu} = \Lambda^{\mu}_{\nu} x^{\nu} \quad \text{such that} \quad \Lambda^{\mu}_{\rho} \eta_{\mu\nu} \Lambda^{\nu}_{\sigma} = \eta_{\rho\sigma}. \tag{2.4}$$

The Lorentz transformation $\mathbf{\Lambda}$ being a 4×4 matrix, it depends on 16 parameters. Because of the 10 constraints provided by the symmetric metric tensor $\boldsymbol{\eta}$, the Lorentz group is in fact $16 - 10 = 6$-dimensional. In addition, it is non-simply connected and involves four connected branches; see figure 2.1 for an illustration of the notions of connected spaces. Indeed equation (2.4) implies that $(\det \mathbf{\Lambda})^2 = 1$, and

$$-\left(\Lambda^0_0\right)^2 + \left(\Lambda^1_0\right)^2 + \left(\Lambda^2_0\right)^2 + \left(\Lambda^3_0\right)^2 = -1.$$

Therefore, we must have $(\Lambda^0_0)^2 \geqslant 1$, that is, $\Lambda^0_0 \geqslant 1$ or $\Lambda^0_0 \leqslant -1$. The four connected components $O(1, 3)^{\uparrow\downarrow}_{\pm}$ correspond to the four cases $\det \mathbf{\Lambda} = \pm 1$ and Λ^0_0 positive ($\uparrow$) or negative ($\downarrow$). They preserve the orientation if $\det \mathbf{\Lambda} = 1$ and the time arrow if $\Lambda^0_0 \geqslant 1$

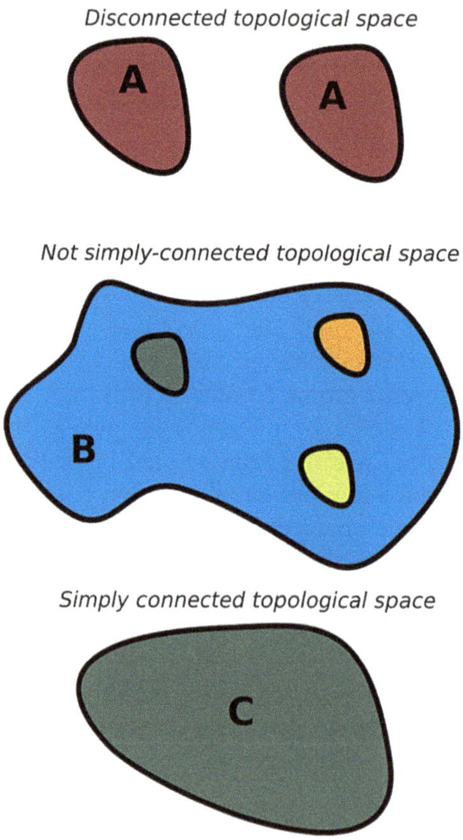

Figure 2.1. Schematic illustration of the notions of connected topological spaces. Space A is disconnected, space B is not simply connected and space C is simply connected. (Figure created by N Schunck.)

The connected subgroup $L_+^\uparrow \equiv SO(1, 3)^\uparrow \equiv O(1, 3)_+^\uparrow$ is called the proper orthochronous Lorentz group; it contains the identity matrix. The other three branches can be constructed from a given $\mathbf{\Lambda} \in L_+^\uparrow$ combined with a space and/or time reflection. For this reason, we will now focus our attention on $L_+^\uparrow$. Each $\mathbf{\Lambda} \in L_+^\uparrow$ can be reconstructed from a Lorentz boost with velocity $\beta = v/c$ in the direction $\boldsymbol{n}$ (with $|\beta| < 1$) together with a spatial rotation $R(\boldsymbol{\alpha}) \in SO(3)$. Explicitly,

$$\mathbf{\Lambda} = \begin{pmatrix} \gamma & \gamma\beta\boldsymbol{n}^T \\ \gamma\beta\boldsymbol{n} & I_3 + (\gamma - 1)\boldsymbol{n}\boldsymbol{n}^T \end{pmatrix} \begin{pmatrix} 1 & 0 \\ 0 & R(\boldsymbol{\alpha}) \end{pmatrix},$$

with $\gamma \equiv \frac{1}{\sqrt{1-\beta^2}}$.

Examples of Lorentz transformations are the rotation of angle θ around the z-axis

$$R_z(\theta) = \begin{pmatrix} 1 & 0 & 0 & 0 \\ 0 & \cos\theta & \sin\theta & 0 \\ 0 & -\sin\theta & \cos\theta & 0 \\ 0 & 0 & 0 & 1 \end{pmatrix} \tag{2.5}$$

or the boost along the z-axis with velocity $v = \tanh\alpha$

$$L_z(\alpha) = \begin{pmatrix} \cosh\alpha & 0 & 0 & -\sinh\alpha \\ 0 & 1 & 0 & 0 \\ 0 & 0 & 1 & 0 \\ -\sinh\alpha & 0 & 0 & \cosh\alpha \end{pmatrix}. \tag{2.6}$$

Two properties will shortly become important in the context of representations:

(i) The Lorentz group is not compact because it contains boosts. Therefore, all unitary representations are infinite-dimensional.

(ii) The Lorentz group is not simply connected because it contains rotations. As a consequence, we need to study the representations of its universal covering group $SL(2, \mathbb{C})$, namely the group formed by complex 2×2 matrices A with $\det A = 1$.

The Poincaré group

The fact that the Lorentz group leaves the pseudonorm $x \cdot x$ of a four-vector invariant is not enough to guarantee that the speed of light is the same in every inertial frame. Instead, we need the line element $dx \cdot dx = dx^\mu \eta_{\mu\nu} dx^\nu = (dt)^2 - (d\boldsymbol{x})^2$ to be invariant [28]. This can be achieved by adding constant space–time translations to the Lorentz transformations,

$$(a, \mathbf{\Lambda}): x \mapsto x' = \mathbf{\Lambda} x + a.$$

The set of transformations $(a, \mathbf{\Lambda})$ form a ten-parameter group, which contains translations, rotations and boosts—it is the Poincaré group, also called the inhomogeneous Lorentz group. The Poincaré group $\mathbb{R}^4 \rtimes O(1, 3)$ has four parameters coming from translations and six coming from Lorentz transformations. In analogy to the Lorentz group, the component which contains the identity $(0, I_4)$, where I_4 is the 4×4 identity matrix, is called $ISO(1, 3)^{\uparrow} \equiv \mathbb{R}^4 \rtimes SO(1, 3)^{\uparrow}$, where I refers to inhomogeneous.

The Lorentz and Poincaré algebras

In the previous sections, we have established some basic properties of the Lorentz and Poincaré groups. These two groups are particular cases of Lie groups.

Let us briefly recall that a Lie group $\mathcal{G}$ is a finite-dimensional, topological group. Its main property is that the application

$$\begin{aligned} \mathcal{G} \times \mathcal{G} &\to \mathcal{G} \\ (x, y) &\mapsto xy \end{aligned}$$

and its inverse are 'smooth', i.e. C^{∞} or in simple terms, infinitely differentiable. In the example of $SO(3)$, the group of rotation, each element of the group is a rotation matrix, which can be characterized by three independent parameters (the rotation angles): the group has dimension 3. The action of two rotations of angles $\boldsymbol{\alpha}$ and $\boldsymbol{\alpha}'$ yields another rotation of angle $\boldsymbol{\alpha} + \boldsymbol{\alpha}''$; see [29–31] for complete, pedagogical introductions to group theory.

The next step is to study their representations, which encode how the elements of the group act in other spaces. In simple terms, a representation of a group $\mathcal{G}$ is a mapping that associates to every element g of $\mathcal{G}$, a linear operator T_g acting in some vector space V that preserves the group operation. For instance, in the case of the $SO(n)$ group of rotations in space, the representations of the group tell us how objects such as square-integrable wave functions or spinors change under a rotation of the coordinate system. The idea here is that the properties of the various structures that are 'physically allowed' under the group of rotations give us more insight into what a rotation actually is.

The determination of the representations of a Lie group is facilitated by the introduction of Lie algebras. Indeed, while we can think of Lie groups as describing in a *global* way the concept of a continuous family of symmetries for physical objects, the associated Lie algebra expresses it in a *local* way. By differentiating the Lie group action, we obtain a Lie algebra action, which is a linearization of the group action. As a linear object, a Lie algebra is often a lot easier to work with than working directly with the corresponding Lie group. Relation among elements of Lie algebras are the same as among elements of Lie groups, since a Lie algebra basically consists of the 'infinitesimal elements' of a Lie group, i.e. the 'elements infinitesimally close to the identity'.

In the special case of a matrix group $\mathcal{G}$, one can show that the associated Lie algebra $\mathfrak{g}$ is formed by all the matrices such that the exponential of that matrix is also an element of $\mathcal{G}$ [32]. In practice, a Lie algebra is a vector space that is also equipped with a bi-linear operation called the Lie bracket, denoted $[x, y]$ for any two vectors x

and y in the Lie algebra. $[x, y]$ is another vector of the Lie algebra. The Lie bracket somewhat measures how non-commutative the underlying Lie group is. In the important example of the Lie algebra of the general linear group $GL(n, \mathbb{C})$ of complex-valued invertible matrices of dimension n, the Lie bracket simply corresponds to the commutator $[A, B] = AB - BA$ of two matrices.

Since Lie algebras are vector spaces, and the Lie bracket is a bi-linear operation, all we really need to characterize the algebra is knowing how a set of basis vectors transforms under the Lie bracket. Such basis vectors are called generators of the group. If we apply the Lie bracket to each pair of generators and write down the resulting vectors as coordinates in the same basis, the set of numbers we obtain are called structure constants. From the generators and the structure constants, we can construct the full Lie algebra. Because of the exponential mapping between the Lie group and the Lie algebra, this implies that generators and structure constants allow us to construct the entire Lie group itself.

In the same manner, the representations of a Lie algebra $\mathfrak{g}$ encode all there is to know about the representations of the Lie group $\mathcal{G}$ it comes from. Casimir operators $\mathcal{C}$ (or invariants) for a Lie algebra play a special role. A Casimir operator is an element of the Lie algebra that commutes with all the generators of the Lie group. If one finds the Casimir invariants of a Lie algebra, then according to a lemma by Schur, these Casimir operators are in any irreducible representation (irrep) proportional to the identity. In other words, their eigenvalues may be used to label the irreps.

For an analogy of what a Lie algebra is with respect to a Lie group, one may think of studying the global motion of a particle by only looking at infinitesimal sections of it. We end up with a differential equation for that motion that encodes all there is to know about it and even allows us to fully determine the whole motion (given some boundary conditions). Note that in standard analysis, infinitesimal elements do not really exist. Technically, a Lie algebra is defined on the tangent space of the Lie group at the identity. Still, the picture of infinitesimal elements is a useful and intuitive way of thinking about this. Formally, a condensed definition of a Lie algebra $\mathfrak{g}$ (in the special case where the Lie group $\mathcal{G}$ is some matrix group) is

$$\mathfrak{g} = \{X \in \mathcal{G} | \forall\, t \in \mathbb{C},\ \exp(tX) \in \mathcal{G}\}.$$

This implies that the elements of the Lie algebra $\mathfrak{g}$ are also elements of $\mathcal{G}$. The construction of the Lie algebra via its generators and structure constants is analogous to the construction of a vector space out of linear combinations of basis elements.

Generators of the Lorentz and Poincaré groups

Let us determine a set of generators for the Lie algebra of the Lorentz group and compute their Lie brackets. Because we are interested in Lorentz transformations infinitesimally close to the identity, we focus first on the $L_+^\uparrow \equiv SO(1, 3)^\uparrow$ subgroup (the only connected branch containing the identity). Transformations of the Lorentz group depend on the parameters of the group, e.g. the angles $\boldsymbol{\alpha}$ of the rotation matrices. The elements Λ_ν^μ of the matrix (in Minkowski space $\mathbb{M}$) of such transformations can thus be viewed as functions of these parameters. These functions are differentiable (this is why they form a Lie group) and we can thus expand them around any point,

$$\Lambda_\nu^\mu(\epsilon) = \delta_\nu^\mu + \epsilon\omega_\nu^\mu + \dots , \tag{2.7}$$

with $\epsilon \ll 1$. For the simple case of a rotation matrix by an angle $\epsilon \ll 1$ around the z-axis, writing

$$R_z(\epsilon) = R_z(0) + \epsilon \frac{\partial R_z}{\partial \theta}\bigg|_{\theta=0} + \mathcal{O}(\epsilon^2)$$

implies that in this case, $\boldsymbol{\omega} \equiv \partial R / \partial \theta|_0$, with $R_z(\theta)$ given by equation (2.5). ω_ν^μ just refer to specific matrix elements of the matrix $\boldsymbol{\omega}$. The condition $\Lambda_\rho^\mu \eta_{\mu\nu} \Lambda_\sigma^\nu = \eta_{\rho\sigma}$ yields

$$\omega^{\mu\nu} + \omega^{\nu\mu} = 0.$$

The matrix ω represents an infinitesimal transformation of $L_+^\uparrow$. It is anti-symmetric and thus characterized by six independent parameters. Likewise, an infinitesimal Poincaré transformation is characterized by the set of matrices $(\alpha^\mu, \omega_\nu^\mu)$.

To determine the corresponding generators of the groups, we look for differential operators expressing how functions $f(x)$ depending on space–time coordinates change under an infinitesimal Poincaré transformation,

$$x'^\lambda \equiv x^\lambda + \delta x^\lambda = x^\lambda + \alpha^\lambda + \omega^{\lambda\nu} x_\nu .$$

We can Taylor-expand the function around point x up to first order, and seek the form

$$f(x^\lambda + \delta x^\lambda) = f(x^\lambda) - i\left(\alpha^\mu P_\mu + \frac{1}{2}\omega^{\mu\nu} J_{\mu\nu}\right) f(x^\lambda),$$

where P_μ and $J_{\mu\nu}$ are differential operators. The calculation is not entirely trivial, but we end up with six generators of the Lorentz transformations represented by the anti-symmetric tensor

$$J_{\mu\nu} = i(x_\mu \partial_\nu - x_\nu \partial_\mu). \tag{2.8}$$

The expressions for the generators can also be deduced from the infinitesimal version of equations (2.5) and (2.6). Writing $R_z(\theta) = I_4 - i\theta J_z + \dots$ for $\theta \ll 1$ and $L_z(\alpha) = I_4 - i\alpha K_z$ for $\alpha \ll 1$, we find

$$J_z = i\begin{pmatrix} 0 & 0 & 0 & 0 \\ 0 & 0 & -1 & 0 \\ 0 & 1 & 0 & 0 \\ 0 & 0 & 0 & 0 \end{pmatrix}$$

and

$$K_z = i\begin{pmatrix} 0 & 0 & 0 & -1 \\ 0 & 0 & 0 & 0 \\ 0 & 0 & 0 & 0 \\ -1 & 0 & 0 & 0 \end{pmatrix}.$$

Defining

$$J_{ij} \equiv \epsilon_{ijk} J^k \qquad K^i \equiv J_{0i},$$

with ϵ_{ijk} the three-dimensional Levi-Civita symbol, a basis for the Lie algebra of the Lorentz group is thus formed by the generators J_x, J_y, J_z of the rotations around the x-, y-, z-axes and the generators K_x, K_y, K_z of the boosts along the same axis. To obtain a basis for the Lie algebra of the Poincaré group, we have to complement the six generators of the Lorentz group by the four generators of space–time translations, the latter being collected in the four-vector

$$P_\mu = -i\partial_\mu.$$

Note that P_0, as the generator of translation in time, is nothing other than the Hamiltonian.

Now that we have a basis for the Lie algebra, we can easily calculate the various Lie brackets among each pair of generators,

$$\begin{aligned} [J_{\mu\nu}, P_\lambda] &= i(\eta_{\nu\lambda}P_\mu - \eta_{\mu\lambda}P_\nu), \\ [J_{\mu\nu}, J_{\lambda\rho}] &= i(\eta_{\nu\lambda}J_{\mu\rho} - \eta_{\mu\lambda}J_{\nu\rho} + \eta_{\mu\rho}J_{\nu\lambda} - \eta_{\nu\rho}J_{\mu\lambda}), \\ [P_\mu, P_\nu] &= 0. \end{aligned}$$

To gain a better understanding of such a structure, we write the Lie brackets in terms of the generators of the Lorentz group first, yielding the structure of the $\mathfrak{so}(1, 3)$ Lorentz algebra

$$[J_i, J_j] = +i\epsilon_{ijk}J_k \qquad [K_i, K_j] = -i\epsilon_{ijk}J_k \qquad [J_i, K_j] = +\epsilon_{ijk}K_k. \tag{2.9}$$

We calculate the remaining Lie brackets to obtain the structure of the $\mathfrak{iso}(1, 3)$ Poincaré algebra

$$\begin{aligned} [J^i, P^0] &= 0 & [K^i, P^0] &= iP^i \\ [J^i, P^j] &= i\epsilon_{ijk}P^K & [K^i, P^j] &= iP^0\delta_{ij}. \end{aligned}$$

These Lie brackets tell us that $J = (J^1, J^2, J^3)$, $K = (K^1, K^2, K^3)$ and $P = (P^1, P^2, P^3)$ transform like the 'usual' three-vector under the rotations of $\mathbb{R}^3$.

We can obtain further insight. Let us form the following linear combination of the Lorentz generators with complex coefficients

$$X_k \equiv \frac{1}{2}(J_k - iK_k) \qquad Y_k \equiv \frac{1}{2}(J_k + iK_k)$$

satisfying

$$[X_j,\ X_k] = i\epsilon_{jkl}X_l \qquad [Y_j,\ Y_k] = i\epsilon_{jkl}Y_l \qquad [X_j,\ Y_k] = 0. \tag{2.10}$$

We say that we work with the complex Lie algebra, $\mathfrak{so}(1,\ 3)_{\mathbb{C}} \equiv \mathfrak{so}(1,\ 3) \times \mathbb{C}$. We recognize two independent familiar $\mathfrak{su}(2)$ sub-algebra generated by X_k and Y_k, respectively, and commuting with one another. Recall that $\mathfrak{su}(2)$ is the Lie algebra associated with the $SU(2)$ group of unitary 2×2 complex-valued matrices, which is a Lie group. Therefore, there is an isomorphism between $\mathfrak{so}(1,\ 3)_{\mathbb{C}}$ and the direct sum of the $\mathfrak{su}(2)$ algebra,

$$\mathfrak{so}(1,\ 3)_{\mathbb{C}} \cong \mathfrak{su}(2) \oplus \mathfrak{su}(2).$$

Because of this isomorphism, we can infer from the properties of the $\mathfrak{su}(2)$ algebra a method to form the Casimir invariants of the $\mathfrak{so}(1,\ 3)_{\mathbb{C}}$ Lie algebra. This is done by forming the quadratic combinations

$$X^2 = X_iX_i \qquad Y^2 = Y_iY_i. \tag{2.11}$$

These matrices indeed commute with all the generators of $\mathfrak{so}(1,\ 3)_{\mathbb{C}}$ as can easily be demonstrated. Their eigenvalues x and y are positive integers of half-integers satisfying

$$X^2|x\rangle = x(x+1)|x\rangle \qquad Y^2|y\rangle = y(y+1)|y\rangle,$$

with $|\ x\rangle$ and $|y\rangle$ the corresponding eigenvectors. They can thus be used to label the irreps of $\mathfrak{so}(1,\ 3)_{\mathbb{C}}$. The introduction of $\pm i$, however, implies that unitary representations of the Lorentz group do not follow in a simple way from those of $SU(2) \times SU(2)$. We therefore have to take a little detour and look at the representations of the $\mathfrak{sl}(2,\ \mathbb{C})$ algebra, i.e. the Lie algebra of the universal cover of the Lorentz group.

Irreducible finite-dimensional representations of $SL(2,\ \mathbb{C})$

There is a simple way to see how $SL(2,\ \mathbb{C})$ and $SO(1,3)^{\uparrow}$ are connected: we associate to any four-vector x^μ the 2×2 Hermitian matrix X such that

$$X = x^\mu\sigma_\mu = x^0 + \boldsymbol{\sigma x} \qquad \det X = x \cdot x = (x^0)^2 - \boldsymbol{x}^2,$$

where the four matrices $\boldsymbol{\sigma}$ form a basis of the $SU(2)$ Lie group. We use the standard convention that $\sigma_0 = I_2$, the identity matrix and $\boldsymbol{\sigma} \equiv (\sigma_x,\ \sigma_y,\ \sigma_z)$ are the usual Pauli matrices. Conversely, we can use the previous relation to express any four-vector of the Minkowski space in terms of a 2×2 Hermitian matrix $X \in SL(2,\ \mathbb{C})$ according to

$$x^\mu = \frac{1}{2}tr(X\sigma_\mu).$$

Any matrix $A \in SL(2,\ \mathbb{C})$ acts on X according to

$$X \mapsto X' = AXA^\dagger,$$

which is indeed Hermitian. Therefore, it can be used to define a new four-vector $x'^\mu = \frac{1}{2}\mathrm{tr}(X'\sigma_\mu)$, the pseudonorm of which is $x' \cdot x' = \det X' = \det X = x \cdot x$. Put differently, the action of any matrix of $SL(2,\ \mathbb{C})$ defines a linear transformation of $\mathbb{M}$ that preserves the Minkowski pseudonorm, i.e. it is a Lorentz transformation.

The matrices associated with the representations of spin j are such that under the action of $A \in SL(2, \mathbb{C})$, we have the transformation

$$|jm\rangle \overset{A}{\rightarrow} |jm'\rangle \mathcal{D}^j_{m'm}$$

We provide here the explicit $(2j + 1)$-dimensional (with j an integer or half-integer) representations $\mathcal{D}^j$ of $SL(2, \mathbb{C})$, namely,

$$\forall A = \begin{pmatrix} a & b \\ c & d \end{pmatrix} \in SL(2, \mathbb{C})$$

$$\mathcal{D}^j_{m,m'}(A) = \sqrt{(j+m)!(j-m)!(j+m')!(j-m')!} \sum_{\substack{n_1,n_2,n_3,n_4 \geqslant 0 \\ n_1+n_2=j+m \\ n_3+n_4=j-m' \\ n_1+n_3=j+m \\ n_2+n_4=j-m'}} \frac{a^{n_1} b^{n_2} c^{n_3} d^{n_4}}{n_1! n_2! n_3! n_4!},$$

with $-j \leqslant m, m' \leqslant j$. We call such a representation $(j,0)$. There exists another non-equivalent $(2j + 1)$-dimensional representation called $(0, j)$ and corresponding to $\mathcal{D}^j((A^\dagger)^{-1})$. Replacing A by $(A^\dagger)^{-1}$ may be interpreted as associating the four-vector x with the matrix $\tilde{X} \equiv x^0 - \boldsymbol{\sigma x}$ instead of the matrix $X \equiv x^0 + \boldsymbol{\sigma x}$.

The most general finite-dimensional irrep of $SL(2, \mathbb{C})$ is then denoted as (j_1, j_2), with j_1 and $j_2 \geqslant 0$ integers or half-integers; it is defined by

$$(j_1, j_2) = (j_1, 0) \otimes (0, j_2), \tag{2.12}$$

with $j_1 \equiv x$ and $j_2 \equiv y$ the eigenvalues of the Casimir operators (2.11). Note that under space inversion, we have

$$x^i \overset{P}{\rightarrow} -x^i \Leftrightarrow \begin{cases} J_i \overset{P}{\rightarrow} +J_i \\ K_i \overset{P}{\rightarrow} -K_i \end{cases} \Leftrightarrow \begin{cases} X_i \overset{P}{\rightarrow} Y_i \\ Y_i \overset{P}{\rightarrow} X_i \end{cases} \Leftrightarrow (j_1, j_2) \overset{P}{\rightarrow} (j_2, j_1). \tag{2.13}$$

Consequently, to be consistent with the full Poincaré symmetry, only representations of the form (j, j) (irreducible) or $(j_1, j_2) \oplus (j_2, j_1)$ (reducible) will be physically acceptable, because they are well-behaved under space reflections.

Representations of the Lorentz group

The representations of the Lorentz group are then the (j_1, j_2) ones. We discuss some of them below in more detail. Tensor representations are the representations (j_1, j_2) where $j_1 + j_2$ is an integer. Some of those are especially relevant:

- (0,0) is the one-dimensional trivial representation. This is how Lorentz scalars transform.
- (1,0) and (0,1) are the representations of self dual two-form fields.
- (1/2,1/2) is the four-dimensional vector representation, specifying how four-vectors transform under Lorentz transformations. The generators are represented by

$$J_k = \frac{1}{2}(\sigma_k \otimes 1 + 1 \otimes \sigma_k) \quad K_k = \frac{1}{2}(\sigma_k \otimes 1 - 1 \otimes \sigma_k).$$

This representation plays a special role because its behavior under a Lorentz transformation involves the Lorentz matrix $\mathbf{\Lambda}$ itself,

$$\left(\frac{1}{2}, \frac{1}{2}\right): \quad x^\mu \overset{\Lambda}{\mapsto} x'^\mu = \Lambda^\mu_\nu x^\nu.$$

It can be used to construct all further (reducible) tensor representations that transform according to

$$T^{\mu\nu\ldots\tau} \overset{\Lambda}{\mapsto} T'^{\mu\nu\ldots\tau} = \underbrace{\Lambda^\mu_\alpha \Lambda^\nu_\beta \ldots \Lambda^\tau_\lambda}_{n \text{ times}} T^{\alpha\beta\ldots\lambda}$$

where $T^{\mu\nu\ldots\tau}$ is a Lorentz tensor of rank n.
- (1,1) denotes the nine-dimensional representation where the symmetric and trace-less 4×4 tensors belong.

Spinor representations correspond to the case where $j_1 + j_2$ is a half-integer. Such representations are *projective* representations, where instead of $\mathcal{D}(\mathbf{\Lambda}')\mathcal{D}(\mathbf{\Lambda}) = \mathcal{D}(\mathbf{\Lambda}'\mathbf{\Lambda})$, one has $\mathcal{D}(\mathbf{\Lambda}')\mathcal{D}(\mathbf{\Lambda}) = e^{i\varphi(\mathbf{\Lambda}', \mathbf{\Lambda})}\mathcal{D}(\mathbf{\Lambda}'\mathbf{\Lambda})$, with a phase φ depending on $\mathbf{\Lambda}$ and $\mathbf{\Lambda}'$. In the case relevant for describing a particle with spin-1/2, we have two non-equivalent two-dimensional representations, namely (1/2,0) and (0,1/2), defined by the infinitesimal generators

$$X_k = \frac{\sigma_k}{2} \quad Y_k = 0 \Leftrightarrow J_k = \frac{\sigma_k}{2} \quad K_k = -i\frac{\sigma_k}{2}$$

for the former and

$$X_k = 0 \quad Y_k = \frac{\sigma_k}{2} \Leftrightarrow J_k = \frac{\sigma_k}{2} \quad K_k = i\frac{\sigma_k}{2}$$

for the latter. Consequently, we need to introduce two different kinds of (two-component) spinors. We will use the generic name ξ for spinors in (1/2, 0) and χ for the ones in (0, 1/2). Throughout this chapter, we assume that a generic matrix $A \in SL(2, \mathbb{C})$ reads

$$A = \begin{pmatrix} a & b \\ c & d \end{pmatrix}.$$

Given such an arbitrary matrix $A \in SL(2, \mathbb{C})$, these two types of (two-component) spinor behave under a Lorentz transformation according to

$$(1/2, 0): \xi = (\xi^\alpha) \overset{\Lambda}{\mapsto} \xi' = A\xi = \begin{pmatrix} a\xi^1 + b\xi^2 \\ c\xi^1 + d\xi^2 \end{pmatrix}$$

and

$$(0,\ 1/2)\colon \chi = (\chi^{\alpha}) \overset{\Lambda}{\mapsto} \chi' = A^{*}\chi = \begin{pmatrix} a^{*}\chi^{1} + b^{*}\chi^{2} \\ c^{*}\chi^{1} + d^{*}\chi^{2} \end{pmatrix}.$$

Note that the anti-symmetric form

$$\forall(\xi,\ \xi'), \qquad (\xi,\ \xi') \equiv \xi^{1}\xi'^{2} - \xi^{2}\xi'^{1} = \xi^{T}(i\sigma_{2})\xi$$

is invariant in (1/2, 0), and likewise

$$\forall(\chi,\ \chi'), \qquad (\chi,\ \chi') \equiv \chi^{1}\chi'^{2} - \chi^{2}\chi'^{1} = \chi^{T}(i\sigma_{2})\chi'$$

defines an invariant alternating form in (0,1/2). One may thus use these forms to lower spinor indices,

$$\begin{array}{llll} (1/2,\ 0)\colon & (\xi,\ \xi') = \xi_{\alpha}\xi'^{\alpha} & \xi_{2} = \xi^{1}; & \xi_{1} = -\xi^{2} \\ (0,\ 1/2)\colon & (\chi,\ \chi') = \chi_{\alpha}\chi'^{\alpha} & \chi_{2} = \chi^{1}; & \chi_{1} = -\chi^{2}. \end{array}$$

If these two spinor types transform in an irreducible way under the Lorentz group $SO(1,\ 3)^{\uparrow}$, equation (2.13) implies, however, that under parity, we have

$$\xi \overset{P}{\to} \chi; \qquad \chi \overset{P}{\to} \xi.$$

Therefore one cannot work with only one type of (two-component) spinor to represent the total Lorentz group. Instead, one must define a four-component spinor ψ called a Dirac spinor

$$\xi \in \left(\frac{1}{2},\ 0\right),\ \chi \in \left(0,\ \frac{1}{2}\right) \Rightarrow \psi \equiv \begin{pmatrix} \xi \\ \chi \end{pmatrix}.$$

A Dirac spinor behaves under a Lorentz transformation $\mathbf{\Lambda} \in SO(1,\ 3)^{\uparrow}$ according to

$$\psi \overset{\Lambda}{\mapsto} \psi' = S(\mathbf{\Lambda})\psi,$$

where

$$S(\mathbf{\Lambda}) \equiv \begin{pmatrix} A & 0 \\ 0 & (A^{\dagger})^{-1} \end{pmatrix},$$

and $A \in SL(2,\ \mathbb{C})$. Under a parity transformation, we obtain

$$\psi \overset{P}{\mapsto} \psi' = \begin{pmatrix} 0 & 1 \\ 1 & 0 \end{pmatrix}\psi.$$

Representations of the Poincaré group

Now that the irreps of the Lorentz group have been constructed, we turn to the Poincaré group. First, we note that according to a theorem of Wigner, the action of proper orthochronous transformations of the Lorentz or Poincaré groups on state vectors of a quantum theory is described by means of unitary representations of these groups, or rather of their universal covers $SL(2, \mathbb{C})$ and $ISL(2, \mathbb{C})$. Unitary representations (of class L^2) of the non-compact group $SL(2, \mathbb{C})$ are necessarily of infinite dimension. The only exception is the trivial representation (0, 0), which describes a state invariant by rotation and by boosts, i.e. the vacuum—which is in fact not of class L^2.

Returning to commutation relations of the Poincaré algebra, one seeks a maximal set of commuting operators—here the four components P^μ of the linear momentum operator. One also considers the Pauli–Lubanski tensor

$$W^\lambda = \frac{1}{2}\epsilon^{\lambda\mu\nu\rho} J_{\mu\nu} P_\rho,$$

where $J_{\mu\nu}$ are defined in equation (2.8). Then,

$$[W^\mu, P^\nu] = 0 \tag{2.14a}$$

$$[W^\mu, W^\nu] = -i\epsilon^{\mu\nu\rho\sigma} W_\rho P_\sigma \tag{2.14b}$$

$$[J_{\mu\nu}, W_\lambda] = i(\eta_{\nu\lambda} W_\mu - \eta_{\mu\lambda} W_\nu). \tag{2.14c}$$

From these relations, one finally shows that $P^2 = P_\mu P^\mu$ and $W^2 = W_\mu W^\mu$ commute with all generators P, J and K. In other words, those are the Casimir operators of the Poincaré algebra. Their eigenvalues, as we will see shortly, are related to the mass of and spin of the system, and may be used to label the irreps.

The Poincaré algebra has many different irreps, but not all of them are physically meaningful. For example, there exist the irreps where $P^2 = m^2 < 0$, with m the mass. Only two types of irreps are relevant for physics, namely those where $P^2 > 0$ and $P^2 = 0$. We refer the interested reader to standard textbooks for the properties of the latter type of irrep (representing zero-mass particles) and focus exclusively on the former type. The irreps where $P^2 = p^2 = m^2 > 0$ with the time-like component of the four-momentum $p^0 > 0$ and $W^2 < 0$ describe massive particles with mass m. The eigenvectors $|p^\mu\rangle$ of P^μ and a *single* component of W^μ (let us call it W) can be chosen as the vectors of the irrep. One then writes

$$\frac{W_\mu}{m} = S_i n_\mu^{(i)} \quad \frac{W^2}{m^2} = -\boldsymbol{S}^2 = -\left(S_1^2 + S_3^2 + S_3^2\right),$$

where $n^{(i)}$ are three four-vectors orthogonal among them and with the four-momentum (hence they are space-like four-vectors), with the norm $(n^{(i)})^2 = -1$. As a consequence of equation (2.14), the S^i satisfy

$$[S_i, S_j] = i\epsilon_{ijk} S_k.$$

We are thus back on the familiar terrain of the irreps of the $\mathfrak{su}(2)$ algebra. Consequently, besides the mass m related to the Casimir invariant P^2, the unitary irreps are characterized by the eigenvalues s of $\boldsymbol{S}^2$ ($\boldsymbol{S}^2|s\rangle = s(s+1)|s\rangle$, with s a positive integer or half-integer), i.e. a spin. Moreover, the $n^{(i)}$ form with the four-momentum p is an orthonormal base called a tetrad $[p] \equiv \{p, n^{(1)}, n^{(2)}, n^{(3)}\}$, assumed to be oriented (det$[p]$ = 1). The vectors $|[p], s, m_s\rangle$ indexed by $[p]$, s and by the eigenvalue m_s of S_3 ($-s \leqslant m_s \leqslant s$) thus form a unitary irrep of the covering $ISL(2, \mathbb{C})$ of the Poincaré group. As expected, we end up with infinite-dimensional irreps since p can take any values on the mass shell $p^2 = m^2 > 0$. The action of the infinitesimal generators on such vectors reads

$$P_\mu|[p], s, m_s\rangle = p_\mu|[p], s, m_s\rangle \tag{2.15a}$$

$$P^2|[p], s, m_s\rangle = p^2\,|[p], s, m_s\rangle = m^2|[p], s, m_s\rangle \tag{2.15b}$$

$$S_3|[p], s, m_s\rangle = m_s|[p], s, m_s\rangle \tag{2.15c}$$

$$(S_1 \pm iS_2)|[p], s, m_s\rangle = \sqrt{s(s+1) - m_s(m_s \pm 1)}\,|[p], s, m_s \pm 1\rangle \tag{2.15d}$$

$$\frac{W^2}{m^2}|[p], s, m_s\rangle = -S^2|[p], s, m_s\rangle = -s(s+1)|[p], s, m_s\rangle. \tag{2.15e}$$

As for the unitary transformations $U(a, A)$ of the covering group $ISL(2, \mathbb{C})$ (associated with the Poincaré transformation (a, Λ)), we have

$$\begin{aligned} U(a, A)|[p], s, m_s\rangle &= U(a, I)U(0, A)|[p], s, m_s\rangle \\ &= e^{i(A.p)a}|[A.\, p], s, m_s'\rangle \mathcal{D}^{s}_{m_s' m_s}([A.\, p]^{-1}A[p]). \end{aligned} \tag{2.16}$$

For the purposes of notation, $[p]$ refers here to an element of $SL(2, \mathbb{C})$ associated with the Lorentz transformation mapping

$$[\mathring{p}] \equiv \{(m, 0, 0, 0); (0, 1, 0, 0); (0, 0, 1, 0); (0, 0, 0, 1)\}$$

onto

$$[p] = \{p, n^{(1)}, n^{(2)}, n^{(3)}\}.$$

Note that $[A.p]^{-1}\,A[p]$ maps $[\mathring{p}]$ onto itself, meaning that the latter is an element of the so-called 'little group', here $SU(2)$.

About the transformations of fields

Under a Lorentz transformation $\Lambda \in O(1, 3)$, objects independent of space–time coordinates transform generically as

$$\varphi_k \overset{\Lambda}{\mapsto} \varphi_k{}' = \mathcal{D}^{(j1, j2)}_{kl}(\Lambda)\varphi_l,$$

where $\mathcal{D}^{(j1, j2)}_{kl}$ are the matrix elements of a given representation (j_1, j_2). When we consider fields $\varphi(x)$, the Lorentz transformation $x' = \Lambda\, x$ must also act on the space–time argument,

$$\varphi_k(x) \overset{\Lambda}{\mapsto} \varphi_k'(x) = D_{kl}^{(j_1,j_2)}(\Lambda)\varphi_l(\Lambda^{-1}x) \Leftrightarrow \varphi_k'(x') = D_{kl}^{(j_1,j_2)}(\Lambda)\varphi_l(x).$$

In Fourier space, the field $\varphi(p)$ corresponding to a massive particle with mass m and spin j transforms according to the irreducible representations of the Poincaré group according to

$$\varphi(p) \overset{(a\Lambda)}{\mapsto} \varphi'(p) = \underbrace{e^{ipa}}_{\substack{\text{Space-time translation}\\ \text{group representations}}} \underbrace{\mathcal{D}^{(j,\,0)}(\Lambda)}_{\text{Lorentz group representations}} \varphi(\Lambda^{-1}p). \tag{2.17}$$

We are done with the properties of the Lorentz and Poincaré groups. We will now see how the transformation law (2.16) constrains the relativistic wave equation verified by a field describing a massive particle with mass m and spin s.

2.1.3 Relativistic wave equations

Because of the second equation of (2.15), any field $\varphi_{m_s}^{s}(p)$ corresponding to the vector $|[p], s, m_s\rangle$ satisfies

$$(p^2 - m^2)\varphi_{m_s}^{s}(p) = 0,$$

or in coordinate-space representation,

$$(\partial_\mu\partial^\mu + m^2)\varphi_{m_s}^{s}(x) = 0.$$

Such an equation is known as the Klein–Gordon equation. A field $\varphi_{m_s}^{s}(p)$ may describe either a spin-0 massive particle that behaves like a Lorentz scalar, or any component of a massive spin-s particle. Note also that $\varphi_{m_s}^{s}(p) = \langle p|[p], s, m_s\rangle$, i.e. it is the momentum space representation of the state vector $|[p], s, m_s\rangle$. In the case of a massive, spin-0, relativistic quantum field—describing, e.g., the σ meson—this equation can be obtained via a variational principle with respect to the action

$$S = \int d^4x\, \mathcal{L} = \frac{1}{2}\int d^4x\left[\partial_\mu\varphi\partial^\mu\varphi - m^2\varphi^2\right].$$

In the general case, the field $\varphi^s(p)$ collecting the $(2s + 1)$ components $\varphi_{m_s}^{s}(p)$ is associated with a massive particle with spin s. The Klein–Gordon equation then applies to each of its components. However, because each component $\varphi_{m_s}^{s}(p)$ belongs to the $\mathcal{D}^{(s,\,0)}(\Lambda)$ representation of the Lorentz group and because the parity operator transforms $\mathcal{D}^{(s,\,0)}(\Lambda)$ into the $\mathcal{D}^{(0,\,s)}(\Lambda)$ representation, the field $P\varphi_{m_s}^{s}(p)$ behaves according to the latter representation under a Lorentz transformation. In this case, we saw in the previous section that we need to construct a field $\psi^s(p)$ belonging to the $\mathcal{D}^s \equiv \mathcal{D}^{(s,\,0)} \oplus \mathcal{D}^{(0,\,s)}$ representation, hence collecting $2(2s + 1)$ components, in order to obtain a well-defined behavior under space reflection. Adapting equation (2.16) to the case of $ISO(1,3)$, we find

$$\psi^s(p) \overset{(a,\,\Lambda)}{\mapsto} \psi'^s(p) = e^{ipa}\mathcal{D}^s(\Lambda)\psi(\Lambda^{-1}p). \tag{2.18}$$

As a drawback, we now have twice as many components of the field than we need for a description of a particle with spin s. The relativistic wave equation is precisely what eliminates the redundant components.

With equation (2.15), we saw that a field $\psi^s(p)$ defined in the orbit $p^2 = m^2$ transforms with respect to the irrep $U^{ms}_{(a\Lambda)}$ (m being the mass) if and only if the field $\psi^s(\mathring{p})$ ($\mathring{p} = (m, 0)$) in the rest frame transforms according to the single representation $\mathcal{D}^{(s,\,0)}$. Note that the distinction between the $(s,0)$ and $(0, s)$ representations has meaning only in the case where the particle has zero-momentum. Indeed, it is under the boost transformations that these two representations differ. Calling π the projector in the $\mathcal{D}^{(s,\,0)}$ representation, such a requirement yields

$$\pi\psi^s(\mathring{p}) = \psi^s(\mathring{p}).$$

Using the transformation law (2.18), the last equation leads to the relation

$$\pi\mathcal{D}^s(\Lambda)\psi^s(p) = \mathcal{D}^s(\Lambda)\psi^s(p); \quad p = \Lambda^{-1}\mathring{p} \tag{2.19}$$

or

$$\pi(p)\psi^s(p) = \psi^s(p), \tag{2.20}$$

where

$$\pi(p) \equiv (\mathcal{D}^s)^{-1}(\Lambda)\pi\mathcal{D}^s(\Lambda). \tag{2.21}$$

Equation (2.20) is the relativistic wave equation for a massive particle with spin s.

The case of a spin-1/2 field—the Dirac equation

In the particular case of a massive spin-1/2 relativistic quantum field such as the nucleon field, we note

$$\psi^{\frac{1}{2}}(p) \equiv \psi(p) = \begin{pmatrix} \xi(p) \\ \chi(p) \end{pmatrix}.$$

Under a Poincaré transformation, each two-component spinor transforms according to

$$\xi(p) \overset{(a\Lambda)}{\mapsto} e^{ipa}\mathcal{D}^{(\frac{1}{2},\,0)}(\Lambda)\xi(\Lambda^{-1}p)$$
$$\chi(p) \overset{(a\Lambda)}{\mapsto} e^{ipa}\mathcal{D}^{(0,\,\frac{1}{2})}(\Lambda)\chi(\Lambda^{-1}p),$$

while the four-component Dirac spinor transforms as

$$\psi(p) \overset{(a\Lambda)}{\mapsto} e^{ipa}\mathcal{D}^{\frac{1}{2}}(\Lambda)\psi(\Lambda^{-1}p),$$

where $\mathcal{D}^{\frac{1}{2}} \equiv \mathcal{D}^{(\frac{1}{2},\,0)} \oplus \mathcal{D}^{(0,\,\frac{1}{2})}$. To remove the two unwanted redundant components, the projection π takes the form

$$\pi = \begin{pmatrix} 1 & 0 & 0 & 0 \\ 0 & 1 & 0 & 0 \\ 0 & 0 & 0 & 0 \\ 0 & 0 & 0 & 0 \end{pmatrix} \equiv \frac{1}{2}(\gamma_0 + I_4).$$

We then deduce from equation (2.21) that the covariant operator $\pi(p)$ takes the form

$$\pi(p) \equiv (\mathcal{D}^{\frac{1}{2}})^{-1}\pi\mathcal{D}^{\frac{1}{2}} = \frac{1}{2m}(\gamma_\mu p^\mu + m),$$

where γ_μ stand for the Dirac matrices. Consequently, the general equation (2.20) reads, in the spin-1/2 case,

$$(\gamma_\mu p^\mu - m)\psi(p) = 0, \tag{2.22}$$

or in coordinate-space representation,

$$(i\gamma_\mu \partial^\mu - m)\psi(x) = 0.$$

Such an equation is known as the Dirac equation and can be obtained from a variational principle with respect to the action

$$S = \int d^4x\mathcal{L} = \int d^4x\bar{\psi}(x)\left[i\gamma_\mu\partial^\mu - m\right]\psi(x).$$

The Dirac matrices introduced above are a set of 4×4 matrices that read

$$\gamma^0 = \begin{pmatrix} 1 & 0 & 0 & 0 \\ 0 & 1 & 0 & 0 \\ 0 & 0 & -1 & 0 \\ 0 & 0 & 0 & -1 \end{pmatrix}, \quad \gamma^1 = \begin{pmatrix} 0 & 0 & 0 & 1 \\ 0 & 0 & 1 & 0 \\ 0 & -1 & 0 & 0 \\ -1 & 0 & 0 & 0 \end{pmatrix},$$

$$\gamma^2 = \begin{pmatrix} 0 & 0 & 0 & -i \\ 0 & 0 & +i & 0 \\ 0 & +i & 0 & 0 \\ -i & 0 & 0 & 0 \end{pmatrix}, \quad \gamma^3 = \begin{pmatrix} 0 & 0 & 1 & 0 \\ 0 & 0 & 0 & -1 \\ -1 & 0 & 0 & 0 \\ 0 & 1 & 0 & 0 \end{pmatrix}.$$

The γ_μ components are obtained as $\gamma_\mu \equiv (\gamma^0, -\gamma^k)$. Another important matrix is γ^5, which is defined by

$$\gamma^5 = i\gamma^0\gamma^1\gamma^2\gamma^3.$$

The case of a spin-1 field—the Proca equation

According to our study of the irreps of the Poincaré group, a massive spin-1 particle, such as the ω and ρ mesons, is represented by a field $\phi^1(p)$ collecting the three components $\phi^1_{m_s}(p)$ with $m_s = \{-1, 0, 1\}$ and behaving under a Poincaré transformation according to

$$U^{m,\,1}_{(a\Lambda)}\phi^1(p) = e^{ipa}\mathcal{D}^{(1,0)}(\Lambda)\phi^1(\Lambda^{-1}p).$$

However, such fields do not transform in a well-defined manner under space inversions. We thus have two possibilities. We can work with the representation $\mathcal{D}^{(1,0)} \oplus \mathcal{D}^{(0,1)}$ as we did in the case of a spin-1/2 particle. We would end up with a field with six components and would project out the three unwanted components, a projection yielding the relativistic wave equation. Alternatively, we can work with the $\mathcal{D}^{(\frac{1}{2},\frac{1}{2})}$ irrep, since

$$\mathcal{D}^{(\frac{1}{2},\frac{1}{2})}|_{\mathrm{SU}(2)} \simeq D^1 + D^0.$$

In other words, the $\mathcal{D}^{(\frac{1}{2},\frac{1}{2})}$ irrep contains a spin-1 particle, and in addition a spin-0 particle that we want to eliminate. Therefore, instead of working with a six-component field and trying to project out three unwanted components, we will work with a four-component field $\phi^\mu(p)$ that behaves like a four-vector, and project out one component, so that it describes a spin-1 particle.

The projector π onto the three-vector space can be written as

$$\pi = \begin{pmatrix} 0 & 0 & 0 & 0 \\ 0 & 1 & 0 & 0 \\ 0 & 0 & 1 & 0 \\ 0 & 0 & 0 & 1 \end{pmatrix} \equiv \frac{1}{2}(\delta_{\mu\nu} - \eta_{\mu\nu}).$$

We then deduce from equation (2.21) that the covariant operator $\pi(p)$ takes the form

$$\pi(p) \equiv (\mathcal{D}^{(\frac{1}{2},\frac{1}{2})})^{-1}\pi\mathcal{D}^{(\frac{1}{2},\frac{1}{2})} = \frac{1}{2}\left(\frac{p_\mu p_\nu}{m^2} - \eta_{\mu\nu}\right).$$

Consequently, the general equation (2.20) in the spin-1 case reads

$$\frac{1}{2}\left(\frac{p_\mu p_\nu}{m^2} - \eta_{\mu\nu}\right)\phi^\mu(p) = \phi_\nu(p).$$

Multiplying both sides with p^ν we obtain

$$p^\nu \phi_\nu(p) = 0,$$

which we can combine with the Klein–Gordon equations satisfied by each component of $\phi^\mu(p)$

$$(p^2 - m^2)\phi_\mu(p) = 0,$$

where m stands for the mass of the spin-1 particle. In the coordinate basis, the last two equations read

$$\partial_\mu\phi^\mu(x) = 0; \quad (\partial_\nu\partial^\nu + m^2)\phi^\mu(x) = 0,$$

which can be cast into the form of a set of first-order equations by setting

$$\begin{cases} B^{\mu\nu} \equiv \partial^\mu\phi^\nu - \partial^\nu\phi^\mu \\ \partial_\mu B^{\mu\nu} + m^2\phi^\nu = 0 \end{cases}.$$

The last equation is known as the Proca equation and can be obtained via a variational principle with respect to the action

$$S = \int d^4x \ \mathcal{L} = \frac{1}{2} \int d^4x \left[m^2 \phi_\mu \phi^\mu - \frac{1}{2} B_{\mu\nu} B^{\mu\nu} \right].$$

We now have all the necessary ingredients to formulate a relativistic description of nuclear systems. Such a description has to implement in one way or another the symmetry properties that QCD exhibits at low energy. Therefore, we first briefly review the symmetry pattern of QCD in the following section.

2.2 Symmetry properties of QCD

QCD is currently accepted as the fundamental theory of the strong interactions which ensure the cohesion of matter. It belongs to the Yang–Mills theories, which are gauge theories based on a (non-Abelian) compact group that provide the conceptual basis underpinning the Standard Model of particle physics. QCD is built on the $SU(3)$ gauge group, where $SU(3)$ stands for the special unitary group in three dimensions, whose elements are the set of complex-valued, unitary 3×3 matrices with determinant one. The form of the QCD Lagrangian is uniquely fixed by a few specific postulates in addition to the general principles of any quantum field theory such as locality, gauge symmetry or the criterion of renormalizability. QCD is based on the gauge color group $SU(N_c)$ ($N_c = 3$ is the number of colors) with colored quarks and gluons as elementary d.o.f.s. They are coupled through the Lagrangian (density)

$$\mathcal{L}_{\mathrm{QCD}} = \bar{q}_f \left(i\gamma^\mu D_\mu - m_f \right) q^f - \frac{1}{4g^2} G^{\mu\nu} G_{\mu\nu}. \tag{2.23}$$

The Dirac field $q_f(x)$ represents a quark with flavor $f = \{u, d, s, c, b, t\}$ (an implicit summation over flavor indices is assumed) and mass m_f. Each $q_f(x)$ belongs to the fundamental representation of the color group $SU(N_c)$. When $N_c = 3$, it is the triplet

$$q^f(x) = \begin{pmatrix} q^f_{\mathrm{red}}(x) \\ q^f_{\mathrm{green}}(x) \\ q^f_{\mathrm{blue}}(x) \end{pmatrix}, \tag{2.24}$$

where each component is a (four-component) Dirac spinor. The term $D_\mu \equiv \partial_\mu + iA_\mu$ is the quark covariant derivative. It involves the spin-1 gluon field $A_\mu \equiv \frac{\lambda_a}{2} A_\mu^a$, $a = 1, \ldots, 8$ collecting the eight color charged gluons. Following [33], our choice for the gauge field A_μ differs from the one usually introduced by

$$A_\mu^{\text{here}} = g A_\mu^{\text{usual}}.$$

This is done in order to make apparent the fact that the factor $1/g$ in equation (2.23) plays the role of a stiffness parameter (it quantifies the energy cost to create curvature in the gauge field). The λ_a stand for the 3×3 Gell-Mann matrices which form a matrix representation of the $SU(3)$ group. Finally, $G_{\mu\nu} \equiv \partial_\mu A_\nu - \partial_\nu A_\mu + i[A_\mu, A_\nu]$ refers to the gluon field strength tensor and g to the quark–gluon coupling constant.

The Lagrangian (2.23) may appear surprising based on what we learned in section 2.1. Indeed, the properties of quarks should in principle be encoded all together in the representation of the (universal cover of the) Poincaré group for a massive spin-1/2 field. The latter yields an invariant characterization of quarks in terms of their mass and spin, such that, only considering quark d.o.f.s, one would expect a Lagrangian of the form

$$\mathcal{L}_0 = \bar{q}(x)(i\gamma^\mu \partial_\mu - m)q(x). \tag{2.25}$$

Such a Lagrangian is far simpler than equation (2.23), and actually fails to account for the overwhelming number of excited states of stable particles found in hadron spectroscopy. The route taken to provide an explanation for the observed patterns was to assume the existence of internal d.o.f.s related to internal symmetries such as, for instance, the $SU(N_f)$ symmetry with N_f the number of flavors, or the $SU(N_c)$ symmetry which explains among others why the observed Δ^{++}, Δ^- and Ω^- baryons can exist without violating the Pauli principle. Introducing such internal symmetries raises new questions about how to articulate these internal dimensions with the usual space–time dimensions. The answer depends on the 'nature' of the symmetry, i.e. global (physical and generally approximate) or local (gauge redundancy), as we will briefly review.

Local and global symmetries

Local and global symmetries are intuitively classified by looking at the group parameters that label the corresponding transformations: if they are independent of the position in space–time, the symmetry is said to be global or rigid. If they are explicitly dependent on space–time coordinates, the symmetry is said to be local. For instance, a 'global' $U(1)$ transformation for a complex field ψ reads

$$\psi \overset{U(1)}{\mapsto} e^{i\theta}\psi,$$

with $\theta \in \mathbb{R}$ a real number, while a 'local' $U(1)$ transformation is of the form

$$\psi \overset{U(1)}{\mapsto} e^{i\theta(x)}\psi,$$

where the parameter $\theta(x)$ is now a function of space–time coordinates. At the same time, this classification does not rigorously determine whether a symmetry of a field theory is gauge or physical [34]. A safe way to discriminate between the gauge or physical nature of a symmetry is to evaluate the associated Noether currents. Physical symmetries have associated Noether currents that lead to non-trivial Noether charges after integration. These charges can be used to parametrize the space of solutions of the classical theory. In contrast, gauge symmetries lead to identically zero Noether charges. They should not be viewed as telling us something fundamental about Nature. Because of the Gauss law, the Hilbert space associated with a quantum system is gauge invariant. In other words, none of the states belonging to the Hilbert space are affected by gauge transformations, so that calling gauge symmetry a symmetry is a misnomer. If a global symmetry is a property of a system, a gauge symmetry is a property of a description of a system, or more precisely a redundancy in the mathematical description of a physical system [35]. Still, gauge redundancies prove extremely useful to make a theory manifestly Lorentz invariant, unitary, local and therefore causal. In this sense they are exact emergent symmetries (they have to be exact otherwise they are not redundancies).

Emergent phenomena are common in condensed-matter physics: global symmetries that emerge in a low-energy limit are always approximate symmetries because they are explicitly violated by operators of higher dimension that are irrelevant in the renormalization group sense. This is very similar to, e.g. the non-conservation of the electron, muon and tau lepton numbers $L_e - L_\mu$ and $L_\mu - L_\tau$ in the standard model of particle physics. Emergent gauge symmetries also provide very useful low-energy descriptions of a system by introducing more d.o.f.s than needed and then 'gluing' them together with gauge fields [36]. Such exact gauge symmetries are emergent indeed, in the sense that they do not have any particular meaning in the microscopic Schrödinger equation for electrons. They typically involve quasiparticles, whose charges can be used to construct all possible representations of the symmetry group. In the context of low-energy effective approaches to QCD, we will run into such an example of emergent gauge symmetry, also referred to as hidden local symmetry [37], with the ρ and ω vector mesons playing the role of the dynamical gauge boson of the theory.

All in all, at the classical level and in the so-called chiral limit where the quarks are considered mass-less, the QCD action

$$S_{\rm QCD}^{m_q=0} = \int d^4x \mathcal{L}_{\rm QCD}^{m_q=0} = \int d^4x \left\{ \bar{q}_f i\gamma^\mu D_\mu q^f - \frac{1}{4g^2} G^{\mu\nu} G_{\mu\nu} \right\} \tag{2.26}$$

exhibits an additional symmetry group $G_{\rm class}$ in addition to Poincaré invariance and the discrete C (charge-conjugation), P (parity) and T (time-reversal) symmetries. $G_{\rm class}$ can be split into the internal symmetry group $G_{\rm int}$ and the dilatation group $R^+_{\rm scale}$:

$$\begin{aligned} G_{\rm class} &\equiv G_{\rm int} \times R^+_{\rm scale} \\ &\equiv SU(3)_c \times SU(N_f)_L \times SU(N_f)_R \times U(1)_B \times U(1)_A \times R^+_{\rm scale}. \end{aligned}$$

We first discuss the symmetries that QCD exhibits at the classical level before addressing how some of these symmetries can be hidden through the mechanism of spontaneous symmetry-breaking (SSB), or broken by the quantum nature of QCD.

2.2.1 Symmetries of QCD at the classical level

The color SU(3) gauge group

Introducing a $SU(3)$ gauge redundancy implies that quarks are ascribed with a more sophisticated structure—that of a *charged* spin-1/2 field: they are represented by a field carrying a color charge as an internal d.o.f.. A quark thus transforms under $SU(3)_c$ according to its fundamental representation

$$q_f(x) \overset{SU(3)_c}{\mapsto} e^{i\alpha_a T_a} q_f(x),$$

with T_a ($a = 1, \ldots, 8$) the generators of the $SU(3)_c$ group. In such a case of a local gauge symmetry, the internal space where the internal symmetry operates and space–time are 'stitched' together through the introduction of a non-trivial structure called a fiber bundle; see figure 2.2. The latter provides the geometrical foundation of (classical) Yang–Mills theory; see e.g. [38] for a detailed review.

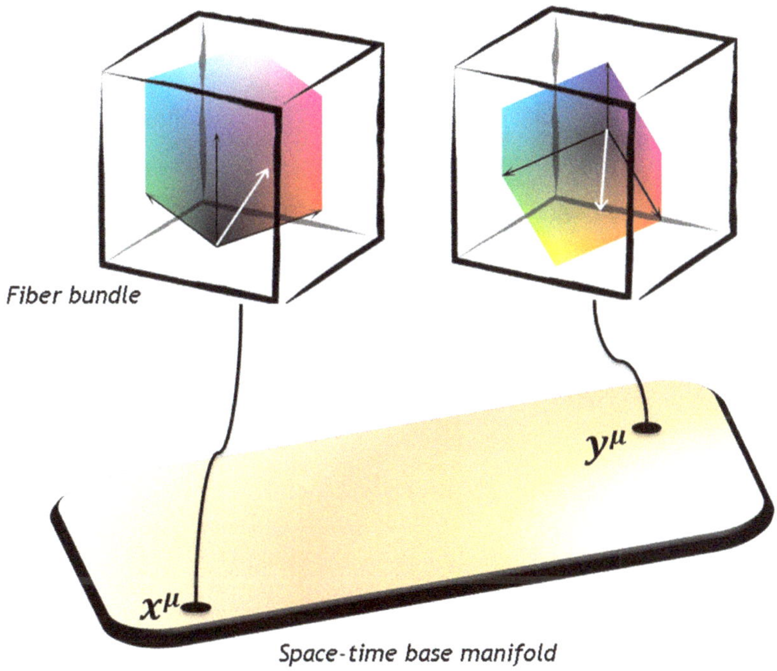

Figure 2.2. A very schematic view of a fiber bundle in the context of QCD. To each four-vector x^μ in the Minkowski space–time, we attach a vector space V_x lying above x. The elements of V_x are three-dimensional vectors that transform according to the fundamental representation of the $SU(3)$ color group. (Figure created by J-P Ebran.)

A theory whose internal space is to be thought of as a bundle over space–time inherits the property to be local, which can be explicitly manifested by substituting the ordinary derivative ∂_μ in the free quark Lagrangian (2.25) by the covariant derivative $D_\mu = \partial_\mu + iA_\mu$ defined in terms of the gauge field A_μ (hence minimally coupled to quark fields), and by promoting the gauge field to the rank of a dynamical d.o.f.. In other words, a (gauge-invariant) kinetic term for the A_μ must be introduced as well. One eventually ends up with the Lagrangian (2.23).

Global internal symmetries of QCD

From hadron spectroscopy, one can infer the existence of several types of quarks (currently six) with different masses that have the same properties with respect to the gluon fields. One distinguishes them with a new quantum number, called flavor: we end up with the quarks *u*, *d*, *s*, *c*, *b* and *t*. Quark (anti-quark) fields are assigned to the N_f-dimensional fundamental representation (its complex conjugate one) with respect to the special unitary group $SU(N_f)$ acting in the internal space of flavors; see e.g. [39] for a recent review of global chiral symmetry.

In the idealized situation where all the quarks share the same mass m, the quark part of the Lagrangian reads

$$\mathcal{L}_q = \bar{q}_f\left(i\gamma^\mu D_\mu - m\right)q^f,$$

where summation of repeated indices is assumed (f runs from 1 to N_f). Such a Lagrangian is invariant under the continuous global transformations of $SU(N_f)$, hence $(N_f^2 - 1)$ Noether conserved currents can be found

$$j_\mu^A(x) = -i\frac{\partial \mathcal{L}_q}{\partial(\partial^\mu q^a)}(T^A)^a_b q^b = \bar{q}_a\gamma_\mu(T^A)^a_b q^b,$$

where the $N_f \times N_f$ Hermitian trace-less T^A matrices are the representatives of the generators of the $SU(N_f)$ group. The T^A matrices hence satisfy the $\mathfrak{su}(N_f)$ algebra $[T^A, T^B] = if_{ABC}T^C$ with A, B, C ranging from 1 to $N_f^2 - 1$ and f_{ABC} the corresponding structure constant. Because of current conservation $\partial^\mu j_\mu^A = 0$, the Noether charges

$$Q^A = \int d^3x\, j_0^A(x)$$

therefore commute with the Hamiltonian of the system

$$[H, Q^A] = 0,$$

so that their eigenvalues (the flavor index) label the quark field. In the real world, the different quarks *u*, *d*, *s*, *c*, *b* and *t* have different masses, so that $SU(N_f)$ is broken. In the $N_f = 2$ and $N_f = 3$ flavor sector however, the relative smallness of the quark masses makes $SU(N_f)$ an approximate global symmetry.

Fermion fields may undergo other types of unitary transformations than those corresponding to flavor symmetry. One example is the set of transformations generated by including the γ_5 Dirac matrix in flavor transformations. Called axial flavor transformations (they change the parity properties of fields) they leave the quark Lagrangian invariant if all the quarks are mass-less,

$$\mathcal{L}_q^{m_q=0} = \bar{q}_f i\gamma^\mu D_\mu q^f - \frac{1}{4g^2} G^{\mu\nu} G_{\mu\nu}. \tag{2.27}$$

In such a case, the conserved currents associated with both flavor and axial flavor transformations read, respectively,

$$j_\mu^A(x) = \bar{q}_a \gamma_\mu (T^A)^a_b q^b \qquad j_{5\mu}^A(x) = \bar{q}_a \gamma_5 \gamma_\mu (T^A)^a_b q^b. \tag{2.28}$$

The corresponding charges

$$Q^A = \int d^3x\, j_0^A(x) \qquad Q_5^A = \int d^3x j_{5,\,0}^A(x) \tag{2.29}$$

satisfy the following algebra

$$[Q^A, Q^B] = if_{ABC} Q^C \quad \left[Q_5^A, Q_5^B\right] = if_{ABC} Q^C \quad \left[Q^A, Q_5^B\right] = if_{ABC} Q_5^C.$$

Note that the axial charges do not form an algebra on their own. The previous algebra can, however, be made more transparent by defining

$$Q_L^A \equiv \frac{1}{2}\left(Q^A - Q_5^A\right) \qquad Q_R^A \equiv \frac{1}{2}\left(Q^A + Q_5^A\right)$$

so that one obtains

$$\left[Q_L^A, Q_L^B\right] = if_{ABC} Q_L^C \quad \left[Q_R^A, Q_R^B\right] = if_{ABC} Q_R^C \quad \left[Q_L^A, Q_R^B\right] = 0.$$

Therefore, the left-handed and right-handed charges are decoupled and operate separately, each of them generating an $SU(N_f)$ group of transformations.

Finally, the Lagrangian (2.27) is also invariant under the $U(1)_B$ and $U(1)_A$ groups. The group $U(1)_B$ corresponds to a global gauge transformation with the same phase for all quark fields, and leads to baryon number B conservation; the group $U(1)_A$ corresponds to global gauge transformations, but with the global phase having the opposite sign for left-handed and right-handed quarks, respectively.

The $U(N_f)_L \times U(N_f)_R \simeq SU(N_f)_L \times SU(N_f)_R \times U(1)_B \times U(1)_A$ group defines the chiral group. Note that for practical reasons, the $SU(N_f)_L \times SU(N_f)_R$ subgroup will also be referred to as the chiral group.

Scale invariance of QCD

At the classical level, the QCD action (2.26) displays another type of invariance, not related to internal d.o.f.s but to another kind of space–time transformation, namely scale invariance R^+_{scale}. The dilatation group R^+_{scale} is part of a larger (15-dimensional) group called the conformal group. A scale transformation with parameter $\lambda \in \mathbb{R}^*$ reads

$$x_\mu \overset{R^+_{\text{scale}}}{\mapsto} x'_\mu = \lambda^{-1} x_\mu$$

and induces a transformation of the fields $\mathbf{\Phi}(x) = \{q(x), A_\mu(x)\}$

$$\mathbf{\Phi}(x) \overset{R^+_{\text{scale}}}{\mapsto} \lambda^\Delta \mathbf{\Phi}(\lambda^{-1} x),$$

with $\Delta = 1$ and 3/2 for the gluon and quark fields, respectively. The generator D of the scale transformation follows from the consideration of infinitesimal transformations, i.e. with parameters of the form $\lambda = 1 + \varepsilon$, $\varepsilon \ll 1$,

$$D = -ix^\mu \partial_\mu.$$

Because the only parameter of the theory, namely the coupling constant g, is dimensionless (in $\hbar = c = 1$ units), the action (2.26) is invariant under such a rescaling, so that the corresponding Noether current

$$J^\mu_D = x_\nu \Theta^{\mu\nu}, \tag{2.30}$$

where $\Theta^{\mu\nu}$ is the (improved [40]) energy–momentum tensor, and is divergence-less and trace-less

$$\partial_\mu J^\mu_D = \Theta^\mu_\mu = 0. \tag{2.31}$$

In other words, at the classical level, the physical content of the theory looks identical whatever the scale of observation λ.

The value of Δ associated with scale transformations can be obtained from simple dimensional analysis considerations. In the system of units with $\hbar = c = 1$, the action

$$S^{m_q=0}_{\text{QCD}} = \int d^4x \; \bar{q}_f i\gamma^\mu D_\mu q^f - \frac{1}{4g^2} G^{\mu\nu} G_{\mu\nu}$$

is dimensionless. Writing all the contributions to the action in units of, say, mass, we obtain

$$\left[\int d^4x\right] = -4 \qquad [\partial_\mu] = [A_\mu] = 1 \qquad [q] = \Delta_q.$$

Because the coupling constant g is dimensionless, we obtain the relation

$$\left[\int d^4x\right] + [\partial_\mu] + 2[q] = 0,$$

that is $\Delta_q = 3/2$.

In summary, at the classical level, the QCD action (2.26) exhibits, in addition to Poincaré invariance and the discrete C (charge-conjugation), P (parity) and T (time-reversal) symmetries, the apparent symmetry

$$G_{\text{class}} \equiv SU(3)_c \times SU(N_f)_L \times SU(N_f)_R \times U(1)_B \times U(1)_A \times R^+_{\text{scale}}.$$

In reality, these symmetries are not always manifest, so that the actual symmetries of QCD are quite different from the ones encoded in G_{class}.

2.2.2 Actual symmetries of QCD

Even if the color $SU(3)$ gauge symmetry plays a fundamental role to build QCD as a manifestly local, Lorentz invariant, unitary (and hence causal) quantum field theory, there is no obvious manifestation of local color gauge symmetry in the low-energy, -temperature and -density regime of QCD because of color confinement [41]. Instead, the relevant symmetry group at this level,

$$G_{\text{class}} \to G_{\text{actual}} \equiv SU(N_f)_L \times SU(N_f)_R \times U(1)_B \times U(1)_A \times R^+_{\text{scale}},$$

as well as the d.o.f.s needed for an EFT, are color-singlet ones, namely hadrons.

Next, chiral symmetry, as a global symmetry, can either be realized in the Wigner–Weyl (WW) scheme or in the Nambu–Goldstone (NG) scheme [42, 43]. In the former, the symmetry of the theory is manifest: its ground state $|0\rangle$ is invariant under the corresponding transformations of the group and (massive) hadrons appear in degenerate parity multiplets. This implies, for instance, that all the Noether charges (2.29) annihilate the ground state

$$Q^A|0\rangle = 0.$$

In the latter realization, the symmetry of the theory is hidden: its ground state does not exhibit the symmetry and hadrons are not in multiplets. However, there appear mass-less, spin-less, bosons, called NG bosons. We refer to the latter manifestation under the name of 'spontaneously broken chiral symmetry'. The concept of SSB translates the fact that stable solutions of equations can manifest lower symmetry than the equations themselves. Note, however, that the term 'broken' is a misnomer insofar as the symmetry is not really broken but rather hidden. In that case, the (axial-vector flavor) generators of the symmetry group do not annihilate the ground state

$$Q_5^A|0\rangle \equiv |\phi^A\rangle \neq 0 \quad A = 1, \ldots, N_f^2 - 1.$$

The SSB mechanism produces states $|\phi^A\rangle$ that share the same quantum properties as their generating axial charges ($|0\rangle$ is assumed to have a positive-parity and no other quantum numbers), i.e. they are pseudo-scalar states. Note that the norm of $|\phi^A\rangle$ is infinite. These modes are not one-particle states, but rather zero-energy limits of a superposition of mass-less, many-particle states. In low-energy nuclear physics, chiral symmetry is realized in the NG way. An explanation for chiral SSB involves the formation in the QCD vacuum of a non-vanishing chiral condensate, that is, a condensate of quark–anti-quark pairs of opposite handedness filling space–time uniformly. Because of it, one can no longer freely rotate left- and right-handed quarks independently since it does not leave the chiral condensate invariant, so that the chiral group $G \equiv SU(N_f)_L \times SU(N_f)_R$ spontaneously breaks into its vector subgroup $H \equiv SU(N_f)_{V=L+R}$. Such an SSB involves NG fields [42, 43] whose number is the number of generators of G that are not also generators of H, i.e. the dimension of the coset G/H (i.e. $N_f^2 - 1$). One can identify the pseudo-scalar NG bosons $\phi \equiv [\pi, K, \eta]$ with the pions in the $N_f = 2$-flavor sector, plus the kaons and an η meson for $N_f = 3$.

> In the chiral limit, the NG modes are mass-less. In the real world, the non-zero mass of quarks, originated at a scale higher than the standard model, explicitly breaks chiral symmetry. When small, e.g. for the light quarks (in the $N_f = 2{,}3$-sectors) relevant to nuclear physics, the explicit breaking of chiral symmetry can be treated in a perturbative way, as exemplified by chiral perturbation theory (χPT).

They couple to the axial current (2.28) according to

$$\langle 0 | j^A_{5\mu} | \phi^B(p) \rangle = i p_\mu f_\phi \delta^{AB},$$

with p the four-momentum of the NG field and f_ϕ its decay constant (from the leptonic decays of pions and kaons [44], one finds ~92 MeV for the former and ~110 MeV for the latter). The ratio $\frac{f_K}{f_\pi} - 1 \simeq 0.2$ measures the extent to which the global flavor $SU(3)$ group is broken.

The symmetry group relevant for building an EFT for low-energy nuclear physics is then

$$G_{\text{actual}} \to G'_{\text{actual}} \equiv SU(N_f)_V \times U(1)_B \times U(1)_A \times R^+_{\text{scale}}.$$

In the next section, we will discuss how in practice one realizes chiral symmetry in a non-linear way (under the form of a non-linear sigma model), based on the coset space G/H (recall that G stands for the chiral group $SU(N_f)_L \times SU(N_f)_R$ and H for its unbroken subgroup $SU(N_f)_V$). As a result, in addition to nucleon d.o.f.s, the corresponding EFT only involves as fundamental d.o.f.s the NG modes associated with the chiral SSB. An alternative viewpoint relies on the fact that a non-linear sigma model based on the manifold G/H is gauge-equivalent to a model having the symmetry $G_{\text{global}} \times H_{\text{local}}$ [37]. After spontaneous breaking of the local symmetry H_{local}, the corresponding gauge boson of the local symmetry H_{local} is identified as the massive vector mesons $V = \{\rho, \omega\}$. In other words, in this alternative linear and gauged representation, the ρ and ω vector mesons play a role as fundamental as the pseudo-scalar NG states of the chiral symmetry in building an EFT [45]. Introducing such heavy fields as relevant d.o.f.s of the EFT forces us to be cautious about the power counting of the theory. However, note that according to the Brown–Rho scaling [46], the masses of the vector mesons tend towards the pseudo-scalar NG modes as the baryonic density of the medium increases, so that the situation may not be as bad as it may seem when working at finite-density.

Symmetry-breaking and anomalies

Symmetries exhibited by QCD at the classical level can be reduced by quantum effects. Such a failure of the symmetry of the theory at the classical level to be also a symmetry of the theory at the quantum level is called an anomaly [47]. At play here are the dynamical features stemming from the structure of QCD as a full-fledged quantum field theory, namely the fact that virtual particle–anti-particles of all types pop in and out of empty space. Among the consequences of this effect, the scale invariance of QCD is totally lost, causing the running of the QCD coupling constant. Indeed, because empty space is full of virtual particle–anti-particle pairs, it behaves as a dynamical medium with dielectric and paramagnetic properties which amount to charge screening and anti-screening. As a consequence, the trace of the energy–momentum tensor (2.31) is no longer zero but instead

$$\Theta^{\mu}_{\mu} = \frac{\beta(\alpha_s)}{4\alpha_s} G^{\mu\nu} G_{\mu\nu}, \tag{2.32}$$

where $\beta(\alpha_s) \equiv \partial\alpha_s/\partial log(\Lambda)$ (with Λ the energy scale at which one looks at the theory) stands for the beta function for the QCD gauge coupling constant $\alpha_s \equiv g^2/4\pi$.

Recent work also suggests that scale invariance may also be 'spontaneously' broken [48, 49], in addition to being explicitly broken by the trace anomaly (2.32). Such an idea is related to the existence of an infrared (IR) fixed-point $\beta(\alpha_{\text{IR}}) = 0$ towards which QCD flows at low energies.

The existence of such a fixed point in the (matter-free) ground state of QCD at a low number of flavors ($N_f = 2{,}3$) remains an open question, even though it appears highly plausible. Such an IR fixed point is even more likely in the QCD ground state at finite baryonic density where it could be generated through strong nuclear correlations as an emergent symmetry, even if it were absent in the matter-free vacuum [50].

The scalar meson σ, which corresponds to $f_0(500)$ in hadronic physics, is identified with the scalar NG boson called the dilaton associated with the spontaneous breaking of scale symmetry [48]. Its mass is generated by explicit symmetry-breaking encoded both in the departure of α_s from the IR fixed point α_{IR} because of a non-zero gluon condensate, and in the non-zero quark mass. These very recent studies emphasize that the non-linear σ meson, which plays a central role in the Walecka model (2.50) of section 2.4 and in the construction of phenomenological Lagrangians, is not only a convenient phenomenological construction, but has a genuine physical origin as the scalar NG boson associated with scale SSB. This will allow us to define an EFT that combines both chiral symmetry and scale symmetry, enabling systematic expansions both in the chiral counting and in the scale counting [51, 52].

2.3 Effective Lagrangians for nuclear systems

Our goal is now to set up an EFT scheme that encompasses as much as possible the physics encoded in the fundamental theory of strong interactions. Put differently, taking the generating functional of QCD

$$Z[J]_{\text{QCD}} = \int \mathcal{D}q\mathcal{D}\bar{q}\mathcal{D}A_\mu e^{i\int d^4x\left[\mathcal{L}_{\text{QCD}}^{m_q=0}+\bar{q}Jq\right]} \tag{2.33}$$

in the presence of the external fields $J(x)$ with

$$J(x) \equiv -s(x) + ip(x)\gamma_5 + \gamma_\mu \nu^\mu(x) + \gamma_\mu a^\mu(x)\gamma_5$$

for the sources of the scalar (s), pseudo-scalar (p), vector (ν^μ) and axial-vector (a^μ) currents, we seek a generating functional of the form

$$Z[J]_{\text{eff}} = \int \mathcal{D}\Phi_{\text{low}} e^{i\int d^4x\mathcal{L}_{\text{eff}}^{\Lambda}(\Phi_{\text{low}},\, J;\partial_\mu\Phi_{\text{low}},\, \ldots)} \tag{2.34}$$

involving relevant low-energy d.o.f. Φ_{low} with dynamics encoded in the effective Lagrangian $\mathcal{L}_{\text{eff}}^{\Lambda}$. The generating functionals (2.33) and (2.34) should yield the same results (to within the working accuracy) for observables computed below the breakdown scale Λ of the EFT. Note that equation (2.33) should include an extra (Fadeev–Popov) term coming with additional functional integrations over anti-commuting Bose fields (called ghost fields) when including a gauge fixing condition.

The choice of a relevant resolution scale Λ conditions the low-energy d.o.f. Φ_{low} to be explicitly considered. Recall that Φ_{low} corresponds to any d.o.f. consistent with the symmetry pattern of QCD in the relevant energy, density and temperature domain. Within the nuclear many-body problem, the occurrence of many different characteristic scales gives rise to the formulation of a variety of possible effective theories, each related one to one another within a web of EFT. Below, we outline two basic strategies to formulate a low-energy EFT of QCD capable of providing a robust and coherent foundation to the phenomenological Lagrangians employed in nuclear CEDFS. The first strategy was proposed by Furnstahl, Serot and Tang [15] and the second one by Rho [51, 52].

2.3.1 Effective Lagrangian à la Furnstahl, Serot and Tang

Relevant degrees of freedom

The effective chiral Lagrangian proposed by Furnstahl, Serot and Tang [15] involves nucleons and pions as the essential d.o.f. (in the $N_f = 2$-sector). In other words, in a diagrammatic expansion, only these fields can appear on external lines with time-like four-momenta.

Recall that at low energies, the local color gauge symmetry is hidden because of color confinement. Therefore it sounds more reasonable to work with hadrons

> rather than quark and gluon fields. Moreover, small momenta compared to Λ only partially probe short-distance physics (e.g. hadrons substructure) that can then be incorporated into the coefficient of the field operators organizing the EFT expansion.

Nucleon fields are indeed the observed fermionic d.o.f.s at low energies, that is, they carry the conserved baryon number B. They are represented by an isospinor-Dirac spinor field

$$\psi(x) = \begin{pmatrix} \psi_n(x) \\ \psi_p(x) \end{pmatrix},$$

where $\psi_n(x)$ and $\psi_p(x)$ stand for the Dirac spinor neutron field and the Dirac spinor proton field, respectively. Although they can emerge as topological solitons, namely skyrmions [53], from a purely pionic Lagrangian, it will prove more convenient to introduce them explicitly by hand.

As pseudo-scalar NG modes associated with the chiral SSB, pions govern the long- and medium-range dynamics of the system. They form an iso-triplet described by the 2×2 special unitary matrix

$$\pi(x) \equiv \vec{\pi}(x) \cdot \frac{1}{2}\vec{\tau} = \frac{1}{2}\begin{pmatrix} \pi^0 & \sqrt{2}\pi^+ \\ \sqrt{2}\pi^- & -\pi^0 \end{pmatrix},$$

with $\vec{\tau}$ the Pauli matrices and the arrow indicating vectors in the isospin space.

The short-range (equivalently high-energy) physics of the system can generically be parametrized by a complete set of terms. A possible representation involves contact terms. As discussed above, alternative representation brings into play non-pseudo-scalar NG bosons such as the $\rho(770)$ meson (isovector–vector field $\rho_\mu(x) \equiv \vec{\rho}_\mu(x) \cdot \frac{1}{2}\vec{\tau}$), the $\omega(782)$ meson (isoscalar–vector field $\omega_\mu(x)$) and the σ meson (isoscalar–scalar field $\sigma(x)$). We used the standard notation $meson(mass_{meson})$.

> The treatment of the sigma field is very similar to that of pion-less EFT where, even if pions are not indispensable, it proves to be much more powerful and predictive to keep pions as effective d.o.f.s in certain cases. Their explicit treatment brings with them their associated low-energy theorems from which one can obtain a highly simplified and efficient description of processes which would require much harder work if the pion were integrated out.

Historically, they were introduced on a phenomenological basis as useful d.o.f.s for parametrizing the medium- and short-range parts of the nucleon–nucleon (NN) interaction (as well as the electromagnetic form factors of hadrons). As discussed in the previous section, recent work suggests that they have in fact a physical origin as gauge bosons and dilaton. In any case, such additional d.o.f.s prove very efficient in capturing the physics encoded in complicated pion loop integrals at the tree level, a property all the more appreciated since the rapid proliferation of the unknown low-energy parameters of the chiral effective Lagrangian at higher chiral orders limits the χPT expansion to low orders (which makes it difficult to address the broad range of nuclear phenomenology).

On the other hand, the introduction of relativistic particles with masses of the order of, or larger than, the breakdown scale Λ_χ, such as nucleons and the NG fields, is known to violate the systematic power counting of the theory [54]. One may employ a heavy-baryon formulation, which involves an expansion of baryon fields in $1/M$ (M being the mass of the heavy baryon), to recover a well-defined power counting [55]. However, this implies that baryons are then treated as non-relativistic d.o.f.s. There are alternative prescriptions that maintain the relativistic nature of baryons while restoring the power counting. They follow from the key observations of Tang [56] that low-energy expansions provided by EFT prove useful only if all of the hard-momentum effects are absorbed into the parameters of the Lagrangian. Since a relativistic treatment of nucleons implies anti-nucleon contributions as hard-momentum effects, absorbing the hard part of the corresponding diagrams into parameters recovers chiral power counting for the remaining soft part. Such a program has been carried out by many teams, e.g. [57], resulting in Lorentz-covariant EFT of QCD with relativistic d.o.f.s with large masses and a systematic power counting regained.

Symmetry-allowed terms

The relevant d.o.f.s being defined, we now seek to form their most general combinations, with the constraint that they preserve the underlying symmetry pattern of QCD, that is, Lorentz covariance, parity conservation, time-reversal and charge-conjugation invariance, (approximate) isospin symmetry as well as (approximate) chiral symmetry. Recall that a symmetry can be realized in two main ways: linear representation (WW) and non-linear realization (NG). Because chiral symmetry belongs to the latter category (i.e. it is spontaneously broken), it is the most tricky symmetry to implement in the effective Lagrangian. Note also that in order for an EFT to be consistent with the fundamental underlying theory, it must reproduce any anomalies of the latter. In QCD, this 't Hooft anomaly matching condition implies that there must be a massless, color-less hadron which couples to the axial current and photon and reproduces exactly the triangle anomaly: hence the pseudo-scalar NG boson.

Consider a symmetry group G spontaneously broken and H a conserved subgroup of G. The corresponding NG field $\phi(x)$ transforms under G as [58]

$$\phi \overset{g\in G}{\mapsto} \phi' = f(g, \phi),$$

with the remarkable property that the non-linear transformation functions f are unique up to field transformations of the NG modes ϕ. First consider the elements $h \in G$ such that $f(h,0) = 0$, i.e. the elements which leave invariant the origin in the ϕ space. They form the unbroken subgroup $H \subset G$. For all other elements g (i.e. $g \in G$ and $g \notin H$), the group property $f(gh,0) = f(g, f(h,0)) = f(g,0)$ with $h \in H$ shows that $f(g,0)$ is an element of the (left) coset space G/H: $\{gH: g \in G\}$. Since the non-linear mapping f between the NG fields $\phi(x)$ and the coset space G/H is invertible, the NG fields provide a parametrization of G/H. In other words, NG fields can be viewed as coordinates of the coset space G/H, such that a field transformation of ϕ is nothing other than a coordinate transformation in G/H.

Non-linear realization of a symmetry

The non-linear realization of a symmetry follows from the geometry of the coset space G/H. Let us first decompose the Lie algebra of the symmetry group G into the generators H_i of the conserved subgroup H ($SU(N_f)_V$ in the chiral case) and the remaining spontaneously broken generators X_a (the axial generators in the chiral case). The Lie algebra reads

$$\left[H_i, H_j\right] = ic_{ijk}H_k \quad [H_i, X_a] = ic_{iab}X_b \quad [X_a, X_b] = ic_{abi}H_i, \tag{2.35}$$

with structure constants c. The first relation of equation (2.35) characterizes the algebra $\mathfrak{h}$ of the subgroup H, while the second relation defines the X_a representation of H. The NG modes ϕ can then parametrize any element ξ of the coset space through the standard form [58]

$$\xi(\phi) = e^{i\phi^a X_a}. \tag{2.36}$$

A symmetry transformation $g \in G$ hence induces in a natural way (by left action) a transformation of $\xi(\phi) \in G/H$,

$$\xi(\phi) \overset{g\in G}{\mapsto} g\xi(\phi) = \xi(\phi')h(\phi, g), \tag{2.37}$$

where the field $h(\phi, g) \in H$, called compensator, reminds the fact that an element of the coset space G/H is only defined up to an H transformation. For $g \in H$ the symmetry is realized in the usual linear way (WW). On the other hand, for $g \in G$ such that $g \notin H$, the symmetry is realized non-linearly, which is signaled by the fact that the compensator field $h(\phi, g)$ depends non-trivially on the NG modes and that the latter transform non-linearly. For practical purposes, one never needs to know the explicit form of the compensator field, but rather the geometry of the coset space. The latter is encoded in the so-called Cartan–Maurer one-form ω_ϕ

$$\omega_\phi\colon \xi(\phi)^{-1}d\xi(\phi) = \Gamma + E = (\Gamma_a + E_a)d\phi^a = (\Gamma^i_a H_i + E^b_a X_b)d\phi^a, \tag{2.38}$$

where we have introduced the vielbein one-form $E(\phi)$ and the natural connection one-form $\Gamma(\phi)$ on coset space. Their transformation under a symmetry $g \in G$ follows from equation (2.37)

$$E \overset{g\in G}{\mapsto} hEh^{-1}, \quad \Gamma \overset{g\in G}{\mapsto} h\Gamma h^{-1} + hdh^{-1}.$$

In particular, the connection $\Gamma(\phi)$ has an important practical application for forming G-invariant combinations involving NG fields and other matter fields. It induces a covariant derivative

$$D\psi \equiv (d + \Gamma)\psi = \left(\partial_\mu + \Gamma_a \frac{\partial \phi^a}{\partial x_\mu}\right)\psi dx^\mu \equiv D_\mu \psi dx^\mu \tag{2.39}$$

supplying the missing ingredient for forming G-invariant terms for the effective Lagrangian.

Indeed, while a generic multi-component matter field $\psi(x)$ transforms in the usual linear way under the conserved subgroup H, i.e. $\psi \overset{h\in H}{\mapsto} \psi' = h_\psi(h)\psi$, where h stands both for the group element and its linear representation in ψ, for a general transformation $g \in G$, one defines

$$\psi \overset{g\in G}{\mapsto} \psi' = h_\psi(\phi, g)\psi$$

in terms of the compensator field $h_\psi(\phi, g)$ in the ψ representation. Because the latter depends on space–time coordinates, the normal derivative $\partial_\mu\psi$ does not transform covariantly (i.e. in the same way as ψ). The remedy is known from gauge theories and the role played by the gauge connection is now taken by the Cartan–Maurer connection $\omega(\phi)$.

In the linear realization case, one has by definition $g \in H$. Therefore, one obtains

$$g\xi(\phi) = e^{i\alpha^k H_k} e^{i\phi^a X_a} = \underbrace{e^{i\alpha^k H_k} e^{i\phi^a X_a} e^{-i\alpha^k H_k}}_{\xi(\phi')=e^{i\phi'^a X_a}} \underbrace{e^{i\alpha^k H_k}}_{h(g,\phi)=h(g)=g},$$

so that the compensator field does not depend on the NG modes and the ϕ themselves transform linearly because of equation (2.35),

$$e^{i\alpha^k H_k} X_a e^{-i\alpha^k H_k} = X_b \mathcal{R}_X(\alpha)_{ba} \Rightarrow \phi'^a = \mathcal{R}_X(\alpha)_{ab}\phi^b,$$

with $\mathcal{R}_X(\alpha)_{ab}$ the X representation of H.

Following the seminal work of Weinberg [59] further developed by Callan, Coleman, Wess and Zumino [58], chiral symmetry is realized in a non-linear way by considering elements of the coset space $\xi(\pi) = (\xi_L(\pi), \xi_R(\pi)) \in G/H$. Here, $G = SU(N_f)_L \times SU(N_f)_R$ is the chiral group, $H = SU(N_f)_{V=L+R}$ its conserved subgroup and π generically stands for the NG fields that describe the chiral SSB in the N_f-flavor sector. The coordinates on the coset space are parametrized by the NG fields, e.g. the standard exponential form (2.36). A chiral transformation $g = (g_L, g_R) \in G$, with $g_{L(R)}$ a global matrix belonging to $SU(N_f)_{L(R)}$, induces a coordinate transformation in the coset space

$$\xi(\pi) \overset{g}{\mapsto} \xi(\pi') = (g_L, g_R)(\xi_L(\pi), \xi_R(\pi))(h^\dagger(\pi, g), h^\dagger(\pi, g)),$$

with $h(\pi, g)$ the corresponding compensator field. The chiral group exhibits an additional symmetry, namely the (outer automorphism) parity, from which it follows that

$$\xi(\pi) \overset{g}{\mapsto} g_R \xi(\pi) h^\dagger(\pi, g) = h(\pi, g)\xi(\pi) g_L^\dagger. \tag{2.40}$$

Such a property allows one to define a field

$$U(\pi) \equiv \xi_R(\pi) \xi_L{}^\dagger(\pi) \tag{2.41}$$

transforming linearly under chiral transformations

$$U(\pi) \overset{g}{\mapsto} g_L U(\pi) g_R^\dagger. \tag{2.42}$$

The standard choice $\xi_R(\phi) = \xi_L{}^\dagger(\phi)$ yields

$$U(\pi) \equiv U(x) \equiv u^2(x) = e^{2i\frac{\pi(x)}{f_\pi}}. \tag{2.43}$$

Because the field $U(x)$ always transforms globally, its derivatives transform in the same way as $U(x)$, so that one can form chirally invariant combinations in terms of the product of U, $U^\dagger$ and their derivatives when it comes to purely pionic interactions.

The action of G on the other d.o.f.s reads [15]

$$\sigma(x) \overset{g}{\mapsto} \sigma(x) \tag{2.44a}$$

$$\omega_\mu(x) \overset{g}{\mapsto} \omega_\mu(x) \tag{2.44b}$$

$$\rho_\mu(x) \overset{g}{\mapsto} h(g, \pi(x)) \rho_\mu(x) h^\dagger(g, \pi(x)) \tag{2.44c}$$

$$\psi(x) \overset{g}{\mapsto} h(g, \pi(x))\psi(x); \quad \bar{\psi}(x) \overset{g}{\to} \bar{\psi}(x) h^\dagger(g, \pi(x)). \tag{2.44d}$$

The isoscalar fields σ and ω are chiral singlets and are thus unaffected by both chiral and isospin transformations. In the special case where g also belongs to $SU(2)_V$, the compensator field $h(\pi(x), g)$ reduces to the element g itself, so that equations (2.44c) and (2.44d) show that ρ meson and nucleon fields transform linearly and globally in accordance with their isospins. However, in the general case $g \in G$, one has to construct their covariant derivative (see equation (2.39)) in order to end up with a chirally invariant effective Lagrangian. To this end, we write the vielbein one-form E and the Cartan–Maurer connection one-form Γ (see equation (2.38)) when the field $U(x)$ is chosen under the form (2.43):

$$E = -\frac{1}{2}(u^\dagger du - udu^\dagger) \equiv -ia_\mu dx^\mu$$

$$\Gamma = \frac{1}{2}(u^\dagger du + udu^\dagger) \equiv iv_\mu^\dagger dx^\mu.$$

The axial a_μ and polar v_μ vectors hence read

$$a_\mu \equiv -\frac{i}{2}\left(u^\dagger \partial_\mu u - u\partial_\mu u^\dagger\right) = a_\mu^\dagger$$
$$\nu_\mu \equiv -\frac{i}{2}\left(u^\dagger \partial_\mu u + u\partial_\mu u^\dagger\right) = \nu_\mu^\dagger$$

and transform homogeneously and non-homogeneously, respectively, under the action of the chiral group:

$$a_\mu(x) \overset{g}{\mapsto} h(\pi, g)a_\mu(x)h^\dagger(\pi, g)$$
$$\nu_\mu(x) \overset{g}{\mapsto} h(\pi, g)(\nu_\mu(x) - \partial_\mu)h^\dagger(\pi, g).$$

The connection ν_μ allows to construct the chirally covariant derivatives for nucleon and ρ meson fields

$$D_\mu\psi \equiv (\partial_\mu + i\nu_\mu)\psi; \quad D_\mu\rho_\nu \equiv \partial_\mu\rho_\nu + i[\nu_\mu, \rho_\nu],$$

which by definition transform under chiral transformations as the nucleon and ρ fields, respectively. To complete the ingredients needed to build the effective chiral Lagrangian, we need the curvature (strength field) tensor

$$\nu_{\mu\nu} \equiv \partial_\mu\nu_\nu - \partial_\nu\nu_\mu + i[\nu_\mu, \nu_\nu] = -i[a_\mu, a_\nu]$$

and the covariant field tensor

$$\rho_{\mu\nu} \equiv D_\mu\rho_\nu - D_\nu\rho_\mu + ig_\rho[\rho_\mu, \rho_\nu],$$

with g_ρ the ρ meson coupling constant.

An infinite number of chirally invariant terms can now be constructed by combining those various ingredients and we need an organizing scheme that tells us which terms are more important than the others in order to make systematic calculations.

Power counting

Ranking such symmetry-allowed terms according to their importance, in other words defining a power-counting scheme, first supposes that we understand how to extract the dimensional scales of each term in the Lagrangian. This can be achieved following Georgi and Manohar's naive dimensional analysis (NDA) [60] and applying naturalness, namely that all appropriately defined, dimensionless couplings are of order unity. This leads us to write the effective Lagrangian in the form

$$\mathcal{L}_{\text{eff}} = \sum_{n,d,p,b}^{\infty} c_{ndpb}\left(\frac{\bar{\psi}\Gamma\psi}{f_\pi^2\Lambda}\right)^{\frac{n}{2}}\left(\frac{\mathcal{D}, m_\pi}{\Lambda}\right)^{d}\left(\frac{\pi(x)}{f_\pi}\right)^{p}\left(\frac{\mathcal{D}, m_\pi}{\Lambda}\right)^{d}$$
$$\times \frac{1}{b!}\left(\frac{\rho(x), \omega(x), \sigma(x)}{f_\pi}\right)^{b} = \sum_{\eta=0}^{\infty} \mathcal{L}^{(\eta)},$$

where c_{ndpb} are dimensionless low-energy constants (LECs) assumed to be natural (i.e. of $\mathcal{O}(1)$), Γ is a product of Dirac matrices, $\mathcal{D}$ is a covariant derivative and m_π is the pion mass (treated as derivatives). The effective Lagrangian is organized in increasing powers of the fields and their derivatives, collected by the index $\eta \equiv \frac{n}{2} + d + b$. Recall that even when we lose a systematic power counting when dealing with a relativistic particle with mass of the order of, or larger than, the breakdown scale Λ [54], prescriptions such as [57] still result in a Lorentz-covariant EFT of QCD with the systematic power counting regained. The effective Lagrangian is also constructed with the constraint that each term is consistent with the electromagnetic gauge symmetry $U(1)_{EM}$.

We are now ready to write the Lorentz-covariant effective Lagrangian at any given order η. We will write it as

$$\mathcal{L}_{\text{eff}}^{(\eta)} = \mathcal{L}_N^{(\eta)} + \mathcal{L}_M^{(\eta)} + \mathcal{L}_{EM}^{(\eta)},$$

where we have separated the nucleon, meson and electromagnetic channels, respectively.

The form of the effective Lagrangian at low order

We will not discuss the electromagnetic channel, which is pretty standard, see e.g. [61]. For $\eta \leqslant 4$ we have for the nucleon part [15]

$$\begin{aligned}\mathcal{L}_N^{(4)} &= \bar{\psi}\Big[\gamma^\mu\big(i\partial_\mu - v_\mu - g_\rho\rho_\mu - g_\omega\omega_\mu\big) + g_A\gamma^\mu\gamma_5 a_\mu - (M - g_\sigma\sigma)\Big]\psi \\ &\quad - \frac{f_\rho g_\rho}{4M}\bar{\psi}\rho_{\mu\nu}\sigma^{\mu\nu}\psi - \frac{f_\omega g_\omega}{4M}\bar{\psi}\omega_{\mu\nu}\sigma^{\mu\nu}\psi - \frac{\kappa_\pi}{M}\bar{\psi}v_{\mu\nu}\sigma^{\mu\nu}\psi,\end{aligned}$$

where $\sigma^{\mu\nu} \equiv 2i[\gamma^\mu, \gamma^\nu]$, $\omega_{\mu\nu} \equiv \partial_\mu\omega_\nu - \partial_\nu\omega_\mu$ is the covariant tensor of the ω meson, $g_A \simeq 1.26$ is the axial coupling constant, g_i, f_i, $i = \{\omega, \rho\}$ are the vector and tensor couplings for ω and ρ mesons [62], g_σ is a Yukawa coupling for the effective scalar field σ and $\kappa_\pi \equiv f_\rho/4$ is the coupling for higher-order πN interaction. For the mesonic part,

$$\begin{aligned}\mathcal{L}_M^{(4)} &= \frac{1}{2}\left(1 + \alpha_1\frac{g_\sigma\sigma}{M}\right)\partial_\mu\sigma\partial^\mu\sigma + \frac{f_\pi^2}{4}\mathrm{tr}\big(\partial_\mu U\partial^\mu U^\dagger\big) \\ &\quad - \frac{1}{2}\mathrm{tr}\big(\rho_{\mu\nu}\rho^{\mu\nu}\big) - \frac{1}{4}\left(1 + \alpha_2\frac{g_\sigma\sigma}{M}\right)\omega_{\mu\nu}\omega^{\mu\nu} - g_{\rho\pi\pi}\frac{2f_\pi^2}{m_\rho^2}\mathrm{tr}\big(\rho_{\mu\nu}v^{\mu\nu}\big) \\ &\quad + \frac{1}{2}\left(1 + \eta_1\frac{g_\sigma\sigma}{M} + \frac{\eta_2}{2}\left(\frac{g_\sigma\sigma}{M}\right)^2\right)m_\omega^2\omega_\mu\omega^\mu + \frac{1}{4!}\zeta_0 g_\omega^2(\omega_\mu\omega^\mu) \\ &\quad + \left(1 + \eta_\rho\frac{g_\sigma\sigma}{M}\right)m_\rho^2\mathrm{tr}\big(\rho_{\mu\nu}\rho^{\mu\nu}\big) - m_\sigma^2\sigma^2\left(1 + \frac{\kappa_3}{3!}\frac{g_\sigma\sigma}{M} + \frac{\kappa_4}{4!}\left(\frac{g_\sigma\sigma}{M}\right)^2\right),\end{aligned}$$

where m_i, $i = \{\sigma, \omega\rho\}$ stands for the σ, ω and ρ mesons masses, $g_\pi\rho\rho$ is the coupling of $\pi\rho\rho$ and $\{\alpha_1, \alpha_2, \eta_1, \eta_2, \eta_\rho, \zeta_0, \kappa_3, \kappa_4\}$ are LECs. Assuming vector-meson dominance, $g_{\pi\rho\rho} = g_\rho$. The trace 'tr' is in the 2×2 isospin space.

At leading order, the effective Lagrangian à la Furnstahl, Serot and Tang has the same form as phenomenological Lagrangians of the non-linear sigma type. However its free parameters are not matched to QCD, nor to two-nucleon scattering data. Rather, a CEDF is deduced from the effective Lagrangian truncated at some order of the chiral expansion and treated at the relativistic mean-field (RMF) level (along the lines of what is detailed in section 2.5). Ascertaining the preservation of the power counting of the theory can be done in an empirical way, by performing many-body calculations for various nuclei based on different chiral orders of the effective Lagrangian and checking the consistency of the radiative corrections. The free parameters of the theory are then adjusted on the usual many-body observables (typically binding energies and radii) of a small set of finite nuclei [63]. In other words, the nucleon interactions are optimized in the presence of the nuclear medium. This is similar to the philosophy of nuclear EDF, and even more similar to the recently defined hybrid approaches using *ab initio*-like many-body methods, however, fueled by chiral potentials partly optimized on many-body observables, namely $NNLO_{sat}$ type potentials [64].

> The free parameters entering the $NNLO_{sat}$ potential are adjusted on both NN scattering data (and deuteron properties) and many-body observables, namely binding energies and charge radii of ^3H, $^{3,4}He$, ^{14}C and ^{16}O, as well as binding energies of $^{22,24,25}O$, computed in a coupled-cluster approach. If such a procedure yields impressive success, note that it is no longer a genuine *ab initio* approach, since a part of the many-body physics is reabsorbed in an uncontrolled way in the NN and three-body (3N) interactions. Neither is it a genuine EFT sine it has not been proven that a well-defined power-counting scheme has been preserved.

2.3.2 Effective Lagrangian à la Rho

Another way of designing an effective Lagrangian that gives a theoretical foundation to the phenomenological Lagrangians underpinning the CEDF approach goes along the lines detailed in [51, 52]. Here, the chiral Lagrangian involves nucleons, pseudo-scalar and scalar NG bosons π and σ, and massive gauge bosons ρ and ω (as well as their equivalents in the $N_f = 3$-sector).

> These ρ, ω and σ d.o.f.s prove convenient to describe the non-vanishing expectation value of the bi-linear nucleon operator (e.g. $\bar{\psi}\psi$ or $\bar{\psi}\gamma^\mu\psi$, with ψ referring to the nucleon spinor) important in nuclear many-body physics. Hence, they can be considered as bosonic collective d.o.f.s in the sense of the Hubbard–Stratonovich transformation [65]. For instance, considering a four-fermion coupling of the current–current type $\mathcal{L}_{\text{int}} = G(\bar{\psi}\psi)^2$, the path integral approach allows one to linearize the four-fermion interaction with the help of Hubbard–Stratonovich transformations. The latter introduces a bosonic auxiliary field $\sigma(x)$ such that $e^{G(\bar{\psi}\psi)^2} = \mathcal{N}\int \mathcal{D}\sigma e^{\frac{\sigma^2}{4G}+\bar{\psi}\sigma\psi}$, with $\mathcal{N}$ a normalization factor.

The corresponding chiral EFT is matched with QCD ground state at 'finite-density'. The sliding of QCD features when going from the matter-free vacuum to the finite-density vacuum can be expressed via, for example, the Brown–Rho scaling [46], so that one can constrain the bare parameters of the EFT based on the finite-density properties of quark and gluon condensates [66–68]. At this stage, one obtains at lowest order a Walecka-like, in-medium, effective Lagrangian [52] characterized by a well-defined power counting. It is possible to further lower the resolution Λ of the effective theory near the Fermi momentum k_F that characterizes nuclear systems. One would then flow from the in-medium chiral EFT to the Laudau Fermi liquid fixed point, although the details to arrive at such an effective Lagrangian remain to be explored (the fact that treating a relativistic Lagrangian along the paradigm of Walecka RMF captures the Fermi liquid fixed-point theory is, however, well accepted; see e.g. [69]).

> The Landau theory of Fermi liquids provides a universal low-energy description of a vast class of interacting fermion systems. It is a fixed point at low-energy in the abstract space of model Hamiltonians in the Wilson renormalization group language. Landau Fermi liquids connect adiabatically and analytically to a non-interacting Fermi gas, their ground state preserves all the symmetries of the system, and their low-energy physics is dominated by Landau fermionic quasi-particles with conserved quantum numbers and whose dynamical properties are renormalized by Landau forward scattering interactions. Strictly speaking, as for many systems in condensed-matter physics, most of the nuclei depart from this fixed point for other stable points corresponding to Hamiltonians with restricted d.o.f.s as compared to the Fermi liquid fixed point via the following possible mechanisms: (i) symmetry-breaking, (ii) the emergence of new symmetry and (iii) non-trivial topology. See [70] for a recent review in the context of condensed-matter physics.

We now construct the effective Lagrangian by considering first the pseudo-scalar and scalar NG d.o.f.s, then the vector mesons introduced via emergent gauge symmetry, and finally the baryonic d.o.f.s.

Nambu–Goldstone fields: the scale-chiral symmetry
As explained above, the σ meson can be considered at the same level of pions and their 'higher flavor' relatives by considering it as the scalar NG boson associated with the scale SSB. This provides the attractive feature that standard χPT can be extended to scale-chiral perturbation theory (χPTσ), which combines both chiral and scale symmetries; see e.g. [48] for details about how we can arrive at a well-defined power counting in this case.

To write the corresponding effective Lagrangian, one introduces the conformal compensator field ζ, a chiral-invariant d.o.f. which can be expressed in terms of the scalar NG boson σ as

$$\zeta(x) = f_\sigma e^{\sigma(x)/f_\sigma},$$

where f_σ is the σ decay constant. The conformal compensator field has mass-dimension $[\zeta] = 1$ and, under dilatation $x \mapsto \lambda^{-1}x$, transforms linearly as

$$\zeta(x) \mapsto \lambda\zeta(\lambda^{-1}x),$$

whereas the dilaton $\sigma(x)$ scales non-linearly,

$$\sigma(x) \mapsto \sigma(\lambda^{-1}x) + f_\sigma ln(\lambda).$$

The pseudo-scalar NG bosons represented by the field $U(x)$ transform under scale transformation as

$$U(x) \mapsto U(\lambda^{-1}x).$$

To the leading scale-chiral order $\mathcal{O}(p^2)$, the effective Lagrangian for scalar and pseudo-scalar NG d.o.f.s hence reads [48]

$$\mathcal{L}_{\chi PT_\sigma}^{LO} = \frac{f_\pi^2}{4}\left(\frac{\zeta}{f_\sigma}\right)^2 \mathrm{Tr}\left(\partial_\mu U \partial^\mu U^\dagger\right) + \frac{1}{2}\partial_\mu\zeta\partial^\mu\zeta + V(\zeta), \tag{2.45}$$

where $V(\zeta)$ is a potential that captures the scale symmetry-breaking, explicit as well as spontaneous.

Vector-meson fields: emergent gauge symmetry
The vector-meson fields $V(x) = (\rho(x), \omega(x))$ can be brought into the theory by realizing that the non-linear sigma model based on the manifold G/H is gauge-equivalent to a model displaying the symmetry $G_{\mathrm{global}} \times H_{\mathrm{local}}$ (called HLS for hidden local symmetry). After spontaneous breaking of the local gauge symmetry H_{local}, the corresponding gauge boson is identified with the massive vector mesons. Note that $H_{\mathrm{local}} = SU(N_f)_V^{\mathrm{local}}$ can be extended to the local gauge symmetries $U(N_f)_V^{\mathrm{local}}$ or $\left[SU(N_f)_V \times U(1)_V\right]_{\mathrm{local}}$. The gauge symmetry comes from the product form (2.41) of the chiral field U [71]; being a product of two local fields brings in infinite redundancies, e.g. in the form of an invariance under the action of $G_{\mathrm{global}} \times H_{\mathrm{local}}$, which leads to the transformation

$$\xi_{L,\,R}(x) \mapsto \xi'_{L,\,R}(x) = h(x)\xi_{L,\,R}(x)g_{L,\,R}^\dagger,$$

where $h(x) \in H_{\mathrm{local}}$ and $g_{L,\,R} \in G$. Contrary to the non-linear realization of chiral symmetry, $h(x) \in H_{\mathrm{local}}$ does not depend on the NG modes. The variables $\xi_{L,\,R}$ can be exponentially parametrized by the NG bosons π and ϑ associated with the spontaneous breaking of G_{global} and H_{local}, respectively,

$$\xi_{L,\,R}(x) = e^{\pm i\frac{\pi(x)}{2f_\pi}} e^{i\frac{\vartheta(x)}{2f_\vartheta}}\, \pi(x) = \pi_a X^a, \quad \vartheta(x) = \vartheta_a H^a,$$

with f_ϑ the decay constant of the $\vartheta(x)$ field, X^a the broken generators of G_{global} and H^a the remaining ones. As before, we associate with the gauge theory the gauge connection one-form $V_\mu(x)$ which defines a covariant derivative $\mathcal{D}_\mu = \partial_\mu - iV_\mu$ and the curvature (or gauge field strength tensor) $V_{\mu\nu}(x) = \partial_\mu V_\nu(x) - \partial_\nu V_\mu(x) - i\left[V_\mu(x), V_\nu(x)\right]$, such that the effective Lagrangian at leading order reads

$$\mathcal{L}_{HLS} = f_\pi^2 \,\mathrm{Tr}(a_{\perp\mu}a_\perp^\mu) + f_\vartheta^2 \,\mathrm{Tr}(a_{\|\mu}a_\|^\mu) - \frac{1}{2g_V^2}\mathrm{Tr}(V_{\mu\nu}V^{\mu\nu}), \tag{2.46}$$

where g_V is the gauge coupling associated with the local gauge group H_{local} and $a_{(\perp,\,\|)\mu}$ are two one-forms defined as

$$a_{\|\mu} \equiv \frac{1}{2i}\left(\mathcal{D}_\mu\xi_R \cdot \xi_R^\dagger + \mathcal{D}_\mu\xi_L \cdot \xi_L^\dagger\right), \quad a_{\perp\mu} \equiv \frac{1}{2i}\left(\mathcal{D}_\mu\xi_R \cdot \xi_R^\dagger - \mathcal{D}_\mu\xi_L\dot{\xi}_L^\dagger\right).$$

The masses of the vector mesons are obtained via the Higgs mechanism, i.e. through the spontaneous breaking of H_{local}. Hence, one takes the unitary gauge $\vartheta(x) = 0$ leading to

$$\xi_L^\dagger = \xi_R \equiv \xi = e^{i\frac{\pi}{2f_\pi}} \qquad U(x) = \xi^2(x), \quad f_\vartheta^2 = 2f_\pi^2$$

and the effective Lagrangian (2.46) reads

$$\mathcal{L}_{HLS} = f_\vartheta^2 \,\mathrm{Tr}\left[(\partial_\mu\xi \cdot \xi^\dagger + \partial_\mu\xi^\dagger \cdot \xi - 2iV_\mu)(\partial^\mu\xi \cdot \xi^\dagger + \partial^\mu\xi^\dagger \cdot \xi - 2iV^\mu)\right] \tag{2.47}$$

from which follows the mass formula $m_V^2 = g_V^2 f_\vartheta^2$.

Baryonic degrees of freedom

Finally, nucleons $\psi(x)$ with mass m can be introduced and coupled to the NG and gauge fields through the Lagrangian

$$\mathcal{L}_\mathrm{B} = \bar{\psi}\left(i\gamma^\mu\mathcal{D}_\mu - \frac{\zeta}{f_\sigma}m + g_A\gamma^\mu\gamma_5 a_\mu\right)\psi \tag{2.48}$$

in the unitary gauge. The total effective Lagrangian $\mathcal{L}_{bsHLS}$ expressed in terms of all the hadronic d.o.f.s hence reads

$$\mathcal{L}_{bsHLS} = \mathcal{L}_\mathrm{M} + \mathcal{L}_B + V(\zeta), \tag{2.49}$$

with $\mathcal{L}_\mathrm{B}$ given in equation (2.48) and $\mathcal{L}_\mathrm{M}$ following from equations (2.45) and (2.47), i.e.

$$\mathcal{L}_\mathrm{M} = f_\pi^2\left(\frac{\zeta}{f_\sigma}\right)^2 \mathrm{Tr}(a_{\perp\mu}a_\perp^\mu) + 2f_\pi^2\left(\frac{\zeta}{f_\sigma}\right)^2 \mathrm{Tr}(a_{\|\mu}a_\|^\mu)$$
$$- \frac{1}{2g_V^2}\mathrm{Tr}(V_{\mu\nu}V^{\mu\nu}) + \frac{1}{2}\partial_\mu\zeta\partial^\mu\zeta.$$

Developing the $a_{(\perp,\,\|)\mu}$ and ζ field at lowest order yields a Lagrangian of the linear sigma type with density-dependent masses, that looks like phenomenological

Lagrangians of the CEDF approach where the density dependence is recast into the coupling constants.

2.4 Phenomenological Lagrangians

The phenomenological Lagrangians underpinning CEDFS are reinterpreted as a downgraded version of a non-renormalizable Lorentz-covariant low-energy effective Lagrangian of QCD, such as the one discussed in the previous section. While they share the symmetry properties of a Lorentz-covariant EFT, phenomenological Lagrangians do not include a proper power counting because of the adjustment of their parameters to many-body observables. In this regard, phenomenological Lagrangians are meant to be used at the 'tree level' only and, like their non-relativistic counterpart, are not characterized by a systematic method to account for higher-order corrections. Such Lagrangians originate from the basic concepts of the Walecka model [8], first proposed by Teller [72, 73], where the Lagrangian reads, in its simplest form

$$\begin{aligned}\mathcal{L}_{\text{Walecka}} = \bar{\psi}\left(i\gamma_\mu\partial^\mu - M - g_\sigma\sigma - g_\omega\gamma_\mu\omega^\mu\right)\psi + \frac{1}{2}\partial_\mu\sigma\partial^\mu\sigma \\ - \frac{1}{2}m_\sigma^2\sigma^2 - \frac{1}{4}\omega^{\mu\nu}\omega_{\mu\nu} + \frac{1}{2}m_\omega^2\omega_\mu\omega^\mu.\end{aligned} \tag{2.50}$$

In practice, the contribution of the ρ meson and the Coulomb interaction have to be included in order to obtain reasonable results. However, the important point is that the Walecka Lagrangian (2.50) indeed appears as a very simple version of the chiral effective Lagrangian derived in the previous section. Because its free parameters are adjusted to many-body observables, it is no longer characterized by a systematic power counting. However, it still retains from the low-energy effective Lagrangians of QCD the consistent realization of the QCD symmetry pattern: it respects Lorentz covariance, parity conservation, time-reversal and charge-conjugation invariance, (approximate) isospin symmetry as well as spontaneously broken (approximate) chiral symmetry.

Within the framework of Walecka-like models, the meson d.o.f.s are to be thought of as effective d.o.f.s in the sense that they parametrize the non-vanishing bi-linear combinations of local nucleon fields (e.g. $\langle\bar{\psi}(x)\psi(x)\rangle$ or $\langle\psi^\dagger(x)\psi(x)\rangle$) in terms of collective bosonic d.o.f.s. Here again, we use the term collective d.o.f.s in the sense of the Hubbard–Stratonovich transformation [65]; see the remark above. The coupling of the collective fields to the nucleon bi-linear forms is interpreted as interactions through meson exchanges. Because these interactions are meant to depict how nucleons are coupled in the medium, the effective meson fields implicitly re-sum the bulk of nucleon correlations. Consequently, the nucleon–meson coupling constants define the phenomenological free parameters of the model to be adjusted on experimental data. Simplicity is the compass that guides the choice of the effective meson fields. One therefore tries to include only as few mesons as possible to reproduce a typical set of nuclear properties, e.g. the isoscalar–scalar σ meson, the isoscalar–vector ω meson, the isovector–scalar ρ meson and the isovector–pseudo-scalar π meson. Together with the electromagnetic four-potential A^μ accounting for Coulomb interaction between protons, phenomenological Lagrangians based of these d.o.f.s are the simplest realization of CEDF. The fact that such a simple approach has already led to a good reproduction of nuclear bulk properties motivated the development of various extensions of the initial Walecka model. In the new few sections, we will review several families of these phenomenological Lagrangians.

The use of phenomenological Lagrangians based on equation (2.50) with the σ, ω, ρ and π mesons (and the Coulomb interaction) is usually referred to as RMF theory, or relativistic Hartree approximation. At this level, the fluctuations for the effective meson fields are not taken into account (i.e. the mesonic fields are not quantized), which leads to a semi-classical theory.

2.4.1 Non-linear Lagrangians

It was soon recognized that an in-medium interaction was crucial if quantitative agreement with experimental data was to be achieved (nuclear surface properties in particular). Historically, the first family of models including an effective medium dependence was the non-linear model, proposed by Boguta and Bodmer [74], in which the medium dependence is included via non-linear self-interactions between the scalar mesons. A typical Lagrangian of the non-linear family reads

$$\begin{aligned}\mathcal{L}_{\rm NL} = \bar{\psi}\left\{i\gamma_\mu\partial^\mu - M - g_\sigma\sigma - g_\omega\gamma_\mu\omega^\mu - g_\rho\gamma_\mu\vec{\rho}^{\,\mu}\cdot\vec{\tau} - e\gamma_\mu A^\mu\frac{1-\tau_3}{2}\right\}\psi \\ + \frac{1}{2}\partial_\mu\sigma\partial^\mu\sigma - U_\sigma[\sigma] - \frac{1}{4}\omega^{\mu\nu}\omega_{\mu\nu} + U_\omega[\omega^\mu] \\ - \frac{1}{4}\vec{\rho}^{\,\mu\nu}\cdot\vec{\rho}_{\mu\nu} + U_\rho[\vec{\rho}^{\,\mu}] - \frac{1}{4}F^{\mu\nu}F_{\mu\nu},\end{aligned} \tag{2.51}$$

with the electromagnetic strength tensor $F_{\mu\nu} \equiv \partial_\mu A_\nu - \partial_\nu A_\mu$ and the non-linear self-coupling terms

$$\begin{aligned}U_\sigma[\sigma] &= \frac{1}{2}m_\sigma^2\sigma^2 + \frac{g_2}{3}\sigma^3 + \frac{g_3}{4}\sigma^4 \\ U_\omega[\omega_\mu] &= \frac{1}{2}m_\omega^2\omega^\mu\omega_\mu + \frac{c_3}{4}(\omega^\mu\omega_\mu)^2 \\ U_\rho[\vec{\rho}_\mu] &= \frac{1}{2}m_\rho^2\vec{\rho}^{\,\mu}\cdot\vec{\rho}_\mu + \frac{d_3}{4}(\vec{\rho}^{\,\mu}\cdot\vec{\rho}_\mu)^2.\end{aligned}$$

The constants g_2, g_3, c_3 and d_3 are free parameters of the model. *A posteriori*, such a phenomenological Lagrangian appears as a particular case of the effective Lagrangian à la Furnstahl, Serot and Tang discussed in the previous section. Note that working with this Lagrangian at the RMF level while in the same time not allowing for parity breaking removes any contribution from the pion field: it carries negative parity so that the corresponding mean field breaks parity at the Hartree level. This is why the pion is usually an omitted d.o.f.. If one were to allow for parity breaking, or to treat explicitly the exchange and pairing terms, or to perform time-dependent calculations, then the following contributions would have to be added in the Lagrangian (2.51)

$$\mathcal{L}_\pi = -\bar{\psi}\frac{f_\pi}{m_\pi}\gamma_5\gamma_\mu\partial^\mu\vec{\pi}\cdot\vec{\tau}\psi + \frac{1}{2}\partial_\mu\vec{\pi}\cdot\partial^\mu\vec{\pi} - \frac{1}{2}m_\pi^2\vec{\pi}\cdot\vec{\pi}.$$

The introduction of non-linear self-coupling terms for the σ effective field already solves some of the issues that are characteristic of the linear Walecka-like model, e.g. the failure to reproduce the incompressibility of nuclear matter, surface properties or deformation of finite nuclei [74]. Non-linearities in the σ potential make the σ effective field dependent on the nuclear density, as if its mass were density-dependent. Such a density dependence of the effective parameters of the model is expected also from more fundamental calculations such as Dirac–Brueckner–Hartree–Fock (DBHF) [75]. The non-linear parameters g_2 and g_3 are adjusted to surface properties of finite nuclei. In many practical cases, this leads to a negative value for the parameter g_3 and produces an unstable theory at large densities. However, in the limit of moderate densities, as found in normal nuclei, the σ field is small enough for the potential $U[\sigma]$ to always be attractive, which results in a reasonable solution [9].

This particular form of the non-linear potential has become standard in realistic applications of RMF theory to nuclei, with NL3 [76] a representative parametrization of this family of Lagrangians. However, note that additional non-linear interaction terms, both in the isoscalar and isovector channels, have been considered [77] in order to better match the density dependence of the vector and scalar potentials found in DBHF calculations.

2.4.2 Density-dependent Lagrangians

First introduced by Brookmann and Toki [75], an alternative way of accounting for the medium dependence of the nucleon interaction is to turn the coupling constants of the Lagrangian into vertex functions of Lorentz scalar bi-linear forms of the nucleon operators. In most applications they depend on the vector density

$$\rho_\mathrm{v} \equiv \sqrt{j_\mu j^\mu},$$

with the nucleon four-current

$$j^\mu = \bar{\psi}\gamma^\mu\psi.$$

Another possible choice would be the scalar density. However, it has been shown that the vector density dependence gives better results for finite nuclei [78]. It is also a more natural choice since the nucleon four-current is a conserved quantity in the Noether sense.

The corresponding Lagrangian density retains linearity and its interaction part reads

$$\mathcal{L}_{\rm int}^{\rm DD} = -\bar{\psi}\Bigg\{g_\sigma(\rho_{\rm v})\sigma + g_\omega(\rho_{\rm v})\gamma_\mu\omega^\mu + g_\rho(\rho_{\rm v})\gamma_\mu\vec{\rho}^\mu\cdot\vec{\tau} + \frac{f_\pi(\rho_{\rm v})}{m_\pi}\gamma_5\gamma_\mu\partial^\mu\vec{\pi}\cdot\vec{\tau} + e\gamma_\mu A^\mu\left(\frac{1-\tau_3}{2}\right)\Bigg\}\psi. \tag{2.52}$$

Such a phenomenological Lagrangian can be related to the effective Lagrangian 'à la Rho' discussed in the previous section. The only difference comes from the fact that the density dependence of the latter is of both intrinsic origin (by virtue of the sliding from the matter-free to the finite-baryon density QCD vacuum) and induced (by virtue of the flow from the chiral effective Lagrangian to the Laudau Fermi liquid fixed point) at the level of the hadron mass. The density dependence of the Lagrangian (2.52) is instead at the level of the nucleon/meson coupling constants. In fact, these two pictures can be related [78]. From DBHF calculations, an in-medium T-matrix $T(1,2)$ can be symbolically written as

$$T(1,2) = g(1)D_m^*(1,2)g(2),$$

where g stands for the free-space (hence density-independent) nucleon/meson coupling and D_m^* for the in-medium (hence density-dependent) meson propagator, obtained from the bare propagator through the standard resumation of the full ladder series. Rewriting the in-medium T-matrix in terms of the bare propagator D_m, one can transfer the medium dependence in the coupling constant

$$T(1,2) = G(1)D_m(1,2)G(2),$$

which defines the vertex functions G in terms of the free-space meson/nucleon vertex through a Bethe–Salpeter-like equation which, diagrammatically, looks like figure 2.3.

As far as the explicit form of the density dependence of the coupling constants is concerned, one chooses an ansatz that can mock up DBHF results, but leaves the actual parameters free to be fitted to properties of nuclear matter and finite nuclei [79, 80]. For the σ and ω mesons, this leads to parametrizing the corresponding coupling constants as follows

$$g_i(\rho_{\rm v}) = g_i(\rho_{\rm sat})f_i(\xi) \quad i = \sigma, \omega, \tag{2.53}$$

where $\rho_{\rm sat}$ is the saturation density and

$$f_i(\xi) = a_i\frac{1 + b_i(\xi + d_i)^2}{1 + c_i(\xi + d_i)^2} \tag{2.54}$$

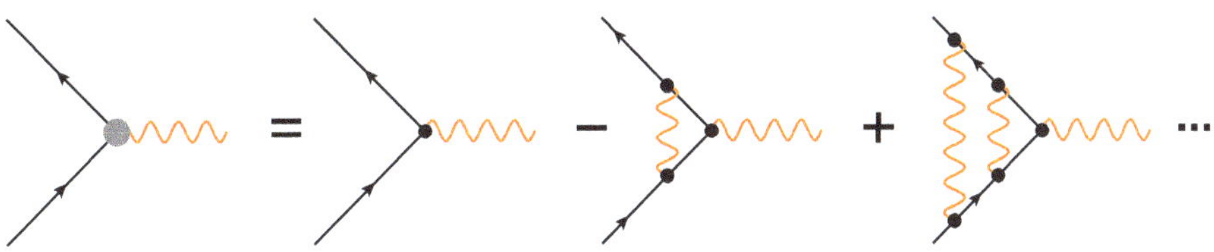

Figure 2.3. Diagrammatic representation of the Bethe–Salpeter equation building the Brueckner ladder to define density-dependent vertex functions out of the free-space ones. (Figure created by J-P Ebran.)

is a function of the dimensionless variable $\xi = \rho_v/\rho_{\rm sat}$. As for the ρ and π mesons, one chooses an exponential density dependence,

$$g_\rho(\rho_v) = g_\rho(0)e^{-a_\rho \xi}$$
$$f_\pi(\rho_v) = f_\pi(0)e^{-a_\pi \xi},$$

with $g_\rho(0)$ and $f_\pi(0)$ standing for the corresponding coupling constants in free space. Typical parametrizations for this family of Lagrangians are DD-ME2 [81] for applications at the RMF level and PKO2 [82] when exchange contributions are explicitly treated.

2.4.3 Point-coupling Lagrangians

A third family of phenomenological Lagrangians replaces the meson exchange in each channel (scalar–isoscalar (S), vector–isoscalar (V), scalar–isovector (TS), and vector–isovector (TV)) by corresponding contact interactions between the nucleons. These point-coupling Lagrangians are closely related to finite-range ones. Both involve expanding the finite-range meson propagators into a zero-range coupling plus gradient corrections.

Anticipating the next section, the Euler–Lagrange equation for the mesonic d.o.f. ϕ_m deduced from a finite-range Lagrangian yields an inhomogeneous Klein–Gordon equation of the type

$$(\Box + m_{\phi_m}^2)\phi_m(x) = s_{\phi_m}(x),$$

with $\Box \equiv \partial_\mu \partial^\mu$ the d'Alembert operator, m_{ϕ_m} the mass of the meson field $\phi_m(x)$ and $s_{\phi_m}(x)$ the source of the meson, typically some nucleon density. A particular solution of the mesonic equation of motion reads

$$\phi_m(x) = \int d^4y \, D_{\phi_m}(x, y) s_{\phi_m}(y),$$

where the propagator $D_{\phi m}$ of the meson field satisfies

$$(\Box + m_{\phi_m}^2) D_{\phi_m}(x, y) = \delta^{(4)}(x, y).$$

Solving this equation yields the usual Yukawa potential

$$D_{\phi_m}(|\boldsymbol{r}_1 - \boldsymbol{r}_2|) \propto \frac{e^{-m_{\phi_m}|\boldsymbol{r}_1 - \boldsymbol{r}_2|}}{m_{\phi_m}|\boldsymbol{r}_1 - \boldsymbol{r}_2|}. \tag{2.55}$$

Apart from the pion, all the mesons involved in CEDF have large masses compared to the typical transferred momenta between nucleons. One can then truncate the Taylor expansion of the propagator at first order,

$$D_{\phi_b}(|\boldsymbol{r}|) = \int \frac{d^3\boldsymbol{k}}{(2\pi)^3} \frac{e^{-i\boldsymbol{k}\cdot\boldsymbol{r}}}{\boldsymbol{k}^2 + m_{\phi_b}^2} = \frac{\delta^{(3)}(\boldsymbol{r})}{m_{\phi_b}^2}[1 + \nabla^2 + ...].$$

Like non-relativistic Skyrme models, finite-range effects are approximated by local derivative terms. The corresponding Lagrangian typically reads [83, 84]

$$\begin{aligned}\mathcal{L}_{\rm PC} &= \bar{\psi}\left(i\gamma_\mu\partial^\mu - m\right)\psi \\ &- \frac{1}{2}\alpha_S(\rho_v)(\bar{\psi}\psi)^2 - \frac{1}{2}\alpha_V(\rho_v)(\bar{\psi}\gamma_\mu\psi)(\bar{\psi}\gamma^\mu\psi) \\ &- \frac{1}{2}\alpha_{TS}(\rho_v)(\bar{\psi}\vec{\tau}\psi)(\bar{\psi}\vec{\tau}\psi) - \frac{1}{2}\alpha_{TV}(\rho_v)(\bar{\psi}\gamma_\mu\vec{\tau}\psi)(\bar{\psi}\gamma^\mu\vec{\tau}\psi) \\ &- \frac{1}{2}\delta_S(\partial_\mu\bar{\psi}\psi)(\partial^\mu\bar{\psi}\psi) - e\bar{\psi}\gamma_\mu A^\mu\frac{1-\tau_3}{2}\psi,\end{aligned} \tag{2.56}$$

where α_i, $i = \{S, V, TS, TV\}$ and δ_S are free parameters to be fitted to nuclear properties. Using equation (2.55), they can be related to the free parameters of a finite-range Lagrangian through

$$\alpha_i = \left(\frac{g_i}{m_i}\right)^2 \qquad \delta_S = \left(\frac{g_\sigma}{m_\sigma}\right)^4.$$

In equation (2.56), the derivative term is the one proportional to δ_S in the last line. It corresponds to the expansion (at first order) of the propagator for the sigma meson only. Similar contributions from other mesons are rarely considered.

2.5 Derivation of the covariant energy density functional

While in non-relativistic approaches, the EDF is either directly parametrized from the one-body density matrix and its various derivatives, or obtained by considering the expectation value of some pseudo-Hamiltonian on a Slater determinant, the CEDF is derived from the Lagrangian density after some intermediate steps. We will illustrate this derivation here in the case of the density-dependent meson exchange family [84, 85]. The derivation for the 'traditional' EDF of the RMF can be found, e.g. in [86]; for Lagrangians with point-couplings, we refer the reader to [87].

2.5.1 Classical equations of motion

The first step to build a covariant EDF is to establish the classical equations of motion from the least-action principle, $\delta S = 0$, with the classical action being given by [28]

$$S = \int d^4x\, \mathcal{L}(x). \tag{2.57}$$

The application of the least-action principle yields a set of (coupled) Euler–Lagrange equations for each d.o.f. of the Lagrangian, i.e. for the nucleon and boson fields.

Nucleonic degrees of freedom
For the nucleon fields, the Euler–Lagrange equation reads

$$\frac{\partial \mathcal{L}}{\partial \bar{\psi}} + \frac{\partial \mathcal{L}}{\partial \rho_{\mathrm{v}}} \frac{\partial \rho_{\mathrm{v}}}{\partial \bar{\psi}} = \partial_{\mu} \frac{\partial \mathcal{L}}{\partial (\partial_{\mu} \bar{\psi})}, \tag{2.58}$$

where we have used Einstein's summation convention (a repeated index implies a summation over that index). The second term on the left-hand side of the previous equation stems from the explicit density dependence of the coupling constants (hence it does not appear in the non-linear parametrization). Equation (2.58) yields the following Dirac equation

$$\left(i\gamma^{\mu}\partial_{\mu} - m - \Sigma\right)\psi(x) = 0,$$

with a nucleon self-energy Σ of the form

$$\Sigma = \Sigma^{S} + \gamma^{\mu}\left(\Sigma_{\mu}^{V} + \Sigma_{\mu}^{R}\right).$$

The Lorentz scalar contribution Σ^{S} reads

$$\Sigma^{S}(x) = g_{\sigma}(\rho_{\mathrm{v}})\sigma(x),$$

the vector one

$$\Sigma_{\mu}^{V}(x) = g_{\omega}(\rho_{\mathrm{v}})\omega_{\mu}(x) + g_{\rho}(\rho_{\mathrm{v}})\vec{\rho}_{\mu}(x)\cdot\vec{\tau} + \frac{f_{\pi}(\rho_{\mathrm{v}})}{m_{\pi}}\gamma_{5}\partial_{\mu}\vec{\pi}(x)\cdot\vec{\tau} + e\frac{1-\tau_{3}}{2}A_{\mu}(x),$$

while the rearrangement one is

$$\begin{aligned}\Sigma_{\mu}^{R}(x) = \frac{j_{\mu}}{\rho_{\mathrm{v}}}\Bigg[&\frac{dg_{\sigma}}{d\rho_{\mathrm{v}}}\bar{\psi}(x)\sigma(x)\psi(x) + \frac{dg_{\omega}}{d\rho_{\mathrm{v}}}\bar{\psi}(x)\gamma_{\nu}\omega^{\nu}(x)\psi(x)\\ &+ \frac{dg_{\rho}}{d\rho_{\mathrm{v}}}\bar{\psi}(x)\gamma_{\nu}\vec{\rho}^{\nu}(x)\cdot\vec{\tau}\psi(x) + \frac{1}{m_{\pi}}\frac{df_{\pi}}{d\rho_{\mathrm{v}}}\bar{\psi}(x)\gamma_{5}\gamma^{\nu}\partial_{\nu}\vec{\pi}(x)\cdot\vec{\tau}\psi(x)\Bigg].\end{aligned}$$

Let us stress here that the treatment of the rearrangement contribution to the self-energy Σ_{μ}^{R} is necessary to satisfy the conservation of the energy–momentum tensor, $\partial_{\mu}T^{\mu\nu} = 0$, with

$$T^{\mu\nu} = -\eta^{\mu\nu}\mathcal{L} + \frac{\partial \mathcal{L}}{\partial(\partial_{\mu}\phi_{i})}\partial^{\nu}\phi_{i},$$

where $\eta^{\mu\nu}$ stands for the Minkowski metric tensor (2.3) while ϕ_i stands for any of the dynamical fields $\phi_i = \{\bar{\psi}, \psi, \sigma, \omega^{\alpha}, \vec{\rho}^{\alpha}, \vec{\pi}, A^{\alpha}\}$.

Bosonic degrees of freedom
For the boson fields, the Euler–Lagrange equations are

$$\frac{\partial \mathcal{L}}{\partial \phi_{b}} = \partial_{\mu}\frac{\partial \mathcal{L}}{\partial(\partial_{\mu}\phi_{b})} \qquad \phi_{b} = \{\sigma, \omega^{\alpha}, \vec{\rho}^{\alpha}, \vec{\pi}, A^{\alpha}\}$$

and yield the following Klein–Gordon equation and the Proca equation

$$
\begin{aligned}
(\Box + m_\sigma^2)\sigma &= -\, g_\sigma(\rho_\mathrm{v})\bar{\psi}\psi \\
\partial_\mu \Omega^{\mu\nu} + m_\omega^2 \omega^\nu &= +\, g_\omega(\rho_\mathrm{v})\bar{\psi}\gamma^\nu\psi \\
\partial_\mu \vec{R}^{\mu\nu} + m_\rho^2 \vec{\rho}^{\,\nu} &= +\, g_\rho(\rho_\mathrm{v})\bar{\psi}\gamma^\nu\vec{\tau}\psi \\
(\Box + m_\pi^2)\vec{\pi} &= +\, \frac{f_\pi}{m_\pi}\partial_\mu(\bar{\psi}\gamma^5\gamma^\mu\vec{\tau}\psi)t \\
\partial_\mu F^{\mu\nu} &= +\, e\bar{\psi}\gamma^\nu\frac{1-\tau_3}{2}\psi
\end{aligned}
$$

Recall that the d'Alembert operator, or d'Alembertian, denoted as $\Box$, is defined by

$$\Box = \partial^\mu \partial_\mu = \frac{1}{c^2}\frac{\partial^2}{\partial t^2} - \Delta.$$

Using the conservation of the four-current $\partial_\mu j^\mu = 0$ and fixing the gauge $\partial_\mu A^\mu = 0$ further simplifies the set of equations above:

$$(\Box + m_\sigma^2)\sigma = -g_\sigma(\rho_\mathrm{v})\bar{\psi}\psi \tag{2.59a}$$

$$(\Box + m_\omega^2)\omega^\mu = +g_\omega(\rho_\mathrm{v})\bar{\psi}\gamma^\mu\psi \tag{2.59b}$$

$$\left(\Box + m_\rho^2\right)\vec{\rho}^{\,\mu} = +g_\rho(\rho_\mathrm{v})\bar{\psi}\gamma^\mu\vec{\tau}\psi \tag{2.59c}$$

$$(\Box + m_\pi^2)\vec{\pi} = +\frac{f_\pi}{m_\pi}\partial_\mu(\bar{\psi}\gamma^5\gamma^\mu\vec{\tau}\psi) \tag{2.59d}$$

$$\Box A^\mu = +e\bar{\psi}\gamma^\mu\frac{1-\tau_3}{2}\psi. \tag{2.59e}$$

These equations will enable some simplifications when deriving the Hamiltonian of the system.

2.5.2 Classical Hamiltonian

An elegant way to quantize a theory based on a classical Lagrangian is by means of the integral functional method. Another possibility is the canonical quantization within a Hamiltonian formulation. The latter hides the explicit covariance of the model, but enables a direct comparison with the non-relativistic case. We then opt for the canonical quantization. Therefore, we first need to derive the classical Hamiltonian of the system.

Recall that in field theory, the Hamiltonian is obtained from the $(\mu, \nu) = (0, 0)$ component of the energy–momentum tensor $T^{\mu\nu}$ [28]. The resulting Hamiltonian can be split into three parts

$$H = H_\mathrm{N} + H_\mathrm{b} + H_\mathrm{int}, \tag{2.60}$$

with the nucleon part given by

$$H_{\rm N} = \int d^3x\, \bar{\psi}(-i\gamma\nabla + m)\psi, \tag{2.61}$$

the boson contribution by

$$\begin{aligned} H_{\rm b} = \frac{1}{2}\int d^3\boldsymbol{x}\Big(&\partial_0\sigma\partial_0\sigma + \nabla\sigma\nabla\sigma + m_\sigma^2\sigma^2 - \Omega^{0i}\partial_0\omega_i + \Omega^{i\mu}\partial_i\omega_\mu \\ &- m_\omega^2\omega_\mu\omega^\mu - \vec{R}^{0i}\cdot\partial_0\vec{\rho}_i + \vec{R}^{i\mu}\cdot\partial_i\vec{\rho}_\mu - m_\rho^2\vec{\rho}_\mu\cdot\vec{\rho}^\mu \\ &+ \partial_0\vec{\pi}\cdot\partial_0\vec{\pi} + \nabla\vec{\pi}\cdot\nabla\vec{\pi} + m_\pi^2\vec{\pi}\cdot\vec{\pi} - F^{0i}\partial_0 A_i + F^{i\mu}\partial_i A_\mu\Big) \end{aligned} \tag{2.62}$$

and the interaction contribution by

$$\begin{aligned} H_{\rm int} = \int d^3\boldsymbol{x}\, \bar{\psi}\Big(& g_\sigma(\rho_{\rm v})\sigma + g_\omega(\rho_{\rm v})\gamma_\mu\omega^\mu + g_\rho(\rho_{\rm v})\gamma_\mu\vec{\rho}^\mu\cdot\vec{\tau} \\ &+ \frac{f_\pi(\rho_{\rm v})}{m_\pi}\gamma_5\gamma\nabla\vec{\pi}\cdot\vec{\tau} + e\gamma_\mu A^\mu\frac{1-\tau_3}{2}\Big)\psi. \end{aligned}$$

Neglecting the time dependence of the effective mesonic d.o.f., i.e. omitting the time-like component of the mesons four-momentum, and using the equations of motion (2.59), simplifies the bosonic contribution (2.62) which reduces to $H_{\rm b} = -H_{\rm int}/2$.

Nucleon interactions mediated by the mesonic d.o.f. are not instantaneous. However, since the transferred momentum through meson exchange is small with respect to the mass of these mesons (this is less true in the case of a pion exchange), one can safely neglect the time dependence of the mesonic fields, which does not affect the Hartree level but amounts to neglecting retardation effects at the Fock level.

Therefore, the total Hamiltonian (2.60) reads

$$\begin{aligned} H = \int d^3\boldsymbol{x}\, \bar{\psi}(-i\gamma\nabla + m)\psi + \frac{1}{2}\int d^3\boldsymbol{x}\, \bar{\psi}\Big(& g_\sigma(\rho_{\rm v})\sigma + g_\omega(\rho_{\rm v})\gamma_\mu\omega^\mu \\ + g_\rho(\rho_{\rm v})\gamma_\mu\vec{\rho}^\mu\cdot\vec{\tau} + \frac{f_\pi(\rho_{\rm v})}{m_\pi}&\gamma_5\gamma\nabla\vec{\pi}\cdot\vec{\tau} + e\gamma_\mu A^\mu\frac{1-\tau_3}{2}\Big)\psi. \end{aligned} \tag{2.63}$$

The form (2.63) is not entirely convenient in comparison to non-relativistic approaches, since it involves both nucleon and mesonic fields. Writing the mesonic and electromagnetic fields ϕ_m as the convolution of their propagator D_{ϕ_m} and the source s_{ϕ_m} entering their equation of motion (2.59),

$$\phi_m(x) = \int d^4y D_{\phi_m}(x, y)s_{\phi_m}(y), \tag{2.64}$$

where the propagators satisfy

$$\left(\Box + m_{\phi_m}^2\right)D_{\phi_m}(x, y) = \delta^{(4)}(x - y),$$

one can eliminate from the total Hamiltonian (2.63) all mesonic and electromagnetic d.o.f.s:

$$
\begin{aligned}
H = & \int d^3x\, \bar{\psi}(-i\gamma\nabla + m)\psi \\
& + \frac{1}{2}\int d^3x d^4y\, \bar{\psi}(x)\bar{\psi}(y)\Gamma_{\phi_m}(x, y)D_{\phi_m}(x, y)\psi(y)\psi(x).
\end{aligned}
\tag{2.65}
$$

The two-body vertices Γ_{ϕ_m} read

$$
\begin{aligned}
\Gamma_\sigma(1, 2) &= -\, g_\sigma(1)g_\sigma(2) \\
\Gamma_\omega(1, 2) &= +\, g_\omega(1)\gamma_\mu(1)g_\omega(2)\gamma^\mu(2) \\
\Gamma_\rho(1, 2) &= +\, g_\rho(1)\gamma_\mu(1)\vec{\tau}(1).\ g_\rho(2)\gamma^\mu(2)\vec{\tau}(2) \\
\Gamma_\pi(1, 2) &= -\, \frac{f_\pi(1)}{m_\pi}\gamma_5(1)\gamma(1)\nabla(1)\vec{\tau}(1) \cdot \frac{f_\pi(2)}{m_\pi}\gamma_5(2)\gamma(2)\nabla(2)\vec{\tau}(2) \\
\Gamma_A(1, 2) &= +\, \frac{e}{4}\gamma_\mu(1)[1 - \tau_3(1)]\gamma^\mu(2)[1 - \tau_3(2)],
\end{aligned}
$$

where (1,2) specifies which part acts on particle 1 or particle 2. The structure of the Hamiltonian (2.65) will be discussed and compared with its non-relativistic counterpart in the following section.

2.5.3 Quantization of the relativistic classical Hamiltonian

The transition to a quantum approach is realized, for example, by imposing the following equal-time anti-commutation and commutation relations. Nucleons are fermions, hence the fields must satisfy

$$
\begin{aligned}
&\left\{\psi_\alpha(\boldsymbol{x}, t), \psi_\beta^\dagger(\boldsymbol{y}, t)\right\} = \delta_{\alpha\beta}\delta^{(3)}(\boldsymbol{x} - \boldsymbol{y}) \\
&\left\{\psi_\alpha(\boldsymbol{x}, t), \psi_\beta(\boldsymbol{y}, t)\right\} = \left\{\psi_\alpha^\dagger(\boldsymbol{x}, t), \psi_\beta^\dagger(\boldsymbol{y}, t)\right\} = 0,
\end{aligned}
$$

while all mesons involved are bosons that must satisfy

$$
\begin{aligned}
&\left[\phi_\mu(\boldsymbol{x}, t), \Pi_\nu(\boldsymbol{y}, t)\right] = i\delta_{\mu\nu}\delta^{(3)}(\boldsymbol{x} - \boldsymbol{y}) \\
&\left[\phi_\mu(\boldsymbol{x}, t), \phi_\mu(\boldsymbol{y}, t)\right] = \left[\Pi_\mu(\boldsymbol{x}, t), \Pi_\nu(\boldsymbol{y}, t)\right] = 0,
\end{aligned}
$$

where α, β stand for spinor indexes and $\Pi_\mu \equiv \partial\mathcal{L}/\partial(\partial_0\phi^\mu)$ is the canonical conjugate variable of ϕ_μ. Focusing on the nucleon d.o.f., one can introduce a one-body basis $\{c_i, d_i\}$ associated with the solutions of the free Dirac equation, such that

$$
\psi(x) = \sum_\alpha \left\{f_\alpha(\boldsymbol{x})e^{-i\epsilon_\alpha t}c_\alpha + g_\alpha(\boldsymbol{x})e^{i\epsilon_{\alpha'} t}d_\alpha^\dagger\right\}
\tag{2.66}
$$

Note that the free Dirac equation yields two kinds of solutions, namely relativistic plane waves with positive energy interpreted as particles, and relativistic plane waves with negative energy interpreted as anti-particles, hence the need for two different sets of creation/annihilation operators. In the above equation, c_α ($d_\alpha^\dagger$) annihilates (creates) a nucleon (anti-nucleon) in the state $|\alpha\rangle$ and the functions $f_\alpha(\boldsymbol{x})$, $g_\alpha(\boldsymbol{x})$ are Dirac spinors. In what is called a no-sea approximation, the contribution coming from the negative-energy states belonging to the Dirac sea is not explicitly taken into account. Such terms correspond to diverging vacuum polarization diagrams associated with an energy scale beyond the one probed by our model. In that sense, the corresponding effects are reabsorbed in the parameters of the model, such that they are implicitly accounted for when adjusting the latter to experimental data; see e.g. [88]. Therefore, we omit the anti-nucleon contribution in equation (2.66), which amounts to dropping the d, $d^\dagger$ anti-nucleon operators. Substituting the nucleon field in the Hamiltonian (2.65) by its quantum version (2.66) in the no-sea prescription yields the Hamiltonian operator

$$\hat{H} = \hat{T} + \sum_m \hat{V}_m,$$

with $\hat{T}$ the usual kinetic energy operator and $\hat{V}_m$ an effective two-body potential,

$$\hat{T} = \sum_{\alpha\beta} t_{\alpha\beta} c_\alpha^\dagger c_\beta \tag{2.67a}$$

$$\hat{V}_m = \frac{1}{2} \sum_{\alpha\beta\gamma\delta} v^m_{\alpha\beta\gamma\delta} c_\alpha^\dagger c_\beta^\dagger c_\delta c_\gamma, \tag{2.67b}$$

where the matrix elements are now given by

$$t_{\alpha\beta} = \int d^3\boldsymbol{x} \bar{f}_\alpha(\boldsymbol{x})(-i\gamma\nabla + m) f_\beta(\boldsymbol{x})$$
$$v^m_{\alpha\beta\gamma\delta} = \int d^3\boldsymbol{x}_1 d^3\boldsymbol{x}_2 \bar{f}_\alpha(\boldsymbol{x}_1)\bar{f}_\beta(\boldsymbol{x}_2)\Gamma_{\phi_m}(1, 2) D_{\phi_m}(\boldsymbol{x}_1, \boldsymbol{x}_2) f_\delta(\boldsymbol{x}_2) f_\gamma(\boldsymbol{x}_1).$$

2.5.4 Derivation of the relativistic energy density functional

From the second-quantized form (2.67) of the Hamiltonian, a relativistic EDF follows as in the non-relativistic case, i.e. introducing a reference product state $|\Phi\rangle$ representing the many-body system. Such a reference state can be taken under the form of a Fermi vacuum, that is, a Slater determinant,

$$|\Phi\rangle = \prod_\alpha a_\alpha^\dagger |-\rangle, \tag{2.68}$$

where $\{a_\alpha^\dagger\}$ are associated with the one-body auxiliary Hartree–Fock (HF) basis yet to be defined and $|-\rangle$ refers to the physical vacuum.

Section 1.1 gives an introduction on the hierarchy of reference states. Chapter 3, in particular section 3.1 contains a detailed presentation of single-reference energy density functional (SR-EDF) techniques, including the definition of density matrices, Bogoliubov transformations, etc. With the introduction of the form (2.67), these techniques can be applied ‘as is’ to the relativistic theory. As we will see below, the main difference to non-relativistic approaches is the physical content of the interaction, or potential, term.

For superfluid systems, an efficient way to account for pairing correlations while retaining a single-particle (s.p.) picture is to take the reference product state $|\Phi\rangle$ under the form of a Bogoliubov vacuum, i.e. a coherent state breaking the $U(1)$ gauge invariance

$$|\Phi(\boldsymbol{g})\rangle = \prod_{\mu} \beta_{\mu}^{(\boldsymbol{g})}|-\rangle,$$

where $\boldsymbol{g}$ is a complex order parameter whose amplitude measures the extent to which the $U(1)$ symmetry is broken while its argument characterizes an orientation in the corresponding internal space, and $\{\beta_{\mu}^{(\boldsymbol{g})}\}$ are ladder quasiparticle operators associated with a one-body auxiliary basis yet to be found and related to the particle basis via

$$\beta_{\mu}^{(\boldsymbol{g})} = \sum_{i}\left(U_{i\mu}^{(\boldsymbol{g})*} c_i + V_{i\mu}^{(\boldsymbol{g})*} c_i^{\dagger}\right)$$
$$\beta_{\mu}^{(\boldsymbol{g})\dagger} = \sum_{i}\left(V_{i\mu}^{(\boldsymbol{g})} c_i + U_{i\mu}^{(\boldsymbol{g})} c_i^{\dagger}\right).$$

The unknown amplitudes U and V define the Bogoliubov transformation.

For the sake of simplicity, we will work within the Fermi vacuum case. We will discuss the two types of relativistic functionals most commonly used in nuclear structure, namely non-local functionals that include explicitly the exchange contributions—the relativistic Hartree–Fock (RHF) approach—and local functionals which exclude the exchange terms and implicitly absorb their effect when adjusting their free parameters to nuclear data—the RMF approach.

We can then normal-order the Hamiltonian (2.67) with respect to the Fermi vacuum (2.68). This involves using the commutation (boson operators) or anti-commutation (fermion operators) to move all creation operators to the left, and all annihilation operators to the right. Here, we find

$$H = \mathcal{E}^{RHF} + H^1 + \mathcal{V}_{\text{res}},$$

where the RHF energy functional $\mathcal{E}^{RHF}$, the one-body operator H^1 and the residual interaction $\mathcal{V}_{\text{res}}$ read

$$\mathcal{E}^{RHF} = \sum_{\alpha\beta} t_{\alpha\beta}\rho_{\beta\alpha} + \sum_{m}\frac{1}{2}\sum_{\alpha\beta\gamma\delta} \bar{v}^{m}_{\alpha\gamma\beta\delta}\rho_{\delta\gamma}\rho_{\beta\alpha}$$
$$H^1 = \sum_{\alpha\beta} t_{\alpha\beta}\colon c_{\alpha}^{\dagger}c_{\beta}\colon + \sum_{m}\frac{1}{2}\sum_{\alpha\beta\gamma\delta} \bar{v}^{m}_{\alpha\gamma\beta\delta}\rho_{\delta\gamma}\colon c_{\alpha}^{\dagger}c_{\beta}\colon$$
$$\mathcal{V}_{\text{res}} = \sum_{m}\sum_{\alpha\beta\gamma\delta} \bar{v}^{m}_{\alpha\beta\gamma\delta}\colon c_{\alpha}^{\dagger}c_{\beta}^{\dagger}c_{\delta}c_{\gamma}\colon$$

In the last set of equations, $\rho_{\beta\alpha} \equiv \langle\Phi|c_\alpha^\dagger c_\beta|\Phi\rangle$ stands for the one-body density matrix while $\bar{v}^m_{\alpha\gamma\beta\delta} \equiv v^m_{\alpha\gamma\beta\delta} - v^m_{\alpha\gamma\delta\beta}$ refers to the anti-symmetrized matrix elements of the potential energy operator.

In the HF basis associated with the ladder operators $a^\dagger a$, where the density matrix reduces to $\rho_{ij} = \delta_{ij}$ (i, j, k... indices label hole states) and $\rho_{ab} = 0$ (a, b, c... indices label particle states), the RHF energy functional reads

$$\mathcal{E}^{RHF}[\mathfrak{T}, \wp^m] = \int d^3x\, \mathfrak{T}(x) + \frac{1}{2}\int d^3x_1 d^3x_2 D_m(x_1, x_2)\wp^m(x_1, x_2)$$

(an implicit summation over m is meant), with the relativistic kinetic density

$$\mathfrak{T}(x) \equiv \sum_i \bar{f}_i(x)(-i\gamma\nabla + m)f_i(x)$$

and the non-local, quartic nucleon functions

$$\wp^m(x_1, x_2) \equiv \sum_{ij} \bar{f}_i(x_1)\bar{f}_j(x_2)\Gamma_m(1, 2)\left[f_j(x_2)f_i(x_1) - f_i(x_2)f_j(x_1)\right].$$

In the RMF case, we omit the explicit contributions of the exchange terms, which yields the functional

$$\mathcal{E}^{RMF}[\mathfrak{T}, \rho^m] = \int d^3x\, \mathfrak{T}(x) + \frac{1}{2}\int d^3x_1 d^3x_2\, D_m(x_1, x_2)\rho^m(x_1)\star\rho^m(x_2),$$

where $\star$ is the usual product but with a phase ± 1 depending on whether the meson is a vector (+1) or a scalar or pseudo-scalar (-1) in the Lorentz sense. The notation $\rho^m(x)$ stands for the various nucleon densities and currents, that is, with $m \equiv \sigma, \omega, \rho, A$,

$$\rho^\sigma(x) \equiv \rho_s(x) = \sum_i \bar{f}_i(x)f_i(x) \tag{2.69a}$$

$$\rho^\omega(x) \equiv j^\mu(x) = \sum_i \bar{f}_i(x)\gamma^\mu f_i(x) = (\rho_v(x), \boldsymbol{j}(x)), \tag{2.69b}$$

$$\rho_v(x) = \sum_i f_i^\dagger(x)f_i(x); \quad \boldsymbol{j}(x) = \sum_i \bar{f}_i(x)\boldsymbol{\gamma} f_i(x) \tag{2.69c}$$

$$\rho^\rho(x) \equiv j_t^\mu(x) = \sum_i \bar{f}_i(x)\gamma^\mu\vec{\tau}f_i(x) = (\rho_{tv}(x), \boldsymbol{j}_t(x)) \tag{2.69d}$$

$$\rho_{tv}(x) = \sum_i f_i^\dagger(x)\vec{\tau}f_i(x); \quad \boldsymbol{j}_t(x) = \sum_i \bar{f}_i(x)\boldsymbol{\gamma}\vec{\tau}f_i(x) \tag{2.69e}$$

$$\rho^A(x) \equiv j_p^\mu(x) = \sum_i \bar{f}_i(x)\gamma^\mu\frac{1-\tau_3}{2}f_i(x) = (\rho_p(x), \boldsymbol{j}(x)), \tag{2.69f}$$

where p means that we restrict the sum over the occupied proton states. Note that the pion is not discussed insofar as its Hartree contribution is zero (when we do not allow for reflection-symmetry-breaking). Using equation (2.64), the RMF functional can be made local at the price of reintroducing the effective meson fields (as well as the electromagnetic four-potential) to be considered as classical objects

$$\mathcal{E}^{RMF}\left[\mathfrak{T}, \rho^m, \phi_m\right] = \int d^3\boldsymbol{x}\left[\mathfrak{T}(\boldsymbol{x}) + \frac{1}{2}\phi_m(\boldsymbol{x})\Gamma_m(\boldsymbol{x})\rho^m(\boldsymbol{x})\right],$$

with $\Gamma_m(\boldsymbol{x})$ the vertex functions

$$\Gamma_\sigma(\boldsymbol{x}) = g_\sigma(\rho_v(\boldsymbol{x})), \qquad \Gamma_\omega(\boldsymbol{x}) = g_\omega(\rho_v(\boldsymbol{x}))\gamma_\mu$$
$$\Gamma_\rho(\boldsymbol{x}) = g_\rho(\rho_v(\boldsymbol{x}))\gamma_\mu\vec{\tau}, \qquad \Gamma_A = \frac{e}{2}\gamma_\mu[1 - \tau_3].$$

As a last step, we make the kinetic part of the classical bosonic d.o.f. appear by proceeding as in the derivation of the classical Hamiltonian, in other words, using the relation $H_b = -H_{\text{int}}/2$:

$$\mathcal{E}^{RMF}\left[\mathfrak{T}, \rho^m, \phi_m\right] = \int d^3\boldsymbol{x}\left(\mathfrak{T}(\boldsymbol{x}) + \phi_m(\boldsymbol{x})\Gamma_m(\boldsymbol{x})\rho^m(\boldsymbol{x})\right) + \mathcal{E}_m\left[\phi_m\right], \tag{2.70}$$

where

$$\mathcal{E}_m\left[\phi_m\right] = \frac{1}{2}\int d^3\boldsymbol{x}\Big(\nabla\sigma\nabla\sigma + m_\sigma^2\sigma^2 - \nabla\omega^\mu\nabla\omega_\mu - m_\omega^2\omega_\mu\omega^\mu$$
$$- \nabla\vec{\rho}^\mu \cdot \nabla\vec{\rho}_\mu - m_\rho^2\vec{\rho}_\mu \cdot \vec{\rho}^\mu - \nabla A^\mu\nabla A_\mu\Big).$$

2.6 Advantages of a relativistic description of nuclear systems

A relativistic description of atomic nuclei first implies promoting the Lorentz group as the relevant symmetry group to describe nuclear phenomenology. In other words, we work with quantities that transform in a well-defined way under Lorentz transformations such as Lorentz scalars, Dirac spinors, or four-vectors. An approach built on such a relativistic foundation proves very powerful to describe nuclear features in both an efficient and physically transparent way.

At first sight, a relativistic treatment seems only to increase unnecessarily the number of d.o.f.s needed to characterize the system. Indeed, nucleons are represented by Dirac spinors, i.e. with twice as many components as a non-relativistic spinor. By the same token, the effective fields parametrizing the interactions between nucleons split into two categories, namely Lorentz scalars on the one hand, and time-like component of four-vectors on the other—a distinction without any counterpart in the non-relativistic realm.

However, it is instead the opposite, i.e. such d.o.f.s provide the minimal number of basic building blocks out of which one can capture many non-trivial features in a condensed form, as a result of their interplay. In that respect, the small number of free parameters ascribed to a relativistic functional—typically 5–8—is a clear indicator of the conciseness of relativistic approaches.

The Dirac equation is the fundamental place where we gather together all these building blocks. It supersedes the non-relativistic Schrödinger equation and provides the foundation for understanding basic features such as the spin–orbit and pseudospin–orbit aspects in nuclei, the saturation of symmetric nuclear matter, the non-trivial medium dependence of the interaction between nucleons, the nuclear clustering phenomenon, etc.

Nuclear clustering

A relativistic framework where nuclear systems are described in terms of the two large fields V and S sheds new light on the phenomenon of nuclear clustering, i.e. the arrangement of nucleons in additional bound sub-structures within the nucleus itself. Clustering effects may coexist with more traditional mean-field effects and have repercussions on the preferred nuclear excitation modes [89–91]. At present, the only comprehensive approach to nuclear structure describing both cluster and mean-field aspects on the same footing, in light as well as in heavier nuclei, is based on the framework of CEDFS [92, 93], see figure 2.4 for the three-dimensional intrinsic densities of the first α self-conjugate nuclei both in their ground and excited states. In particular, the fact that the S and V fields both determine the central and spin–orbit potentials provides additional constraints on the depth of the confining potential [94], which is typically at least 10 MeV deeper than its non-relativistic counterparts. This depth is one of the necessary conditions for clustering to develop [94–96], see figure 2.5.

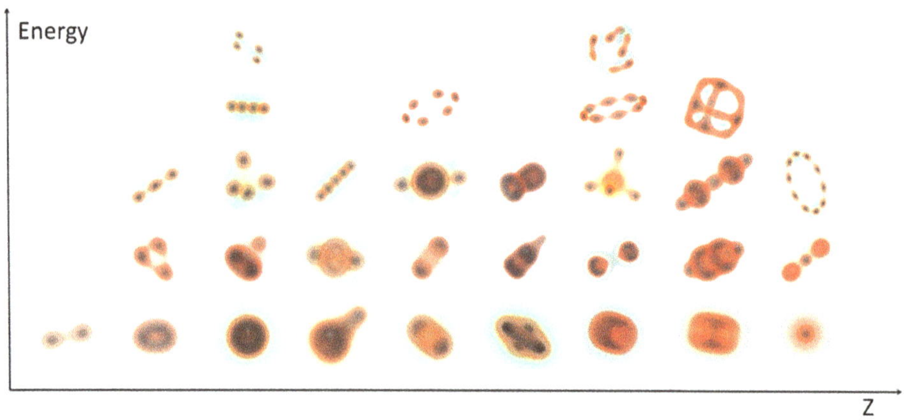

Figure 2.4. Intrinsic densities in the 3D plane of the first α self-conjugate nuclei from Be to Ca both in their ground and excited states, solutions of a CEDF calculation at the RMF level with the DD-ME2 parametrization set. (Figure created by J-P Ebran.)

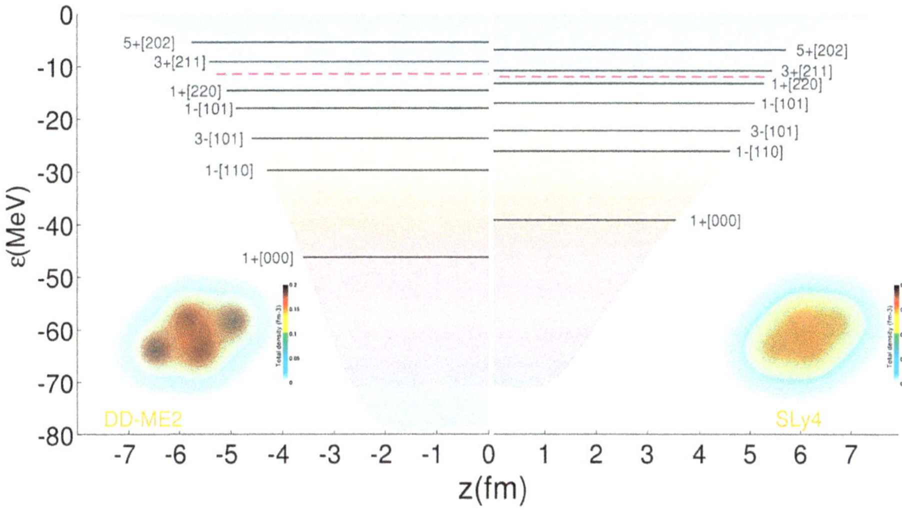

Figure 2.5. Comparison of the relativistic DD-ME2 (left) and the non-relativistic SLy4 (right) neutron s.p. spectrum (labeled with Nilsson quantum numbers) of ^{20}Ne in its ground state. The shape of the self-consistent confining potential is also displayed along the axial (Oz) axis, which is deeper in the relativistic calculation. The corresponding intrinsic density in the 3D plane is shown in both cases, with clear signs of nucleon localization due to the depth of the potential in the relativistic case. (Figure created by J-P Ebran.)

The interplay between the large and the small components of the nucleon spinor, the vector and scalar nucleon self-energies, or the vector and scalar densities can be made more manifest by processing the non-relativistic limit of the one-particle Dirac equation, where two types of relativistic corrections are revealed, namely kinematical ones that can always be seamlessly included in any non-relativistic formulation of the nuclear many-body problem, and dynamical ones that are far more subtle [97]. This final section aims at addressing these different facets, starting with the one-particle Dirac equation together with its non-relativistic limit.

2.6.1 One-particle Dirac equation

Working at the RMF level and considering time-even systems simplify the dynamical d.o.f.s of the theory. Indeed, in this case, only the time-like component of the various four-vectors ($\omega \equiv \omega^0$, $\boldsymbol{\rho} \equiv \boldsymbol{\rho}^0$, $A \equiv A^0$, $\rho_v \equiv j^0$ and $\boldsymbol{\rho}_t \equiv \boldsymbol{j}_t^0$) are *a priori* non-zero, and only the third components of isovectors ($\rho \equiv \rho_3$ and $\rho_{tv} \equiv j_{t3}^0$) survive. The RMF energy functional (2.70) then reads

$$\mathcal{E}^{RMF}[\mathfrak{K}, \rho_s, \rho_v, \rho_{tv}, \sigma, \omega, \rho, A] = \int d^3\boldsymbol{x}\Big\{ \mathfrak{K} + g_\sigma\sigma\rho_s + g_\omega\omega\rho_v + g_\rho\rho\rho_{tv} + eA\rho_{v,p} + \frac{1}{2}\nabla\sigma\nabla\sigma + \frac{1}{2}m_\sigma^2\sigma^2 - \frac{1}{2}\nabla\omega\nabla\omega - \frac{1}{2}m_\omega^2\omega^2 - \frac{1}{2}\nabla\rho\cdot\nabla\rho - \frac{1}{2}m_\rho^2\rho^2 - \frac{1}{2}\nabla A\nabla A\Big\}. \tag{2.71}$$

Let us first focus on the nucleon d.o.f. The minimization of the RMF functional (2.71) with respect to the nucleon Dirac spinor

$$f_i(\boldsymbol{r}) \equiv \begin{pmatrix} \varphi_i(\boldsymbol{r}) \\ \xi_i(\boldsymbol{r}) \end{pmatrix}$$

$\bar{f}_i(\boldsymbol{r})$, under the constraint that the latter is normalized, yields the one-particle Dirac equation

$$[-i\boldsymbol{\alpha} \cdot \nabla + \beta(m + S) + V]f_i(\boldsymbol{r}) = E_i f_i(\boldsymbol{r}), \tag{2.72}$$

where $\boldsymbol{\alpha}$ and β stand for the combination of the Dirac matrices

$$\boldsymbol{\alpha} \equiv \gamma_0 \boldsymbol{\gamma} = \begin{pmatrix} 0 & \boldsymbol{\sigma} \\ \boldsymbol{\sigma} & 0 \end{pmatrix} \qquad \beta \equiv \gamma_0 = \begin{pmatrix} I_2 & 0 \\ 0 & -I_2 \end{pmatrix}.$$

$\boldsymbol{\sigma}$ collects the three Pauli matrices, the Lagrange multiplier E_i corresponds to the energy of the nucleon $|i\rangle$, and V and S refer to the vector and scalar nucleon self-energy

$$S(\boldsymbol{r}) = g_\sigma(\rho_v(\boldsymbol{r}))\sigma(\boldsymbol{r}),$$
$$V(\boldsymbol{r}) = g_\omega(\rho_v(\boldsymbol{r}))\omega(\boldsymbol{r}) + g_\rho(\rho_v(\boldsymbol{r}))\rho(\boldsymbol{r})\tau^3 + eA\frac{1 - \tau^3}{2} + \Sigma_R(\boldsymbol{r}),$$

where $\Sigma_R(\boldsymbol{r}) = \frac{dg_\sigma}{d\rho_{\mathrm{v}}}\sigma(\boldsymbol{r})\rho_{\mathrm{s}}(\boldsymbol{r}) + \frac{dg_\omega}{d\rho_{\mathrm{v}}}\omega(\boldsymbol{r})\rho_{\mathrm{v}}(\boldsymbol{r}) + \frac{dg_\rho}{d\rho_{\mathrm{v}}}\rho(\boldsymbol{r})\rho_{\mathrm{tv}}(\boldsymbol{r})$ stands for the rearrangement term. Introducing the structure of the Dirac spinor in terms of its large and small components, the Dirac equation (2.72) becomes

$$\begin{pmatrix} m + S + V & -i\boldsymbol{\sigma} \cdot \nabla \\ -i\boldsymbol{\sigma} \cdot \nabla & -m - S + V \end{pmatrix} \begin{pmatrix} \varphi_i \\ \chi_i \end{pmatrix} = E_i \begin{pmatrix} \varphi_i \\ \chi_i \end{pmatrix}. \tag{2.73}$$

To gain more insight into the physics content of the Dirac equation, let us decompose the energy E_i of the nucleon in state $|i\rangle$ into the sum of a non-relativistic contribution ϵ_i and the mass energy m,

$$E_i = \epsilon_i + m.$$

We can now write equation (2.73) in the form

$$\begin{pmatrix} S + V & -i\boldsymbol{\sigma} \cdot \nabla \\ -i\boldsymbol{\sigma} \cdot \nabla & V - S - 2m \end{pmatrix} \begin{pmatrix} \varphi_i \\ \chi_i \end{pmatrix} = \epsilon_i \begin{pmatrix} \varphi_i \\ \chi_i \end{pmatrix}. \tag{2.74}$$

Figure 2.6 gives a visual representation of the typical energy scales associated with the s.p. solutions to the Dirac equation (2.74). Far from the origin of the nucleus, the potentials V and S vanish: the Dirac equation (2.74) represent a system of non-interacting nucleons. Its solutions are of the type (2.66), one with positive energy describing a free nucleon, the other with a negative energy describing a free anti-nucleon. The two energy *continua* are separated by a gap equal to twice the nucleon mass.

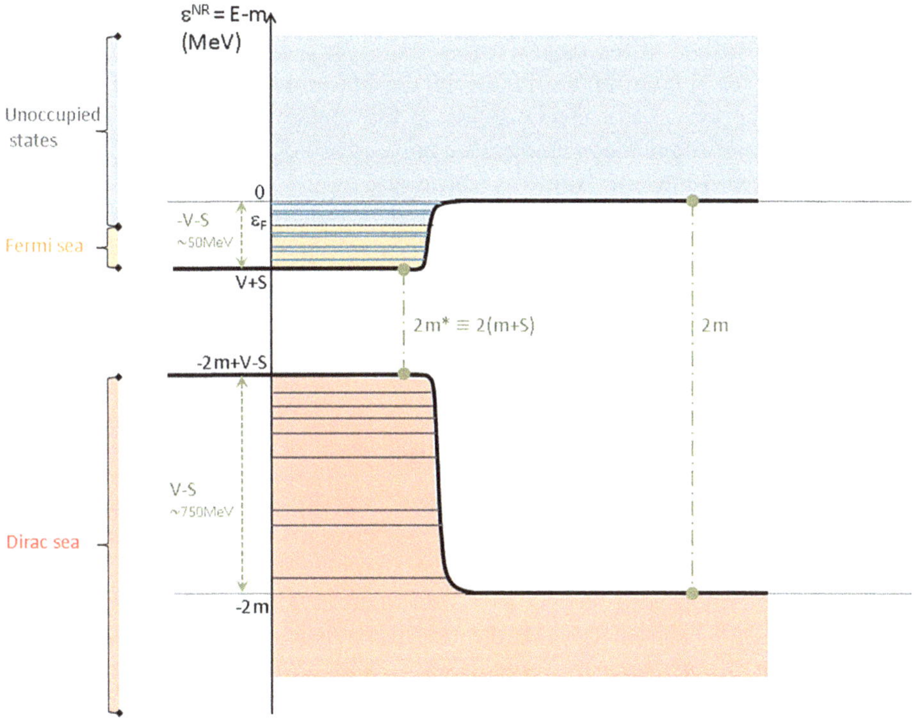

Figure 2.6. Schematic illustration of the S and V relativistic potential features as a function of the radial distance from the origin of the nucleus, and their relation with nucleon and anti-nucleon single-particle properties. (Figure created by J-P Ebran.)

> Recall that V is large and positive, while S is large and negative [63]. Non-relativistic approaches only encode energy scales of the order of the depth of the average nuclear potential, which is precisely the same order of magnitude as $V + S$.

In the nuclear interior, the relativistic potentials S and V both introduce a relevant scale of the order of the QCD mass scale. Their combinations under the form $V + S$ and $V - S$ display a Woods–Saxon-like shape [63]. The large component is mostly sensitive to $V + S \sim 50$ MeV, while the small component is driven by the additive combination $V - \mathrm{S} \sim 750$ MeV. These nucleon self-energies lower the energy gap between the nucleon and anti-nucleon states to about twice the so-called Dirac mass $m^* \equiv m + S \sim 540$ MeV. The Dirac mass should not be confused with the effective mass $\tilde{M} \equiv m + \frac{1}{2}[S(\boldsymbol{r}) - V(\boldsymbol{r})] \sim 565\,\mathrm{MeV}$ that we will introduce later in the context of the non-relativistic limit of the Dirac equation. At the same time, this mechanism creates attractive finite potentials for both the nucleon and the anti-nucleon, which can now occupy discrete energy levels.

Note that within the no-sea prescription discussed in the previous section, excitations stemming from the presence of anti-nucleons filling the Dirac sea are not explicitly accounted for. Indeed, in an EFT viewpoint their characteristic energies lie over the breakdown scale of the model; the corresponding effects can thus safely be absorbed in the local parameters of the model (this is the reason why we dropped the set of creation/annihilation $\{d^{\dagger}d\}$ of anti-nucleon states). However, the impact of the presence of the Dirac sea is fully taken into account through the small component of the nucleon spinor, the norm of which quantifies the probability of extracting an anti-nucleon from the Dirac sea.

We will see later that it is precisely the small component that maintains a distinction between the vector density $\rho_V = \sum_i \bar{f}_i \gamma^0 f_i = \sum_i (|\varphi_i|^2 + |\chi_i|^2)$ and the scalar density $\rho_S = \sum_i \bar{f}_i f_i = \sum_i (|\varphi_i|^2 - |\chi_i|^2)$. It is also the quantum numbers associated with the small component that label pseudospin–orbit partners; see section 2.6.5.

2.6.2 Non-relativistic limit of the Dirac equation

The standard way to take the non-relativistic limit in a relativistic system is the Foldy–Wouthuysen unitary transformation [98, 99]. However, it cannot be applied in nuclear physics because in that case, the small parameter entering the transformation is not small with respect to 1. We then have to look for another small parameter that can be used to take the non-relativistic limit. Note that the non-relativistic limit of the nuclear Dirac Hamiltonian has also been studied in the framework of the similarity renormalization group [100].

A Schrödinger-like equation

Before discussing the parameters relevant for a non-relativistic expansion, let us remark that non-relativistic equations apply to two-component spinors, not four-component ones like the solutions to the Dirac equation. However, it is straightforward to use the Dirac equation and obtain an equation for the large component only (which is a two-component spinor). We achieve this by rewriting equation (2.74) in terms of the large component only (a two-component spinor) by eliminating the small components via its second equation

$$\chi_i = \frac{-i}{2m + \epsilon_i + S - V} \boldsymbol{\sigma} \cdot \nabla \varphi_i \tag{2.75}$$

leading to

$$\left\{ \boldsymbol{\sigma} \cdot \nabla \frac{-1}{2\tilde{M}(\boldsymbol{r}) + \epsilon_i} \boldsymbol{\sigma} \cdot \nabla + V + S \right\} \varphi_i = \epsilon_i \varphi_i, \tag{2.76}$$

where $\tilde{M}(\boldsymbol{r})$ stands for the nucleon effective mass

$$\tilde{M}(\boldsymbol{r}) \equiv m + \frac{1}{2}[S(\boldsymbol{r}) - V(\boldsymbol{r})]. \tag{2.77}$$

We have thus obtained a Schrödinger-like equation (2.76) which is strictly equivalent to the original Dirac equation (2.72). However, the large component $\varphi_i(\boldsymbol{r})$ alone is not properly normalized. Therefore we need to work a little more to obtain an expression suitable for a non-relativistic expansion. In the following, it will be convenient to introduce the quantity

$$\mathcal{B} \equiv \frac{1}{2\tilde{M}(\boldsymbol{r}) + \epsilon_i}.$$

With this definition, the relation (2.75) between the small and large component thus becomes

$$\chi_i = -i\mathcal{B}\boldsymbol{\sigma} \cdot \nabla\varphi_i. \tag{2.78}$$

Effective Hamiltonians

Non-relativistic effective Hamiltonians can be introduced generically through the following kernels,

$$\mathfrak{H}_D = \int d^3\boldsymbol{r}\ \phi_i^\dagger(h_{\text{eff}} - \epsilon_i{}')\phi_i, \tag{2.79}$$

where $\phi_i(\boldsymbol{r})$ is a normalized two-component spinor. Let us introduce the following kernel,

$$\mathfrak{H}_D \equiv \int d^3\boldsymbol{r}\ f_i^\dagger(h_D - \epsilon_i)f_i, \tag{2.80}$$

where h_D is the one-particle Dirac Hamiltonian

$$h_D \equiv \begin{pmatrix} S + V & -i\boldsymbol{\sigma} \cdot \nabla \\ -i\boldsymbol{\sigma} \cdot \nabla & -2m - S + V \end{pmatrix} \tag{2.81}$$

and f_i is the full four-component Dirac spinor. In terms of the large component of that spinor, the kernel (2.80) reads

$$\mathfrak{H}_D = \int d^3\boldsymbol{r}\ \varphi_i^\dagger\left\{\boldsymbol{\sigma} \cdot \nabla\frac{-1}{2\tilde{M}(\boldsymbol{r}) + \epsilon_i}\boldsymbol{\sigma} \cdot \nabla + V + S - \epsilon_i\right\}\varphi_i.$$

Using the fact that the total nucleon spinor is normalized

$$\int d^3\boldsymbol{r}\ \bar{f}_i\gamma^0 f_i = \int d^3\boldsymbol{r}\left(\varphi_i^\dagger\varphi_i + \chi_i^\dagger\chi_i\right) = 1$$

and using equation (2.78), we obtain

$$\int d^3\boldsymbol{r}\ \varphi_i^\dagger(1 + \boldsymbol{\sigma} \cdot \nabla'\mathcal{B}^2\boldsymbol{\sigma} \cdot \nabla)\varphi_i = 1,$$

where ∇' means that the gradient operator acts on the left. Let us define

$$\hat{\mathcal{N}} = 1 + \boldsymbol{\sigma} \cdot \nabla'\mathcal{B}^2\boldsymbol{\sigma} \cdot \nabla.$$

We can now construct a normalized two-component spinor from the large component of the nucleon Dirac spinor according to

$$\phi_i = \hat{\mathcal{N}}^{\frac{1}{2}}\varphi_i; \quad \varphi_i = \hat{\mathcal{N}}^{-\frac{1}{2}}\phi_i.$$

By identification with equation (2.79), the effective Hamiltonian $\hat{h}_{\text{eff}}$ reads

$$\hat{h}_{\text{eff}} = \hat{\mathcal{N}}^{-\frac{1}{2}}[-\boldsymbol{\sigma}\cdot\nabla\mathcal{B}\boldsymbol{\sigma}\cdot\nabla + V + S]\hat{\mathcal{N}}^{-\frac{1}{2}}. \tag{2.82}$$

Such an effective Hamiltonian is suitable for a non-relativistic expansion, and we can now go back to the problem of identifying the relevant parameters to carry out such an expansion.

We draw the attention of the reader to some kind of formal analogy between this transformation and the definition of orthonormal collective wave functions in the generator coordinate method (GCM); see equation (3.88) for the discrete version and section 6.4.1 for the continuous version.

The non-relativistic limit

Single-particle energies ϵ_i typically range between −10 and +5 MeV near the Fermi level of nuclear systems, and are almost never smaller than ~ − 50 MeV. At the same time, the typical values of the self-energies are $S \sim -400$ MeV and $V \sim +\,350$ MeV, and the effective mass $\tilde{M}$ is of the order of ~600 MeV. These estimates suggest taking the variable $\epsilon_i/2\tilde{M} \sim 1/2\tilde{M}$ as our small parameter. We thus define

$$\mathcal{B}_0 \equiv \frac{1}{2\tilde{M}(\boldsymbol{r})}.$$

Up to the second order in $\mathcal{B}_0$, we find

$$\mathcal{B} \approx \mathcal{B}_0(1 - \epsilon_i\mathcal{B}_0) \approx \mathcal{B}_0(1 - \mathcal{B}_0)$$
$$\hat{\mathcal{N}}^{-\frac{1}{2}} \approx 1 - \frac{1}{2}\boldsymbol{\sigma}\cdot\nabla'\mathcal{B}_0^2\boldsymbol{\sigma}\cdot\nabla$$

leading to

$$\begin{aligned}\hat{h}_{\text{eff}} = {} & S + V - \boldsymbol{\sigma}\cdot\nabla\mathcal{B}_0\boldsymbol{\sigma}\cdot\nabla + \boldsymbol{\sigma}\cdot\nabla\mathcal{B}_0^2\boldsymbol{\sigma}\cdot\nabla \\ & - \frac{1}{2}\boldsymbol{\sigma}\cdot\nabla'\mathcal{B}_0^2\boldsymbol{\sigma}\cdot\nabla(S + V) - \frac{1}{2}(S + V)\boldsymbol{\sigma}\cdot\nabla'\mathcal{B}_0^2\boldsymbol{\sigma}\cdot\nabla.\end{aligned} \tag{2.83}$$

Instead of giving the full demonstration that yields to equation (2.84), we will give a few pointers and intermediate steps. First, one needs to make use of the following pair of properties of the Pauli matrices,

$$(\boldsymbol{\sigma} \cdot \mathbf{A})(\boldsymbol{\sigma} \cdot \mathbf{B}) = \mathbf{A} \cdot \mathbf{B} + i\boldsymbol{\sigma} \cdot (\mathbf{A} \times \mathbf{B})$$
$$(\boldsymbol{\sigma} \cdot \nabla)(\boldsymbol{\sigma} \cdot \nabla) = \nabla^2.$$

We then note that

$$\nabla \mathcal{B}_0 = \nabla \frac{1}{2\tilde{M}(\boldsymbol{r})} = -\frac{1}{2\tilde{M}^2} \nabla \tilde{M} = \mathcal{B}_0^2 \nabla (V - S).$$

Finally, we use

$$\boldsymbol{\sigma} \cdot \nabla \mathcal{B}_0^2 \boldsymbol{\sigma} \cdot \nabla \approx \mathcal{B}_0^2 \nabla^2.$$

This expression can be transformed into a more familiar form by working on the last four terms. After some algebra, we find up to second order in $\mathcal{B}_0$,

$$\begin{aligned}\hat{h}_{\text{eff}} = S + V - \mathcal{B}_0(1 - \mathcal{B}_0)\nabla^2 - \mathcal{B}_0^2 \nabla(V - S) \cdot \nabla - i\mathcal{B}_0^2 \boldsymbol{\sigma} \cdot \nabla(V - S) \times \nabla \\ - \frac{1}{2} \nabla' \mathcal{B}_0^2 [\nabla(S + V)] - i\boldsymbol{\sigma} \cdot \nabla' \times \mathcal{B}_0^2 [\nabla(S + V)] \\ - \frac{1}{2}(S + V) \nabla' \mathcal{B}_0^2 \nabla - i\boldsymbol{\sigma} \cdot \nabla' \times \mathcal{B}_0^2 [\nabla(S + V)].\end{aligned} \tag{2.84}$$

Components of the effective Hamiltonian

To facilitate the comparison with standard, non-relativistic forms of the Hamiltonian, we can write the last equation in the compact form

$$\hat{h}_{\text{eff}} = \hat{t} + \hat{v}_{\text{cent}} + \hat{v}_{\text{so}} + \hat{v}_{\text{corr}}. \tag{2.85}$$

We first note that the kinetic energy operator does not involve the bare nucleon mass, but the effective mass $\tilde{M}(\boldsymbol{r})$, since

$$\hat{t} = -\mathcal{B}_0(1 - \mathcal{B}_0)\nabla^2 = -\frac{1}{2\tilde{M}(\boldsymbol{r})}\left(1 - \frac{1}{2\tilde{M}(\boldsymbol{r})}\right)\nabla^2.$$

The central potential $\hat{v}_{\text{cent}}$ is simply

$$\hat{v}_{\text{cent}} = S + V.$$

It is of the order of ~ -50 MeV because of the near cancellation of the scalar and vector self-energies. It is also small compared to the mass energy of a nucleon at rest ($\sim$1 GeV), which implies that nuclear kinematics can safely be considered non-relativistic. A spin–orbit potential $\hat{v}_{\text{so}}$ naturally emerges from the constructive combination of the nucleon self-energies. It appears as a relativistic correction since it is proportional to $1/\tilde{M}^2$. Its original expression,

$$\hat{v}_{\text{so}} = -i\mathcal{B}_0^2 \boldsymbol{\sigma} \cdot \nabla(V - S) \times \nabla,$$

can be recast into

$$\hat{v}_{\text{so}} = \frac{1}{4M^2}\nabla V_{ls} \cdot \boldsymbol{p} \times \boldsymbol{\sigma} \tag{2.86}$$

if one poses

$$V_{ls} = \frac{M}{\tilde{M}(\boldsymbol{r})}(V - S). \tag{2.87}$$

If the system is described in a spherical basis, the gradient operator takes a simple form and equation (2.86) becomes

$$\hat{v}_{\text{so}} = \frac{1}{2r\tilde{M}^2(\boldsymbol{r})}\frac{d}{dr}(V - S)\boldsymbol{l} \cdot \boldsymbol{s}, \tag{2.88}$$

with $\boldsymbol{l} = \boldsymbol{r} \times \boldsymbol{p}$ and $\boldsymbol{s} = \frac{\sigma}{2}$. Finally, we find an additional correction $\hat{v}_{\text{corr}}$ which has no non-relativistic counterpart

$$\hat{v}_{\text{corr}} = -\frac{1}{4\tilde{M}^2(\boldsymbol{r})}\nabla(V - S) \cdot \nabla - \frac{1}{2}\nabla'\mathcal{B}_0^2[\nabla(S + V)] - i\boldsymbol{\sigma} \cdot \nabla' \times \mathcal{B}_0^2[\nabla(S + V)]$$
$$- \frac{1}{2}(S + V)\nabla'\mathcal{B}_0^2\nabla - i\boldsymbol{\sigma} \cdot \nabla' \times \mathcal{B}_0^2[\nabla(S + V)].$$

The physical content of the Dirac equation can already be addressed at this stage. However, it is even more instructive to remap each of these terms as density-dependent kernels. In other words, we wish to rewrite $\hat{h}_{\text{eff}}$ in the form $\hat{h}_{\text{eff}} = \hat{t} + \hat{v}[\rho]$, with ρ a non-relativistic local density. Doing so requires taking two additional steps. We first need to express each of these potentials as functionals of the relativistic densities and currents (2.69). Second, we need to introduce a local, non-relativistic density.

The local density form of the self-energies

Let us first express the two self-energies $\hat{S}$ and $\hat{V}$ as functionals of the various relativistic densities and currents (2.69). Since the typical momenta transferred between nucleons are small with respect to the meson masses, it is reasonable to treat the exchange of mesons between nucleons as zero-range interactions. In this paragraph, the pion is not included in the discussion. Starting with the scalar self-energy

$$S(\boldsymbol{r}) = g_\sigma(\rho_v(\boldsymbol{r}))\sigma(\boldsymbol{r}),$$

we recall that the σ field is a solution of the inhomogeneous Klein–Gordon equation

$$[-\nabla^2 + m_\sigma^2]\sigma = -g_\sigma \rho_s \tag{2.89}$$

and that its propagator reads

$$D_\sigma(\boldsymbol{r}_1, \boldsymbol{r}_2) = \int \frac{d^3\boldsymbol{k}}{(2\pi)^3}\frac{e^{i\boldsymbol{k}\cdot(\boldsymbol{r}_1-\boldsymbol{r}_2)}}{\boldsymbol{k}^2 + m_\sigma^2}. \tag{2.90}$$

A solution of equation (2.89) can then be written in the generic form

$$\sigma(\boldsymbol{r}) = -\int d^3\boldsymbol{r}'\ D_\sigma(\boldsymbol{r}, \boldsymbol{r}')g_\sigma[\rho_v(\boldsymbol{r}')]\rho_s(\boldsymbol{r}'). \tag{2.91}$$

Expanding equation (2.90) at the zeroth order in k^2/m_σ^2,

$$D_\sigma(\boldsymbol{r}_1, \boldsymbol{r}_2) = \frac{1}{m_\sigma^2}\int \frac{d^3\boldsymbol{k}}{(2\pi)^3}e^{i\boldsymbol{k}\cdot(\boldsymbol{r}_1-\boldsymbol{r}_2)} = \frac{1}{m_\sigma^2}\delta^{(3)}(\boldsymbol{r}_1 - \boldsymbol{r}_2), \tag{2.92}$$

yields

$$\sigma(\boldsymbol{r}) = -\frac{g_\sigma(\rho_v(\boldsymbol{r}))}{m_\sigma^2}\rho_s(\boldsymbol{r}). \tag{2.93}$$

Note that considering the next order is necessary to obtain sensible results, and yields a gradient term (surface term) correcting the zero-range approximation. For the sake of simplicity, we do not consider these extra terms. Setting $\alpha_i \equiv (\frac{g_i(\rho_v(\boldsymbol{r}))}{m_i})^2$ for $i = \{\sigma, \omega, \rho\}$, we then deduce an expression for the scalar self-energy

$$S(\boldsymbol{r}) = -\alpha_\sigma\rho_s(\boldsymbol{r}). \tag{2.94}$$

Proceeding in a similar way for the vector self-energy, we end up with

$$\begin{aligned} V(\boldsymbol{r}) = {} & \alpha_\omega\rho_v(\boldsymbol{r}) + \tau_3\alpha_\rho\rho_{tv}(\boldsymbol{r}) + eA\frac{1-\tau^3}{2} \\ & + \underbrace{\frac{1}{2}\left[-\frac{d\alpha_\sigma}{d\rho_{\text{v}}}\rho_{\text{s}}^2(\boldsymbol{r}) + \frac{d\alpha_\omega}{d\rho_{\text{v}}}\rho_{\text{v}}^2(\boldsymbol{r}) + \frac{d\alpha_\rho}{d\rho_{\text{v}}}\rho_{\text{tv}}^2(\boldsymbol{r})\right]}_{\Sigma_R(\boldsymbol{r})}. \end{aligned} \tag{2.95}$$

Thereafter, we will omit the Coulomb contribution since its physical content is similar to that of non-relativistic approaches.

Non-relativistic local density approximation

The next step is to express the various densities in terms of the non-relativistic local density $\rho_B(\boldsymbol{r}) \equiv \sum_i\phi_i^\dagger(\boldsymbol{r})\phi_i(\boldsymbol{r})$. The case of the vector density ρ_v is trivial

$$\rho_v = \sum_i \bar{f}_i\gamma^0 f_i = \sum_i\left[\varphi_i^\dagger\varphi_i + \chi_i^\dagger\chi_i\right] = \sum_i \phi_i^\dagger\phi_i \equiv \rho_B.$$

In a relativistic formulation, it is the vector density ρ_v that is the closest equivalent to the 'traditional' density ρ of non-relativistic approaches. Note that ρ_v is not a vector in itself, it is only-derived from vector-like quantities. In terms of the non-relativistic neutron and proton densities, we have

$$\rho_v(\boldsymbol{r}) = \rho_B(\boldsymbol{r}) = \rho_B^{(n)}(\boldsymbol{r}) + \rho_B^{(p)}(\boldsymbol{r}).$$

Likewise, the isovector density reads

$$\rho_\tau(\boldsymbol{r}) = \rho_B^{(n)}(\boldsymbol{r}) - \rho_B^{(p)}(\boldsymbol{r}) \equiv \rho_{tB}(\boldsymbol{r}).$$

The scalar density ρ_s case is more subtle,

$$\rho_s = \sum_i \left[\varphi_i^\dagger \varphi_i - \chi_i^\dagger \chi_i\right] = \sum_i \left[\phi_i^\dagger \phi_i - 2\left(\phi_i^\dagger \boldsymbol{\sigma} \cdot \nabla'\right)\mathcal{B}_0^2\left(\boldsymbol{\sigma} \cdot \nabla \phi_i\right)\right] + \mathcal{O}\left(\mathcal{B}_0^4\right).$$

Using properties of the Pauli matrices, we find

$$\rho_s(\boldsymbol{r}) = \rho_B(\boldsymbol{r}) - 2\mathcal{B}_0^2[\tau(\boldsymbol{r}) - \nabla \cdot \boldsymbol{J}(\boldsymbol{r})], \tag{2.96}$$

where

$$\tau(\boldsymbol{r}) \equiv \sum_i \left(\nabla \phi_i^\dagger\right) \cdot \left(\nabla \phi_i\right)$$

is the non-relativistic nucleon kinetic density and

$$\boldsymbol{J} \equiv \sum_i \left(\nabla \phi_i^\dagger\right) \cdot (\nabla \times \sigma)\phi_i$$

is the vector part of the spin-current density; compare to equation (1.22) in the non-relativistic context. Going back to the scalar and vector self-energies, we deduce that

$$S(\boldsymbol{r}) = -\alpha_\sigma\left(\rho_B(\boldsymbol{r}) - 2\mathcal{B}_0^2[\tau(\boldsymbol{r}) - \nabla \cdot \boldsymbol{J}(\boldsymbol{r})]\right), \tag{2.97}$$

while the vector self-energy (without the Coulomb contribution) reads

$$\begin{aligned} V(\boldsymbol{r}) = \alpha_\omega \rho_B(\boldsymbol{r}) + \tau_3 \alpha_\rho \rho_{tB}(\boldsymbol{r}) + \frac{1}{2}\Bigg[\frac{d\alpha_\omega}{d\rho_B}\rho_B^2(\boldsymbol{r}) + \frac{d\alpha_\rho}{d\rho_B}\rho_{tB}^2(\boldsymbol{r}) \\ - \frac{d\alpha_\sigma}{d\rho_B}\left\{\rho_B(\boldsymbol{r}) - 2\mathcal{B}_0^2[\tau(\boldsymbol{r}) - \nabla \cdot \boldsymbol{J}(\boldsymbol{r})]\right\}^2\Bigg], \end{aligned} \tag{2.98}$$

whose various combinations yield a fully non-relativistic expression of the ingredients entering the non-relativistic effective Hamiltonian (2.85). We now have all the necessary information to discuss the benefits from a relativistic description of nuclear systems.

2.6.3 Saturation of symmetric nuclear matter

The saturation mechanism of symmetric nuclear matter (SNM) may be the best illustration of the benefits of a relativistic approach, as it allows a fine analysis of several, different types of relativistic corrections. We recall that the main feature of SNM is the existence of a minimum in the curve of the binding energy versus density of the system, called the empirical saturation point and characterized by $E^{\mathrm{NM}} \approx 16 \pm 1\mathrm{MeV}$ and $\rho = \rho_{\mathrm{sat}} \approx 0.16 \pm 0.01\ \mathrm{fm}^{-3}$.

> Experimental measurements of nuclear radii and binding energies reveal that the radii are proportional $A^{1/3}$, A being the mass number. This implies a nearly constant

average nucleon density $\rho = A/V \simeq 0.16\ \text{fm}^{-3}$. At the same time, the binding energy per nucleon is also nearly constant and roughly equal to 8 MeV. Such saturation properties are closely related to the nature of the nuclear force sustaining nucleons in atomic nuclei, particularly its short-range nature and the existence of a repulsive core.

Non-relativistic calculations of nuclear matter

In non-relativistic approaches, two ingredients are fundamental in order to reproduce the empirical saturation point. First, short-range correlations mainly induced by the strongly repulsive nature of the NN interaction at short distance must be included. Given a NN interaction in free space, a consistent way of incorporating such correlations is either to soften the interaction ($V_{\text{low k}}$ or similarity renormalization group (SRG)) and treat it within many-body perturbation theory (MBPT), or employ a Brueckner–Hartree–Fock (BHF) approach involving the bare NN interaction with a hard core.

In the first two cases, the inclusion of short-range correlations does produce a saturation point. However, not a single parametrization of the bare NN interaction yields this saturation point close to the empirical one: this is the famous problem of the Coester line [101]. In order to reproduce the empirical saturation point, the treatment of a second fundamental ingredient is needed, namely many-body contributions (three- and four-body terms) that make infinite nuclear matter (INM) stiffer at high densities; see figure 2.7.

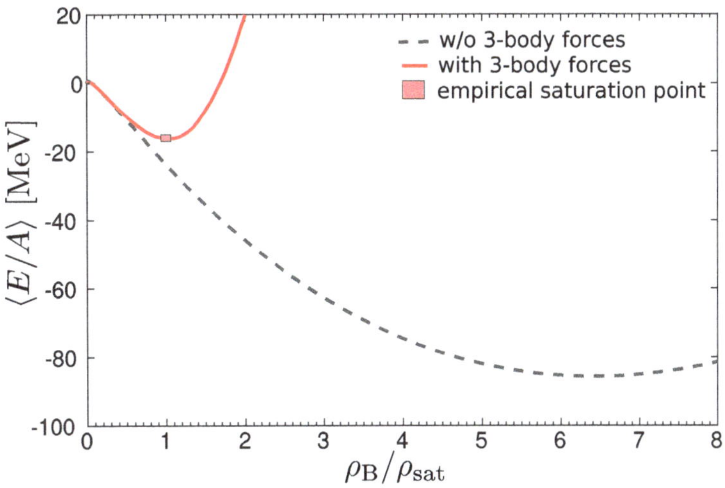

Figure 2.7. Binding energy versus density in SNM. Without the explicit treatment of many-body contributions to the nucleon interactions (dotted line), the curve totally misses the empirical saturation point. The results shown here are illustrative, yet representative of any parametrization of a bare NN force. Only if 3N forces are included can the empirical saturation point be reproduced satisfactorily (plain line). (Figure created by J-P Ebran.)

A third option is to introduce an effective in-medium pseudo-potential such as the Skyrme and Gogny one, that implicitly resums the bulk of nucleon correlations. In this case, the density-dependent term ρ^{α}, with α often a non-integer, captures not only the short-range nucleon correlations that play a key role to obtain saturation, but also the averaged many-body contributions. However, as discussed in section 3.2 of chapter 3 such non-analytical density-dependent terms bring about pathologies related to the violation of the Pauli exclusion principle.

A relativistic description of symmetric nuclear matter

How a relativistic approach achieves saturation of SNM is particularly enlightening to understand the advantages of adopting a relativistic language to describe nuclear properties. In this paragraph, we will use the simplest implementation of a CEDF (i.e. the RMF level) based on the most elementary Walecka-like Lagrangian, that is a Lagrangian of the linear sigma type whose only bosonic d.o.f.s are the the σ and ω effective meson fields and without any medium dependence of the sort (neither non-linear mesonic self-interaction terms nor explicit density dependence in the coupling constants). Such a basic approach already leads to the correct empirical saturation point of SNM. We can understand this surprising success as the effect of both kinematical and dynamical relativistic corrections.

Let us start with the Dirac equation and Klein–Gordon equations following our simple relativistic approach and expressed in terms of the ingredients defined above

$$\{-i\boldsymbol{\alpha}\nabla + \beta[m + S] + V\}f_i = e_i f_i \tag{2.99a}$$

$$[-\nabla^2 + m_\sigma^2]\sigma = -g_\sigma \rho_s \tag{2.99b}$$

$$[-\nabla^2 + m_\omega^2]\omega = +g_\omega \rho_v, \tag{2.99c}$$

with $S = g_\sigma \sigma$ and $V = g_\omega \omega$ the scalar and vector self-energies, respectively.

In the special case of nuclear matter, the invariance under translations justifies working in the momentum representation. This yields a nucleon spinor in the form of a relativistic plane wave

$$f_i \to f_{\boldsymbol{p},s,\tau}(\boldsymbol{r}) \equiv e^{i\boldsymbol{p}\boldsymbol{r}} u(\boldsymbol{p}, s)\chi_\tau^{1/2}, \tag{2.100}$$

with $u(\boldsymbol{p}, s)$ a Dirac spinor yet to be found and $\chi_\tau^{1/2}$ a bi-spinor in the isospin space. By the same token, the effective meson fields do not depend on space–time coordinates, so that the equations of the model read

$$\{\boldsymbol{\alpha p} + \beta[m + S(\boldsymbol{p})] + V(\boldsymbol{p})\}u(\boldsymbol{p}, s) = e(\boldsymbol{p})u(\boldsymbol{p}, s) \tag{2.101a}$$

$$m_\sigma^2 \sigma(\boldsymbol{p}) = -g_\sigma \rho_s(\boldsymbol{p}) \tag{2.101b}$$

$$m_\omega^2 \omega(\boldsymbol{p}) = +g_\omega \rho_v(\boldsymbol{p}). \tag{2.101c}$$

The scalar and vector nucleon self-energies thus read

$$S(\boldsymbol{p}) \equiv S = -\alpha_\sigma \rho_s \qquad V(\boldsymbol{p}) \equiv V = +\alpha_\omega \rho_\nu,$$

where $\alpha_i \equiv (g_i/m_i)^2$ for $i = \{\sigma, \omega\}$. Setting $m^*(\boldsymbol{p}) = m + S(\boldsymbol{p}) \equiv m^*$ and defining $e^*(\boldsymbol{p}) \equiv e(\boldsymbol{p}) - V$, the Dirac equation (2.101a) takes the form

$$(\boldsymbol{\alpha p} + \beta m^*)u(\boldsymbol{p}, s) = e^*(\boldsymbol{p})u(\boldsymbol{p}, s).$$

Recall that the free-particle Dirac equation (2.22) that we presented in section 2.1.3 was originally written

$$(\gamma_\mu p^\mu - m)\psi(p) = 0.$$

Introducing

$$\beta = \gamma^0 \quad \alpha^i = \beta\gamma^i \quad (i = 1, 2, 3)$$

allows us to rewrite its static version in the form

$$(\boldsymbol{\alpha} \cdot \boldsymbol{p} + \beta m)\psi(p) = 0.$$

Adopting the form (2.100), its normalized solutions $u^0(\boldsymbol{p}, s)$ ($\nu^0(\boldsymbol{p}, s)$) with positive (negative) energies thus read

$$u^0(\boldsymbol{p}, s) = \sqrt{\frac{e_p + m}{2e_p}} \begin{pmatrix} 1 \\ \dfrac{\sigma p}{e_p + m} \end{pmatrix} \chi_s^{1/2},$$

$$\nu^0(\boldsymbol{p}, s) = \sqrt{\frac{e_p + m}{2e_p}} \begin{pmatrix} \dfrac{\sigma p}{e_p + m} \\ 1 \end{pmatrix} \chi_s^{1/2},$$

$$e_p \equiv \sqrt{\boldsymbol{p}^2 + m^2}.$$

In other words, it has the exact same form as the free-particle Dirac equation (2.22), except that the bare quantities m and e are simply replaced by the dressed quantities m^* and e^*. The corresponding solutions follow from the free-case solutions substituting the bare quantities with the dressed ones. For example, the positive-energy solution reads

$$u^*(\boldsymbol{p}, s) = \sqrt{\frac{e_p^* + m^*}{2e_p^*}} \begin{pmatrix} 1 \\ \dfrac{\sigma p}{e_p^* + m^*} \end{pmatrix} \chi_s^{1/2}, \qquad e_p^* \equiv \sqrt{\boldsymbol{p}^2 + m^{*2}}.$$

The vector and scalar densities thus read

$$\rho_\nu = \sum_{\text{isospin}} \sum_{\text{spin}} \int_0^{k_F} \frac{d^3\boldsymbol{k}}{(2\pi^3)} \bar{u}^* \gamma^0 u^* = \frac{2}{\pi^2} \int_0^{k_F} dk k^2 = \frac{2}{3\pi^2} k_F^3$$

$$\rho_s = \sum_{\text{isospin}} \sum_{\text{spin}} \int_0^{k_F} \frac{d^3\boldsymbol{k}}{(2\pi^3)} \bar{u}^* u^* = \frac{2}{\pi^2} \int_0^{k_F} dk k^2 \frac{m^*(k)}{\sqrt{k^2 + m^{*2}(k)}}.$$

Note that the scalar density is smaller than the vector one due to the factor m^*/e^*, which is an effect of Lorentz contraction. Therefore, the contribution of rapidly moving nucleons to the scalar source is significantly reduced. More importantly, the involved form of the scalar density induces a transcendental self-consistency equation for the scalar self-energy (or equivalently the Dirac mass m^*),

$$m^*(k) = m - \alpha_\sigma \frac{2}{\pi^2} \int_0^{k_F} dk k^2 \frac{m^*(k)}{\sqrt{k^2 + m^{*2}(k)}},$$

that must be solved at each value of the Fermi momentum k_F; see figure 2.8.

This illustrates the non-perturbative nature of the physics encompassed within the Lorentz scalar sector, which will shortly be rendered more manifest by considering several types of particular cases. Before doing this, let us write the total binding energy of SNM in our simplified CEDF model

$$\begin{aligned} \left\langle \frac{E}{A} \right\rangle &= E_\omega + E_\sigma + E_{\text{kin}} \\ &= \frac{1}{\rho_v} \left[\frac{1}{2} \alpha_\omega \rho_v^2 - \frac{1}{2} \alpha_\sigma \rho_s^2 + \frac{2}{\pi^2} \int_0^{k_F} d^k k^2 \frac{k^2 + m m^*}{\sqrt{k^2 + m^{*2}}} \right]. \end{aligned} \tag{2.102}$$

This quantity, plotted in figure 2.9 (solid line) against the ratio ρ_B/ρ_{sat} where $\rho_B \equiv \rho_v$ is the non-relativistic nucleon density and $\rho\text{sat} = 0.16\ \text{fm}^{-3}$, reproduces the empirical saturation point for $m^2\alpha_\sigma = 357.4$ and $m^2\alpha_\omega = 273.8$ [102].

Kinematical and dynamical relativistic corrections

To understand the reason for such a success, we first investigate the full non-relativistic limit, where, according to the previous section, at the zeroth order,

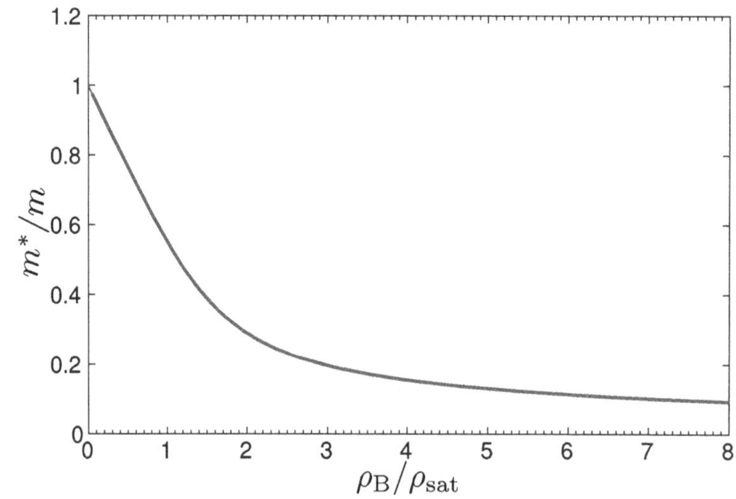

Figure 2.8. Ratio of the effective nucleon mass m^* over the nucleon bare mass m as a function of the density of the system normalized by its saturation density $\rho_{\text{sat}} = 0.16\ \text{fm}^{-3}$. (Figure created by J-P Ebran.)

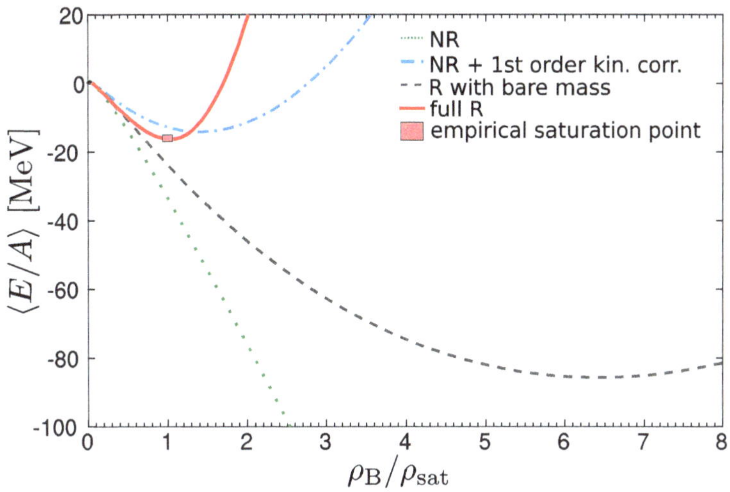

Figure 2.9. Binding energy of SNM against $\rho/\rho_{\rm sat}$ computed at four different levels: the full relativistic calculation (solid line) that reproduces the empirical saturation point; a non-relativistic limit at zeroth order (dotted line) that loses the saturation property; a non-relativistic calculation with first-order 'kinematical' corrections (dash-dotted line) that saturates but too far away from the empirical saturation point; and a relativistic calculation where the dressed quantities are replaced by their bare expression (dashed line). (Figure created by J-P Ebran.)

$\rho_s = \rho_v \equiv \rho_B$, i.e. we lose the distinction between a Lorentz scalar and the time-like component of the four-vector, and the non-relativistic nucleon bi-spinor $\varphi(\boldsymbol{p})$ satisfies a Schrödinger equation

$$\left(\frac{p^2}{2m} + \hat{v}_{\rm cent}\right)\varphi(\boldsymbol{p}) = e(\boldsymbol{p})\varphi(\boldsymbol{p}), \tag{2.103}$$

where

$$\hat{v}_{\rm cent} = V + S = (\alpha_\omega - \alpha_\sigma)\rho_B.$$

Note that at the lowest order, it is the 'bare' mass m that is involved in equation (2.103), not the 'dressed' mass m^* or the 'effective' mass $\tilde{M}$. In addition, contrary to the relativistic case, where the solutions of the free nucleon equation of motion are relativistic plane waves dressed by medium effects, the non-relativistic solutions of equation (2.103) are simple plane waves $e^{i\boldsymbol{p}\boldsymbol{r}}u^*(\boldsymbol{p}, s)\chi_\tau^{1/2} \xrightarrow{NR} e^{i\boldsymbol{p}\boldsymbol{r}}\chi_s^{1/2}\chi_\tau^{1/2}$ (as in the non-relativistic free problem). The corresponding binding energy reads

$$\begin{aligned}\left\langle\frac{E}{A}\right\rangle^{NR} &= E_\omega^{NR} + E_\sigma^{NR} + E_{\rm kin}^{NR} \\ &= \frac{1}{\rho_B}\left[\frac{\alpha_\omega - \alpha_\sigma}{2}\rho_B^2 + \frac{2}{\pi^2}\int_0^{k_{\rm F}} dk k^2\left(\frac{k^2}{2m} + m\right)\right]\end{aligned}$$

and is plotted as a dotted line in figure 2.9. Therefore, losing the distinction between the Lorentz scalars and the time-like component of the four-vectors thus translates into losing the minimum in the equation of state (EoS) of SNM. In the non-relativistic limit, the potential contribution involves the nucleon density ρ_B weighted by the negative factor $\alpha_\omega - \alpha_\sigma$ hence a monotonic behavior with increasing density.

The next step is to restore a distinction between the Lorentz scalars and the time-like component of the four-vectors. We first focus on the so-called kinematical relativistic corrections, that is, corrections in powers of $1/m$, with m the bare nucleon mass. We start with equation (2.96) adapted to the case of nuclear matter: the spin–orbit current is zero. Recall that the 'small' parameter $\mathcal{B}_0$ that we use for our non-relativistic expansion is

$$\mathcal{B}_0 = \frac{1}{2\tilde{M}} = \frac{1}{2m + S - V}.$$

Since $S - V$ is of the order of -50 MeV while the bare nucleon mass is of the order of 1 GeV, we can expand $\mathcal{B}_0^2$ up to first order in $1/m$. This leads to

$$\rho_s \simeq \rho_B - \frac{2}{m^2}\tau_{\mathrm{F}},$$

with

$$\tau_{\mathrm{F}} = \frac{1}{\pi^2}\int_0^{k_{\mathrm{F}}} k^2 k^2 dk$$

so that

$$\rho_s \simeq \rho_B - \frac{2}{5\pi^2 m^2}k_{\mathrm{F}}^5.$$

Using this new expression of ρ_s in equation (2.102) and setting $\rho_v = \rho_B$ yields

$$\left\langle \frac{E}{A} \right\rangle^{NR+\mathrm{corr}} = \frac{1}{\rho_B}\left[\frac{1}{2}\alpha_\omega \rho_B^2 - \frac{1}{2}\alpha_\sigma\left(\rho_B - \frac{2}{m^2}\tau_{\mathrm{F}}\right)^2 + \frac{2}{\pi^2}\int_0^{k_{\mathrm{F}}} dk k^2\left(\frac{k^2}{2m} + m\right)\right],$$

the dashed-dotted line of figure 2.9. In other words, saturation is recovered, but quite off the empirical point. As the Fermi momentum k_{F} increases, the kinetic density $\tau_{\mathrm{F}} \sim k_{\mathrm{F}}^5$ grows faster than the nucleon density $\rho_B \sim k_{\mathrm{F}}^3$, hence the occurrence of a minimum.

In order to obtain a more realistic, stiffer binding energy curve at high densities, we need to incorporate the physics of a many-body contribution to the nucleon interaction. This is exactly what dressing the nucleon mass and s.p. energy achieve. To see this, we can 'undress' the solution of the Dirac equation in nuclear matter: $u^*(\boldsymbol{p}, s) \overset{\text{undressing}}{\rightarrow} u^0(\boldsymbol{p}, s)$. The corresponding binding energy is shown by the dashed line in figure 2.9. Because the distinction between ρ_s and ρ_v is maintained, we obtain a saturation point. However, the undressing of both the mass and the s.p. energy yields an inconsistent treatment of the kinetic energy property that is no longer affected by the Lorentz contraction. This is the reason why the saturation point is even in lower agreement with the empirical one as compared to the non-relativistic case with the first relativistic corrections of kinematical type.

We thus see that the effect of dressing quantities in a relativistic framework leads to similar quantitative results as the inclusion of 3N forces in non-relativistic calculations. The additional correlations incorporated when going from the bare to dressed quantities are called dynamical. We can obtain a better understanding of these correlations by decomposing the dressed spinor u^* on the basis formed by the free spinors u^0 and ν^0, namely

$$u^*(\boldsymbol{p}, s) = \frac{1}{\sqrt{1 + \zeta^2(\boldsymbol{p})}}\left[u^0(\boldsymbol{p}, s) + \zeta(\boldsymbol{p})\sum_{s'}\langle s|\boldsymbol{\sigma p}|s'\rangle \nu^0(-\boldsymbol{p}, s')\right]$$

where the parameter

$$\zeta(\boldsymbol{p}) = -\frac{pS}{2mm^*}$$

is roughly 0.1 in our simple model. Based on such an expansion, the dynamical correlations involve an anti-nucleon component ν^0. At the first non-zeroth order in $\zeta(\boldsymbol{p})$, the dynamical correction to the vector self-energy V thus reads

$$\delta V|_{\rm dyn} = \frac{p^2}{M}\left(\frac{S}{M}\right)^2$$

corresponding to the Feynmann diagram of figure 2.10. Put differently, such a dynamical correction is topologically equivalent to a 3N term, hence the effect of making nuclear matter binding energy stiffer at high densities.

Therefore, a simple relativistic approach without any medium dependence encompasses in a very effective way non-trivial correlations of both kinematical

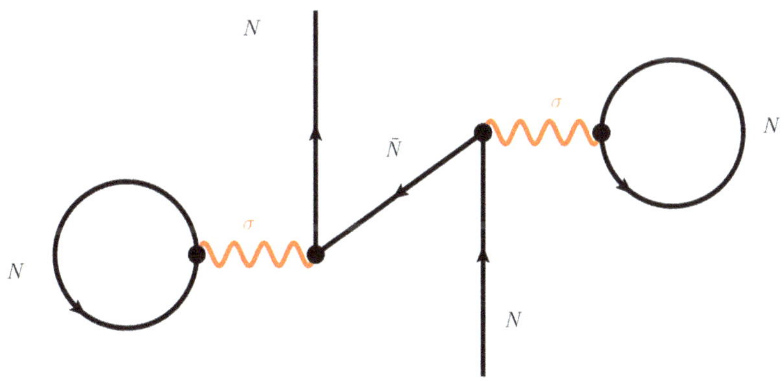

Figure 2.10. The so-called Z-diagram representing the first non-zero order of relativistic dynamical corrections to the nucleon self-energy. (Figure created by J-P Ebran.)

and dynamical types. This implies that we can describe nuclear saturation very economically. Working with four-component Dirac spinors allows one to maintain a difference between the vector density ρ_v and the scalar density ρ_s. Whenever the nucleon density increases, the scalar and vector nucleon self-energies start to grow, so that the energy gap between the Fermi and Dirac seas shrinks; recall figure 2.6. It implies that the probability to extract an anti-nucleon from the Dirac sea, measured by the small component $|\chi_i|^2$, is enhanced. In other words, the increase of the nucleon density that first yields the increase of the scalar self-energy S eventually leads to a decrease of the scalar density, which is the source of the scalar self-energy S, stabilizing the nuclear system.

2.6.4 Spin–orbit properties

There is no need to reaffirm that the spin–orbit coupling plays a crucial role in nuclear structure, and that its inclusion in the effective single-nucleon potential is not only essential to reproduce the empirical magic numbers around the valley of β-stability [103, 104], but also provides the microscopic foundation for the rotational features of the excitation spectra of light nuclei, understood in a many-particle spherical shell-model approach in terms of the Elliott and Harvey SU(3) coupling scheme [105–107].

We have seen in section 2.6.2 how the spin–orbit potential appears after taking the non-relativistic limit of the Dirac equation; see section 2.6.2. The fact that the spin–orbit potential emerges automatically with the proper strength as the constructive combination of the scalar and vector nucleon self-energies is commonly used as an argument in favor of a relativistic formulation of the nuclear many-body problem; see e.g. [9].

Less noticed is the fact that the typical spin–orbit energy splitting between spin–orbit partners is of the same order of magnitude as the average energy spacing between single-particle levels due to the confining potential. Here again, a relativistic approach enables us to understand such a feature [108]. To illustrate this, we use the expression (2.88) of the spin–orbit potential in spherical symmetry and, for the sake of simplicity, neglect the effect of the ρ meson and the explicit density dependence of the coupling constants and set the constant values $V \equiv \alpha_\omega \rho_{\text{sat}}$, $S \equiv -\alpha_\sigma \rho_{\text{sat}}$. We have

$$\hat{v}_{\text{so}} \simeq \frac{\alpha_\omega + \alpha_\sigma}{2m^2\left[1 - \frac{V-S}{2m}\right]^2} \frac{1}{r}\frac{d}{dr}\rho(r)\boldsymbol{l} \cdot \boldsymbol{s}.$$

Using $\langle \frac{1}{r}\frac{d}{dr}\rho(r) \rangle \sim -\frac{1}{R^5}$ with R the radius of the system, we obtain

$$\hat{v}_{\text{so}} \simeq \frac{V-S}{2m^2R^2\left[1 - \frac{V-S}{2m}\right]^2}\boldsymbol{l} \cdot \boldsymbol{s}.$$

We deduce the typical energy splitting δ^{so} between two spin–orbit partners with an angular quantum number l

$$\delta^{\mathrm{so}} \simeq \frac{(V - S)\left(l + \frac{1}{2}\right)}{2m^2R^2\left[1 - \frac{V - S}{2m}\right]^2}. \tag{2.104}$$

On the other hand, the typical energy spacing δ^{cent} between s.p. levels due to the confining potential is well approximated by the harmonic oscillator case [108]

$$\delta^{\mathrm{cent}} \simeq \frac{1}{R}\sqrt{\frac{-2(S + V)}{M}},$$

with $-(S + V)$ the depth of the confining potential. After introducing the dimensionless quantity

$$\eta \equiv \frac{m}{V - S}$$

we find that the ratio of the spin–orbit splitting versus the principal energy splitting reads $|\delta^{\mathrm{so}}|/|\delta^{\mathrm{cent}}| = K\mathcal{R}$, with the coefficients K and $\mathcal{R}$ defined as

$$K \equiv \frac{l + \frac{1}{2}}{2R\sqrt{-2m(S + V)}}, \qquad \mathcal{R} \equiv \left| \frac{1}{\eta\left[1 - \frac{1}{2\eta}\right]^2} \right|.$$

K typically ranges from 0.5 to 1. Hence, the essential feature of the spin–orbit coupling, that is, the characteristic energy gap between spin–orbit partners and its sign (i.e. which of the two levels goes up and which goes down) is governed by the parameter η. This parameter is the ratio between the bare mass of the fermionic d.o.f. and the constructive combination $V - S$ of the vector and scalar self-energies. In nuclear systems, the accidental feature $|V - S| \approx m$ entails that the ratio $\mathcal{R}$ is of order 1. Therefore, the relativistic formulations allow one to understand why the spin–orbit splitting in nuclei is of the same order of magnitude as the principal energy spacing: it is an emergent property related to the features of the scalar and vector nucleon self-energies compared to the nucleon mass, that is, to the typical energy gap across the QCD non-perturbative vacuum which encodes the physics of the chiral SSB.

Interestingly, the expression for $\mathcal{R}$, which assumes short-range interactions between the fermionic d.o.f.s, still holds for spring-like interactions such as those quarkonia experience, as well as for interactions that follow $1/r$ as Coulomb systems. The conclusions pertaining to the effect of the spin–orbit coupling on the s.p. spectrum can thus be extended to the cases of quarkonia, atoms, or even hypernuclei; see figure 2.11.

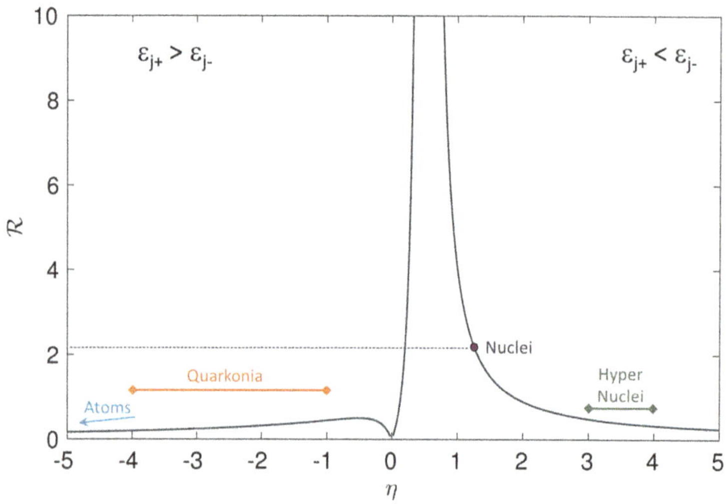

Figure 2.11. Ratio $\mathcal{R}$ with respect to the parameter η. Particular cases are highlighted: nuclei, hypernuclei, quarkonia and atoms. If we note $j_\pm = \ell \pm \frac{1}{2}$, then the sign of the spin–orbit coupling determines whether the s. p. level ϵ_{j+} is higher/lower than the level ϵ_{j-}. (Figure created by J-P Ebran.)

2.6.5 Pseudospin symmetry

The near equality $V \simeq -S$ not only justifies the non-relativistic kinematics for nucleons in nuclear systems, but is also the source of the approximate realization of pseudospin symmetry in atomic nuclei [10]. The PSS symmetry is a theoretical construction designed to explain why two spherical s.p. states with quantum numbers $(n, l, j = l + 1/2)$ and $(n - 1, l + 2, j = l + 3/2)$ are nearly degenerate [109, 110]. When relabeled with what is called pseudo-radial and pseudo-orbital quantum numbers $(\tilde{n} = n - 1, \tilde{l} = l + 1)$, two such levels appear as pseudospin–orbit partners $j = \tilde{l} \pm 1/2$. In a non-relativistic approach, the PSS symmetry is an empirical construct. However, like the spin–orbit case, it is naturally explained as a property of the one-particle Dirac Hamiltonian [10, 111]: the pseudo-orbital angular momentum $\tilde{l}$ is nothing other than the orbital angular momentum of the small component of the Dirac spinor.

Let us define the following 4×4 matrices,

$$S_i \equiv \begin{pmatrix} s_i & 0 \\ 0 & \breve{s}_i \end{pmatrix} \quad \breve{S}_i \equiv \begin{pmatrix} \breve{s}_i & 0 \\ 0 & s_i \end{pmatrix} \quad i = \{x, y, z\},$$

where $s_i \equiv \sigma_i/2$ and

$$\breve{s}_i \equiv \left(\frac{\boldsymbol{\sigma} \cdot \boldsymbol{p}}{|\boldsymbol{p}|}\right)^\dagger s_i \left(\frac{\boldsymbol{\sigma} \cdot \boldsymbol{p}}{|\boldsymbol{p}|}\right).$$

One notices that the matrices verify

$$[S_i, S_j] = \epsilon_{ijk} S_k, \qquad [\breve{S}_i, \breve{S}_j] = \epsilon_{ijk} \breve{S}_k.$$

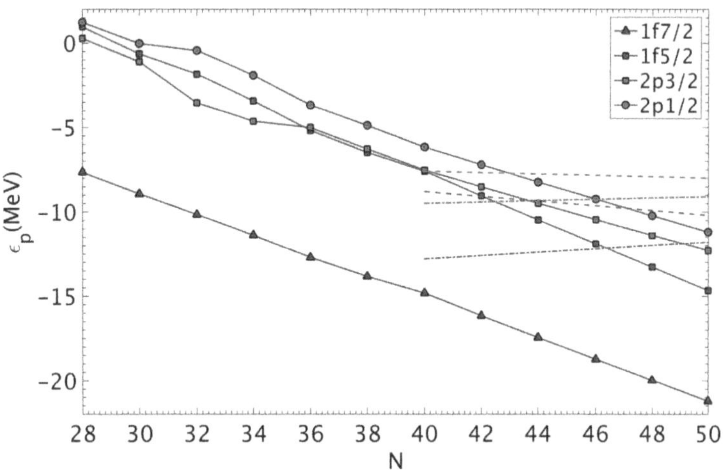

Figure 2.12. Proton single-particle levels in the Ni isotopic chain calculated within a CEDF at the RMF level with the $NL3$ parametrization set. The dotted lines without markers correspond to Otsuka's predictions [114]. (Figure created by J-P Ebran.)

In other words, $\boldsymbol{S}$ and $\check{S}$ generate separately a $\mathfrak{su}(2)$ algebra associated with the $SU(2)_{\text{spin}}$ and $SU(2)_{\text{pseudospin}}$ groups, respectively. If $V - S$ (for the spin symmetry case) and $V + S$ (for the PSS case) do not depend on space coordinates, then their commutator with the one-particle Dirac Hamiltonian (2.81) yields

$$[S_i, h_D] = 0; \qquad [\check{S}_i, h_D] = 0.$$

The spin–orbit potential is proportional to $\nabla(V - S)$, and hence provides an estimate of the breaking of the spin symmetry. Likewise, a pseudospin–orbit potential can be defined by expressing the Dirac equation in terms of the small component of the nucleon spinor instead of its large component [112]. Such a potential is proportional to $\nabla(V + S)$, hence it measures the breaking of PSS.

Because $V + S$ is inherently small in actual nuclei, contrary to $V - S$, the PSS is approximately realized, leading to the near degeneracy of the pseudospin–orbit partners $j = \tilde{l} \pm 1/2$. A relativistic description of the s.p. shell structure based on the one-particle Dirac Hamiltonian hence naturally accounts for the empirical PSS and its dependence on the depth of the confining potential, the surface diffusivity, or the neutron-to-proton ratio [113].

Pseudospin symmetry and shape coexistence

The PSS symmetry also provides a mechanism to explain shape coexistence in the region $Z \simeq 28$ and $N \simeq 50$ that is far more convincing than Otsuka's tensor mechanism [114]. In this region, Otsuka's tensor mechanism is usually invoked to account for the crossing of the 1f5/2 and 2p3/2 states, the former dropping below the latter as the neutron number increases; see figure 2.12. However, Otsuka's tensor coupling also implies that the π 1f7/2 orbital should increase relative to its partner π 1f5/2, thus an erosion of the Z = 28 magic number around N = 50. This erosion has been discarded by recent experimental campaigns such as [115]. In contrast, the PSS provides a consistent description of the evolution of these three proton s.p. states. Since 1f5/2 and 2p3/2 are pseudospin–orbit partners, the PSS not only explains their near degeneracy but also the evolution of the pseudospin–orbit splitting with increasing isospin, namely, the fact that the pseudospin–orbit partner $j = \tilde{l} + 1/2$ goes lower than $j = \tilde{l} - 1/2$. In addition, the 1f7/2 state does not belong to a pseudospin–orbit doublet, so that it evolves according to the central and spin–orbit features as the neutron number increases. In fact, the PSS symmetry is consistent with the persisting nature of the Z = 28 magic number even in exotic nuclei.

References

[1] Chen W-C and Piekarewicz J 2014 Building relativistic mean field models for finite nuclei and neutron stars *Phys. Rev.* C **90** 044305

[2] Baran V *et al* 2005 Reaction dynamics with exotic nuclei *Phys. Rep.* **410** 335–466

[3] Jenkins E and Manohar A V 1991 Baryon chiral perturbation theory using a heavy fermion Lagrangian *Phys. Lett.* B **255** 558–62

[4] Weinberg S 1996 What is quantum field theory, and what did we think it is? *Conceptual Foundations of Quantum Field Theory, Proceedings, Symposium and Workshop, Boston, MA*1–3 March pp 241–51 arXiv: hep-th/9702027

[5] Weinberg S 2016 Effective field theory, past and future *Int. J. Mod.Phys.* A **31** 1630007

[6] Georgi H 1993 Effective field theory *Ann. Rev. Nucl. Part. Sci.* **43** 209–52

[7] Shankar R 1999 Effective field theory in condensed matter physics *Conceptual Foundations of Quantum Field Theory* ed T Y Cao (Cambridge: Cambridge University Press) pp 47–55

[8] Serot B D and Walecka J D 1986 The relativistic nuclear many body problem *Adv. Nucl. Phys.* **16** 1–327

[9] Ring P 1996 Relativistic mean field theory in finite nuclei *Prog. Part. Nucl. Phys.* **37** 193–263

[10] Ginocchio J N 1997 Pseudospin as a relativistic symmetry *Phys. Rev. Lett.* **78** 436–9

[11] Nambu Y and Jona-Lasinio G 1961 Dynamical model of elementary particles based on an analogy with superconductivity. II *Phys. Rev.* **124** 246–54

[12] Muroya S *et al* 2003 Lattice QCD at finite density—an introductory review *Prog. Theor. Phys.* **110** 615–68

[13] Furnstahl R J, Griegel D K and Cohen T D 1992 QCD sum rules for nucleons in nuclear matter *Phys. Rev.* C **46** 1507–27

[14] Finelli P *et al* 2003 Nuclear many-body dynamics constrained by QCD and chiral symmetry *Eur. Phys. J.* A **17** 573–8

[15] Furnstahl R J, Serot B D and Tang H-B 1997 A chiral effective Lagrangian for nuclei *Nucl. Phys.* A **615** 441–82

[16] Engel Y M *et al* 1975 Time-dependent Hartree–Fock theory with Skyrme's interaction *Nucl. Phys.* A **249** 215–38

[17] Dobaczewski J and Dudek J 1995 Time-odd components in the mean field of rotating superdeformed nuclei *Phys. Rev.* C **52** 1827–39

[18] Dobaczewski J and Dudek J 1996 Time-odd components in the rotating mean field and identical bands *Acta Phys. Pol.* B **27** 45

[19] Bender M *et al* 2002 Gamow–Teller strength and the spin–isospin coupling constants of the Skyrme energy functional *Phys. Rev.* C **65** 064308

[20] Carlsson B G, Dobaczewski J and Kortelainen M 2008 Local nuclear energy density functional at next-to-next-to-next-to-leading order *Phys. Rev.* C **78** 044326

[21] Schweber S S 2005 *An Introduction to Relativistic Quantum Field Theory* 2nd edn (New York: Dover)

[22] Weinberg S 1995 *The Quantum Theory of Fields*vol 1 (Cambridge: Cambridge University Press)

[23] Peskin M E and Schroeder D V 1995 *An Introduction to Quantum Field Theory* (Boulder, CO: Westview Press)

[24] Itzykson C and Zuber J-B 1980 *Quantum Field Theory* (New York: McGraw-Hill)

[25] Galilei G 2001 *Dialogue Concerning the Two Chief World Systems: Ptolemaic and Copernican* (New York: The Modern Library)

[26] Lee A R and Kalotas T M 1975 Lorentz transformations from the first postulate *Am. J. Phys.* **43** 434–7

[27] Lévy-Leblond J-M 1976 One more derivation of the Lorentz transformation *Am. J. Phys.* **44** 271–7

[28] Landau L D and Lifshitz E M 1980 *The Classical Theory of Fields* 4th edn (Oxford: Butterworth-Heinemann)

[29] Cornwell J F 1984 *Group Theory in Physics*vol 1 (New York: Academic)

[30] Cornwell J F 1984 *Group Theory in Physics*vol 2 (New York: Academic)

[31] Hamermesh M 1989 *Group Theory and Its Application to Physical Problems* (New York: Dover)

[32] Hall B C 2000 *Lie Groups. Lie Algebras and Representations* (Berlin: Springer)

[33] Wilczek F 2002 QCD in extreme conditions *Theoretical Physics at the End of the Twentieth Century: Lecture Notes of the CRM Summer School, Banff, Alberta* ed Y Saint-Aubin and L Vinet (New York: Springer) pp 567–636

[34] Barceló C *et al* 2016 From physical symmetries to emergent gauge symmetries *J. High Energy Phys.* **10** 84

[35] Martin C 2003 On continuous symmetries and the foundations of modern physics *Symmetries in Physics: Philosophical Reflections* ed K Brading and E Castellani (Cambridge: Cambridge University Press)

[36] Lee P A 2008 From high temperature superconductivity to quantum spin liquid: progress in strong correlation physics *Rep. Prog. Phys.* **71** 012501

[37] Bando M, Kugo T and Yamawaki K 1988 Nonlinear realization and hidden local symmetries *Phys. Rep.* **164** 217–314

[38] Nakahara M 2003 *Geometry, Topology and Physics* 2nd edn (London: Taylor and Francis)

[39] Sazdjian H 2017 Introduction to chiral symmetry in QCD *EPJ Web Conf.* **137** 02001

[40] Forger M and Römer H 2004 Currents and the energy–momentum tensor in classical field theory: a fresh look at an old problem *Ann. Phys.* **309** 306–89

[41] Gross D J 1996 Asymptotic freedom, confinement and QCD *History of Original Ideas and Basic Discoveries in Particle Physics* (Boston, MA: Springer) pp 75–99

[42] Yoichiro N 1960 Quasi-particles and gauge invariance in the theory of superconductivity *Phys. Rev.* **117** 648–63
[43] Goldstone J 1961 Field theories with superconductor solutions *Nuovo Cim.* **19** 154–64
[44] Olive K A *et al* 2014 Review of particle physics *Chin. Phys.* **C38** 090001
[45] Holt J W, Rho M and Weise W 2016 Chiral symmetry and effective field theories for hadronic, nuclear and stellar matter *Phys. Rep.* **621** 2–75
[46] Brown G E and Rho M 1991 Scaling effective Lagrangians in a dense medium *Phys. Rev. Lett.* **66** 2720–3
[47] Bardeen W A 1969 Anomalous ward identities in spinor field theories *Phys. Rev.* **184** 1848–59
[48] Crewther R J and Tunstall L C 2015 $\Delta I = 1/2$ rule for kaon decays derived from QCD infrared fixed point *Phys. Rev.* D **91** 034016
[49] Golterman M and Shamir Y 2016 Low-energy effective action for pions and a dilatonic meson *Phys. Rev.* D **94** 054502
[50] Beane S R and van Kolck U 1994 The dilated chiral quark model *Phys. Lett.* B **328** 137–42
[51] Li Y-L, Ma Y-L and Rho M 2017 Chiral-scale effective theory including a dilatonic meson *Phys. Rev.* D **95** 114011
[52] Paeng W-G *et al* 2017 Scale-invariant hidden local symmetry, topology change, and dense baryonic matter. II *Phys. Rev.* D **96** 014031
[53] Skyrme T H R 1962 A unified field theory of mesons and baryons *Nucl. Phys.* **31** 556–69
[54] Gasser J, Sainio M E and Svarc A 1988 Nucleons with chiral loops *Nucl. Phys.* B **307** 779–853
[55] Bernard V *et al* 1992 Chiral structure of the nucleon *Nucl. Phys.* B **388** 315–45
[56] Hua-Bin T 1996 A new approach to chiral perturbation theory for matter fields arXiv: hep-ph/9607436
[57] Lehmann D and Prézeau G 2001 Effective field theory dimensional regularization *Phys. Rev.* D **65** 016001
[58] Coleman S, Wess J and Zumino B 1969 Structure of phenomenological Lagrangians *Phys. Rev.* **177** 2239–47
[59] Weinberg S 1968 Nonlinear realizations of chiral symmetry *Phys. Rev.* **166** 1568–77
[60] Manohar A and Georgi H 1984 Chiral quarks and the non-relativistic quark model *Nucl. Phys.* B **234** 189–212
[61] Serot B and Zhang X 2012 Electroweak interactions in a chiral effective Lagrangian for nuclei *Advances in Quantum Field Theory* ed S Ketov (Rijeka: InTech) ch 4
[62] Machleidt R 1989 The meson theory of nuclear forces and nuclear structure *Advances in Nuclear Physics* ed J W Negele and E Vogt (Boston, MA: Springer) pp 189–376
[63] Walecka J D 2004 *Theoretical Nuclear and Subnuclear Physics* 2nd edn (London: Imperial College Press)
[64] Ekström A *et al* 2015 Accurate nuclear radii and binding energies from a chiral interaction *Phys. Rev.* C **91** 051301
[65] Hubbard J 1959 Calculation of partition functions *Phys. Rev. Lett.* **3** 77–8
[66] Vretenar D and Weise W 2004 Exploring the nucleus in the context of low-energy QCD *Extended Density Functionals in Nuclear Structure Physics* ed G A Lalazissis, P Ring and D Vretenar (Berlin: Springer) pp 65–117
[67] Finelli P *et al* 2006 Relativistic nuclear energy density functional constrained by low-energy QCD *Nucl. Phys.* A **770** 1–31

[68] Mannque R 2017 Pinning down the axial-vector coupling constant in nuclei and dense matter, arXiv: 1705.10864
[69] Matsui T 1981 Fermi-liquid properties of nuclear matter in a relativistic mean-field theory *Nucl. Phys.* A **370** 365–88
[70] Su Y-H and Lu H-T 2017 Breakdown of Landau Fermi liquid theory: restrictions on the degrees of freedom of quantum electrons *Front. Phys.* **13** 137103
[71] Harada M and Yamawaki K 2003 Hidden local symmetry at loop: a new perspective of composite gauge boson and chiral phase transition *Phys. Rep.* **381** 1–233
[72] Johnson M H and Teller E 1955 Classical field theory of nuclear forces *Phys. Rev.* **98** 783–7
[73] Duerr H-P and Teller E 1956 Interaction of antiprotons with nuclear fields *Phys. Rev.* **101** 494–5
[74] Boguta J and Bodmer A R 1977 Relativistic calculation of nuclear matter and the nuclear surface *Nucl. Phys.* A **292** 413–28
[75] Brockmann R and Toki H 1992 Relativistic density-dependent Hartree approach for finite nuclei *Phys. Rev. Lett.* **68** 3408–11
[76] Lalazissis G A, König J and Ring P 1997 New parametrization for the Lagrangian density of relativistic mean field theory *Phys. Rev.* C **55** 540–3
[77] Vretenar D *et al* 2005 Relativistic Hartree–Bogoliubov theory: static and dynamic aspects of exotic nuclear structure *Phys. Rep.* **409** 101–259
[78] Fuchs C, Lenske H and Wolter H H 1995 Density dependent hadron field theory *Phys. Rev.* C **52** 3043–60
[79] de Jong F and Lenske H 1998 Relativistic Brueckner–Hartree–Fock calculations with explicit intermediate negative energy states *Phys. Rev.* C **58** 890–9
[80] Typel S and Wolter H H 1999 Relativistic mean field calculations with density-dependent meson–nucleon coupling *Nucl. Phys.* A **656** 331–64
[81] Lalazissis G A *et al* 2005 New relativistic mean-field interaction with density-dependent meson–nucleon couplings *Phys. Rev.* C **71** 024312
[82] Long W-H, Giai N V and Meng J 2006 Density-dependent relativistic Hartree–Fock approach *Phys. Lett.* B **640** 150–4
[83] Rusnak J J and Furnstahl R J 1997 Relativistic point-coupling models as effective theories of nuclei *Nucl. Phys.* A **627** 495–521
[84] Nikšić T, Vretenar D and Ring P 2011 Relativistic nuclear energy density functionals: mean-field and beyond *Prog. Part. Nucl. Phys.* **66** 519
[85] Ebran J P 2010 *Description relativiste de l'état fondamental des noyaux atomiques par l'approche du champ moyen auto-cohérent, incluant la déformation et la superfluidité Thesis Université* Paris Sud-Paris
[86] Kucharek H and Ring P 1991 Relativistic field theory of superfluidity in nuclei *Z. Phys.* A **339** 23
[87] Sulaksono A *et al* 2003 Mapping exchange in relativistic Hartree–Fock *Ann. Phys.* **306** 36
[88] Furnstahl R J 2004 Next generation relativistic models *Extended Density Functionals in Nuclear Structure Physics* ed G A Lalazissis, P Ring and D Vretenar (Berlin: Springer) pp 1–29
[89] von Oertzen W, Freer M and Kanada-Enyo Y 2006 Nuclear clusters and nuclear molecules *Phys. Rep.* **432** 43–113
[90] Kanada-En'yo Y, Kimura M and Horiuchi H 2003 Antisymmetrized molecular dynamics: a new insight into the structure of nuclei *Compt. Rend. Phys.* **4** 497–520
[91] Marević P *et al* 2018 Quadrupole and octupole collectivity and cluster structures in neon isotopes *Phys. Rev.* C **97** 024334

[92] Ebran J-P *et al* 2014 Density functional theory studies of cluster states in nuclei *Phys. Rev.* C **90** 054329

[93] Ebran J-P *et al* 2017 Localization and clustering in atomic nuclei *J. Phys. G: Nucl. Part. Phys.* **44** 103001

[94] Ebran J-P *et al* 2012 How atomic nuclei cluster *Nature* **487** 341

[95] Ebran J-P *et al* 2013 Localization and clustering in the nuclear Fermi liquid *Phys. Rev.* C **87** 044307

[96] Ebran J-P *et al* 2014 Cluster–liquid transition in finite, saturated fermionic systems *Phys. Rev.* C **89** 031303

[97] Bouyssy A *et al* 1987 Relativistic description of nuclear systems in the Hartree–Fock approximation *Phys. Rev.* C **36** 380–401

[98] Foldy L L and Wouthuysen S A 1950 On the Dirac theory of spin 1/2 particles and its non-relativistic limit *Phys. Rev* **78** 29–36

[99] Acharya R and Sudarshan E C G 1960 "Front" description in relativistic quantum mechanics *J. Math. Phys.* **1** 532–6

[100] Jian-You G 2012 Exploration of relativistic symmetry by the similarity renormalization group *Phys. Rev.* C **85** 021302

[101] Coester F *et al* 1970 Variation in nuclear-matter binding energies with phase-shift-equivalent two-body potentials *Phys. Rev.* C **1** 769–76

[102] Serot B D 1992 Quantum hadrodynamics *Rep. Prog. Phys.* **55** 1855

[103] Otto Haxel J, Jensen H D and Suess H E 1949 On the "magic numbers" in nuclear structure *Phys. Rev.* **75** 6–1766

[104] Mayer M G 1949 On closed shells in nuclei. II *Phys. Rev.* **75** 1969–70

[105] Elliott J P 1958 Collective motion in the nuclear shell model. I. Classification schemes for states of mixed configurations *Proc. R. Soc. Lond.* A **245** 128–45

[106] Elliott J P 1958 Collective motion in the nuclear shell model II. The introduction of intrinsic wave-functions *Proc. R. Soc. Lond.* A **245** 562–81

[107] Elliott J P 1963 Collective motion in the nuclear shell model III. The calculation of spectra *Proc. R. Soc. Lond.* A **272** 557–77

[108] Ebran J-P *et al* 2016 Spin–orbit coupling rule in bound fermion systems *J. Phys. G: Nucl. Part. Phys.* **43** 085101

[109] Hecht K T and Adler A 1969 Generalized seniority for favored $I + 0$ pairs in mixed configurations *Nucl. Phys.* A **137** 129–43

[110] Arima A, Harvey M and Shimizu K 1969 Pseudo LS coupling and pseudo SU_3 coupling schemes *Phys. Lett.* B **30**(8) 517–22

[111] Meng J *et al* 1999 Pseudospin symmetry in Zr and Sn isotopes from the proton drip line to the neutron drip line *Phys. Rev.* C **59** 154–63

[112] Liang H, Meng J and Zhou S-G 2015 Hidden pseudospin and spin symmetries and their origins in atomic nuclei *Phys. Rep.* **570** 1–84

[113] Alberto P *et al* 2001 Isospin asymmetry in the pseudospin dynamical symmetry *Phys. Rev. Lett.* **86** 5015–18

[114] Otsuka T, Matsuo T and Abe D 2006 Mean field with tensor force and shell structure of exotic nuclei *Phys. Rev. Lett.* **97** 162501

[115] Olivier L *et al* 2017 Persistence of the $Z = 28$ shell gap around ^{78}Ni: first spectroscopy of ^{79}Cu *Phys. Rev. Lett.* **119** 192501

Chapter 3

Single-reference and multi-reference formulations

Michael Bender, Nicolas Schunck, Jean-Paul Ebran and Thomas Duguet

The single-reference energy density functional (SR-EDF) and multi-reference energy density functional (MR-EDF) are the modern names for what used to be called the self-consistent mean-field theory and beyond-mean-field theory, respectively [1]. These names have their origin in the realization that the techniques used in nuclear physics since at least the late 1950s share many similarities with the concepts of density functional theory (DFT) for electronic many-body systems. In particular, the energy functional itself does not have to be derived from an actual interaction (or pseudo-potential), since in practice it encapsulates many-body correlations that go well beyond what a realistic nuclear potential would provide at the Hartree–Fock (HF) level, in analogy with the exchange-correlation energy of atomic physicists. In addition, it was realized in the 2000s that the generalization of projection techniques or the generator coordinate method (GCM) to energy functionals not strictly derived from a density-independent Hamiltonian was ambiguous at best, and impossible at worst. These are some of the reasons that led to the change of names.

The SR-EDF approach is based on the diagonal densities (3.1), that are defined as expectation values of combinations of creation and annihilation operators on a product state. It relies on the standard (static) Wick theorem to facilitate calculations of expectation values of operators. We will only present the time-independent, zero-temperature version of the Wick theorem in section 3.1.2. In chapter 7, we will also make use of its generalization to statistical ensembles. The SR-EDF framework thus encompasses the HF, Bardeen–Cooper–Schrieffer (BCS) and Hartree–Fock–Bogoliubov (HFB) theories, which should be regarded now as merely a set of mathematical techniques to handle various types of product states. For example, the HF theory is, mathematically, nothing other than a change of single-particle (s.p.) basis that provides the 'best' s.p. basis for a given functional, that is, the basis where the energy of the ground state is the lowest.

doi:10.1088/2053-2563/aae0edch3

The MR-EDF approach is based on the off-diagonal, or transition densities (1.17), which are defined from two different product states $|\Phi^{(g)}\rangle$ and $|\Phi^{(g')}\rangle$. It relies on the generalized Wick theorem, which gives expectation values of operators in terms of transition densities. The MR-EDF encompasses standard projection techniques and the GCM.

In this chapter, we give a short overview of both the SR-EDF and MR-EDF approaches. We have tried to focus our presentation on those concepts that have not been discussed in great detail in existing textbooks, or that have only recently been discovered. For example, we give a formal presentation of the concept of symmetry-breaking, discuss the problem of computing norm kernels in MR-EDF, etc. Note that several other chapters contain material that is also relevant to the following two sections. In particular, section 6.4 of chapter 6 is entirely devoted to the particular case of the Gaussian overlap approximation (GOA) of the GCM, while chapter 8 discusses many practical aspects of implementing the SR-EDF theory.

3.1 Single-reference implementation of nuclear energy density functionals

The first level of description associated with the horizontal expansion discussed in the introduction involves a unique symmetry-unrestricted vacuum: the one associated with an auxiliary independent-particle system with the lowest energy. Such a step corresponds to the SR-EDF implementation of the energy density functional (EDF) approach, sometimes referred to as the self-consistent mean-field level. We will focus our presentation on the HFB theory, which is the *de facto* best compromise between simplicity (of implementation, calculation) and predictive power. For a macroscopic system, that is, a system for which one can consistently take the thermodynamical limit, a symmetry-unrestricted state such as the kind described within the HFB theory is admissible as an effective ground state since the relaxation of the system from the wave packet state to the exact symmetrical ground state becomes exceedingly long as $N \to \infty$ (N being the number of particles): quantum fluctuations associated with the relevant order parameters are strongly suppressed by the 'macroscopic heaviness' of the system; see [2; p 44]. In a finite system, however, these fluctuations are not negligible and tend to restore the broken symmetries. This is why the SR-EDF is necessarily an intermediate step only. The proper treatment of many-body observables requires *in fine* an explicit restoration of broken symmetries; see section 3.2.

The single-reference implementation of the nuclear EDF exclusively retains the information contained in the diagonal part $E[\boldsymbol{g}, \boldsymbol{g}]$ of the kernel (1.16). The latter is a functional of the one-body density matrix and the anomalous density,

$$\rho_{ij}^{g,g} \equiv \frac{\langle \Phi(\boldsymbol{g})| c_j^\dagger c_i \, |\Phi(\boldsymbol{g})\rangle}{\langle \Phi(\boldsymbol{g})|\Phi(\boldsymbol{g})\rangle} \tag{3.1a}$$

$$\kappa_{ij}^{g,g} \equiv \frac{\langle \Phi(g)|c_j c_i\,|\Phi(g)\rangle}{\langle \Phi(g)|\Phi(g)\rangle} \tag{3.1b}$$

$$\kappa_{ij}^{g,g*} \equiv \frac{\langle \Phi(g)|c_i^\dagger c_j^\dagger|\Phi(g)\rangle}{\langle \Phi(g)|\Phi(g)\rangle}, \tag{3.1c}$$

with $\{c^\dagger, c\}$ being creation and annihilation operators associated with an arbitrary s.p. basis of $\mathcal{H}_1$, and $|\Phi(g)\rangle$ a symmetry-unrestricted vacuum. This vacuum can, in principle, break as many symmetries of the nuclear Hamiltonian as needed to lower the energy of the system.

3.1.1 Spontaneous symmetry-breaking

Before we begin, let us recall that G is said to be a symmetry group of the Hamiltonian $\hat{H}$ if the latter commutes with all the unitary operators $\hat{U}(g)$ associated with the elements g of G, that is, $[\hat{H}, \hat{U}(g)] = 0$, $\forall g \in G$. Section 2.1 of chapter 2 gives an introduction to Lie groups, Lie algebras and their representations in the context of relativistic quantum field theory, which is based on Minkowski space–time. The reader interested in this branch of mathematics may consult the textbooks [3–6] for an introduction.

The symmetry group G of the nuclear Hamiltonian reads

$$G \equiv T(3) \rtimes SO(3) \times I \times U(1)_N \times U(1)_Z \times T, \tag{3.2}$$

where:

- $T(3) \simeq (\mathbb{R}^3, +)$ is the Abelian, non-compact (however, locally compact) Lie group of space translations. Its Casimir operator is $\hat{\boldsymbol{P}}^2$, where $\hat{\boldsymbol{P}}$ is the linear momentum of the center-of-mass (c.o.m.). Its infinitesimal generators are the components of $\hat{\boldsymbol{P}}$. They define the exponential map $R_{\boldsymbol{P}}(\boldsymbol{a}) = e^{-\frac{i}{\hbar}\boldsymbol{a}\cdot\hat{\boldsymbol{P}}}$ (with $\boldsymbol{a} \in \mathbb{R}^3$) and the one-dimensional irreps $S^{\boldsymbol{Q}}(\boldsymbol{a}) = e^{-\frac{i}{\hbar}\boldsymbol{a}\cdot\boldsymbol{Q}}$ (plane waves) with $\boldsymbol{Q} \in \mathbb{R}^3$ the eigenvalues of $\boldsymbol{P}$. In nuclear SR-EDF methods, a translationally invariant reference state can only be constructed for the description of infinite homogeneous nuclear matter. For finite systems such as atomic nuclei, translational invariance is always broken by the reference state being localized in space, which mixes states with finite c.o.m. momentum into the quasiparticle (q.p.) vacua.
- $SO(3)$ is the non-Abelian, compact Lie group of space rotations. Its infinitesimal generators $\hat{\boldsymbol{J}}$ (total angular momentum) define the exponential map $R_J(\boldsymbol{\Omega}) = e^{-i\alpha\hat{J}_z}e^{-i\beta\hat{J}_y}e^{-i\gamma\hat{J}_z}$ where $\boldsymbol{\Omega} = (\alpha, \beta, \gamma)$ is the set of Euler angles, $(\alpha, \beta, \gamma) \in ([0, 4\pi] \times [0, \pi] \times [0, 2\pi])$. The eigenvalues $J(J+1)$ (with $2J \in \mathbb{N}$) of the Casimir operator $\hat{\boldsymbol{J}}^2$ label the $(2J+1)$-dimensional irreducible representations (irreps) $S^J(\boldsymbol{\Omega}) = D^J_{MK}(\boldsymbol{\Omega})$ of the group, where $(2M, 2K) \in \mathbb{Z}^2$ and $-2J \leqslant 2M, 2K \leqslant 2J$. These functions are called Wigner matrices [7]. It is possible to construct q.p. vacua describing finite nuclei that are eigenstates of

angular momentum, most importantly spherically symmetric states with $J = 0$. For a few magic nuclei, such states even give the lowest binding energy when calculated at the SR-EDF level. For the majority of nuclei and nuclear phenomena, however, q.p. vacua that break rotational symmetry have to be considered. Its breaking mixes states with different total angular momentum and/or third component of angular moment into the q.p. vacua, which is then signaled by non-zero expectation values of spherical tensor operators of finite rank, most importantly multipole moments $\langle r^{\lambda} Y_{\lambda}^{\mu} \rangle$. Rotational invariance, however, is often not completely broken, for example when the q.p. vacua retain axial symmetry, i.e. invariance under an arbitrary rotation about a single specific axis, or signature symmetry, i.e. invariance under a rotation by π about a specific axis.

We will discuss symmetries such as the signature and simplex in more detail in section 8.1.1 of chapter 8, which is devoted to the numerical implementation of the SR-EDF method. Indeed, symmetries are very useful tools to accelerate calculations by reducing the computational complexity of the problem.

- I is the Abelian, finite and discrete group of reflection symmetry $\hat{P}$. Its generators read $\{1, \Pi\}$ and the eigenvalues $p = \pm 1$ of the Casimir operator Π label the one-dimensional irreps $S^p = D^p(I)$. Its breaking is unambiguously signaled by parity-odd operators taking non-zero expectation values, for example multipole moments $\langle r^{\lambda} Y_{\lambda}^{\mu} \rangle$ with odd λ. In the majority of cases of interest, breaking of parity is also accompanied by breaking of rotational invariance. It is worth noting that in parity-conserving SR-EDF calculations it is possible to construct q.p. vacua for either positive or negative parity eigenvalues, although states with negative parity are then necessarily non-collective q.p. excitations.
- $U(1)_X$, $\hat{X} = \{\hat{N}, \hat{Z}\}$ is the Abelian, compact, global gauge group (a Lie group). The infinitesimal operator $\hat{X}$ defines the exponential mapping $R_X(\varphi) = e^{i\varphi\hat{X}}$ with $\varphi \in [0, 2\pi]$ and the eigenvalues $x \in \mathbb{Z}$ of the Casimir operator $\hat{X}^2$ label the one-dimensional irreps $S^x(\varphi) = e^{ix\varphi}$. In nuclear EDF methods, the breaking of $U(1)_X$ mixes states with different total particle number of the species X into the reference states, which is signaled by the anomalous density of that nucleon species taking non-zero values. A subgroup of $U(1)_X$ worth mentioning is the number parity group I_X, an Abelian, finite and discrete group with the same properties as the parity group I. The latter remains conserved by Bogoliubov q.p. vacua (at zero temperature).

Additional discussion of the symmetry-breaking at the SR-EDF level can be found in section 3.1.5, where we briefly recall how the various ingredients of the HFB theory transform under symmetry operations, and in chapter 8, where we discuss the consequences on the implementation of HFB solvers of breaking or conserving specific symmetries.

As discussed in the introduction, opening the variational space to q.p. vacua with symmetry lower than the symmetry of the underlying Hamiltonian, thence describing a state with higher order, allows us to account for non-dynamical nucleon correlations in an independent-particle picture. The appearance of order parameters is governed by bifurcations that result from the non-linearity of the mean-field equations, leading to the concept of emergent phenomena. The original problem of particle–hole (p.h.) quasi-degeneracies occurring in the symmetry-conserving picture is removed when symmetries are spontaneously broken. The spontaneous breaking of continuous symmetries also gives rise to another kind of degeneracies, i.e. a manifold of degenerate Goldstone modes that are involved at the MR-EDF level when restoring the broken symmetries. Note that if spontaneous symmetry-breaking (SSB) is a very inexpensive way to explicitly treat static collective correlations, it also implies that the connection between computed quantities and experimental observables is not always direct—at least as long as the broken symmetries are not restored; see section 3.2.3.

3.1.2 Densities for a quasiparticle vacuum

Perhaps the most common realization of the SR-EDF approximation is the one described by the HFB theory. This theory has been described in great detail, in particular in various review articles [8, 9] and in textbooks [1, 10]. In this section, we only wish to recall some of the most important features of this theory.

The Bogoliubov transformation

The nuclear many-body state is represented by an HFB coherent state that is a product state of quasiparticle annihilation operators

$$|\Phi(\boldsymbol{g})\rangle \equiv |\Phi_{\mathrm{HFB}}\rangle \equiv \prod_{\lambda} \beta_{\lambda}\, |-\rangle. \tag{3.3}$$

In this expression $|-\rangle$ refers to the particle vacuum $c_j|-\rangle = 0$. For the discussion of the properties of a given q.p. vacuum, the index $\boldsymbol{g}$ labeling different q.p. vacua will be dropped whenever possible. This state is a vacuum

$$\beta_{\mu}|\Phi\rangle = 0 \tag{3.4}$$

for the entire basis of q.p. operators $\{\beta, \beta^{\dagger}\}$ obtained from the s.p. operators $\{c, c^{\dagger}\}$ through a Bogoliubov transformation

$$\beta_{\mu} = \sum_{n}\left(U^{*}_{n\mu}c_n + V^{*}_{n\mu}c^{\dagger}_n\right) \tag{3.5a}$$

$$\beta^{\dagger}_{\mu} = \sum_{n}\left(V_{n\mu}c_n + U_{n\mu}c^{\dagger}_n\right) \tag{3.5b}$$

between the s.p. and q.p. basis. As we will see below when discussing blocked states, it is sometimes necessary to redefine the q.p. vacuum by omitting specific q.p.

operators from its definition (3.3). In matrix form, the Bogoliubov transformation can be written as

$$\begin{pmatrix} \beta \\ \beta^\dagger \end{pmatrix} = \begin{pmatrix} U^\dagger & V^\dagger \\ V^T & U^T \end{pmatrix} \begin{pmatrix} c \\ c^\dagger \end{pmatrix} = \mathcal{W}^\dagger \begin{pmatrix} c \\ c^\dagger \end{pmatrix}, \tag{3.6}$$

where $\mathcal{W}$ is called the Bogoliubov matrix. Ensuring that Fermionic anti-commutation relations are fulfilled by the two sets of operators $\{c, c^\dagger\}$ and $\{\beta, \beta^\dagger\}$ makes $\mathcal{W}$ unitary,

$$\mathcal{W}^\dagger \mathcal{W} = \mathcal{W}\mathcal{W}^\dagger = 1. \tag{3.7}$$

This leads to the set of relations

$$U^\dagger U + V^\dagger V = (U^\dagger U + V^\dagger V)^* = 1 \tag{3.8a}$$

$$V^T U + U^T V = (V^T U + U^T V)^* = 0 \tag{3.8b}$$

$$VV^\dagger + U^* U^T = (VV^\dagger + U^* U^T)^* = 1 \tag{3.8c}$$

$$UV^\dagger + V^* U^T = (UV^\dagger + V^* U^T)^* = 0 \tag{3.8d}$$

between the U and V matrices.

HFB spinors in coordinate space

The Bogoliubov transformation of equation (1.13) is written in configuration space: the matrices U and V give the coefficient of the decomposition of the q. p. μ in the s.p. basis. We can use the transformation rules (1.2) to write the Bogoliubov transformation in terms of field operators,

$$\beta_\mu^{(g)} = \sum_i U_{i\mu}^{(g)*} \int dx\, \psi_i^*(x) c_x + \sum_i V_{i\mu}^{(g)*} \int dx\, \psi_i(x) c_x^\dagger$$

$$\beta_\mu^{(g)\dagger} = \sum_i V_{i\mu}^{(g)} \int dx\, \psi_i^*(x) c_x + \sum_i U_{i\mu}^{(g)} \int dx\, \psi_i(x) c_x^\dagger.$$

Let us introduce the q.p. spinor fields

$$U_\mu^{(g)}(x) = \sum_i U_{i\mu}^{(g)} \psi_i(x) \tag{3.9a}$$

$$V_\mu^{(g)}(x) = \sum_i V_{i\mu}^{(g)*} \psi_i(x). \tag{3.9b}$$

The Bogoliubov transformation thus becomes

$$\beta_\mu^{(g)} = \int dx\, U_\mu^{(g)*}(x) c_x + \int dx\, V_\mu^{(g)*}(x) c_x^\dagger$$
$$\beta_\mu^{(g)\dagger} = \int dx\, V_\mu^{(g)}(x) c_x + \int dx\, U_\mu^{(g)}(x) c_x^\dagger.$$

This set of equations gives the q.p. ladder operators to create and annihilate a quasiparticle in state μ.

While both are product states, there are several important differences between the definition of a Bogoliubov q.p. vacuum (3.3) and the definition of a Slater determinant:

1. The unitarity of $\mathcal{W}$ implies that U and V in equation (3.6) are square matrices, meaning that the Bogoliubov transformation provides as many q.p. annihilation operators β_μ as there are creation *and* annihilation operators in the underlying s.p. basis.
2. As the Bogoliubov q.p. state (3.3) is the vacuum of *all* q.p. annihilation operators (3.4), the product over λ in equation (3.3) necessarily runs over as many terms as there are single-particle states in the s.p. basis $(c, c^\dagger)$. With this, the number of factors in the product (3.3) depends on the size of the basis in which the q.p. operators are represented, and is not a physical parameter characteristic of the system to be described.

   In contrast, an A-body Slater determinant is always the product of exactly A s.p. creation operators acting on the particle vacuum, be it written in the natural basis $(a, a^\dagger)$ of the Slater determinant or in an arbitrary basis $(c, c^\dagger)$:

   $$\prod_{\mu=1}^{A} \hat{a}_\mu^\dagger |-\rangle = \prod_{\mu=1}^{A} \left[\sum_k D_{\mu k} \hat{c}_k^\dagger \right] |-\rangle,$$

   where $D_{\mu k}$ is a transformation that develops the A natural s.p. states of the Slater determinant into the full s.p. basis that has $|-\rangle$ as a vacuum.

3. The quasiparticle vacuum as defined through equation (3.3) is not normalized. As we will see below, this does not cause any practical problem, since observables calculated for a state $|\Phi\rangle$ do not depend on its normalization.

Wick theorem and density matrices

Whenever the vacuum state entering equation (3.1) is a product state such as a Slater determinant or a Bogoliubov vacuum of the kind (3.3), calculations of expectation values of operators on the vacuum are greatly facilitated by the use of the Wick theorem. We refer the reader to [10] for an in-depth presentation of the general Wick theorem and its demonstration. Here, we only wish to recall (one particular) formulation of the theorem by J Dobaczewski [11].

Contractions and self-contractions

Consider some arbitrary many-body reference state $|\Phi\rangle$ (not necessarily normalized) and an operator $\hat{A}$ that acts in Fock space. We can always expand $\hat{A}$ as follows

$$\hat{A} = \hat{A}_0 + \hat{A}_- + \hat{A}_+, \tag{3.10}$$

with $\hat{A}_0$ a constant, and $\hat{A}_-|\Phi\rangle = 0$ and $\langle\Phi|\hat{A}_+ = 0$. In fact, if we introduce the projector $\hat{P}_\Phi$ on the state $|\Phi\rangle$, it is easy to show that

$$\begin{aligned}
\hat{A}_0 &= \frac{\langle\Phi|\hat{A}|\Phi\rangle}{\langle\Phi|\Phi\rangle} \\
\hat{A}_- &= \left(\hat{A} - \frac{\langle\Phi|\hat{A}|\Phi\rangle}{\langle\Phi|\Phi\rangle}\right)(1-\hat{P}_\Phi)^{\cdot} \\
\hat{A}_+ &= (1-\hat{P}_\Phi)\hat{A}\hat{P}_\Phi
\end{aligned}$$

It follows from these definitions that the expectation value of a product of operators $\hat{A}$ and $\hat{B}$ on $|\Phi\rangle$ is given by

$$\frac{\langle\Phi|\hat{A}\hat{B}|\Phi\rangle}{\langle\Phi|\Phi\rangle} = \frac{\langle\Phi|\hat{A}|\Phi\rangle\langle\Phi|\hat{B}|\Phi\rangle}{\langle\Phi|\Phi\rangle^2} + \frac{\langle\Phi|\hat{A}_-\hat{B}_+|\Phi\rangle}{\langle\Phi|\Phi\rangle}.$$

We define the contraction of the product of operators $\hat{A}\hat{B}$ and the self-contraction of the operator $\hat{A}$ as

$$\overline{AB} = \frac{\langle\Phi|\hat{A}_-\hat{B} + \hat{B}\hat{A}_-|\Phi\rangle}{\langle\Phi|\Phi\rangle} \qquad \overline{A} = \frac{\langle\Phi|\hat{A}|\Phi\rangle}{\langle\Phi|\Phi\rangle} = \hat{A}_0$$

Note that when $|\Phi\rangle$ is a product state, we do not really need the second term proportional to $\hat{B}\hat{A}_-$ in the definition of the contraction. Another (perhaps more familiar) definition of the contraction follows simply from the expansion (3.10). By definition, we have

$$AB = \frac{\langle\Phi|\hat{A}\hat{B}|\Phi\rangle}{\langle\Phi|\Phi\rangle} + \frac{\langle\Phi|\hat{A}|\Phi\rangle}{\langle\Phi|\Phi\rangle}\frac{\langle\Phi|\hat{B}|\Phi\rangle}{\langle\Phi|\Phi\rangle}.$$

It is essential to note that, by virtue of their definition, contractions are intrinsically dependent on a given reference state $|\Phi\rangle$.

As an example, let us calculate the contraction $\overline{c_j^\dagger c_i}$ associated with a Slater determinant defined with the same s.p. creation operators,

$$|\Phi\rangle = \prod_{k=1}^{A} c^{\dagger}_{\mu_k}|0\rangle.$$

In this particular example, the self-contractions $c_{j,0} = c^{\dagger}_{j,0} = 0$, since this would corresponds to overlaps between two Slater determinants with different particle numbers (see the definition of self-contraction). In addition, we suppose the Slater determinant properly normalized. Therefore,

$$\overline{c^{\dagger}_j c_i} = \langle\Phi|c^{\dagger}_{j,-}c_{i,+}\,|\Phi\rangle$$

Let us evaluate the term $c^{\dagger}_{j,-}$ for a given index j and a given reference state $|\Phi\rangle$. We note that if $j \in \{\mu_k\}_{k=1,\ldots,A}$, then $c^{\dagger}_j|\Phi\rangle = 0$. If $j \notin \{\mu_k\}_{k=1,\ldots,A}$, we can only have $c^{\dagger}_{j,-} = 0$. Proceeding similarly for $c_{i,+}$, we find the two sets of definitions

$$c^{\dagger}_{j,-} = \begin{cases} c^{\dagger}_j & j \in \{\mu_k\}_{k=1,\ldots,A} \\ 0 & \text{otherwise} \end{cases} \qquad c_{i,+} = \begin{cases} c_i & i \in \{\mu_k\}_{k=1,\ldots,A} \\ 0 & \text{otherwise} \end{cases}.$$

Given these results is then straightforward to compute the contraction and find

$$\overline{c^{\dagger}_j c_i} = \sum_{k=1}^{A} \delta_{j\mu_k}\delta_{i\mu_k}.$$

Theorem 1 (Wick theorem). *Given the expectation value $\langle\Phi|\hat{A}_1\ldots\hat{A}_n|\Phi\rangle$, if all mutual contractions of pairs of operators in the product are numbers, then the average value of the product of these operators equals the linear combination of products of all possible contractions and auto-contractions,*

$$\begin{aligned}\langle\Phi|\hat{A}_1\ldots\hat{A}_n\,|\Phi\rangle &= \overline{A_1}\langle\Phi|\hat{A}_2\ldots\hat{A}_n\,|\Phi\rangle + \overline{A_1A_2}\langle\Phi|\hat{A}_3\ldots\hat{A}_n\,|\Phi\rangle \\ &\quad - \overline{A_1A_3}\langle\Phi|\hat{A}_2\hat{A}_4\ldots\hat{A}_n\,|\Phi\rangle + \ldots \\ &\quad + (-1)^{n-2}\overline{A_1A_n}\langle\Phi|\hat{A}_2\ldots\hat{A}_{n-1}\,|\Phi\rangle \\ &\quad + \ldots.\end{aligned} \tag{3.11}$$

The sign of each time equals $(-1)^m$, where m is the number of permutations of operators needed to obtain the 'right' operator in the contraction outside the remaining matrix element.

The standard Wick theorem as presented here can only be used when bra and ket correspond to the same reference state. However, this formulation can easily be generalized to transition matrix elements between two non-orthogonal state $|\Phi_1\rangle$ and $|\Phi_2\rangle$. The decomposition of an arbitrary operator $\hat{A}$ still reads $\hat{A} = \hat{A}_0 + \hat{A}_+ + \hat{A}_-$, only the various operators are now

$$\hat{A}_0 = \frac{\langle\Phi_2|\hat{A}|\Phi_1\rangle}{\langle\Phi_2|\Phi_1\rangle}$$

$$\hat{A}_- = \left(\hat{A} - \frac{\langle\Phi_2|\hat{A}|\Phi_1\rangle}{\langle\Phi_2|\Phi_1\rangle}\right)(1-\hat{P})$$

$$\hat{A}_+ = (1-\hat{P}_\Phi)\hat{A}\hat{P},$$

with the projector now given by $\hat{P} = |\Phi_1\rangle\langle\Phi_2|/\langle\Phi_2|\Phi_1\rangle$. The contraction and self-contraction in this case become

$$\overbracket{AB} = \frac{\langle\Phi_2|\hat{A}\hat{B}|\Phi_1\rangle}{\langle\Phi_2|\Phi_1\rangle} - \frac{\langle\Phi_2|\hat{A}|\Phi_1\rangle\langle\Phi_2|\hat{B}|\Phi_1\rangle}{\langle\Phi_2|\Phi_1\rangle^2}$$

$$\overbracket{A} = \frac{\langle\Phi_2|\hat{A}|\Phi_1\rangle}{\langle\Phi_2|\Phi_1\rangle}.$$

Product states such as Slater determinants or Bogoliubov q.p. vacua belong, specifically, to the class of many-body states for which all mutual contractions are numbers. The interest of this theorem is thus clear: expectation values of even the most complicated n-body operators on such product states will come down to enumerating all possible combinations of *pairs* of creation and/or annihilation operators, and use the set of relations (3.12). Once these elementary contractions have been computed (which depends on the reference state), everything becomes quite simple.

Because of their definition (3.1), we see that the one-body density matrix and the anomalous density are nothing other than contractions

$$\rho_{ij} = \overbracket{c_j^\dagger c_i} \tag{3.12a}$$

$$\kappa_{ij} = \overbracket{c_j c_i} \tag{3.12b}$$

$$(1-\rho)^*_{ij} = \overbracket{c_j c_i^\dagger} \tag{3.12c}$$

$$\kappa^*_{ij} = \overbracket{c_i^\dagger c_j^\dagger}. \tag{3.12d}$$

Note that the order of the indices is different in κ^*. Let us go back to the definition of the one-body density matrix in equation (3.1)

$$\rho_{kl} = \frac{\langle\Phi|c_l^\dagger c_k|\Phi\rangle}{\langle\Phi|\Phi\rangle} \tag{3.13}$$

and compute it when $|\Phi\rangle$ is the (non-normalized) q.p. vacuum (3.3). We can express this matrix in the quasiparticle representation by transforming the c operators into β operators thanks to the transformation (3.6). We obtain

$$\frac{\langle\Phi|c_l^\dagger c_k\,|\Phi\rangle}{\langle\Phi|\Phi\rangle} = \frac{1}{\langle\Phi|\Phi\rangle}\langle\Phi|\sum_{\mu\nu}\Big(V_{l\mu}U_{k\nu}\,\beta_\mu\beta_\nu + V_{l\mu}V_{k\nu}^*\,\beta_\mu\beta_\nu^\dagger$$
$$+\,U_{l\mu}^*U_{k\nu}\,\beta_\mu^\dagger\beta_\nu + U_{l\mu}^*V_{k\nu}^*\,\beta_\mu^\dagger\beta_\nu^\dagger\Big)|\Phi\rangle.$$

The denominator $\langle\Phi|\Phi\rangle$ is the product of the two normalization factors necessary to normalize the q.p. vacuum (3.3) in the bra and in the ket. The two terms that have a β_ν to the right will give 0, since $|\Phi\rangle$ is by definition a vacuum for q.p.s. The term with two q.p. creation operators will also vanish, since it is nothing other than the 'bra' associated with $\beta_\mu\beta_\nu|\Phi\rangle$, which is zero. There remains

$$\sum_{\mu\nu}V_{l\mu}V_{k\nu}^*\,\frac{\langle\Phi|\beta_\mu\beta_\nu^\dagger|\Phi\rangle}{\langle\Phi|\Phi\rangle} = \sum_{\mu\nu}V_{l\mu}V_{k\nu}^*\frac{\langle\Phi|(\delta_{\mu\nu}-\beta_\nu^\dagger\beta_\mu)|\Phi\rangle}{\langle\Phi|\Phi\rangle}$$
$$= \sum_{\mu}V_{l\mu}V_{k\mu}^*\,\frac{\langle\Phi|\Phi\rangle}{\langle\Phi|\Phi\rangle}.$$

We thus arrive at the result

$$\rho = V^*V^T. \tag{3.14}$$

The calculation of the anomalous density as defined in equation (3.1)

$$\kappa_{kl} = \frac{\langle\Phi|c_l c_k\,|\Phi\rangle}{\langle\Phi|\Phi\rangle} \tag{3.15}$$

can be performed in exactly the same way, yielding

$$\kappa = V^*U^T. \tag{3.16}$$

In fact, all the information about a system of independent q.p.s is contained in the one-body normal and anomalous densities ρ and κ defined by equation (3.1). In the HFB theory, these densities can be put into a compact form called the Valatin generalized density matrix $\mathcal{R}$

$$\mathcal{R} \equiv \begin{pmatrix}\rho & \kappa\\ -\kappa^* & 1-\rho^*\end{pmatrix} = \begin{pmatrix}V^*V^T & V^*U^T\\ U^*V^T & U^*U^T\end{pmatrix}. \tag{3.17}$$

Using equations (3.14), (3.16) and the unitarity of the Bogoliubov transformation (3.8) it can be easily shown that the generalized density is idempotent, that is,

$$\mathcal{R}^2 = \mathcal{R}. \tag{3.18}$$

This leads to the following relations between ρ and κ

$$\rho\rho - \kappa\kappa^* = \rho$$
$$\rho\kappa - \kappa\rho^* = 0$$

and the complex conjugate relations. From relations (3.12), it can also be easily seen that ρ is Hermitian, while κ and κ^* are skew-symmetric

$$\rho^*_{\mu\nu} = \rho_{\nu\mu} \qquad \kappa_{\mu\nu} = -\kappa_{\nu\mu} \qquad \kappa^*_{\mu\nu} = -\kappa^*_{\nu\mu} \tag{3.19}$$

from which it directly follows that the generalized density matrix is Hermitian

$$\mathcal{R}^\dagger = \mathcal{R}. \tag{3.20}$$

For the formal analysis of the Bogoliubov transformation, it is useful to introduce a matrix γ [10],

$$\gamma \equiv \begin{pmatrix} 0 & 1 \\ 1 & 0 \end{pmatrix}, \tag{3.21}$$

with $\gamma\gamma = \gamma\gamma^T = \gamma^T\gamma = \gamma^T\gamma^T = 1$ that transforms 2×2 block matrices and the corresponding column and row vectors of Bogoliubov theory as

$$\begin{pmatrix} 0 & 1 \\ 1 & 0 \end{pmatrix}\begin{pmatrix} A & B \\ C & D \end{pmatrix}\begin{pmatrix} 0 & 1 \\ 1 & 0 \end{pmatrix} = \begin{pmatrix} D & C \\ B & A \end{pmatrix} \tag{3.22a}$$

$$\begin{pmatrix} 0 & 1 \\ 1 & 0 \end{pmatrix}\begin{pmatrix} a \\ b \end{pmatrix} = \begin{pmatrix} b \\ a \end{pmatrix} \tag{3.22b}$$

$$(a, b)\begin{pmatrix} 0 & 1 \\ 1 & 0 \end{pmatrix} = (b, a). \tag{3.22c}$$

From this follows that the Bogoliubov transformation (3.6) has the symmetry

$$\gamma\mathcal{W}\gamma = \mathcal{W}^* \qquad \gamma\mathcal{W}^\dagger\gamma = \mathcal{W}^T. \tag{3.23}$$

From the relations (3.12) and the skew-symmetry of κ^* (3.19) it follows that the transposed generalized density matrix $\mathcal{R}$ (3.17) is the matrix of contractions

$$\mathcal{R}^T = \begin{pmatrix} \overline{c^\dagger c} & \overline{c^\dagger c^\dagger} \\ \overline{cc} & \overline{cc^\dagger} \end{pmatrix} = \frac{1}{\langle\Phi|\Phi\rangle}\langle\Phi|\begin{pmatrix} c^\dagger c & c^\dagger c^\dagger \\ cc & cc^\dagger \end{pmatrix}|\Phi\rangle. \tag{3.24}$$

The matrix in the expectation value on the rhs of this expression can be transformed to the q.p. basis by multiplying from the left and right with $\gamma^2 = 1$, using equation (3.22), factorizing the matrix into the product of a column and a row vector, and using the Bogoliubov transformation (3.6) between the two bases as well as its symmetry (3.23):

$$\begin{pmatrix} c^\dagger c & c^\dagger c^\dagger \\ cc & cc^\dagger \end{pmatrix} = \gamma \begin{pmatrix} cc^\dagger & cc \\ c^\dagger c^\dagger & c^\dagger c \end{pmatrix} \gamma = \gamma \begin{pmatrix} c \\ c^\dagger \end{pmatrix} (c^\dagger, c)\, \gamma$$
$$= \gamma\, \mathcal{W} \begin{pmatrix} \beta \\ \beta^\dagger \end{pmatrix} (\beta^\dagger, \beta)\, \mathcal{W}^\dagger \gamma$$
$$= \gamma\, \mathcal{W} \gamma \begin{pmatrix} \beta^\dagger \beta & \beta^\dagger \beta^\dagger \\ \beta\beta & \beta\beta^\dagger \end{pmatrix} \gamma\, \mathcal{W}^\dagger \gamma$$
$$= \mathcal{W}^* \begin{pmatrix} \beta^\dagger \beta & \beta^\dagger \beta^\dagger \\ \beta\beta & \beta\beta^\dagger \end{pmatrix} \mathcal{W}^T.$$

To evaluate now the rhs of equation (3.24), we take the normalized expectation value of all operators in this matrix, which yields the matrix of contractions of the q.p. operators

$$\begin{aligned} \mathcal{R}^T &= \frac{1}{\langle \Phi | \Phi \rangle} \langle \Phi | \mathcal{W}^* \begin{pmatrix} \beta^\dagger \beta & \beta^\dagger \beta^\dagger \\ \beta\beta & \beta\beta^\dagger \end{pmatrix} \mathcal{W}^T | \Phi \rangle \\ &= \mathcal{W}^* \begin{pmatrix} \overline{\beta^\dagger \beta} & \overline{\beta^\dagger \beta^\dagger} \\ \overline{\beta\beta} & \overline{\beta\beta^\dagger} \end{pmatrix} \mathcal{W}^T = \mathcal{W}^* \tilde{\mathcal{R}}^T \mathcal{W}^T, \end{aligned} \tag{3.25}$$

where $\tilde{\mathcal{R}}$ is the generalized density matrix in the quasiparticle basis. This relation has two important consequences. First, the transformation between the generalized density matrix $\mathcal{R}$ in the s.p. and q.p. bases is simply provided by the Bogoliubov transformation

$$\tilde{\mathcal{R}} = \mathcal{W}^\dagger \mathcal{R}\, W. \tag{3.26}$$

Here and throughout this book, we will use a tilde sign to refer to matrices in the q.p. basis. Second, with respect to their vacuum, all contractions of the q.p. operators are zero except for $\overline{\beta_\mu \beta_\nu^\dagger} = \delta_{\mu\nu}$. Therefore, in the quasiparticle basis the generalized density matrix $\tilde{\mathcal{R}}$ is diagonal with eigenvalues 0 and 1. This follows from its idempotence $\mathcal{R}^2 = \mathcal{R}$, which is not affected by unitary transformations. We find

$$\tilde{\mathcal{R}} = \begin{pmatrix} 0 & 0 \\ 0 & 1 \end{pmatrix}. \tag{3.27}$$

Note that any unitary transformation of q.p. operators that only mixes creation operators among themselves and annihilation operators among themselves yields a q.p. basis that still diagonalizes $\tilde{\mathcal{R}}$.

Thouless theorem

Given a Bogoliubov q.p. vacuum $|\Phi_0\rangle$ and corresponding q.p. operators $(\beta, \beta^\dagger)$, any other q.p. vacuum $|\Phi_1\rangle$ that is non-orthogonal, $\langle \Phi_0 | \Phi_1 \rangle \neq 0$, can be expressed in the form

$$|\Phi_1\rangle = \langle\Phi_0\,|\Phi_1\rangle \exp\left[\frac{1}{2}\sum_{\mu\mu'} Z_{\mu\mu'}\,\beta^\dagger_\mu\,\beta^\dagger_{\mu'}\right]|\Phi_0\rangle, \tag{3.28}$$

which is known as the Thouless theorem [1, 10, 12]. Note that neither q.p. vacuum has to be normalized. The possible difference in normalization factors is absorbed by the overlap $\langle\Phi_0|\Phi_1\rangle$.

We give here only a brief introduction to the Thouless theorem. In addition to its interest for formal developments, the Thouless theorem is at the heart of the gradient method to solve the HFB equation. We will thus expand our discussion of the theorem in section 8.3.2 where we present this powerful technique.

The Thouless matrix is a skew-symmetric $Z = -Z^T$ tensor of rank-2 given through

$$Z = V^* U^{*-1}, \tag{3.29}$$

where the matrices U and V are determined by the Bogoliubov transformations W_0 and W_1 that define the two q.p. vacua

$$U = U_0^\dagger U_1 + V_0^\dagger V_1 \tag{3.30a}$$

$$V = V_0^T U_1 + U_0^T V_1. \tag{3.30b}$$

As single-particle states can be interpreted as a trivial case of q.p. states with vacuum $|-\rangle$, a useful corollary of equation (3.28) is that any q.p. vacuum that is non-orthogonal to $|-\rangle$ can be written as

$$|\Phi\rangle = \langle -|\Phi\rangle \exp\left[\frac{1}{2}\sum_{kl} Z_{kl}\; c_k^\dagger\; c_l^\dagger\right]|-\rangle, \tag{3.31}$$

where the U and V matrices entering Z in equation (3.29) are now directly the Bogoliubov matrices defining $|\Phi\rangle$ through equation (3.6).

3.1.3 HFB equation

In an SR-EDF approach, one is interested in determining the q.p. vacuum that gives the lowest energy. For that purpose, we first introduce a convenient one-body basis associated with the ladder operators $\{c^\dagger, c\}$ to express the effective nuclear Hamiltonian in second quantized form

$$\begin{aligned}\hat{H} &= \sum_{ij} t_{ij} c_i^\dagger c_j + \frac{1}{2}\sum_{ijkl} v_{ijkl} c_i^\dagger c_j^\dagger c_l c_k \\ &= \sum_{ij} t_{ij} c_i^\dagger c_j + \frac{1}{4}\sum_{ijkl} \bar{v}_{ijkl} c_i^\dagger c_j^\dagger c_l c_k,\end{aligned} \tag{3.32}$$

where in the second line we introduced the anti-symmetrized matrix elements of the two-body interaction

$$\bar{v}_{ijkl} \equiv v_{ijkl} - v_{ijlk} = v_{ijkl} - v_{jikl}, \tag{3.33}$$

with the property

$$\bar{v}_{ijkl} = -\bar{v}_{ijlk}. \tag{3.34}$$

Their use allows for compact notation in formal manipulations. The generalization to a Hamiltonian also containing genuine three-body and higher interactions is straightforward, but leads to cumbersome expressions. In this section, we will also assume that $\hat{H}$ is a genuine Hamiltonian, that is, the vertex v_{ijkl} does not depend on the density but is instead the matrix element of a true operator. Although this assumption allows for rigorous mathematical expressions, it is not necessarily very convenient since, in practice, nearly all the most popular energy functionals are not derived form a genuine Hamiltonian, but from a pseudo-Hamiltonian with a density-dependent term, $\bar{v}_{ijkl} \equiv \bar{v}_{ijkl}[\rho]$. In fact, if the vertex depends on the density, $v_{ijkl} \equiv v_{ijkl}[\rho]$, one is, strictly speaking, not allowed to write expressions such as equation (3.32). This is a simple consequence of the fact that the density is given by equation (3.12), i.e. it depends on the reference state $|\Phi\rangle$ [13]. Whenever needed, we will thus complement our presentation with vignettes to emphasize how formulas should be modified to account for such density dependencies.

Hamiltonian in the quasiparticle basis

In section 1.1.1 of chapter 1, we recalled the Fock space representation of arbitrary operators. Equation (1.5) was obtained by introducing an s.p. basis of the Hilbert space and using ladder operators $(c, c^\dagger)$ to express the action of an arbitrary operator in Fock space. The form (3.32) for the effective nuclear Hamiltonian is a particular case. When working with Bogoliubov vacua of the type of (3.3), it is in fact advantageous to substitute the s.p. ladder operators by q. p. operators thanks to the Bogoliubov transformation (3.5). For a two-body operator, the calculation leads to the following general expression [1, 8],

$$\hat{H} = \hat{H}^0 + \hat{H}^{11} + \hat{H}^{20} + \hat{H}^{22} + \hat{H}^{31} + \hat{H}^{40}, \tag{3.35}$$

where

$$\begin{aligned}
\hat{H}^0 &= E_{\text{HFB}} \\
\hat{H}^{11} &= \sum_{\mu\nu} H^{11}_{\mu\nu}\beta^\dagger_\mu\beta_\nu \\
\hat{H}^{20} &= \frac{1}{2}\sum_{\mu\nu}\left(H^{20}_{\mu\nu}\beta^\dagger_\mu\beta^\dagger_\nu + H^{20*}_{\mu\nu}\beta_\nu\beta_\mu\right) \\
\hat{H}^{22} &= \frac{1}{4}\sum_{\mu\nu\gamma\delta} H^{22}_{\mu\nu\gamma\delta}\beta^\dagger_\mu\beta^\dagger_\nu\beta_\delta\beta_\gamma + \text{h.c.} \\
\hat{H}^{31} &= \sum_{\mu\nu\gamma\delta} H^{31}_{\mu\nu\gamma\delta}\beta^\dagger_\mu\beta^\dagger_\nu\beta^\dagger_\gamma\beta_\delta + \text{h.c.} \\
\hat{H}^{40} &= \sum_{\mu\nu\gamma\delta} H^{40}_{\mu\nu\gamma\delta}\beta^\dagger_\mu\beta^\dagger_\nu\beta^\dagger_\gamma\beta^\dagger_\delta + \text{h.c..}
\end{aligned}$$

where 'h.c.' refers to the Hermitian conjugate operation

$$H^{31}_{\mu\nu\gamma\delta}\beta^\dagger_\mu\beta^\dagger_\nu\beta^\dagger_\gamma\beta_\delta \rightarrow H^{31*}_{\mu\nu\gamma\delta}\beta^\dagger_\delta\beta_\gamma\beta_\nu\beta_\mu.$$

Note that there can be different conventions for the objects $H^{22}_{\mu\nu\gamma\delta}$, $H^{31}_{\mu\nu\gamma\delta}$ and $H^{40}_{\mu\nu\gamma\delta}$. Here we follow the definitions of [1]. Among the matrix elements entering these expressions, the most commonly used are

$$H^{11}_{\mu\nu} = (U^\dagger h U)_{\mu\nu} - (V^\dagger h^T V)_{\mu\nu} + (U^\dagger \Delta V)_{\mu\nu} - (V^\dagger \Delta^* U)_{\mu\nu}$$
$$H^{20}_{\mu\nu} = (U^\dagger h V^*)_{\mu\nu} - (V^\dagger h^T U^*)_{\mu\nu} + (U^\dagger \Delta U^*)_{\mu\nu} - (V^\dagger \Delta^* V^*)_{\mu\nu}.$$

The advantage of this representation is two-fold. First, one can calculate expectation values of operators thanks to the Wick theorem applied with contractions of q.p. operators. Second, it makes it relatively easy to evaluate expectation values of operators such as $\hat{H}\beta^\dagger_\mu\beta^\dagger_\nu$, $\beta_\mu\beta_\nu\hat{H}$, etc.

Let us calculate the expectation value of $\hat{H}\beta^\dagger_\mu\beta^\dagger_\nu$ on the HFB reference state. Since $|\Phi\rangle$ is a vacuum for all q.p. operators, the only term in equation (3.35) that will give a non-zero contribution is the one that contains exactly two annihilation operators, that is, the term proportional to $H^{20*}_{\mu\nu}$, so that

$$\langle \hat{H}\beta^\dagger_\mu\beta^\dagger_\nu\rangle = \frac{1}{2}\sum_{\gamma\delta} H^{20*}_{\delta\gamma}\left(\overbrace{\beta_\gamma\beta_\delta\beta^\dagger_\mu\beta^\dagger_\nu} + \overbrace{\beta_\gamma\beta_\delta\beta^\dagger_\mu\beta^\dagger_\nu}\right) = \frac{1}{2}\sum_{\gamma\delta} H^{20*}_{\delta\gamma}(-\delta_{\gamma\mu}\delta_{\delta\nu} + \delta_{\gamma\nu}\delta_{\delta\mu}) = \frac{1}{2}(H^{20*}_{\mu\nu} - H^{20*}_{\nu\mu}) = H^{20*}_{\mu\nu},$$

where the last equality simply comes from the fact that H^{20} is skew-symmetric, as can be easily verified from its definition.

Total energy at the HFB approximation

Let us write down the expectation value of $\hat{H}$ on the HFB vacuum (3.3). We find first

$$\mathcal{E} = \sum_{ij} t_{ij}\frac{\langle\Phi|c^\dagger_i c_j|\Phi\rangle}{\langle\Phi|\Phi\rangle} + \frac{1}{4}\sum_{ijkl}\bar{v}_{ijkl}\frac{\langle\Phi|c^\dagger_i c^\dagger_j c_l c_k|\Phi\rangle}{\langle\Phi|\Phi\rangle}.$$

Introducing the contractions and using the Wick theorem, we obtain

$$\mathcal{E} = \sum_{ij} t_{ij}\overbrace{c^\dagger_i c_j} + \frac{1}{4}\sum_{ijkl}\bar{v}_{ijkl}\left(\overbrace{c^\dagger_i c^\dagger_j}\,\overbrace{c_l c_k} - \overbrace{c^\dagger_i c^\dagger_j c_l c_k} + \overbrace{c^\dagger_i c^\dagger_j c_l c_k}\right).$$

Replacing contractions by their expressions in terms of the one-body density matrix and anomalous density we obtain

$$\mathcal{E} = \sum_{ij} t_{ij}\rho_{ji} + \frac{1}{4}\sum_{ijkl}\bar{v}_{ijkl}\left(\kappa^*_{ij}\kappa_{kl} - \rho_{li}\rho_{kj} + \rho_{ki}\rho_{lj}\right).$$

We now use the properties (3.34) of the anti-symmetrized matrix elements (3.34) to obtain

$$\mathcal{E} = \sum_{ij} t_{ij}\rho_{ji} + \frac{1}{2}\sum_{ijkl} \bar{v}_{ijkl}\rho_{lj}\rho_{ki} + \frac{1}{4}\sum_{ijkl} \bar{v}_{ijkl}\kappa_{ij}^{*}\kappa_{kl}. \tag{3.36}$$

Effective pseudo-Hamiltonian

The expression (3.36) for the energy was derived from the expectation value of the pseudo-Hamiltonian (3.32). In the spirit of EDF approaches, one often uses different vertices in the p.h. and particle–particle (p.p.) channels and/or modifies, approximates, or even drops some exchange terms as discussed, e.g. in chapter 1. In this case (3.36) should be modified to read

$$\mathcal{E} = \sum_{ij} t_{ij}\rho_{ji} + \frac{1}{2}\sum_{ijkl} \bar{v}_{ijkl}^{\rho\rho}[\rho]\,\rho_{lj}\rho_{ki} + \frac{1}{4}\sum_{ijkl} \bar{v}_{ijkl}^{\kappa\kappa}[\rho]\,\kappa_{ij}^{*}\kappa_{kl}. \tag{3.37}$$

Also, we have assumed implicitly that the vertices might only depend on the density, but one should not exclude a dependence on the anomalous density, even though this has never really been considered in the literature.

Variation of the HFB energy

The one-body generalized density matrix $\mathcal{R}$, i.e. the one-body normal and anomalous density matrices ρ and κ, or equivalently the Bogoliubov amplitudes U and V, are determined self-consistently from the variation of the HFB energy $\mathcal{E}$ under the constraints that (i) the average particle number corresponds to the actual number of particles and (ii) the many-body state remains an independent q.p. state of the form (3.3). In other words, the minimization reads

$$\delta\left(\mathcal{E}[\mathcal{R}] - \frac{1}{2}\lambda_{\mathrm{F}}[\mathrm{tr}(\rho) + \mathrm{tr}(\rho^{*})] - \mathrm{tr}(\Lambda[\mathcal{R}^{2} - \mathcal{R}])\right) = 0, \tag{3.38}$$

with λ_{F} being a Lagrange multiplier that is adjusted such that the trace of the density matrix equals the particle number $\mathrm{tr}(\rho) = \mathrm{tr}(\rho^{*}) = N_0$, while Λ is a matrix of Lagrange multipliers enforcing that the generalized density matrix remains idempotent (3.18) during the variation. The Lagrange multiplier λ_{F} is usually called the Fermi energy, or chemical potential, and equals $\partial\mathcal{E}/\partial N$. Because of the properties (3.19) of ρ (Hermitian) κ and κ^{*} (skew-symmetric), the irreducible set of independent variational variables is $\rho_{\mu\nu}$, $\rho_{\mu\nu}^{*}$, $\kappa_{\mu\nu}$ and $\kappa_{\mu\nu}^{*}$ for all $\nu < \mu$, as well as $\rho_{\mu\mu}$, for all μ. The variation of the HFB energy thus takes the form

$$\delta\mathcal{E} = \sum_{\nu\leqslant\mu}\left(\frac{\delta\mathcal{E}}{\delta\rho_{\mu\nu}}\delta\rho_{\mu\nu} + \frac{\delta\mathcal{E}}{\delta\rho^*_{\mu\nu}}\delta\rho^*_{\mu\nu} + \frac{\delta\mathcal{E}}{\delta\kappa_{\mu\nu}}\delta\kappa_{\mu\nu} + \frac{\delta\mathcal{E}}{\delta\kappa^*_{\mu\nu}}\delta\kappa^*_{\mu\nu}\right)$$
$$\equiv\frac{1}{2}\sum_{\mu\nu}\left(h_{\nu\mu}\delta\rho_{\mu\nu} + h^*_{\nu\mu}\delta\rho^*_{\mu\nu} - \Delta^*_{\nu\mu}\delta\kappa_{\mu\nu} - \Delta_{\nu\mu}\delta\kappa^*_{\mu\nu}\right)$$
$$=\frac{1}{2}\text{tr}(h\delta\rho + h^*\delta\rho^* - \Delta^*\delta\kappa - \Delta\,\delta\kappa^*),$$

where we introduced for $\nu \leqslant \mu$,

$$h_{\nu\mu} \equiv \frac{\delta\mathcal{E}}{\delta\rho_{\mu\nu}} = h^*_{\mu\nu} \qquad h^*_{\nu\mu} \equiv \frac{\delta\mathcal{E}}{\delta\rho^*_{\mu\nu}} = h_{\mu\nu} \tag{3.39a}$$

$$\Delta_{\mu\nu} \equiv \frac{\delta\mathcal{E}}{\delta\kappa^*_{\mu\nu}} = -\Delta_{\nu\mu} \qquad \Delta^*_{\mu\nu} \equiv \frac{\delta\mathcal{E}}{\delta\kappa_{\mu\nu}} = -\Delta^*_{\nu\mu}. \tag{3.39b}$$

By running summations over all indices $\nu \leqslant \mu$, we formally account for the fact that $\delta\rho^*_{\mu\mu} = \delta\kappa_{\mu\mu} = \delta\kappa^*_{\mu\mu} = 0$, the first one because the diagonal elements of ρ are real numbers, the other two because the diagonal elements of κ and κ^* are zero anyway. The term $h_{\mu\nu}$ is called the mean field, or HF potential. The term $\Delta_{\mu\nu}$ is called the pairing field. In analogy to the one-body density matrix and anomalous density, the mean field $h_{\nu\mu}$ is Hermitian, whereas the pairing field $\Delta_{\mu\nu}$ is skew-symmetric. The variation of the energy can be further compacted after introducing the generalized HFB matrix $\mathcal{H}$ defined by

$$\mathcal{H} \equiv \begin{pmatrix} h & \Delta \\ -\Delta^* & -h^* \end{pmatrix}. \tag{3.40}$$

The HFB matrix is Hermitian, which follows directly from the definitions (3.39). This leads to

$$\delta\mathcal{E} = \frac{1}{2}\text{tr}(\mathcal{H}\delta\mathcal{R})$$

The variation of the constraint on particle number leads to

$$\delta(\lambda_{\text{F}}\ \text{tr}(\rho)) = \lambda_{\text{F}}\ \text{tr}(\delta\rho) \qquad \delta(\lambda_{\text{F}}\ \text{tr}(\rho^*)) = \lambda_{\text{F}}\ \text{tr}(\delta\rho^*),$$

while the variation of the second constraint gives

$$\delta\text{tr}(\Lambda[\mathcal{R}^2 - \mathcal{R}]) = \sum_{\mu\leqslant\nu}\frac{\delta}{\delta\mathcal{R}_{\mu\nu}}\sum_{\alpha\beta}\Lambda_{\alpha\beta}\left[\sum_{\gamma}\mathcal{R}_{\beta\gamma}\mathcal{R}_{\gamma\alpha} - \mathcal{R}_{\beta\alpha}\right]\delta\mathcal{R}_{\mu\nu}$$
$$= \text{tr}\{(\mathcal{R}\Lambda + \Lambda\mathcal{R} - \Lambda)\delta\mathcal{R}\}.$$

Combining each of these three terms, the minimization (3.38) reads

$$\text{tr}\{(\mathcal{H}' - \mathcal{R}\Lambda - \Lambda\mathcal{R} + \Lambda)\} = 0, \tag{3.41}$$

with

$$\mathcal{H}' \equiv \begin{pmatrix} h-\lambda_{\mathrm{F}} & \Delta \\ -\Delta^* & -h^*+\lambda_{\mathrm{F}} \end{pmatrix}. \tag{3.42}$$

Since equation (3.41) has to be true for an arbitrary variation $\delta\mathcal{R}$, we find the condition

$$\mathcal{H}'-\mathcal{R}\Lambda-\Lambda\mathcal{R}+\Lambda=0.$$

We obtain two independent equations by multiplying the lhs by $\mathcal{R}$ from the left and from the right. Subtracting the two and using the idempotency of $\mathcal{R}$, we find the final form of the HFB equation,

$$[\mathcal{H}', \mathcal{R}]=0. \tag{3.43}$$

This matrix equation can be solved easily by finding the q.p. basis that diagonalizes simultaneously the matrices of $\mathcal{H}'$ and $\mathcal{R}$. As will be discussed in more detail in chapter 8, section 8.3.1, this basis is found by solving the HFB eigenvalue problem

$$\mathcal{H}'\begin{pmatrix} U \\ V \end{pmatrix}_\mu = E_\mu \begin{pmatrix} U \\ V \end{pmatrix}_\mu, \tag{3.44}$$

with E_μ being the q.p. energy.

Here, we have obtained equation (3.43) by requiring that the energy computed at the HFB approximation is a minimum with respect to variations of $\mathcal{R}$. In section 8.3.2 of chapter 8, we will present an alternative method to directly minimize the HFB energy, based on a generalization of the gradient method and the Thouless parametrization of HFB vacua.

The Bloch–Messiah–Zumino theorem

A unitary matrix $\mathcal{W}$ of the particular form (3.6) can always be decomposed into a sequence of three transformations

$$\mathcal{W}=\mathcal{D}\bar{\mathcal{W}}\mathcal{C}=\begin{pmatrix} D & 0 \\ 0 & D^* \end{pmatrix}\begin{pmatrix} \bar{U} & \bar{V} \\ \bar{V} & \bar{U} \end{pmatrix}\begin{pmatrix} C & 0 \\ 0 & C^* \end{pmatrix}, \tag{3.45}$$

where C and D are unitary transformations and $\bar{U}$ and $\bar{V}$ are the real sub-matrices of a Bogoliubov transformation that takes the very simple canonical form

$$\bar{U}=\begin{pmatrix} 0 & & & & 0 \\ & \ddots & & & \\ & & u_k & 0 & \\ & & 0 & u_{\bar{k}} & \\ & & & & \ddots \\ 0 & & & & 1 \end{pmatrix} \quad \bar{V}=\begin{pmatrix} 1 & & & & 0 \\ & \ddots & & & \\ & & 0 & v_k & \\ & & v_{\bar{k}} & 0 & \\ & & & & \ddots \\ 0 & & & & 0 \end{pmatrix}. \tag{3.46}$$

These definitions imply that $\bar{U}_{kk} = u_k$, $\bar{U}_{\bar{k}\bar{k}} = u_{\bar{k}}$, $\bar{V}_{k\bar{k}} = v_k$, $\bar{V}_{\bar{k}k} = v_{\bar{k}}$. This is known as the Bloch–Messiah–Zumino theorem [1, 14, 15], which takes us from an arbitrary basis of s.p. operators $(c, c^\dagger)$ to the arbitrary basis of q.p. operators $(\beta, \beta^\dagger)$ in three consecutive steps,

$$\mathcal{W}^\dagger\colon \{c\} \to \{\beta\} = \{c\} \overset{\mathcal{D}^\dagger}{\to} \{a\} \overset{\bar{\mathcal{W}}^T \equiv (\bar{U}^T, \bar{V}^T)}{\to} \{\alpha\} \overset{\mathcal{C}^\dagger}{\to} \{\beta\},$$

where:

- $\mathcal{D}^\dagger$ is a transformation among the s.p. states, leading to the single-particle basis $(a, a^\dagger)$. This transformation diagonalizes the one-body density matrix ρ and puts the anomalous density κ into a canonical form analogous to $\bar{V}$. It defines the 'canonical' basis of HFB theory, which we will discuss further in the next section.
- $\bar{\mathcal{W}}^T \equiv (\bar{U}^T, \bar{V}^T)$ is a transformation from the canonical s.p. basis $(a, a^\dagger)$ to a q.p. basis $(\alpha, \alpha^\dagger)$ that is the special Bogoliubov transformation of the BCS theory.
- $\mathcal{C}^\dagger$ is a transformation of q.p. states among themselves, leading to the q.p. basis $(\beta, \beta^\dagger)$.

For the following discussion, it is important to note that the q.p. vacuum is invariant (possibly up to a phase factor) under the basis change associated with the $\mathcal{C}$ transformation of the Bloch–Messiah–Zumino theorem, as such a transformation only mixes q.p. annihilation (creation) operators among themselves, $\beta = C^\dagger \alpha$, $\beta^\dagger = C^T \alpha^\dagger$. In order to obtain another q.p. basis associated with a different vacuum $|\Phi'\rangle$, one would need to invoke a unitary transformation between the two bases that mixes q.p. creation with q.p. annihilation operators. Here, since the q.p. vacuum $|\Phi\rangle$ is invariant, it is a vacuum both for the operators β and for the operators α

$$\prod_\lambda \beta_\lambda \,|-\rangle \sim \prod_\mu \alpha_\mu \,|-\rangle. \tag{3.47}$$

This interesting property allows for calculating observables and analyzing properties of the q.p. vacuum in the intermediate bases defined through the Bloch–Messiah–Zumino theorem, which are easier to handle thanks to the special form of the matrices $\bar{U}$ and $\bar{V}$.

The canonical basis

As already mentioned above, in the canonical single-particle basis, as defined by the Bloch–Messiah–Zumino theorem, the one-body density matrix and anomalous density take the particularly simple form

$$\rho = \begin{pmatrix} 1 & & & & \\ & \ddots & & & \\ & & v_k^2 & & \\ & & & v_{\bar{k}}^2 & \\ & & & & \ddots & \\ & & & & & 0 \end{pmatrix} \quad \kappa = \begin{pmatrix} 0 & & & & & 0 \\ & \ddots & & & & \\ & & 0 & u_k v_k & & \\ & & u_{\bar{k}} v_{\bar{k}} & 0 & & \\ & & & & \ddots & \\ 0 & & & & & 0 \end{pmatrix}. \tag{3.48}$$

The s.p. states associated with the basis which diagonalizes the density matrix fall into three different categories:

1. There is a set $\{I^f\}$ of fully occupied states that we will label with indices f. These states correspond to the eigenvalue $\rho_{ff} = 1$ and can be identified by having a matrix element $\bar{V}_{ff}$ with value 1 on the diagonal of $\bar{V}$. These do not participate in pairing correlations and consequently the corresponding matrix elements of κ are zero.
2. There is a set $\{I^b\}$ of completely empty states that we will label with indices b. These correspond to the eigenvalue $\rho_{bb} = 0$ and can be identified by having a matrix element $\bar{U}_{bb}$ with value 1 on the diagonal of $\bar{U}$. These also do not participate in pairing correlations, hence do not have non-zero matrix elements of κ.
3. There is a set $\{I^p\}$ of paired single-particle states which all come in pairs with degenerate eigenvalues $\rho_{kk} = \rho_{\bar{k}\bar{k}} = \nu_k^2 = \nu_{\bar{k}}^2$ that take values $0 < \nu_k^2 < 1$. The single-particle states with indices k and $\bar{k}$ are called *conjugate pairs*, and the anomalous density $\kappa_{k\bar{k}} = (\bar{V}\bar{U})_{k\bar{k}} = \nu_k u_{\bar{k}} = -\kappa_{\bar{k}k}$ has only matrix elements between two such states. There might be additional degeneracies because of symmetries of the single-particle basis, but the number of paired states with same occupation is always even.

In the canonical single-particle basis, the normalized q.p. vacuum takes the particularly simple and well-known form of a BCS state

$$|\Phi_{\text{BCS}}\rangle = \prod_{k\in\{I^p\}>0} \left(u_k + \nu_k\, a_k^\dagger a_{\bar{k}}^\dagger\right) \prod_{f\in\{I^f\}} a_f^\dagger|-\rangle. \tag{3.49}$$

The shorthand $k > 0$ in a product or a sum over paired single-particle states in the canonical basis indicates that the index k only runs over the labels without a bar. Requiring that equation (3.49) be normalized implies that $u_k^2 + \nu_k^2 = 1$.

There are several possible phase conventions for the matrix elements of $\bar{U}$ and $\bar{V}$ that satisfy the relations mentioned above. The most common one is $u_k = u_{\bar{k}} \geqslant 0$ and $\nu_{\bar{k}} = -\nu_k$. However, the choice of which matrix element of $\nu_{\bar{k}}$ or ν_k actually takes a negative value cannot be made independently of the choice made for the relative phases of the single-particle states, since their combined phase has to match the phase convention of the q.p. states.

The corresponding quasiparticle annihilation operators obtained after the special Bogoliubov transformation of the Bloch–Messiah–Zumino theorem have the simple form

$$\begin{aligned} \alpha_k &= u_k a_k + \nu_{\bar{k}} a_{\bar{k}}^\dagger \\ \alpha_{\bar{k}} &= u_{\bar{k}} a_{\bar{k}} + \nu_k a_k^\dagger \\ \alpha_f &= a_f^\dagger \\ \alpha_b &= a_b. \end{aligned} \tag{3.50}$$

The two conjugate s.p. states always have the same eigenvalue $\rho_{kk} = \rho_{\bar{k}\bar{k}}$ of the density matrix. In a general, symmetry-unrestricted HFB calculation, however, there is no *a priori* reason that the absolute value of the expectation value of any one-body operator is the same for the two states that form such a conjugate pair. It is only when the q.p. vacuum is invariant under time-reversal that the two s.p. states in a conjugate pair are connected by a symmetry transformation, namely time-reversal: $\Psi_{\bar{k}}(\boldsymbol{r}) = \pm\hat{T}\ \Psi_k(\boldsymbol{r})$ and $\Psi_k(\boldsymbol{r}) = \mp\hat{T}\ \Psi_{\bar{k}}(\boldsymbol{r})$. And only then are the expectation values of one-body operators the same (possibly up to a sign) for both states in a conjugate pair.

The use of the canonical basis greatly simplifies the evaluation of many-body expectation values since the number of sums over basis states is greatly reduced. Also, many formal results are much easier to interpret. An example is the dispersion of particle number

$$\langle(\Delta\hat{N})^2\rangle \equiv \langle\Phi_{\rm BCS}\,|\hat{N}^2|\Phi_{\rm BCS}\rangle - \langle\Phi_{\rm BCS}\,|\hat{N}|\Phi_{\rm BCS}\rangle^2 = 4\sum_{k\in\{I^p\}>0} u_k^2 v_k^2,$$

that provides an indicator for the pairing correlations present in a q.p. vacuum.

The BCS form (3.49) of the q.p. vacuum can easily be brought to a Thouless-like form

$$\begin{aligned} |\Phi_{\rm BCS}\rangle &= \prod_{k\in\{I^p\}>0} u_k\left(1 + \frac{v_k}{u_k}\,a_k^\dagger a_{\bar{k}}^\dagger\right)\prod_{f\in\{I^f\}} a_f^\dagger\,|-\rangle \\ &= \left[\prod_{k\in\{I^p\}>0} u_k\right]\exp\left[\sum_{k\in\{I^p\}>0} Z_{k\bar{k}}\,a_k^\dagger\,a_{\bar{k}}^\dagger\right]\prod_{f\in\{I^f\}} a_f^\dagger\,|-\rangle, \end{aligned} \tag{3.51}$$

with $Z_{k\bar{k}} = v_k/u_k$. From this expression it is evident that a q.p. vacuum can be described as a Thouless state built on the s.p. vacuum only when there are no fully occupied s.p. states, i.e. when the cardinality of $\{I^f\}$ is zero. Similarly, the link between the q.p. vacuum (3.3) and its BCS form (3.49) can be established. We find

$$\begin{aligned} |\Phi_{\rm BCS}\rangle &= \prod_{k\in\{I^p\}>0}\left(u_k + v_k\,a_k^\dagger a_{\bar{k}}^\dagger\right)\prod_{f\in\{I^f\}} a_f^\dagger\,|-\rangle \\ &= \prod_{k\in\{I^p\}>0}\left(u_k\,a_k a_k^\dagger - v_{\bar{k}}\,a_k^\dagger a_{\bar{k}}^\dagger\right)\prod_{f\in\{I^f\}} a_f^\dagger\,|-\rangle \\ &= \prod_{k\in\{I^p\}>0}\frac{1}{v_k}\left(u_k a_k + v_{\bar{k}} a_{\bar{k}}^\dagger\right)v_k a_k^\dagger\prod_{f\in\{I^f\}} a_f^\dagger\,|-\rangle \\ &= \prod_{k\in\{I^p\}>0}\frac{1}{v_k}\left(u_k a_k + v_{\bar{k}} a_{\bar{k}}^\dagger\right)\left(u_{\bar{k}} a_{\bar{k}} + v_k a_k^\dagger\right)\prod_{f\in\{I^f\}} a_f^\dagger\,|-\rangle \\ &= \prod_{k\in\{I^p\}>0}\frac{1}{v_k}\,\alpha_k\alpha_{\bar{k}}\prod_{f\in\{I^f\}}\alpha_f\,|-\rangle. \end{aligned} \tag{3.52}$$

In these manipulations, we used the commutators $a_k a_k^\dagger = 1 - a_k^\dagger a_k$, $a_k^\dagger a_{\bar{k}}^\dagger = -a_{\bar{k}}^\dagger a_k^\dagger$, as well as $v_k = -v_{\bar{k}}$ along the way. The final result can be rewritten as

$$|\Phi\rangle = (-)^p \prod_{k \in \{I^p\} > 0} \alpha_k \alpha_{\bar{k}} \prod_{f \in \{I^f\}} \alpha_f \prod_{b \in \{I^b\}} \alpha_b \, |-\rangle \tag{3.53}$$

and differs from the q.p. vacuum on the rhs of equation (3.47) in three respects. First, equation (3.53) contains a phase factor $(-)^p$, where p is the number of permutations necessary to bring the q.p. operator into that order. Second, the normalization of the BCS state yields a factor $1/v_k$ for each pair of operators $\alpha_k \alpha_{\bar{k}}$ of paired quasiparticles on the rhs of equation (3.52). Third, and most importantly, the definition of the BCS state (3.49) does not contain a product over the annihilation operators α_b, which raises the question if $|\Phi_{\rm BCS}\rangle$ is a vacuum of the corresponding creation operators $\alpha_b^\dagger$.

From expressions (3.52) and (3.53) it is evident that a q.p. vacuum defined through (3.3) is not normalized. For a q.p. vacuum whose canonical basis only contains filled and paired s.p. states with finite occupation $0 < v_k^2 \leqslant 1$, one finds $\langle \Phi | \Phi \rangle = \prod_{k \in \{I^p\} > 0} v_k^2$, which means that the normalization factor $1/\sqrt{\langle \Phi | \Phi \rangle}$ can take an arbitrarily huge value. When the canonical basis of the q.p. vacuum also contains empty s.p. states, then the q.p. vacuum cannot be normalized anymore.

From the definition (3.50) of the annihilation operators α_b follows $\alpha_b|-\rangle = a_b|-\rangle = 0$, meaning that α_b are basis vectors of the null space of the Bogoliubov transformation. The corresponding q.p. wave functions do not contribute to observables of the q.p. vacuum, but they are needed to completely define the q.p. basis and with that contribute to any vertical expansion built on the q.p. vacuum.

In spite of the absence of α_b operators in its definition (3.49), the BCS state is nevertheless the vacuum of the corresponding creation operators $\alpha_b^\dagger = a_b^\dagger$, as these are also creation operators with respect to the s.p. vacuum $|-\rangle$. To drop the α_b operators from the definition (3.49) of $|\Phi_{\rm BCS}\rangle$ is necessary for the normalization of the BCS state.

Number parity

Bogoliubov q.p. vacua $|\Phi\rangle$ have a symmetry called number parity

$$e^{-i\pi\hat{N}}|\Phi\rangle = \pi_N \, |\Phi\rangle = \pm|\Phi\rangle \tag{3.54}$$

related to the phase they take when rotating them by π in gauge space. The number parity of a Bogoliubov q.p. state $|\Phi\rangle$ is independent on the mean particle number $\langle \Phi | \hat{N} | \Phi \rangle / \langle \Phi | \Phi \rangle$ that the state has been constrained to when solving the HFB equations (3.44). It is not affected by changing the basis the q. p. vacuum is expressed in. Like most other operators, number parity is most easily evaluated in the canonical basis. Using the relation $e^{-i\varphi\hat{N}} c_k^\dagger e^{i\varphi\hat{N}} = c_k^\dagger e^{-i\varphi}$ for gauge-space rotation about an arbitrary angle, it can be easily seen that each factor $(u_k + v_k a_k^\dagger a_{\bar{k}}^\dagger)$ in equation (3.49) gives a factor +1 when rotating by π. The number parity of a given q.p. vacuum is therefore solely determined by the number of fully occupied single-particle states in the canonical basis. More precisely, $\pi_N = (-1)^r$, where r is the cardinality of the set $\{I^f\}$. Fully paired quasiparticle vacua always have the number parity $\pi_N = +1$, blocked one-quasiparticle states the number parity $\pi_N = -1$, etc.

For the interpretation of number parity it is important to note that eigenstates $|\Psi^{N_0}\rangle$ of the particle-number operator, $\hat{N}\,|\Psi^{N_0}\rangle = N_0\,|\Psi^{N_0}\rangle$ are also trivially eigenstates of number parity

$$e^{-i\pi\hat{N}}|\Psi^{N_0}\rangle = (-1)^{N_0}|\Psi^{N_0}\rangle.$$

Bogoliubov q.p. vacua with number parity +1 therefore have the correct structure to describe systems with even particle number, whereas those with number parity −1 have the correct structure to describe systems with odd particle number.

Selection of quasiparticle states

Using equation (3.22), one can show that the matrix γ defined in equation (3.21) transforms the HFB matrix (3.42) into

$$\gamma\,\mathcal{H}'\,\gamma = -\mathcal{H}'^*. \tag{3.55}$$

Transforming the HFB equation (3.44) using the above relation and equation (3.22) yields

$$0 = \gamma[\mathcal{H}' - E]\gamma\,\gamma\begin{pmatrix} U \\ V \end{pmatrix} = -[\mathcal{H}'^* + E]\begin{pmatrix} V \\ U \end{pmatrix}.$$

Taking the complex conjugate of this equation, it can be recast into

$$\mathcal{H}'\begin{pmatrix} V^* \\ U^* \end{pmatrix} = -E\begin{pmatrix} V^* \\ U^* \end{pmatrix}.$$

The inherent symmetry (3.55) of the HFB Hamiltonian has the consequence that for every eigenstate with eigenvalue E there is a second eigenstate with eigenvalue of opposite sign $-E$ that is obtained by simultaneous exchange of upper and lower components of the quasiparticle wave function and complex conjugation. We will label these states with indices μ and $\breve{\mu}$

$$\Psi_\mu = \begin{pmatrix} U \\ V \end{pmatrix}_\mu \qquad \Psi_{\breve{\mu}} = \begin{pmatrix} V^* \\ U^* \end{pmatrix}_\mu. \tag{3.56}$$

Thanks to relations (3.8) between the U and the V matrices imposed by the unitarity of the Bogoliubov transformation, we see that these two solutions correspond to different eigenvalues of the generalized density matrix (3.17)

$$\mathcal{R}\,\Psi_{\breve{\mu}} = 1\,\Psi_{\breve{\mu}} \qquad \mathcal{R}\,\Psi_{\mu} = 0. \tag{3.57}$$

Following [16], these will be called filled and empty q.p. states. The adjectives 'filled' and 'empty' have here a meaning similar to the solutions of the Dirac equation, where a filled state in the Dirac sea is not available as degree of freedom unless it is 'emptied' by creating a hole; see section 2.6.1. From equation (3.5) it follows that the creation and annihilation operators associated with these two types of solutions of the HFB equation obey the relations

$$\beta_{\breve{\mu}} = \beta_{\mu}^{\dagger} \qquad \beta_{\breve{\mu}}^{\dagger} = \beta_{\mu}. \tag{3.58}$$

This has the important consequence that acting with the creation operator $\beta_{\breve{\mu}}^{\dagger}$ for a filled q.p. state on the q.p. vacuum gives zero

$$\beta_{\breve{\mu}}^{\dagger}|\Phi\rangle = \beta_{\mu}\,|\Phi\rangle = 0. \tag{3.59}$$

Hence, the q.p. vacuum (3.3) is the vacuum of 'empty' q.p. states, i.e. the ones that are eigenstates of the generalized density matrix $\mathcal{R}$ with eigenvalue zero, but not the 'filled' ones.

Denoting the size of the single-particle basis $(c, c^{\dagger})$ with Ω, then the Boguliubov transformation (3.5) defines also Ω q.p. creation and annihilation operators. As a consequence, the definition of the q.p. vacuum (3.3) is a product also containing Ω operators β_{μ}. Because of its doubled dimension, however, the HFB equation (3.44) has 2Ω solutions, half of which are consequently not entering the definition of the q. p. vacuum (3.3).

Which q.p. states are filled and which ones are empty is, however, a matter of choice. Indeed, replacing β_{μ} by $\beta_{\breve{\mu}}$ in the definition of the q.p. vacuum will modify the corresponding generalized density matrix (3.17) such that with respect to the new $\mathcal{R}$ the empty state becomes a filled one and vice versa. That q.p. states are empty or filled is not a result of solving the HFB equation, but the result of a choice made when building the q.p. vacuum, and with that $\mathcal{R}$, from the eigenstates of the HFB Hamiltonian. Through self-consistency, the HFB Hamiltonian will of course be different for each of the possible choices. For the stable iterative solution of the HFB equations it is therefore necessary to find a procedure that ensures that a consistent choice can be systematically made.

The sign of the corresponding quasiparticle energies $\pm E_{\mu}$ is irrelevant for the identification of empty and filled states. In particular, there is no *a priori* reason that it has to be the same for all empty (filled) states. In practice, however, one is usually interested in q.p. vacua that approximate the nuclear ground state or a low-lying excitation. This leads to the question of how to identify configurations that will most probably lead to the largest possible binding energy when building the generalized density matrix $\mathcal{R}$ without having to calculate the total binding energy for each possibility.

It can easily be seen that the H^{11}, H^{20} and H^{02} parts of the q.p. representation of the many-body Hamiltonian introduced above can be written as matrix elements of the HFB Hamiltonian $\mathcal{H}'$ (3.42), which in the basis of its eigenvectors (3.56) take the simple form

$$H^{11}_{\mu\nu} = \Psi^{\dagger}_{\mu}\,\mathcal{H}'\,\Psi_{\nu} = E_{\mu}\delta_{\mu\nu} \tag{3.60}$$

$$H^{20}_{\mu\nu} = \Psi^{\dagger}_{\mu}\,\mathcal{H}'\,\Psi_{\breve{\nu}} = 0 \tag{3.61}$$

$$H^{02}_{\mu\nu} = \Psi^{\dagger}_{\breve{\mu}}\,\mathcal{H}'\,\Psi_{\nu} = 0, \tag{3.62}$$

where in the last two relations we used that q.p. states with indices μ and $\breve{\nu}$ as defined above are always orthogonal, including the case $\breve{\nu} = \breve{\mu}$. Expressions (3.61) and (3.62) imply that the diagonalization of $\mathcal{H}'$ suppresses any coupling between the q.p. vacuum built from its eigenstates and all possible two-q.p. excitations.

Neglecting the possible contributions from H^{31} and H^{13}, from the expression for H^{11} (3.60) it follows that a perturbative one-q.p. excitation $\beta^{\dagger}_{\mu}|\Phi\rangle$ on the q.p. vacuum changes the expectation value of the total energy by E_{μ}. This estimate suggests that the most bound q.p. vacuum that can be built from the eigenstates of $\mathcal{H}'$ corresponds to the configuration where all q.p. states with negative E_{μ} are filled, such that only q. p. states with positive E_{μ} can be excited. From this follows that the excited configurations built from the same q.p. basis have at least one empty state with negative E_{μ}.

The lowest possible q.p. vacuum, however, is not always the one that provides the ground state of the system of interest, as it might not have the correct quantum numbers. The most relevant for the present discussion is number parity, which distinguishes between q.p. vacua representing even and odd systems.

Figure 3.1 presents an example of a situation where also q.p. states with an negative eigenvalue have to be considered when building the q.p. vacuum of an even–even nucleus. It shows the spectrum of neutron q.p. energies E_{μ} as the function of rotational frequency obtained from a series of cranked HFB calculations of ^{24}Mg. In such a calculation, adding a constraint $-\omega\hat{J}_{\mu}$ on the mean value of a component of the angular-momentum vector $\hat{J}_{\mu}$ allows for studying the behavior of the nucleus at rapid collective rotation. The q.p. spectrum is clearly symmetric around the $E = 0$ line. In the calculation, the selection of the filled and empty q.p. states has been made such that the q.p. vacua have the same number parity +1 of a configuration of an even–even nucleus at all rotational frequencies. At rotational frequency $\omega = 0$, time-reversal is dynamically conserved and all q.p. levels have a two-fold Kramers degeneracy. At finite rotational frequency, time-reversal invariance is lifted by the constraint on angular momentum. At rotational frequency around $\omega \approx 2.75$ MeV $\hbar^{-1}$, two negative parity levels cross at $E = 0$, such that at higher rotational frequency ω there is an 'empty' q.p. level with negative q.p. energy and a 'filled' one with positive q.p. energy. Choosing the q.p. operators β_{μ} that build the q.p. vacuum only from q.p. states with positive E_{μ}. would lead at $\omega \approx 3$ MeV $\hbar^{-1}$ to a

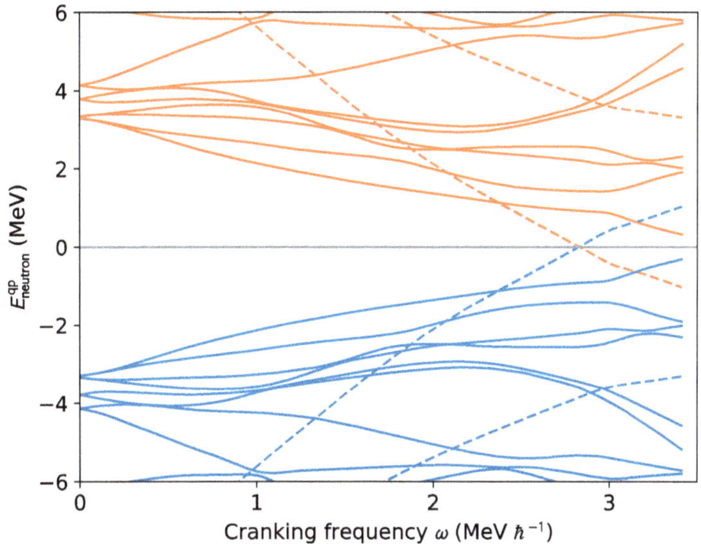

Figure 3.1. Spectrum of neutron q.p. levels in a cranked HFB calculation of ^{24}Mg as a function of rotational frequency ω. 'Empty' q.p. levels are drawn in orange and 'filled' ones in blue, where the number parity of the q.p. vacuum is the same at all rotational frequencies. Solid (dashed) lines indicate the conserved positive (negative) parity of the q.p. levels in this calculation. (Figure created by W Ryssens.)

configuration that has a negative number parity and thus describes a rotating state in an odd nucleus instead of ^{24}Mg.

Blocking

We have seen above in the analysis of q.p. vacua in the canonical basis (3.52) that having canonical s.p. levels with eigenvalue $\rho_{bb} = 0$ requires that the corresponding q.p. annihilation operator α_b be absent from the definition of the q.p. vacuum in its BCS form (3.52). There are two reasons why such a situation may arise

1. Pairs of conjugate s.p. states might just happen to have occupation zero, i.e. they are the $v_k^2 \to 0$ limit of a conjugate pair in the BCS representation (3.49) of the q.p. vacuum $(u_k + v_k\, a_k^\dagger a_{\bar{k}}^\dagger) \to 1$.
2. Pairs of conjugate s.p. states might be 'broken', meaning that one q.p. out of a conjugate pair becomes a filled state, the other one a blocked state, $\alpha_k \alpha_{\bar{k}} \to \alpha_f \alpha_b$.

The latter case is generated by acting with a q.p. creation operator on a paired state and introduces what one calls the blocking approximation. For the sake of simplifying notations, assume that the reference q.p. vacuum is completely paired. In addition, we will assume that the q.p. vacuum is normalized according to equation (3.52), neglecting, however, a possible phase. Creating a q.p. excitation on top of this normalized vacuum expressed in the canonical basis leads to

$$\alpha_\mu^\dagger \prod_\lambda \frac{1}{\sqrt{v_\lambda}} \alpha_\lambda \,|-\rangle = (-)^p \alpha_\mu^\dagger \frac{1}{v_\mu} \alpha_\mu \alpha_{\bar\mu} \prod_{\lambda \neq \mu\bar\mu} \frac{1}{\sqrt{v_\lambda}} \alpha_\lambda \,|-\rangle$$
$$= a_\mu^\dagger (-)^p \prod_{\lambda \neq \mu\bar\mu} \frac{1}{\sqrt{v_\lambda}} \alpha_\lambda \,|-\rangle.$$

Note that the first operator in the final expression for the product defining the new vacuum (in terms of operators of the old vacuum) is a *particle* operator, not a q.p. operator. To obtain this result, we have used the property that $a_\mu^\dagger a_\mu^\dagger|-\rangle = 0$ and

$$\alpha_\mu^\dagger \alpha_\mu \alpha_{\bar\mu} \,|-\rangle = \left(u_\mu a_\mu^\dagger + v_{\bar\mu} a_{\bar\mu}\right)\left(u_\mu a_\mu + v_{\bar\mu} a_{\bar\mu}^\dagger\right)\left(u_{\bar\mu} a_{\bar\mu} + v_\mu a_\mu^\dagger\right)|-\rangle$$
$$= \left(u_\mu^2\, v_\mu\, a_\mu^\dagger + v_{\bar\mu}^2\, v_\mu\, a_\mu^\dagger\right)|-\rangle = v_\mu\, a_\mu^\dagger|-\rangle.$$

The phase factor $(-)^p$ depends on the number p of permutations necessary to bring the q.p. operators into suitable order.

These expressions imply, not surprisingly, that the canonical bases of the blocked and the original (non-blocked) state are different. Indeed, using the identification of q.p. operators (3.50) suggests that blocking converts the pair of q.p. operators α_μ and $\alpha_{\bar\mu}$ into a different pair of q.p. operators α_b and α_f. With this notation, the (normalized) blocked state can be rewritten in its own canonical basis as

$$\alpha_\mu^\dagger \prod_\lambda \frac{1}{\sqrt{v_\lambda}} \alpha_\lambda \,|-\rangle \to (-)^p \alpha_f \prod_{\lambda \neq \mu\bar\mu} \frac{1}{\sqrt{v_\lambda}} \alpha_\lambda \,|-\rangle. \tag{3.63}$$

Compared to the original vacuum, there is one q.p. operator fewer than before. The one which has disappeared is $\alpha_\mu \to \alpha_b$, which has the property $\alpha_b|-\rangle = a_b|-\rangle = 0$. The occupation of this state in the new canonical basis is zero $\rho_{bb} = 0$. The operator $\alpha_{\bar\mu} \to \alpha_f$ has also changed its structure such that it corresponds to a filled s.p. state in the new canonical basis $\rho_{ff} = 1$.

With that, a pair of conjugate s.p. states in the canonical basis has indeed been broken. As the number of fully occupied s.p. states in the canonical basis changes by one, the new q.p. vacuum has a different number parity than the old one. A one-q.p. state built on a state that describes a system with even particle number therefore describes a system with odd particle number. The same procedure can be repeated on other pairs to generate blocked broken-pair two-q.p. states, three-q.p. states. etc. Each time the number parity changes its sign.

As elsewhere, we assume here in the formal discussion of the HFB equation that there is only one abstract particle species, while an actual nucleus is composed of two particle species. An odd–odd nucleus is described by a simultaneous one-q.p. excitation of the q.p. vacuum describing neutrons and one-q.p. excitation of the q.p. vacuum describing protons.

There also is a possibility to create a two-q.p. excitation of the same pair. After cumbersome algebra, the resulting normalized two-q.p. state can be expressed in the canonical s.p. basis of the original state as

$$\alpha_\mu^\dagger \alpha_{\bar\mu}^\dagger \prod_\lambda \alpha_\lambda |-\rangle \propto \left(-v_\mu + u_\mu\, a_k^\dagger a_{\bar k}^\dagger\right) \prod_{\substack{k\in\{IP\}>0 \\ k\neq\mu,\bar\mu}} \left(u_k + v_k\, a_k^\dagger a_{\bar k}^\dagger\right)|-\rangle, \tag{3.64}$$

where we omit normalization and phase factors on the lhs. Such a state is sometimes called a 'decorrelated pair' state [17]. The blocked single-particle states in the canonical basis are still paired, but their occupations are different from those in the original q.p. vacuum.

Above, we have discussed the formal properties of blocked states in the canonical basis, which differs from the basis that diagonalizes the HFB Hamiltonian by the $\mathcal{C}$ transformation of the Bloch–Messiah–Zumino theorem. Blocked configurations can of course also be constructed directly in the HFB basis [8, 16, 18–21], but the expressions are less transparent. In the HFB basis, blocking a quasiparticle with index b modifies the one-body density matrix and anomalous density according to

$$\rho_{kl}^{(b)} = (V^* V^T)_{kl} - V_{kb}^* V_{lb} + U_{kb} U_{lb}^*$$
$$\kappa_{kl}^{(b)} = (V^* U^T)_{kl} - V_{kb}^* U_{lb} + U_{kb} V_{lb}^*,$$

which simply corresponds to the substitution of the q.p. wave function with index b by the q.p. wave function with index $\breve{b}$ as defined above. When blocking a q.p. with index b, the resulting state is a vacuum for the operators $(\beta_1, \ldots, \beta_{b-1}, \beta_b^\dagger, \beta_{b+1}, \ldots)$. These operators correspond to a new Bogoliubov transformation for which the column b of the (U, V) matrix has to be substituted with the column b of the (V^*, U^*) matrix, see the discussion of equation (3.56).

In general, it cannot be expected that blocked q.p. vacua built on a self-consistent solution of the HFB equation (3.44) remain a solution of the HFB equation. There are several reasons for this. First, the blocked state will in general have a slightly different average particle number than the state it was constructed from. More importantly, breaking a pair inevitably breaks time-reversal invariance. If the original q.p. vacuum was time-reversal invariant, then the contributions from time-odd mean fields in h lift the Kramers degeneracy and also add particular contributions to the pairing field Δ. This means that in order to construct a blocked state, the HFB equations have to be solved specifically for the blocked configuration taking into account all polarization and rearrangement effects induced by the blocking.

Solving the HFB equations for blocked quasiparticle states

Because of the added complexity of having to break time-reversal and other symmetries, to handle time-odd densities and fields, and to converge the self-consistent calculations for several configurations in order to find the one of interest, there have been many fewer large-scale HFB calculations of odd and odd–odd nuclei than for even–even systems. A technical complication specific to calculating blocked states is the need to tag the targeted configuration such that it can be unambiguously followed through the iterations of the self-consistent problem. Also, one has to pay attention not to accidentally end up with a configuration that has the wrong number parity [16].

A detailed analysis of the description of odd nuclei with Slater determinants and Bogoliubov q.p. vacua is given in [22]. The various contributions to the odd–even staggering of nuclear masses in the presence of pairing correlations have been analyzed in [23, 24]. Examples of systematic calculations of blocked states in odd nuclei can be found in [21, 25–28]. For a discussion of the impact of imposing symmetries on blocked one-q.p. states when solving the HFB equations, see [21]. A detailed analysis of the polarization effects when blocking a quasiparticle was presented in [29].

A widely used approximation to fully self-consistent blocked HFB calculations is the equal-filling approximation (EFA) [30], which consists in a statistical mixing of different q.p. vacua for which the time-odd contributions to the HFB Hamiltonian cancel out. In the canonical basis, it simply consists in setting $\rho_{ee} = \rho_{\bar{e}\bar{e}} = 1/2$ and $\kappa_{e\bar{e}} = \kappa_{\bar{e}e} = 0$ for the equally filled canonical basis state. The main advantage of the EFA is that it conserves time-reversal symmetry. Detailed benchmarks showed that for ground-state properties at least, the impact of time-odd terms is small and can often be neglected [31].

3.1.4 Stability condition

As mentioned already, the solution to the HFB equation (3.43) determines the one-body density matrix ρ and anomalous density κ (or equivalently the generalized density $\mathcal{R}$) such that the energy $\mathcal{E}[\rho, \kappa, \kappa^*]$ is minimal. Equivalently, the HFB solution determines the matrix of the Bogoliubov transformation, hence the q.p. operators and the product state (3.3). However, when we derived equation (3.43), we simply computed the variations of the HFB energy: the fact that $\delta\mathcal{E} = 0$ does not imply that the energy is a minimum, only that it is an extremum. In this section, we will use what one may call a generalization of the Taylor expansion to determine a criterion for the solution of the HFB equation to be a minimum (stability condition). Such an expansion will prove ubiquitous throughout the book, since we will see that it involves the quasiparticle random-phase approximation (QRPA) matrix. The QRPA and linear response theory are the subject of chapter 5, but the QRPA matrix is also an important building block of theories of large amplitude collective motion presented in chapter 6.

In our expansion of the total energy near the HFB minimum, we will parametrize the perturbations of the generalized density to carry out a Taylor expansion (up to second order). When discussing the gradient method in section 8.3.2, we will perform a similar expansion, this time by parametrizing directly the many-body product state thanks to the Thouless theorem.

Unitary transformation of the generalized density
Given the generalized density $\mathcal{R} \equiv \mathcal{R}^{(0)}$ given by equation (3.17) that satisfies the HFB equation, we want to parametrize 'small variations' of this matrix. To this end, let us choose a unitary transformation of the generalized density $\mathcal{R}$ in the following form

$$\mathcal{R} = e^{i\chi}\mathcal{R}^{(0)}e^{-i\chi}, \tag{3.65}$$

with the following notation,

$$\mathcal{R}^{(0)} = \begin{pmatrix} \rho^{(0)} & \kappa^{(0)} \\ -\kappa^{(0)*} & 1-\rho^{(0)*} \end{pmatrix},$$

for the generalized density matrix associated with the Bogoliubov vacuum (3.3). The operator $\hat{\chi}$ is a one-body operator, which can thus be written in any arbitrary s.p. basis of the Hilbert space as

$$\hat{\chi} = \sum_{ij} \chi_{ij} c_i^\dagger c_j.$$

We further assume that $||\chi_{ij}|| \ll 1$, i.e. the operator is a 'small' perturbation. Introducing the expansion of the generalized density as a function of the perturbation up to second order, we find

$$\mathcal{R} = \mathcal{R}^{(0)} + \mathcal{R}^{(1)} + \mathcal{R}^{(2)} + \mathcal{O}(\chi^3). \tag{3.66}$$

We now impose that both $\mathcal{R}$ and $\mathcal{R}^{(0)}$ are projectors with trace equal to particle number. In other words, they both correspond to q.p. vacua. Under these conditions, it is straightforward to show that the first- and second-order term of the expansion read

$$\mathcal{R}^{(1)} = i[\chi, \mathcal{R}^{(0)}] \tag{3.67a}$$

$$\mathcal{R}^{(2)} = \frac{1}{2}[[\chi, \mathcal{R}^{(0)}], \chi]. \tag{3.67b}$$

This equation is an operatorial equation—it is valid in any particular basis. In particular, we can express it either in an arbitrary s.p. basis or in the q.p. basis that diagonalizes the HFB Hamiltonian. Let us indicate with a tilde sign all matrices in the q.p. basis. Since $\mathcal{R}^{(0)}$ is a solution to the HFB equation (3.43), we have

$$\tilde{\mathcal{R}}^{(0)} = \begin{pmatrix} 0 & 0 \\ 0 & 1 \end{pmatrix}, \qquad \tilde{\mathcal{H}}^{(0)} = \begin{pmatrix} E & 0 \\ 0 & -E \end{pmatrix}. \tag{3.68}$$

To set the notations, we will write the matrix χ of the perturbation in the q.p. basis

$$\tilde{\chi} = \begin{pmatrix} \chi^{11} & \chi^{12} \\ \chi^{21} & \chi^{22} \end{pmatrix}. \tag{3.69}$$

Energy at second order and the stability matrix

Let us turn to the problem of the stability of the HFB minimum. To this end, we seek to express the energy of the system up to second order in perturbation of the generalized density using the expansion (3.66) based on the unitary transformation (3.65). We first note that the HFB energy can be written most generally

$$\mathcal{E}_{\mathrm{HFB}} = \frac{1}{4}\mathrm{tr}[(\mathcal{H} + \mathcal{T})\mathcal{S}], \tag{3.70}$$

with the following definitions (in the double s.p. basis),

$$\mathcal{H} = \begin{pmatrix} h & \Delta \\ -\Delta^* & -h^* \end{pmatrix} \quad \mathcal{T} = \begin{pmatrix} t & 0 \\ 0 & -t^* \end{pmatrix} \quad \mathcal{S} = \begin{pmatrix} \rho & \kappa \\ -\kappa^* & -\rho^* \end{pmatrix}.$$

The fields h and Δ are defined by equation (3.39) and the densities ρ and κ by equations (3.13) and (3.15), respectively; see also equations (3.14) and (3.16) for the expression of these densities in terms of the U and V matrices of the Bogoliubov transformation. We can also introduce the matrices $\mathcal{S}^{(n)}$ and $\mathcal{K}^{(n)}$ in the s.p. basis,

$$\mathcal{S}^{(n)} = \begin{pmatrix} \rho^{(n)} & \kappa^{(n)} \\ -\kappa^{(n)*} & -\rho^{(n)*} \end{pmatrix} \quad \mathcal{K}^{(n)} = \begin{pmatrix} \Gamma^{(n)} & \Delta^{(n)} \\ -\Delta^{(n)*} & -\Gamma^{(n)*} \end{pmatrix}, \tag{3.71}$$

which will correspond, respectively, to the densities and fields up to nth-order in powers of χ. We note that $\mathcal{R} = \mathcal{R}^{(0)} + \mathcal{S}^{(1)} + \mathcal{S}^{(2)}$. In other words, $\mathcal{R}^{(1)} \equiv \mathcal{S}^{(1)}$ and $\mathcal{R}^{(2)} \equiv \mathcal{S}^{(2)}$. Let us now expand the energy (3.70) up to second order in powers of χ,

$$\begin{aligned} \mathcal{E}[\mathcal{R}] &= \frac{1}{4}\mathrm{tr}[(\mathcal{H}^{(0)} + \mathcal{K}^{(1)} + \mathcal{K}^{(2)} + \mathcal{T})(\mathcal{S}^{(0)} + \mathcal{S}^{(1)} + \mathcal{S}^{(2)})] \\ &= \frac{1}{4}\mathrm{tr}[(\mathcal{H}^{(0)} + \mathcal{T})\mathcal{S}^{(0)}] + \frac{1}{4}\mathrm{tr}[\mathcal{H}^{(0)}\mathcal{S}^{(1)} + (\mathcal{K}^{(1)}\mathcal{S}^{(0)} + \mathcal{T}\mathcal{S}^{(1)})] \\ &\quad + \frac{1}{4}\mathrm{tr}[\mathcal{H}^{(0)}\mathcal{S}^{(2)} + (\mathcal{K}^{(2)}\mathcal{S}^{(0)} + \mathcal{T}\mathcal{S}^{(2)})] + \frac{1}{4}\mathrm{tr}[\mathcal{K}^{(1)}\mathcal{S}^{(1)}]. \end{aligned}$$

We can easily show that

$$\mathrm{tr}[\mathcal{H}^{(0)}\mathcal{S}^{(n)}] = \mathrm{tr}[(\mathcal{K}^{(n)}\mathcal{S}^{(0)} + \mathcal{T}\mathcal{S}^{(n)})].$$

This is a simple consequence of the fact that

$$\begin{aligned} \mathrm{tr}[h^{(0)}\rho^{(n)}] &= \sum_{ij}\left(t_{ij} + \frac{1}{2}\sum_{kl}\bar{v}_{ikjl}\rho_{lk}^{(0)}\right)\rho_{ji}^{(n)} \\ &= \sum_{ij} t_{ij}\rho_{ji}^{(n)} + \sum_{kl}\frac{1}{2}\sum_{ij}\bar{v}_{ikjl}\rho_{ji}^{(n)}\rho_{lk}^{(0)} \\ &= \mathrm{tr}[t\rho^{(n)} + \Gamma^{(n)}\rho^{(0)}] \end{aligned}$$

and similarly for $\mathrm{tr}[\Delta^{(n)}\kappa^{(0)*}]$. If the energy functional cannot be derived from an effective pseudo-potential $\hat{v}$, the previous equality still holds. Therefore, we arrive at

$$\mathcal{E}[\mathcal{R}] = \mathcal{E}_{\rm HFB} + \frac{1}{2}\text{tr}(\mathcal{H}^{(0)}\mathcal{S}^{(1)}) + \frac{1}{2}\text{tr}(\mathcal{H}^{(0)}\mathcal{S}^{(2)}) + \frac{1}{4}\text{tr}(\mathcal{K}^{(1)}\mathcal{S}^{(1)}). \tag{3.72}$$

We then use the relations (3.67a) and (3.67b) to compute these traces. Using the cyclic invariance of the trace and the fact that the HFB equation is satisfied at order 0 (by definition), we immediately find that

$$\text{tr}\big(\mathcal{H}^{(0)}\mathcal{S}^{(1)}\big) = i\text{tr}\big(\mathcal{H}^{(0)}[\chi, \mathcal{R}^{(0)}]\big) = i\text{tr}\big([\mathcal{R}^{(0)}, \mathcal{H}^{(0)}]\chi\big) = 0.$$

The calculation of the second trace is a little more involved. We can use again the properties of the commutators to write

$$\text{tr}(\mathcal{H}^{(0)}\mathcal{S}^{(2)}) = \frac{1}{2}\text{tr}\big([\chi, \mathcal{H}^{(0)}][\chi, \mathcal{R}^{(0)}]\big).$$

Each of these two commutators is most easily computed in the q.p. basis. We are free to choose the most convenient basis, since traces are independent of the basis. A straightforward calculation leads to

$$\text{tr}\big([\tilde{\chi}, \tilde{\mathcal{H}}^{(0)}][\tilde{\chi}, \tilde{\mathcal{R}}^{(0)}]\big) = \big(\chi_{ij}^{12*}, \chi_{ij}^{12}\big)\begin{pmatrix} (E_i + E_j)\delta_{ik}\delta_{jl} & 0 \\ 0 & (E_i + E_j)\delta_{ik}\delta_{jl} \end{pmatrix}\begin{pmatrix} \chi_{kl}^{12} \\ \chi_{kl}^{12*} \end{pmatrix},$$

where we have used a linearized notation for the matrix elements of χ^{12} and its complex conjugate. Recall that a linearized matrix simply consists in stacking all matrix elements of an $n \times n$ matrix into a vector of size n^2.

For the last term, we will again take advantage of the properties of the q.p. basis. Recall that $\mathcal{R}^{(1)} \equiv \mathcal{S}^{(1)}$ is defined by equation (3.67a). As a consequence, it is easy to show that the matrix $\tilde{\mathcal{R}}^{(1)}$ is off-block diagonal, that is,

$$\tilde{\mathcal{R}}^{(1)} = \begin{pmatrix} 0 & R^{12} \\ R^{21} & 0 \end{pmatrix} = i\begin{pmatrix} 0 & \chi^{12} \\ \chi^{21} & 0 \end{pmatrix}.$$

Therefore,

$$\text{tr}(\mathcal{K}^{(1)}\mathcal{S}^{(1)}) = -i\text{tr}(H^{12}\chi^{12\dagger}) + i\text{tr}(H^{12\dagger}\chi^{12}).$$

The calculation of the matrix H^{12} is cumbersome but does not pose any difficulty. The key is to take advantage of (i) the relation (3.67a) between $\mathcal{R}^{(1)}$ and χ, (ii) the definition of the matrix $\mathcal{R}^{(n)}$ in terms of the densities $\rho^{(n)}$ and $\kappa^{(n)}$, and (iii) the particular block structure of some of the matrices in the q.p. basis and the fact that one goes from the s.p. basis to the q.p. basis through the Bogoliubov transformation (which is fixed). Given these elements, we can first express $\mathcal{R}^{(1)}$ in terms of the matrix $\tilde{\chi}$, insert the resulting expressions of $\rho^{(1)}$ and $\kappa^{(1)}$ into the definition of $\mathcal{K}^{(1)}$ (see equation (3.71)) and transform back to the q.p. basis. We arrive at

$$\begin{aligned}\text{tr}(\mathcal{K}^{(1)}\mathcal{S}^{(1)}) = {} & [A_{ij\mu\nu} - (E_i + E_j)\delta_{i\mu}\delta_{j\nu}]\chi_{\mu\nu}^{12}\chi_{ij}^{12*} + B_{ij\mu\nu}\chi_{\mu\nu}^{12*}\chi_{ij}^{12} \\ & + [A_{ji\mu\nu}^{*} - (E_j + E_j)\delta_{j\mu}\delta_{i\nu}]\chi_{\mu\nu}^{12*}\chi_{ji}^{12} + B_{ji\mu\nu}^{*}\chi_{\mu\nu}^{12*}\chi_{ji}^{12}.\end{aligned}$$

Taking advantage of the fact that all indices are summation indices and can thus be interchanged consistently, we can put this trace into the matrix form

$$\mathrm{tr}(\mathcal{K}^{(1)}\mathcal{S}^{(1)}) = \left(\chi_{ij}^{12*}, \chi_{ij}^{12}\right)\begin{pmatrix} A_{ij\mu\nu} - (E_i + E_j)\delta_{i\mu}\delta_{j\nu} & B_{ij\mu\nu} \\ B_{ij\mu\nu}^{*} & A_{ij\mu\nu}^{*} - (E_j + E_j)\delta_{j\mu}\delta_{i\nu} \end{pmatrix}\begin{pmatrix} \chi_{\mu}^{12} \\ \chi_{\nu}^{12*} \end{pmatrix}.$$

Putting everything together, we see that the total energy can be conveniently and succinctly written as

$$\mathcal{E}[\mathcal{R}] = \mathcal{E}_{\mathrm{HFB}} + \frac{1}{4}(\chi^{12*}, \chi^{12})\begin{pmatrix} A & B \\ B^{*} & A^{*} \end{pmatrix}\begin{pmatrix} \chi^{12} \\ \chi^{12*} \end{pmatrix},$$

with the QRPA matrix given by

$$\mathcal{M} = \begin{pmatrix} A & B \\ B^{*} & A^{*} \end{pmatrix},$$

with the blocks A and B in their full glory,

$$\begin{aligned} A_{ij\mu\nu} = {} & (E_i + E_j)\delta_{i\mu}\delta_{j\nu} \\ & + U_{i\alpha}^{\dagger}V_{\beta j}^{*}\bar{v}_{\alpha k\beta l}U_{l\mu}V_{\nu k}^{T} - V_{i\alpha}^{\dagger}V_{\beta j}^{*}\bar{v}_{\alpha\beta kl}^{*}V_{k\nu}V_{\mu l}^{T} \\ & + U_{i\alpha}^{\dagger}U_{\beta j}^{*}\bar{v}_{\alpha\beta kl}U_{k\mu}U_{\nu l}^{T} - V_{i\alpha}^{\dagger}U_{\beta j}^{*}\bar{v}_{\alpha k\beta l}^{*}V_{l\nu}U_{\mu k}^{T} \end{aligned} \tag{3.73a}$$

$$\begin{aligned} B_{ij\mu\nu} = {} & - U_{i\alpha}^{\dagger}V_{\beta j}^{*}\bar{v}_{\alpha k\beta l}V_{l\nu}^{*}U_{\mu k}^{\dagger} + V_{i\alpha}^{\dagger}V_{\beta j}^{*}\bar{v}_{\alpha\beta kl}^{*}U_{k\mu}^{*}U_{\nu l}^{\dagger} \\ & - U_{i\alpha}^{\dagger}U_{\beta j}^{*}\bar{v}_{\alpha\beta kl}V_{k\nu}^{*}V_{\mu l}^{\dagger} + V_{i\alpha}^{\dagger}U_{\beta j}^{*}\bar{v}_{\alpha k\beta l}^{*}U_{l\mu}^{*}V_{\nu k}^{\dagger}. \end{aligned} \tag{3.73b}$$

The same QRPA matrix is also derived in section 5.4.3 in the context of the linear response theory following similar arguments.

Stability condition for effective pseudo-potential

Our derivation was based on a proper, density-independent potential $\hat{v}$. If the latter is an effective density-dependent pseudo-potential of the type $\hat{v} \equiv \hat{v}[\rho]$, additional terms have to be taken into account when expanding HFB matrix as a function of the perturbed density. The most general form of such an expansion involves writing, at first order in $\mathcal{R}$,

$$\mathcal{H} = \mathcal{H}^{(0)} + \mathcal{H}^{(1)} = \mathcal{H}^{(0)} + \frac{\delta\mathcal{H}}{\delta\mathcal{R}}\delta\mathcal{R} \equiv \mathcal{H}^{(0)} + \frac{\delta\mathcal{H}}{\delta\mathcal{R}}\mathcal{R}^{(1)}.$$

Therefore,

$$\mathcal{H}_{ij}^{(1)} = \sum_{a\leq b}\frac{\delta\mathcal{H}_{ij}}{\delta\mathcal{R}_{ab}}\mathcal{R}_{ab}^{(1)}.$$

This implies that

$$\mathcal{H}_{ij}^{(1)} = \sum_{a\leq b} \frac{\delta \mathcal{H}_{ij}}{\delta \rho_{ab}} \rho_{ab}^{(1)} + \sum_{a\leq b} \frac{\delta \mathcal{H}_{ij}}{\delta \kappa_{ab}} \kappa_{ab}^{(1)} + \sum_{a\leq b} \frac{\delta \mathcal{H}_{ij}}{\delta \kappa_{ab}^{*}} \kappa_{ab}^{(1)*} + \sum_{a\leq b} \frac{\delta \mathcal{H}_{ij}}{\delta \rho_{ab}^{*}} \rho_{ab}^{(1)*}.$$

One should also remember equations (3.39) and (3.40), which imply that

$$\mathcal{H} = \begin{pmatrix} \dfrac{\partial \mathcal{E}}{\partial \rho_{nm}} & \dfrac{\partial \mathcal{E}}{\partial \kappa_{mn}^{*}} \\ \dfrac{\partial \mathcal{E}}{\partial \kappa_{mn}} & \dfrac{\partial \mathcal{E}}{\partial \rho_{nm}^{*}} \end{pmatrix}.$$

Taking into account the properties of ρ (which also holds for $\rho^{(1)}$), we find after some elementary algebra,

$$\Gamma_{\alpha\beta}^{(1)} = \sum_{ab} \frac{\partial^2 \mathcal{E}}{\partial \rho_{\beta\alpha} \partial \rho_{ab}} \rho_{ab}^{(1)} + \frac{1}{2} \sum_{ab} \frac{\partial^2 \mathcal{E}}{\partial \rho_{\beta\alpha} \partial \kappa_{ab}} \kappa_{ab}^{(1)} + \frac{1}{2} \sum_{ab} \frac{\partial^2 \mathcal{E}}{\partial \rho_{\beta\alpha} \partial \kappa_{ab}^{*}} \kappa_{ab}^{(1)*}$$

$$\Delta_{\alpha\beta}^{(1)} = \sum_{ab} \frac{\partial^2 \mathcal{E}}{\partial \kappa_{\alpha\beta}^{*} \partial \rho_{ab}} \rho_{ab}^{(1)} + \frac{1}{2} \sum_{ab} \frac{\partial^2 \mathcal{E}}{\partial \kappa_{\alpha\beta}^{*} \partial \kappa_{ab}} \kappa_{ab}^{(1)}.$$

From these expressions, we verify that when the energy functional is built from a true potential, we recover

$$\Gamma_{\alpha\beta}^{(1)} = \sum_{ab} \frac{\partial^2 \mathcal{E}}{\partial \rho_{\beta\alpha} \partial \rho_{ab}} \rho_{ab}^{(1)} = \sum_{ab} \bar{v}_{ab\beta a} \rho_{ab}^{(1)}$$

$$\Delta_{\alpha\beta}^{(1)} = \frac{1}{2} \sum_{ab} \frac{\partial^2 \mathcal{E}}{\partial \kappa_{\alpha\beta}^{*} \partial \kappa_{ab}} \kappa_{ab}^{(1)} = \frac{1}{2} \sum_{ab} \bar{v}_{\alpha\beta ab} \kappa_{ab}^{(1)}.$$

3.1.5 Self-consistent symmetries

To discuss self-consistent symmetries, we will assume that we work with an effective pseudo-Hamiltonian characterized by a pseudo-potential $\hat{v}$. We consider symmetry operations $\hat{S}$ that can be either unitary, $\hat{S}\hat{S}^{\dagger} = 1$, or anti-unitary. Formally, the symmetry operation transforms s.p. states as

$$|i'\rangle = \hat{S}|i\rangle = \sum_{a} S_{ia} \, |i\rangle.$$

It is assumed that the effective pseudo-potential conserves whatever symmetry we are considering (signature, simplex, parity, time-reversal, etc).

Transformation of densities and fields
The one-body density matrix ρ_{ij} can be viewed as the matrix elements of a linear operator in the s.p. basis

$$\rho_{ij} = \langle i|\hat{\rho}|j\rangle.$$

The transformation rule of ρ_{ij} is therefore the same as that of a matrix, i.e.

$$\rho'_{ij} = \langle i'|\hat{\rho}|j'\rangle = \langle i|\hat{S}^{\dagger}\hat{\rho}\hat{S}|j\rangle,$$

and therefore, in matrix form,

$$\rho' = S^{\dagger}\rho S. \tag{3.74}$$

Transformation properties 'propagate' to the mean field. Indeed, the symmetry-transformed mean field is

$$\Gamma'_{nm} = \sum_{ij} \bar{v}'_{injm}\rho'_{ji},$$

with

$$\bar{v}'_{injm} = \langle i'n'|\bar{v}|j'm'\rangle = \sum_{abcd} S^{\dagger}_{ia}S^{\dagger}_{nb}\bar{v}_{abcd}S_{cj}S_{dm}$$

$$\rho'_{ji} = \sum_{kl} S^{\dagger}_{jk}\rho_{kl}S_{li}.$$

It is straightforward to show that the transformation rule of the mean field is therefore

$$\Gamma' = S^{\dagger}\Gamma S. \tag{3.75}$$

The anomalous density κ is *not* a one-body operator. Interpreting the HFB vacuum $|\Phi\rangle$ as a superposition of Slater determinants corresponding to even particle number +, $N-2$, N, $N+2$, …, and following the definition (3.15) of κ_{ij}, we can interpret it as a sum of overlaps between $(N-2)$- and N-particle states. It is therefore best expressed as

$$\kappa_{ij} = \langle ij|\kappa\rangle,$$

where $|\kappa\rangle$ is a (two-body) correlated state in Fock space. The transformation rule of κ_{ij} under a unitary transformation of the s.p. state is thus

$$\kappa'_{ij} = \langle ij|\hat{S}^{\dagger}\hat{S}^{\dagger}|\kappa\rangle.$$

Explicitly, we can introduce resolution of the identity for two-body states, $\sum_{kl}|kl\rangle\langle kl| = 1$, and we obtain

$$\kappa'_{ij} = \sum_{kl} S^{\dagger}_{ik}\kappa_{kl}S^{*}_{lj} \Rightarrow \kappa' = S^{\dagger}\kappa S^{*}. \tag{3.76}$$

Like the anomalous density the pairing field Δ cannot be interpreted as the matrix of some linear operator. From its definition, we obtain

$$\Delta_{mn} = \frac{1}{2}\sum_{ij} \bar{v}_{mnij}\kappa_{ij} = \frac{1}{2}\sum_{ij} \langle mn|\hat{v}|ij\rangle\langle ij|\kappa\rangle = \frac{1}{2}\langle mn|\hat{v}\kappa\rangle. \tag{3.77}$$

Since we have seen that $|\kappa\rangle$ is somewhat analogous to a two-body state, so is $\hat{v}|\kappa\rangle$. The transformation rule of the pairing field can thus be formally established by writing

$$\Delta'_{mn} = \frac{1}{2}\sum_{ij} \bar{v}'_{injm}\kappa'_{ij},$$

with $\kappa'_{ij} = \langle ij|\hat{S}^{\dagger}\hat{S}^{\dagger}\kappa\rangle$, where from it follows that

$$\Delta'_{mn} = \sum_{kl} S^{\dagger}_{mk}\Delta_{kl}S^{*}_{ln} \Rightarrow \Delta' = S^{\dagger}\Delta S^{*}. \tag{3.78}$$

The two-body density matrix, or two-particle density matrix [32, 33], can be defined in analogy with equation (3.13). We give here its expression in coordinate-space representation,

$$\gamma(x_1, x_2, x'_1, x'_2) = \frac{\langle\Phi|c^{\dagger}_{x'_2}c^{\dagger}_{x'_1}c_{x_2}c_{x_1}|\Phi\rangle}{\langle\Phi|\Phi\rangle},$$

where $|\Phi\rangle$ is the Bogoliubov vacuum. After applying the Wick theorem, we can obtain the following relations between the local, two-body density matrix, the one-body density matrix and the anomalous density:

$$\gamma(\mathbf{r}_1, \mathbf{r}_2) = \rho(\mathbf{r}_2, \mathbf{r}_1)\rho(\mathbf{r}_1, \mathbf{r}_2) - \rho(\mathbf{r}_1)\rho(\mathbf{r}_2) - \kappa^{*}(\mathbf{r}_1, \mathbf{r}_2)\kappa(\mathbf{r}_2, \mathbf{r}_1).$$

Since the two-body density matrix contains all two-body correlations in a multi-fermion system, we see that the anomalous density somehow describes those correlations that cannot be reduced to one-body types [32]. Note that these considerations only apply if $|\Phi\rangle$ is a product state.

Overall, the density matrix ρ_{ij}, mean field Γ_{ij} and therefore the s.p. Hamiltonian h_{ij} transform under a unitary transformation $\hat{S}$ like the matrix of linear operators. In contrast, the anomalous density κ_{ij} and pairing field Δ_{ij} transform differently.

Symmetries in the quasiparticle basis

For the matrix S of a given unitary transformation $\hat{S}$ in the s.p. basis, we can construct the matrix $\mathcal{S}$ of the same operator in the 'doubled' s.p. space. This matrix reads

$$\mathcal{S} = \begin{pmatrix} S & 0 \\ 0 & S^{*} \end{pmatrix}.$$

We note that the matrix of a unitary transformation in the double basis has a different structure than that of a Hermitian operator, see the remark in section 8.3.1 in the discussion of constrained calculations. The form of this matrix is needed to satisfy the various transformation rules established before for the mean field and pairing field. It is straightforward to show that the HFB matrix transforms as

$$\mathcal{H}' = \mathcal{S}^{\dagger}\mathcal{H}\mathcal{S}.$$

If S corresponds to a conserved symmetry, then $\mathcal{H}' = \mathcal{H}$, and therefore $\mathcal{H}$ and $\mathcal{S}$ commute. This implies that there exists a basis where both matrices are diagonal. Since U and V are eigenvectors of the HFB matrix, they must also be eigenvectors of $\mathcal{S}$,

$$\mathcal{S}\begin{pmatrix} U_\mu \\ V_\mu \end{pmatrix} = e^{i\varphi_\mu}\begin{pmatrix} U_\mu \\ V_\mu \end{pmatrix}.$$

The form $e^{i\varphi_\mu}$ of the eigenvalue comes from the unitarity of the symmetry operation. Gathering all eigenvectors in a matrix, we obtain

$$\begin{aligned} SU &= e^{i\varphi}U, \\ S^{*}V &= e^{i\varphi}V. \end{aligned}$$

This result is important, because it implies that the matrices of U and V do not behave like actual matrices under a symmetry operation, but rather like vectors. In other words, one cannot associate them with some linear operator, the transformation rules of which would define a certain block structure in a certain basis.

3.2 Multi-reference implementation of nuclear energy density functionals

For finite-size systems quantum fluctuations are large. Put differently, when $\hat{H}_0(\boldsymbol{g})$ breaks a symmetry of $\hat{H}$ in the partitioning (1.15), the component of $\hat{\mathcal{V}}_{\text{res}}$ that compensates such a symmetry-violation cannot be neglected. It is then necessary to perform a subsequent post-mean-field step that restores the broken symmetries and hence the linearity of the many-body Schrödinger equation. This is the MR-EDF implementation of nuclear EDF, where the many-body state reads as equation (1.14) and obeys all the original symmetries of the Hamiltonian. More precisely, for each value of the amplitude g of the order parameter, the degenerate modes with different phases φ_g are mixed with coefficients related to the structure of the symmetry group. We will briefly outline some important aspects of the MR-EDF in this section. Let us stress that the effects of the mean-field symmetry-breaking do survive the symmetry restoration and still manifest themselves in the ground-state properties and in the excitation spectra. In addition, for systems easily deformable along a collective coordinate g, another contribution of the residual interaction becomes non-negligible, the one responsible for collective mean-field oscillations. Asking for the total energy of the system to be minimal for states of the form (1.14) yields the Hill–Wheeler–Griffin equations to be solved in order to obtain the coefficients entering in the horizontal expansion.

The multi-reference implementation of the energy density functional method is usually formulated in the context of what is known as the GCM. It provides a very general framework to treat correlations through the mixing of mean-field states (i.e. the generator states). It has been developed to achieve what we dubbed above as the horizontal expansion. However, techniques pertaining to vertical expansions such as the random-phase approximation (RPA) or QRPA, second RPA, etc, discussed in chapter 5, can, at least formally, also be expressed as special cases of the GCM.

3.2.1 General formalism

The starting point for MR-EDF calculations is a set of, in general, non-orthogonal A-body wave functions $|\Phi(\boldsymbol{q})\rangle$ that depend on a set of coordinates $\boldsymbol{q} \equiv (q_1, \dots, q_N)$. These might be collective shape degrees of freedom, labels of single-particle excitations, the rotation angles associated with symmetry restoration, or a combination of all of the above. The GCM wave functions $|\Psi_\mu\rangle$ that can be constructed in that basis are written as

$$|\Psi_\mu\rangle = \int d^N\boldsymbol{q}\, f_\mu(\boldsymbol{q})|\Phi(\boldsymbol{q})\rangle, \tag{3.79}$$

where $d^N\boldsymbol{q} = dq_1 \times \dots \times dq_N$ is the volume element in the N-dimensional collective space. In practice, the states $|\Phi(\boldsymbol{q})\rangle$ are (necessarily) non-orthogonal, which has to be taken care of.

The notation that we use to denote the vector of collective variables might erroneously suggest that $\boldsymbol{q}$ refers to multipole moments. As discussed in section 3.1.1, multipole moments correspond to the amplitude $||g||$ of the order parameter characterizing the loss of rotational invariance. The formalism presented here is more general: $\boldsymbol{q}$ can also refer to the phase of the order parameter, e.g. the Euler rotation angles or the Gauge angle for particle number.

Variational principle

In equation (3.79), the weight functions $f_\mu(\boldsymbol{q})$ are the unknowns. In the discretized version of the GCM ansatz, (3.82), it is the values of the weight functions at the lattice points q_k that are the unknowns. They can be determined by solving a generalized eigenvalue problem obtained by requiring that the energy be stationary under arbitrary infinitesimal variations of $f_\mu(\boldsymbol{q})$. If the energy is calculated from an effective pseudo-Hamiltonian, that is,

$$E_\mu = \frac{\langle\Psi_\mu|\hat{H}|\Psi_\mu\rangle}{\langle\Psi_\mu|\Psi_\mu\rangle},$$

the variational principle reads

$$\delta E_\mu = \frac{\langle \delta\Psi_\mu |(\hat{H}-E_\mu)|\Psi_\mu\rangle}{\langle \Psi_\mu |\Psi_\mu\rangle} + \frac{\langle \Psi_\mu |(\hat{H}-E_\mu)|\delta\Psi_\mu\rangle}{\langle \Psi_\mu |\Psi_\mu\rangle},$$

where

$$\delta|\Psi_\mu\rangle = \frac{\delta}{\delta f_\mu}|\psi_\mu\rangle\, \delta f_\mu = \delta f_\mu \int d\boldsymbol{q}\, |\Phi_\mu(\boldsymbol{q})\rangle.$$

Variations of the 'bra', $\delta\langle\Psi_\mu|$, take a similar form but involve the complex conjugate of the function $f_\mu(\boldsymbol{q})$. One can easily show that the variational principle leads to the generalized eigenvalue problem [34]

$$\int d\boldsymbol{q}\big[\mathcal{H}(\boldsymbol{q}', \boldsymbol{q}) - E_\mu \mathcal{N}(\boldsymbol{q}', \boldsymbol{q})\big] f_\mu(\boldsymbol{q}) = 0, \tag{3.80}$$

where E_μ are the energies of the collective states corresponding to the weight functions $f_\mu(\boldsymbol{q})$. This is known in the literature as the Hill–Wheeler–Griffin equation. The integrand is in principle a function of $\boldsymbol{q}'$; the variational principle states that this function must be identically zero. In order to solve the Hill–Wheeler–Griffin equation, one has to calculate the norm kernel and energy kernel

$$\mathcal{N}(\boldsymbol{q}, \boldsymbol{q}') = \langle \Phi(\boldsymbol{q})|\Phi(\boldsymbol{q}')\rangle \tag{3.81a}$$

$$\mathcal{H}(\boldsymbol{q}, \boldsymbol{q}') = \langle \Phi(\boldsymbol{q})|\hat{H}|\Phi(\boldsymbol{q}')\rangle. \tag{3.81b}$$

Note that the time-odd terms in the EDF almost always contribute to the energy kernel, even when the states $|\Phi(\boldsymbol{q})\rangle$ and $|\Phi(\boldsymbol{q}')\rangle$ are both time-reversal invariant [35].

Discretization of the collective variables

In an EDF framework, the reference states $|\Phi(\boldsymbol{q})\rangle$ entering equation (3.79) are the solution of the self-consistent HFB equation, for which in general an analytical expression as a function of the generating coordinates $\boldsymbol{q}$ does not exist. In order to numerically solve the general Hill–Wheeler–Griffin equation in such a basis, the integral is discretized and we are thus led to consider the following ansatz for $|\Psi_\mu\rangle$:

$$|\Psi_\mu\rangle = \sum_k f_\mu(\boldsymbol{q}_k)\, |\Phi(\boldsymbol{q}_k)\rangle \tag{3.82}$$

Recall that in the general case considered here, $\boldsymbol{q}$ is a vector that contains more than one variable. Each of these takes values on some discrete mesh, $q_i = q_{i,1}, \dots, q_{i,n_i}$. Even in this general case, the Hill–Wheeler–Griffin equation can be cast into a generalized eigenvalue problem, since one stacks the vectors (on the mesh) of each collective variable one after the other to form a large vector of length $N \times \prod_{i=1}^{N} n_i$.

Starting from equation (3.80), we easily find the discretized Hill–Wheeler–Griffin equation

$$\sum_{l}\left[\mathcal{H}_{kl}-E_{\mu}\,\mathcal{N}_{kl}\right]f_{l,\mu}=0\quad\forall k,\tag{3.83}$$

where we have noted $f_{l,\mu}\equiv f_{\mu}(q_l)$, $\mathcal{H}_{kl}\equiv\mathcal{H}(q_k,\,q_l)$ and $\mathcal{N}_{kl}\equiv\mathcal{N}(q_k,\,q_l)$. The discretized weight functions satisfy the generalized condition of orthonormality

$$\sum_{kl}f_{k,\mu}\,\mathcal{N}_{kl}f_{l,\nu}=\delta_{\mu\nu}$$

and consequently cannot be interpreted as wave functions. Having solved equation (3.83), the expectation values and transition matrix elements of all operators can be simply calculated from the matrix elements between the reference states as

$$\langle\Psi_{\mu}\,|\hat{O}|\Psi_{\nu}\rangle=\sum_{kl}f_{k,\mu}\,\langle\Phi(\boldsymbol{q}_k)|\hat{O}|\Phi(\boldsymbol{q}_l)\rangle f_{l,\nu}\,.\tag{3.84}$$

Collective Schrödinger equation

Provided there exist matrices $\mathcal{N}_{kl}^{1/2}\equiv(\mathcal{N}^{1/2})_{kl}$ and $\mathcal{N}_{kl}^{-1/2}\equiv(\mathcal{N}^{-1/2})_{kl}$ (for the construction of these matrices see equation (3.96) below) that verify the properties

$$\sum_{m}\mathcal{N}_{km}^{1/2}\,\mathcal{N}_{ml}^{1/2}=\mathcal{N}_{kl}\tag{3.85a}$$

$$\sum_{m}\mathcal{N}_{km}^{-1/2}\mathcal{N}_{ml}^{1/2}=\delta_{kl}\tag{3.85b}$$

$$\sum_{mn}\mathcal{N}_{km}^{-1/2}\mathcal{N}_{mn}\mathcal{N}_{nl}^{-1/2}=\delta_{kl},\tag{3.85c}$$

the generalized eigenvalue problem (3.83) can be formally transformed into a standard eigenvalue problem [35]

$$\sum_{l}\tilde{\mathcal{H}}_{kl}g_{l,\mu}=E_{\mu}g_{k,\mu},\tag{3.86}$$

with the transformed Hamiltonian kernel

$$\tilde{\mathcal{H}}_{kl}=\sum_{mn}\mathcal{N}_{km}^{-1/2}\mathcal{H}_{mn}\mathcal{N}_{nl}^{-1/2}\tag{3.87}$$

and corresponding orthonormal collective wave functions

$$g_{k,\mu}=\sum_{l}\mathcal{N}_{kl}^{1/2}f_{l,\mu},\tag{3.88}$$

that are more useful for the interpretation of the outcome of a GCM calculation than the weight functions $f_{l,\mu}$. It has to be noted, however, that the square of a collective wave function $g_{k,\mu}$ does not represent the probability of finding the reference state $|\Phi(\boldsymbol{q}_l)\rangle$ in the GCM state $|\Psi_\mu\rangle$. The reason is that the reference states $|\Phi(\boldsymbol{q}_l)\rangle$ are non-orthogonal, such that each of them is also contained in all others.

Having solved equation (3.86), all observables can then be calculated in terms of the collective wave functions from the transformed operator kernels

$$\tilde{O}_{kl} = \sum_{mn} \mathcal{N}^{-1/2}_{mk} \langle \Phi_m | \hat{O} | \Phi_n \rangle \mathcal{N}^{-1/2}_{nl} \tag{3.89}$$

as

$$\langle \Psi_\mu | \hat{O} | \Psi_{\mu'} \rangle = \sum_{kl} g_{k,\mu} \tilde{O}_{kl} g_{l,\mu'}.$$

Elimination of redundant states

The existence of the matrix $\mathcal{N}^{-1/2}$ in equation (3.87) requires that the norm kernel matrix $\mathcal{N}$ does not have zero eigenvalues. The presence of the latter signals that the set of reference states $|\Phi(\boldsymbol{q}_l)\rangle$ is over-complete, containing redundant states that can be expressed as a linear combination of other reference states. The likelihood of having redundant states in a GCM calculation increases with the density of discretization points $\boldsymbol{q}_l$. Redundant states can also appear systematically in the projection of states with some residual symmetries [36, 37]. These redundant states can be removed without altering observables. In fact, the numerical solution of the discretized generalized eigenvalue problem (3.83) requires that they are eliminated. In order to do so, the norm kernel matrix has to be diagonalized first

$$\sum_l \mathcal{N}_{kl} u_{l,\nu} = \lambda_\nu u_{k,\nu}.$$

As the norm kernel $\mathcal{N}_{kl}$ is a Hermitian, positive, semi-definite matrix, its eigenvalues are real and larger than, or equal to, zero, $\lambda_\nu = \lambda_\nu^* \geqslant 0$. Its eigenstates form an orthonormal basis

$$\sum_k u_{k,\nu} u_{k,\nu'} = \delta_{\nu\nu'} \qquad \sum_\mu u_{k,\nu} u_{l,\nu} = \delta_{kl},$$

Limiting oneself to the eigenstates u_ν with non-zero eigenvalues, it is now possible to construct (rectangular) matrices

$$\bar{\mathcal{N}}^{1/2}_{k\nu} = \sqrt{\lambda_\nu} u_{k,\nu}$$
$$\bar{\mathcal{N}}^{-1/2}_{\nu k} = \frac{1}{\sqrt{\lambda_\nu}} u_{k,\nu},$$

with the properties

$$\sum_{k} \bar{\mathcal{N}}^{1/2}_{k\nu} \bar{\mathcal{N}}^{1/2}_{k\nu'} = \lambda_\nu \, \delta_{\nu\nu'} \tag{3.90a}$$

$$\sum_{\nu} \bar{\mathcal{N}}^{1/2}_{k\nu} \bar{\mathcal{N}}^{1/2}_{l\nu} = \mathcal{N}_{kl} \tag{3.90b}$$

$$\sum_{\nu} \bar{\mathcal{N}}^{-1/2}_{\nu k} \bar{\mathcal{N}}^{1/2}_{l\nu} = \delta_{kl} \tag{3.90c}$$

$$\sum_{kl} \bar{\mathcal{N}}^{-1/2}_{\nu k} \mathcal{N}_{kl} \bar{\mathcal{N}}^{-1/2}_{\nu' l} = \delta_{\nu\nu'}. \tag{3.90d}$$

These matrices differ from $\mathcal{N}^{1/2}_{kl}$ and $\mathcal{N}^{1/2}_{kl}$ introduced above (3.90). These connect the basis of the reference states $|\Phi(\boldsymbol{q}_k)\rangle$ with the basis of norm eigenstates u_ν. With these matrices, the generalized eigenvalue problem (3.83) can again be transformed into a standard eigenvalue problem, now expressed in the (possibly smaller) basis of eigenstates of the norm kernel with non-zero eigenvalue

$$\sum_{\nu'} \bar{\mathcal{H}}_{\nu\nu'} \bar{g}_{\nu',\mu} = E_\mu \bar{g}_{\nu,\mu}, \tag{3.91}$$

with the transformed Hamiltonian kernel

$$\bar{\mathcal{H}}_{\nu\nu'} = \sum_{k} \bar{\mathcal{N}}^{-1/2}_{\nu k} \mathcal{H}_{kl} \bar{\mathcal{N}}^{-1/2}_{\nu' l} \tag{3.92}$$

and collective wave functions

$$\bar{g}_{\nu,\mu} = \sum_{k} \bar{\mathcal{N}}^{1/2}_{\nu k} f_{k,\mu}\,. \tag{3.93}$$

In terms of the collective wave functions, all observables can again be calculated from transformed operator kernels

$$\bar{\mathcal{O}}_{\nu\nu'} = \sum_{kl} \bar{\mathcal{N}}^{-1/2}_{k\nu} \langle \Phi_k \,| \hat{O} | \Phi_l \rangle \bar{\mathcal{N}}^{-1/2}_{l\nu'} \tag{3.94}$$

and transformed collective wave functions as

$$\langle \Psi_\mu \,| \hat{O} | \Psi_{\mu'} \rangle = \sum_{\nu\nu'} \bar{g}_{\nu,\mu} \, \bar{\mathcal{O}}_{\nu\nu'} \bar{g}_{\nu',\mu'}. \tag{3.95}$$

Alternatively, one can invert relation (3.93) and calculate observables from the resulting $f_{k,\mu}$ through equation (3.84).

Because of numerical inaccuracies that accumulate when calculating the norm kernel matrix and diagonalizing it, the redundant states will not be eigenstates of the norm kernel with an eigenvalue that is *exactly* zero, but a small number that is possibly even negative. As a consequence, in order to eliminate redundant states one has to introduce a cutoff, below which all eigenstates of the norm are removed, at the price of equations

(3.91), (3.93) and (3.94) becoming approximations and the possibility of unintentionally cutting physical states that have a small norm eigenvalue. From a numerical point of view, it is often advantageous to remove a reference state $|\Phi(\boldsymbol{q}_k)\rangle$ that only adds a near-zero norm eigenvalue and which does not significantly contribute to observables of low-lying states from the entire GCM calculation, instead of keeping it and eliminating the corresponding norm eigenstate. A careful and detailed analysis of the problem of choosing a cutoff in the norm eigenspace in a GCM for a model system that can also be solved by direct diagonalization is given in [38].

If all eigenvalues of the norm kernel matrix are non-zero, then one can use its eigenstates to construct directly the matrices $\mathcal{N}^{1/2}$ and $\mathcal{N}^{-1/2}$ of equations (3.85) in the basis of the original reference states

$$\mathcal{N}^{1/2}_{km} = \sum_{\nu} \sqrt{\lambda_\nu}\, u_{k,\nu} u_{m,\nu} \tag{3.96a}$$

$$\mathcal{N}^{-1/2}_{km} = \sum_{\nu} \frac{1}{\sqrt{\lambda_\nu}} u_{k,\nu} u_{m,\nu} \tag{3.96b}$$

and solve the Hill–Wheeler–Griffin equation (3.86) directly in this basis.

Real generator coordinates

In most, if not all, existing applications of the GCM to configuration mixing of stationary states that use an MR-EDF to calculate the Hamiltonian kernel, the reference states $|\Phi(\boldsymbol{q}_k)\rangle$ are chosen such that the norm and Hamiltonian kernels are real, which implies that the relative phases between the reference states $|\Phi(\boldsymbol{q}_k)\rangle$, and between the reference states and the GCM states $|\Psi_\mu\rangle$, are real. In that case, the norm kernel $\mathcal{N}_{kl}$ can be diagonalized with a (real) orthogonal transformation $\mathcal{Q}^T \mathcal{N} \mathcal{Q}$, such that the eigenstates of the norm $u_{k,\nu} = \mathcal{Q}_{k\nu}$ are real, as are $\bar{\mathcal{N}}^{-1/2}$, $\bar{\mathcal{N}}^{1/2}$, the weight functions $f_\mu(\boldsymbol{q}_k)$ and the collective wave functions $g_\mu(\boldsymbol{q}_k)$. This assumption has been made throughout the above presentation of the formalism.

However, in applications of MR-EDF to restore continuous symmetries, which we will discuss later, one necessarily has to work with complex weight functions, norm and energy kernels. Fortunately, the complex weight functions are determined in this case by the properties of the symmetry group such that the procedure described above does not have to be extended to complex kernels. In addition, when mixing projected reference states in a subsequent GCM, then the conditions discussed above with real kernels (between projected reference states) are recovered.

In contrast, in the time-dependent extension of the GCM [39, 40], $f_\mu(\boldsymbol{q}_k)$ cannot be chosen to be real anymore. The same also holds for a GCM where also the conjugate momenta of the $\boldsymbol{q}_k$ are used as additional generator coordinates [41]. To the best of our knowledge, such calculations have not yet been carried out in an MR-EDF approach.

Collective space

The GCM states (3.79) are entirely defined by the functions $f_\mu(\boldsymbol{q})$ acting on the collective coordinates. The idea behind the concept of a collective space is to map the complete Hilbert space into the space of functions acting on $\boldsymbol{q}$. Moreover, we would like to translate the action of an arbitrary operator $\hat{F}$ on $|\Psi_\mu\rangle$ into the action of a collective operator $\hat{\mathcal{F}}$ on the collective wave functions $g_\mu(\boldsymbol{q})$ associated with $|\Psi_\mu\rangle$. We will show here how to build an example of such a collective mapping. This section generalizes the notions already introduced in section 3.2.1 to the case of continuous GCM, that is, when the value of the collective variables are not discretized. Let us denote by $\mathcal{F}$ the collective kernel of an operator $\hat{F}$ of the Hilbert space defined as

$$\mathcal{F}(\boldsymbol{q}, \boldsymbol{q}') = \langle \phi_{\boldsymbol{q}} | \hat{F} | \phi_{\boldsymbol{q}'} \rangle.$$

The norm kernel $\mathcal{N}$ and the energy kernel $\mathcal{H}$, which play a major role in the derivation of the collective equation of motion, are particular examples of such kernels for the identity and Hamiltonian operators, respectively. Let us recall here the expressions (3.81)

$$\mathcal{N}(\boldsymbol{q}, \boldsymbol{q}') = \langle \phi_{\boldsymbol{q}} | \phi_{\boldsymbol{q}'} \rangle \tag{3.97a}$$

$$\mathcal{H}(\boldsymbol{q}, \boldsymbol{q}') = \langle \phi_{\boldsymbol{q}} | \hat{H} | \phi_{\boldsymbol{q}'} \rangle. \tag{3.97b}$$

Given an arbitrary kernel $\mathcal{F}$ we can define a linear operator (denoted $\hat{\mathcal{F}}$) acting on the collective functions $g_\mu(\boldsymbol{q})$ as

$$(\hat{\mathcal{F}} g_\mu)(\boldsymbol{q}) = \int_{\boldsymbol{q}'} \mathcal{F}(\boldsymbol{q}, \boldsymbol{q}')\, g_\mu(\boldsymbol{q}') d\boldsymbol{q}'.$$

The product of two operators $\hat{\mathcal{F}}$ and $\hat{\mathcal{G}}$ corresponds to the composition of $\hat{\mathcal{F}}$ and $\hat{\mathcal{G}}$,

$$(\hat{\mathcal{F}}\hat{\mathcal{G}} g_\mu)(\boldsymbol{q}) = \int_{\boldsymbol{q}''} \int_{\boldsymbol{q}'} \mathcal{F}(\boldsymbol{q}, \boldsymbol{q}'')\, \mathcal{G}(\boldsymbol{q}'', \boldsymbol{q}')\, g_\mu(\boldsymbol{q}') d\boldsymbol{q}' d\boldsymbol{q}''. \tag{3.98}$$

The composition rule (3.98) allows us to introduce the square root of the operator $\hat{\mathcal{F}}$, which we denote by $\hat{\mathcal{F}}^{1/2}$ (if it exists). One needs to be careful here: while $\mathcal{F}^{1/2}(\boldsymbol{q}, \boldsymbol{q}'')$ truly corresponds to taking the square root of the value of the scalar function $\mathcal{F}(\boldsymbol{q}, \boldsymbol{q}'')$, this is not the case for $\hat{\mathcal{F}}^{1/2}$, which is merely a notation. The square root of the operator is defined through the property

$$\begin{aligned}(\hat{\mathcal{F}} g_\mu)(\boldsymbol{q}) &= \left(\hat{\mathcal{F}}^{1/2}\hat{\mathcal{F}}^{1/2} g_\mu\right)(\boldsymbol{q}) \\ &= \int_{\boldsymbol{q}''} \int_{\boldsymbol{q}'} \mathcal{F}^{1/2}(\boldsymbol{q}, \boldsymbol{q}'')\, \mathcal{F}^{1/2}(\boldsymbol{q}'', \boldsymbol{q}')\, g_\mu(\boldsymbol{q}') d\boldsymbol{q}'\, d\boldsymbol{q}''.\end{aligned}$$

This gives us an important equality

$$\int_{q'} \mathcal{F}(q, q')g_\mu(q')dq' = \int_{q''}\int_{q'} \mathcal{F}^{1/2}(q, q'')\, \mathcal{F}^{1/2}(q'', q')\, g_\mu(q')dq'\, dq''. \quad (3.99)$$

With the same composition rule (3.98), we can introduce the operator $\hat{\mathcal{F}}^{-1}$, the inverse (if it exists) of $\hat{\mathcal{F}}$. Denoting by $\hat{\mathcal{I}}$ the identity operator in the collective space, we have

$$\left(\hat{\mathcal{I}}g_\mu\right)(q) = \left(\hat{\mathcal{F}}^{-1}\hat{\mathcal{F}}g_\mu\right)(q) = \left(\hat{\mathcal{F}}\hat{\mathcal{F}}^{-1}g_\mu\right)(q) = g_\mu(q).$$

We complete the construction of our collective space by defining a scalar product (.|.) acting on collective functions through the relation

$$(g_\mu\,|h_\nu) = \int_q g^*(q)_\mu h_\nu(q)dq. \quad (3.100)$$

Here, the expression (.|.) is used for the scalar product in the collective space. The same notation is sometimes used to denote the regular, non anti-symmetrized, scalar product in Fock space. Given the GCM ansatz (3.79) for the many-body wave function, the collective wave functions $g_\mu(q)$ associated with the GCM state are the functions $g_\mu(q) = (\hat{\mathcal{N}}^{1/2}f_\mu)(q)$, that is,

$$g_\mu(q) = \int_{q'} \mathcal{N}^{1/2}(q, q')f_\mu(q')dq'. \quad (3.101)$$

These collective wave functions are built in order to mimic the wave function of a system whose dynamical variables would be the collective variables. This definition, along with the scalar product (3.100) implies a set of intuitive properties in the collective space. First, we note that if $|\Psi_\mu\rangle$ is a normalized GCM state, then g_μ are orthonormal according to

$$(g_\mu\,|g_\nu) = \int_q g_\mu(q)g_\nu^*(q)\,dq = \delta_{\mu\nu}.$$

If the GCM states are orthonormalized, then $\langle\Psi_\mu|\Psi_\nu\rangle = \delta_{\mu\nu}$. Inserting an ansatz (3.79) into the GCM yields

$$\langle\Psi_\mu\,|\Psi_\nu\rangle = \int dq f_\mu^*(q) \int dq' f_\nu(q')\mathcal{N}(q, q').$$

We then insert equation (3.99) to replace the second integral over q', and after noting that $\mathcal{N}(q, q') = \mathcal{N}^*(q', q)$, this gives us the orthonormality of the collective wave functions.

The key feature of the collective space, however, is that for any operator $\hat{F}$ in the Hilbert space we can associate a collective operator $\hat{\mathcal{F}}_{\text{coll}}$ acting in the collective subspace and *having the same matrix elements*. This collective operator is defined as

$$\hat{\mathcal{F}}_{\text{coll}} = \hat{\mathcal{N}}^{-1/2}\hat{\mathcal{F}}\hat{\mathcal{N}}^{-1/2}$$

and yields the equality

$$\forall(\mu\nu): \quad \langle\Psi_\mu|\hat{F}|\Psi_\nu\rangle = (g_\mu|\hat{\mathcal{F}}_{\text{coll}}|g_\nu), \tag{3.102}$$

where g_μ and g_ν are the collective wave functions associated with $|\Psi_\mu\rangle$ and $|\Phi_\nu\rangle$, respectively.

Note that the existence of the operator $\hat{\mathcal{N}}^{-1/2}$ is not guaranteed. This operator verifies by definition $\hat{\mathcal{N}}^{-1/2}\hat{\mathcal{N}}^{1/2} = \hat{\mathcal{N}}^{1/2}\hat{\mathcal{N}}^{-1/2} = \hat{\mathcal{I}}$, where $\hat{\mathcal{I}}$ is the identity operator in the collective space and $\hat{\mathcal{N}}^{1/2}$ is the notation for the operator that verifies $\hat{\mathcal{N}}^{1/2}\hat{\mathcal{N}}^{1/2} = \hat{\mathcal{N}}$.

Scope of the MR-EDF approach

The MR-EDF formalism can be used to achieve several goals

1. First, it can be used to restore broken symmetries, which provides correlation energies, a spectrum of excited states and transition matrix elements between them that respect the selection rules of the corresponding symmetry. In this case, the set of states $|\Phi(\boldsymbol{q})\rangle$ is constructed by acting with the elements of the corresponding symmetry group on a mean-field state; see the discussion in section 3.1.1. As a reminder, these symmetry groups are the group of spatial rotations for angular-momentum projection, global $U(1)$ gauge rotation for particle-number projection and space inversion for parity projection. For Abelian groups such as global gauge symmetry and parity, the weight functions $f_k(\boldsymbol{q})$ are already entirely determined by the matrix representation of the group. For non-Abelian groups such as rotation, however, $f_k(\boldsymbol{q})$ are not entirely determined by the group such that a generalized eigenvalue problem (3.80) has to be solved in order to orthogonalize irreps with the same restored quantum number [37].
2. Second, the ansatz (3.79) can be used to describe configuration mixing within a set of states $|\Phi(\boldsymbol{q})\rangle$ with different shapes obtained from constrained SR-EDF calculations, where the value of the constraints serve as generator coordinates $\boldsymbol{q}$. Physics cases of interest are shape vibrations in soft nuclei and the study of shape coexistence. As mean-field states with different deformations are in general non-orthogonal, the states corresponding to different minima in the potential energy surface of a nucleus should be mixed in order to compare to experiments. Typical shape degrees of freedom considered as generator coordinates in such studies are the axial or full triaxial quadrupole moment [42–46] and the octupole moment [44, 47–51], but others such as the hexadecapole moment, the amplitude of pairing correlations [52–55], or the intrinsic angular momentum as obtained with a cranking constraint can be used as well [56].
3. Third, the ansatz (3.79) can also be used to mix states that differ in some s.p. degrees of freedom, for example, blocked configurations in odd nuclei [37, 57]. Like states with different deformation, different blocked HFB states

with the same conserved quantum numbers are in general not orthogonal, such that they should be mixed before comparing them to experiments.

All of the above can of course be combined. We will come back to that below.

3.2.2 Calculation of overlaps and kernels

The GCM kernels of an arbitrary operator between two different, but non-orthogonal q.p. vacua $|\Phi(\boldsymbol{q})\rangle$ and $|\Phi(\boldsymbol{q}')\rangle$ can be expressed with the help of the generalized Wick theorem [58]. As the name indicates, this theorem generalizes the Wick theorem presented in section 3.1.2 to the calculation of transition matrix elements between two different product states. For a one-body operator $\hat{O}$ expressed in some s.p. basis $(c, c^\dagger)$, one obtains

$$\begin{aligned}\langle\Phi(\boldsymbol{q}')|\hat{O}|\Phi(\boldsymbol{q})\rangle &= \sum_{ij} O_{ji}\frac{\langle\Phi(\boldsymbol{q}')|c_j^\dagger c_i\,|\Phi(\boldsymbol{q})\rangle}{\langle\Phi(\boldsymbol{q}')|\Phi(\boldsymbol{q})\rangle\rangle}\langle\Phi(\boldsymbol{q}')|\Phi(\boldsymbol{q})\rangle \\ &= \sum_{ij} O_{ji}\,\rho_{ij}^{q'q}\,\langle\Phi(\boldsymbol{q}')|\Phi(\boldsymbol{q})\rangle,\end{aligned}$$

with $\rho_{ij}^{q'q}$ being the transition density matrix defined above in equation (1.17). The kernel of a two-body operator $\hat{V}$ involves all the contractions; one can simply use equation (3.11) with the bra and ket being $\langle\Phi(\boldsymbol{q}')|$ and $|\Phi(\boldsymbol{q})\rangle$,

$$\begin{aligned}\langle\Phi(\boldsymbol{q}')|\hat{V}|\Phi(\boldsymbol{q})\rangle &= \sum_{ijmn}\bar{v}_{ijnm}\frac{\langle\Phi(\boldsymbol{q}')|c_i^\dagger c_j^\dagger c_m c_n\,|\Phi(\boldsymbol{q})\rangle}{\langle\Phi(\boldsymbol{q}')|\Phi(\boldsymbol{q})\rangle}\langle\Phi(\boldsymbol{q}')|\Phi(\boldsymbol{q})\rangle \\ &= \sum_{ijmn}\bar{v}_{ijnm}\Big[\rho_{ni}^{q'q}\,\rho_{mj}^{q'q}-\rho_{mi}^{q'q}\,\rho_{nj}^{q'q}+\kappa_{ij}^{qq'*}\,\kappa_{nm}^{q'q}\Big]\langle\Phi(\boldsymbol{q}')|\Phi(\boldsymbol{q})\rangle.\end{aligned}$$

Compared to a one-body operator, one also needs the two anomalous transition density matrices $\kappa_{ij}^{qq'*}$ and $\kappa_{nm}^{q'q}$ defined in equation (1.17), which in general are independent of one another. Similar expressions hold for three-body operators, etc. In practice, the only difference to the standard Wick theorem is that the usual normal and anomalous density matrix elements are replaced by the homologous transition density matrices, and that the entire expression is then multiplied once by the overlap. From the formulas above, it should become clear that the two main ingredients of a GCM calculation are (i) the norm kernel $\mathcal{N}(\boldsymbol{q}, \boldsymbol{q}')$ and (ii) the transition densities (1.17). Once these quantities are known, calculating the ingredients of the Hill–Wheeler–Griffin equation becomes straightforward.

Calculation of the norm kernel

The two Bogoliubov q.p. states $|\Phi(\boldsymbol{q})\rangle$ and $|\Phi(\boldsymbol{q}')\rangle$ involved in the norm kernel are each defined through a set of q.p. operators

$$\alpha_i\,|\Phi(\boldsymbol{q}')\rangle = 0 \qquad \beta_j\,|\Phi(\boldsymbol{q})\rangle = 0.$$

When expressed in a common basis, the two sets of q.p. operators are connected by a Bogoliubov transformation that can be written as

$$\begin{pmatrix} \alpha \\ \alpha^\dagger \end{pmatrix} = \begin{pmatrix} (D^{-1})^* & -E \\ -E^* & D^{-1} \end{pmatrix} \begin{pmatrix} \beta \\ \beta^\dagger \end{pmatrix}. \tag{3.103}$$

The fact that the transformation between the two q.p. sets of operators is a Bogoliubov transformation implies, among other things, that the matrix of the transformation is unitary. It has also the same formal block structure; compare with the Hermitian conjugate of the matrix in equation (3.6). It has been shown [58] that the only non-vanishing contractions between both sets of q.p. operators are

$$\frac{\langle \Phi(\boldsymbol{q}')|\alpha_i \alpha_j^\dagger|\Phi(\boldsymbol{q})\rangle}{\langle \Phi(\boldsymbol{q}')|\Phi(\boldsymbol{q})\rangle} = \delta_{ij} \tag{3.104a}$$

$$\frac{\langle \Phi(\boldsymbol{q}')|\alpha_i \alpha_j \,|\Phi(\boldsymbol{q})\rangle}{\langle \Phi(\boldsymbol{q}')|\Phi(\boldsymbol{q})\rangle} = (ED)_{ij} \tag{3.104b}$$

$$\frac{\langle \Phi(\boldsymbol{q}')|\beta_i \beta_j^\dagger|\Phi(\boldsymbol{q})\rangle}{\langle \Phi(\boldsymbol{q}')|\Phi(\boldsymbol{q})\rangle} = \delta_{ij} \tag{3.104c}$$

$$\frac{\langle \Phi(\boldsymbol{q}')|\beta_i^\dagger \beta_j^\dagger|\Phi(\boldsymbol{q})\rangle}{\langle \Phi(\boldsymbol{q}')|\Phi(\boldsymbol{q})\rangle} = (DE^*)_{ij} \tag{3.104d}$$

$$\frac{\langle \Phi(\boldsymbol{q}')|\alpha_i \beta_j^\dagger|\Phi(\boldsymbol{q})\rangle}{\langle \Phi(\boldsymbol{q}')|\Phi(\boldsymbol{q})\rangle} = (D^T)_{ij}, \tag{3.104e}$$

while all other possible contractions between the α, $\alpha^\dagger$, β and $\beta^\dagger$ are zero.

Using the above notation, the expression for the norm kernel, which was first given by Onishi [59] reads

$$\mathcal{N}(\boldsymbol{q}, \boldsymbol{q}') = \langle \Phi(\boldsymbol{q}')|\Phi(\boldsymbol{q})\rangle = \sqrt{\det D^{-1}}. \tag{3.105}$$

This leaves its sign undetermined. The overlap $\mathcal{N}(\boldsymbol{q}, \boldsymbol{q}')$ is in general complex, in which case the Onishi formula fixes the overlap up to a phase of π. In some simple GCM calculations [60], the matrix D in equation (3.103) might remain block diagonal with pairwise identical blocks, such that $\det D^{-1}$ becomes the product of squares of determinants of the sub-matrices of D, which in turn fixes the sign of the square root in equation (3.105).

In pure particle-number projection of a q.p. vacuum, the sign of the overlap is also directly accessible as it can be evaluated with the *standard* Wick theorem when going to the canonical basis of the non-rotated q.p. vacuum [61–63].

In most cases of interest, in particular when combining several generator coordinates, the determination of the sign of the overlap requires additional effort. One possibility is to use the Onishi formula (3.105) and to follow the evolution of its sign through a Taylor expansion when the kernel can be connected in small steps to a known reference overlap [64, 65]. This, however, becomes very complicated, even impossible, in multi-dimensional calculations and when breaking time-reversal invariance.

A first alternative to the Onishi formula (3.105) that determines the overlap with its sign was presented by Neergård and Wüst in [66]. It requires the diagonalization of a matrix constructed from the Thouless matrices $Z_{ij}^{\boldsymbol{q}}$ and $Z_{ij}^{\boldsymbol{q}'}$ connecting each of the q.p. vacua $|\Phi(\boldsymbol{q})\rangle$ and $|\Phi(\boldsymbol{q}')\rangle$ to a common basis in the *particle* vacuum $|-\rangle$, see sections 3.1.2 and 8.3.2:

$$|\Phi(\boldsymbol{q})\rangle = \exp\left[\frac{1}{2}\sum_{ij} Z_{ij}^{\boldsymbol{q}} c_i^\dagger c_j^\dagger\right]|-\rangle \tag{3.106a}$$

$$|\Phi(\boldsymbol{q}')\rangle = \exp\left[\frac{1}{2}\sum_{ij} Z_{ij}^{\boldsymbol{q}'} c_i^\dagger c_j^\dagger\right]|-\rangle. \tag{3.106b}$$

In this representation, the Onishi formula reads [1]

$$\langle\Phi(\boldsymbol{q}')|\Phi(\boldsymbol{q})\rangle = \sqrt{\det\left[1 + (Z^{\boldsymbol{q}'})^*(Z^{\boldsymbol{q}})^T\right]}. \tag{3.107}$$

It can be shown that the eigenvalues λ_i of the $M \times M$ matrix $(Z^{\boldsymbol{q}'})^*(Z^{\boldsymbol{q}})^T$ are pairwise degenerate. The overlap is then obtained as the product over half of the eigenvalues, one out of each pair [66, 67],

$$\langle\Phi(\boldsymbol{q}')|\Phi(\boldsymbol{q})\rangle = \prod_{i=1}^{M/2} (1 + \lambda_i).$$

In realistic applications the use of this scheme can become cumbersome and has therefore not often been implemented, a notable exception being calculations by the Madrid group [44, 68]. The reason is that a fully occupied s.p. state in a given q.p. vacuum $|\Phi(\boldsymbol{q})\rangle$ leads to $\langle\Phi(\boldsymbol{q})|-\rangle = 0$, which then inhibits the straightforward connection of both states via the Thouless theorem, see the discussion of equation (3.51) in section 3.1.3. Instead, the contribution of such (non-paired) s.p. states has to be treated separately.

A second alternative to the Onishi formula that also directly determines the overlap with its sign was proposed by Robledo [69] in terms of the Pfaffian of a skew-symmetric $2M \times 2M$ matrix, which again is obtained from the Thouless matrices $Z^{\boldsymbol{q}}$ and $Z^{\boldsymbol{q}'}$ defined through equation (3.106)

$$\langle\Phi(\boldsymbol{q}')|\Phi(\boldsymbol{q})\rangle = (-)^{M(M+1)/2}\mathrm{pf}\begin{pmatrix} Z^{\boldsymbol{q}'} & -1 \\ 1 & -Z^{\boldsymbol{q}} \end{pmatrix}. \tag{3.108}$$

Like the determinant, the Pfaffian is a characteristic polynomial of a matrix that is, however, only defined for skew-symmetric ones. It has the intriguing property that $(\mathrm{pf}A)^2 = \det A$, which establishes the connection between equations (3.108) and (3.107). Further properties of Pfaffians that are frequently needed for formal manipulations can be found, for example, in [69].

Fully occupied s.p. states in either one of the two q.p. vacua do again lead to complications [70, 71]. In the original derivation of [69], the Pfaffian emerged when mapping the q.p. vacua on fermionic coherent states, whose overlap is then evaluated with the help of Grassmann algebra. More recently it has been pointed out that a Pfaffian expression is also the natural consequence of bringing the q.p. operators into a particular order when evaluating their matrix elements [71, 72], which gives a much more direct access to formal manipulations. The connection between the various schemes to calculate the overlap is analyzed in [67]. All schemes to calculate the overlap with the help of Pfaffians mentioned so far rely on expressing both q.p. vacua in a common s.p. basis. An alternative that can be also applied when $|\Phi(\boldsymbol{q})\rangle$ and $|\Phi(\boldsymbol{q}')\rangle$ are each expressed in a different s.p. basis has been constructed in [73]. Such an extension is for example necessary when expressing each q.p. state in its canonical s.p. basis.

Multi-reference techniques for *ab initio* Hamiltonians

The need for the formulas discussed in this section has come mostly for applications of the MR-EDF framework, e.g. in GCM calculations of configuration mixing or symmetry restoration. Until very recently, these techniques were used in conjunction with generator states $|\Phi(\boldsymbol{q})\rangle$ of the HF or HFB type, that is, product states. However, it is possible to envision the dissemination of these techniques to *ab initio* theory. Indeed, there have been recent attempts to apply many-body techniques such as coupled-clusters [74] or in-medium similarity renormalization groups (IM-SRGs) [75] on top of symmetry-breaking reference states such as HFB vacua [76, 77]. The next step would be to extend this to GCM-mixed states. In this case, the various formulas discussed here do not apply since the many-body states in the coupled-cluster or IM-SRG techniques are not product states. For this reason another completely different approach to the calculation of the overlap, without the need to calculate a determinant or pfaffian, has recently been proposed in [78].

Calculation of the transition densities

There are many possibilities for the practical implementation of the calculation of operator matrix elements with the generalized Wick theorem, which are usually chosen to suit the choice of underlying basis. Most authors calculate the normal and anomalous transition density matrices in a common s.p. basis. The most basic strategy relates the normal and anomalous transition density matrices to the Thouless matrices of equation (3.106) through [1]

$$\rho^{q'q} = -\, Z^{q}(1 - Z^{q'*}Z^{q})^{-1}Z^{q'*}$$
$$\kappa^{q'q} = Z^{q}(1 - Z^{q'*}Z^{q})^{-1}$$
$$\kappa^{qq'*} = (1 - Z^{q'*}Z^{q})^{-1}Z^{q'*}.$$

This is, however, again limited to the case where $\langle \Phi(\boldsymbol{q}')|-\rangle \neq 0$ and $\langle \Phi(\boldsymbol{q})|-\rangle \neq 0$. The treatment of the general case where the normal and anomalous transition density matrices are expressed through the matrices D and E of equation (3.103) is described, for example, in [79].

A very efficient way to calculate the normal and anomalous transition density matrices consists in using the s.p. creation operators in the canonical s.p. basis of the 'bra' in a many-body kernel, and the annihilation operators of the canonical s.p. basis of the 'ket' [65, 80, 81]. This corresponds to working with a non-unitary Bogoliubov transformation that relates the two sets of q.p. operators α and β. This scheme also naturally connects to the way in which kernels are evaluated for Slater determinants.

Relevance of pairing correlations in the reference states

Using paired Bogoliubov q.p. states instead of Slater determinants as reference states in an MR-EDF calculation of configuration-mixing type is in many cases a necessity. Assume two different normalized N-body Slater determinants, each represented in the respective natural basis

$$|\Phi(\boldsymbol{q}')\rangle = \prod_{l=1}^{N} a_l^{\dagger}|-\rangle \qquad |\Phi(\boldsymbol{q})\rangle = \prod_{r=1}^{N} b_r^{\dagger}|-\rangle.$$

The $N \times N$ matrix of overlaps between the single-particle states

$$\mathcal{R}_{lr} = \langle -|a_l b_r^{\dagger}|-\rangle \tag{3.109}$$

for $l, r \leqslant N$ contains all the information needed to calculate the overlap of the two Slater determinants

$$\langle \Phi(\boldsymbol{q}')|\Phi(\boldsymbol{q})\rangle = \det \mathcal{R} \tag{3.110}$$

and also all contractions entering the operator kernels calculated with the generalized Wick theorem, provided that the two Slater determinants are non-orthogonal [82, 83],

$$\rho_{rl}^{q'q} = \frac{\langle \Phi(\boldsymbol{q}')|a_l^{\dagger} b_r\,|\Phi(\boldsymbol{q})\rangle}{\langle \Phi(\boldsymbol{q}')|\Phi(\boldsymbol{q})\rangle} = (\mathcal{R}^{-1})_{rl},$$

with $r, l \leqslant N$. These simple relations have a number of important consequences for applications. If the two Slater determinants $|\Phi(\boldsymbol{q})\rangle$ and $|\Phi(\boldsymbol{q}')\rangle$ share a common symmetry, then the single-particle states in both respective natural bases can be chosen to have the corresponding quantum number. The matrix $\mathcal{R}_{lr}$ then becomes block diagonal with non-zero matrix elements only between single-particle states having the same quantum number. When the number of single-particle states of any given quantum number is different in the two Slater determinants, then the determinant of $\mathcal{R}$, and with that the overlap (3.110), is zero. As a consequence, such Slater determinants cannot be mixed with the ansatz (3.79).

We recall that a vanishing overlap between two Slater determinants does not imply that the corresponding operator matrix elements are zero. In fact, the majority of configuration-mixing techniques develop the correlated state into a basis of orthogonal Slater determinants. The matrix elements of the Hamiltonian and other operators then have to be evaluated with the standard Wick theorem.

For the present discussion, the most important example of a conserved symmetry that introduces such a block structure into the matrix $\mathcal{R}$ of equation (3.109) is parity. Indeed, when working with parity-conserving reference states, the number of single-particle states of given parity below the Fermi energy usually changes very quickly with quadrupole deformation. Under such conditions, a GCM calculation (3.79) requires the use of Bogoliubov q.p. vacua, all of which have to be paired, as reference states. This also explains why the reference states for such calculations are usually constructed, either with the Lipkin–Nogami scheme or with a variation after projection (VAP) on particle number; both procedures ensure the presence of pairing correlations in the reference states.

The only possibility to safely work with Slater determinants as reference states in a configuration-mixing-type MR-EDF calculation is to enforce the breaking of a sufficient number of symmetries in all reference states such that the matrix $\mathcal{R}$ does not become singular for any pair of reference states entering the calculation. This is different for projection of Slater determinants, however, where the number of single-particle states that conserve a symmetry common to all relative orientations of the reference states is always the same in both natural bases.

3.2.3 Projection of broken symmetries

Projection of broken symmetries is a particular case of the MR-EDF formalism. To outline the principal ideas, we introduce a schematic representation $\hat{P}^S$ of the projection operators corresponding to a continuous symmetry group $\mathcal{G}$ of the nuclear Hamiltonian $\hat{H}$. Starting from a symmetry-breaking reference state $|\Phi(\varphi_0)\rangle$ at some arbitrary orientation φ_0, a symmetry-conserving many-body state with symmetry quantum number S is obtained as

$$|\Psi^S\rangle = \hat{P}^S|\Phi\rangle = \int d\varphi\, f^S(\varphi)\, \hat{R}(\varphi)|\Phi(\varphi_0)\rangle, \tag{3.111}$$

where $\hat{R}(\varphi)$ is the unitary symmetry transformation, which changes the orientation of a state by φ, and $f^S(\varphi)$ is a weight function, which is assumed to contain normalization factors such that $\langle \Psi^S | \Psi^S \rangle = 1$. For discrete symmetries, the integral becomes a sum instead. For its numerical evaluation, the integral over the continuous group parameter φ is usually also discretized. The superposition of states described by equation (3.111) is a special case of the GCM ansatz, where the multi-reference basis is obtained by unitary transformations from a single state. Because $[\hat{R}(\varphi), \hat{H}] = 0$ and the unitarity of the symmetry transformation $\hat{R}(\varphi)\,\hat{R}^\dagger(\varphi) = 1$, all possible orientations of $|\Phi(\varphi)\rangle$ are degenerate, $\langle \Phi(\varphi_0) | \hat{R}^\dagger(\varphi)\,\hat{H}\,\hat{R}(\varphi) | \Phi(\varphi_0) \rangle = \langle \Phi(\varphi_0) | \hat{H} | \Phi(\varphi_0) \rangle$, such that the orientation φ_0 can indeed be freely chosen; see figure 3.2.

The energy of the symmetry-conserving state (3.111) is obtained as

$$
\begin{aligned}
E_{\text{proj}} &= \langle \Psi^S | \hat{H} | \Psi^S \rangle \\
&= \iint d\varphi\, d\varphi'\, f^{S*}(\varphi')\, f^S(\varphi)\, \langle \Phi(\varphi_0) | \hat{R}^\dagger(\varphi')\, \hat{H}\, \hat{R}(\varphi) | \Phi(\varphi_0) \rangle \\
&= \langle \Phi(\varphi_0) | \hat{H}\, \hat{P}^S | \Phi(\varphi_0) \rangle \\
&= \int d\varphi\, f^S(\varphi)\, \langle \Phi(\varphi_0) | \hat{H}\, \hat{R}(\varphi) | \Phi(\varphi_0) \rangle,
\end{aligned}
$$

where in the second line we have used that the projection operator on symmetry groups of the Hamiltonian commutes with the Hamiltonian, such that the double integral can be replaced by a single integral. Introducing $\varphi_1 \equiv \varphi_0 + \varphi$ for the orientation of the rotated state, the energy can be rewritten as

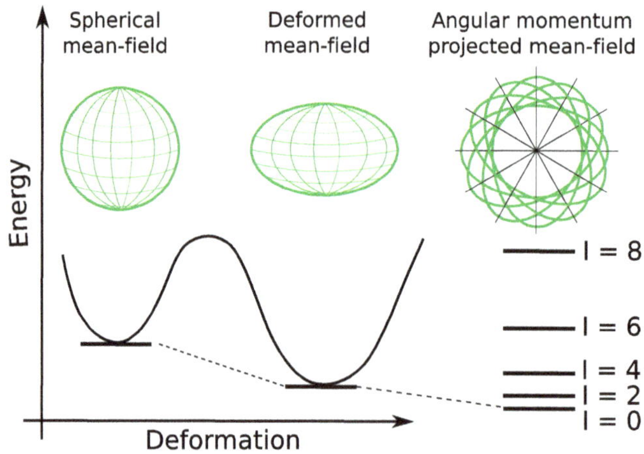

Figure 3.2. Schematic illustration of the effect of projection in the case of angular momentum. We start from a symmetry-conserving, i.e. spherical, solution which, as discussed in section 1.1, is a poor approximation of the ground state except for a few doubly closed shell nuclei. Symmetry-breaking (deformation) lowers the energy. Projecting on angular momentum implies mixing Euler-rotated solutions, which restores angular-momentum quantum numbers. (Figure created by N Schunck.)

$$
\begin{aligned}
E_{\text{proj}} &= \int d(\varphi_1-\varphi_0) f^S(\varphi_1-\varphi_0) \langle \Phi(\varphi_0)|\hat{H}\,|\Phi(\varphi_1)\rangle \\
&= \int d(\varphi_1-\varphi_0) f^S(\varphi_1-\varphi_0) \mathcal{E}[\varphi_0, \varphi_1].
\end{aligned}
\tag{3.112a}
$$

It is customary, although not necessary, to set $\varphi_0 = 0$. The projected energy is obtained from the integral over energy kernels $\mathcal{E}[\varphi_0, \varphi_1]$ (1.16) taken at all possible *relative* orientations of the reference states. The last equality establishes the direct connection to the general MR-EDF approach.

Energy kernels and symmetry groups

In the discussion of the projected energy above, we have assumed an underlying Hamiltonian $\hat{H}$. When the energy functional is not derived from a Hamiltonian but simply defined through the energy kernel $\mathcal{E}[g', g]$ of equation (1.18), the rigorous derivations of the formulas given below involves more advanced notions of group theory. Parameterizing the symmetry group $\mathcal{G}$ again by $g \equiv \varphi$ and assuming for simplicity that it has only one-dimensional irreps labeled by $m \in \mathbb{Z}$, any function of φ can be expanded as

$$
f(\varphi) = \sum_m f_m e^{im\varphi}. \tag{3.113}
$$

In this case, the projection consists in expanding the energy and norm kernels on the basis of irrep functions of the group. The coefficients of the expansion are obtained by using the orthogonality of the irrep with respect to the integration measure associated with the parameters φ of the group. A complete and more rigorous presentation of these techniques can be found in [84].

The use of the generalized Wick theorem allows transforming these non-diagonal kernels into functionals of the transition densities. Just like the GCM, the bottleneck is the calculation of these densities. In the following, we will give a few elements of projection techniques for the two most important ones: particle number and angular momentum. Additional details can be found in the references we provide. We will not cover projection on linear momentum, parity or isospin. Projection on eigenstates of linear momentum is discussed in [83, 85–87]. The case of parity restoration is treated in great detail in [88]. The formalism for isospin projection, which is particularly relevant for the study of nuclei close to the $N = Z$ line and calculations of β-decay, is presented in [89–91]. Until now it has been restricted to HF reference states.

Since isospin symmetry is broken by the Coulomb interaction, it is not a symmetry group of the nuclear Hamiltonian. There is nevertheless an interest in projecting on good isospin in order to separate non-physical aspects of SSB introduced by the use of simple product states in the EDF approach from the physical symmetry-breaking introduced by the Hamiltonian. The idea is to project the symmetry-breaking Slater determinants first on good isospin, and then to diagonalize the Hamiltonian, which in isospin space is a tensor operator of rank-2 instead of a scalar, within the many-body basis of states with good isospin.

Particle-number projection

A very detailed presentation of particle-number projection can be found in [61, 92, 93]. Complementary information is given in [62, 63, 94], especially with regard to possible inconsistencies of standard formulas for density-dependent pseudo-potentials. In the following, we sketch the most relevant properties of the particle-number projection operator and how particle-number projection can be implemented.

The projector $\hat{P}^N$ associated with the restoration of the global gauge invariance of the eigenstates of the particle-number operator reads

$$\hat{P}^{N_0} = \frac{1}{2\pi}\int_0^{2\pi} d\varphi e^{-\mathrm{i}\varphi(\hat{N}-N_0)}. \tag{3.114}$$

It is straightforward to verify that it is a true projector in the mathematical sense with the properties

$$\hat{P}^{N_1}\hat{P}^{N_0} = \delta_{N_1 N_0}\,\hat{P}^{N_0} \qquad \left(\hat{P}^{N_0}\right)^{\dagger} = \hat{P}^{N_0}.$$

From these relations it follows that for the calculation of projected matrix elements of operators $\hat{O}$ that commute with the particle-number operator, $[\hat{O}, \hat{N}] = 0$, it is sufficient to project either the bra or the ket

$$\langle\Phi|\left(\hat{P}^{N_0}\right)^{\dagger}\,\hat{O}\,\hat{P}^{N_0}|\Phi'\rangle = \langle\Phi|\hat{P}^{N_0}\,\hat{O}|\Phi'\rangle = \langle\Phi|\hat{O}\,\hat{P}^{N_0}|\Phi'\rangle.$$

By definition, only operators that change particle number do not commute with $\hat{N}$. Such operators are usually associated with nuclear decay and reaction modes that transform a nucleus into a different one and are beyond the scope of this section. It can be easily shown that a q.p. vacuum can only be projected on eigenstates of $\hat{N}$ that have the same number parity π_N as defined above, irrespective of the mean particle number that the q.p. vacuum has been constrained to with the Lagrange multiplier λ_{F} in the HFB equation.

Numerical representation of particle-number projection

There are several possibilities to evaluate the integral over the gauge angle φ in equation (3.114). The most direct strategy is to discretize it. Taking advantage of the number parity, see above, it can be shown that the integration interval in equation (3.114) can be reduced to $[0, \pi]$. A simple prescription based on a M-point trapezoidal quadrature [95]

$$\hat{P}^{N_0} \to \frac{1}{\pi}\int_0^{\pi} d\varphi e^{-\mathrm{i}\varphi(\hat{N}-N_0)} \approx \frac{1}{M_\varphi}\sum_{m=1}^{M_\varphi} e^{-\mathrm{i}\pi\frac{m-1}{M_\varphi}(\hat{N}-N_0)}.$$

turns out to be very efficient and converges already to a very high precision with very few discretization points, which can typically be chosen to be on the order of at most 15 when aiming at the dominant components contained in a given q.p. vacuum. This procedure can also be combined with other symmetry restorations and the calculation of particle-number-projected operator kernels for a subsequent GCM calculation.

In MR-EDF calculations limited to the gauge angle of particle-number projection as the only collective coordinate, matrix elements can also be evaluated with recurrence relations [96], or through a mapping on a contour integral in the complex plane which is then evaluated with the residue theorem [62, 63]. The latter representation is particularly instructive for the formal analysis of the corresponding MR-EDF.

Using the generalized Wick theorem, we find that the norm kernel and energy kernel depend on the following set of transition densities

$$\rho_{\mu\nu}^{0\varphi} = \frac{\langle\Phi|c_\nu^\dagger c_\mu e^{i\varphi\hat{N}}|\Phi\rangle}{\langle\Phi|e^{i\varphi\hat{N}}|\Phi\rangle} \qquad \kappa_{\mu\nu}^{0\varphi} = \frac{\langle\Phi|c_\nu c_\mu e^{i\varphi\hat{N}}|\Phi\rangle}{\langle\Phi|e^{i\varphi\hat{N}}|\Phi\rangle} \qquad \kappa_{\mu\nu}^{\varphi 0*} = \frac{\langle\Phi|c_\mu^\dagger c_\nu^\dagger e^{i\varphi\hat{N}}|\Phi\rangle}{\langle\Phi|e^{i\varphi\hat{N}}|\Phi\rangle}.$$

These are nothing other than the adaptation of equation (1.17) to the particular case where $|\Phi^{(g')}\rangle \equiv |\Phi\rangle$ is an HFB reference state and $|\Phi^{(g)}\rangle$ is the HFB state rotated in gauge space. The expressions for these densities in the canonical basis can be found in [61–63]. In an arbitrary quasiparticle basis one finds [61]

$$\rho_{\mu\nu}^{0\varphi} = (\mathcal{V}^*(\varphi)V^T)_{\mu\nu} \qquad \kappa_{\mu\nu}^{0\varphi} = (\mathcal{V}^*(\varphi)U^T)_{\mu\nu} \qquad \kappa_{\mu\nu}^{\varphi 0*} = (V\mathcal{U}^\dagger(\varphi))_{\mu\nu}.$$

Note that $\kappa_{ij}^{\varphi 0*}$ is not the complex conjugate of $\kappa_{ij}^{0\varphi}$. In these expressions, we have set

$$\mathcal{U}(\varphi) = U + V^*\chi^*(\varphi)$$
$$\mathcal{V}(\varphi) = V + U^*\chi^*(\varphi),$$

where (U, V) are the sub-matrices of the Bogoliubov transformation (3.6) associated with the HFB state $|\Phi\rangle$, while $\chi(\varphi)$ is the contraction (3.104)

$$\chi_{\mu\nu}(\varphi) = \frac{\langle\Phi|\beta_\nu\beta_\mu\, e^{i\varphi\hat{N}}|\Phi\rangle}{\langle\Phi|e^{i\varphi\hat{N}}|\Phi\rangle} = (E(\varphi)D(\varphi))_{\nu\mu},$$

where

$$D^{-1}(\varphi) = \cos(\varphi) - i\sin(\varphi)\, N^{11}$$
$$E^*(\varphi) = i\sin(\varphi)\, N^{20},$$

with $N^{11} = U^\dagger U - V^\dagger V$ and $N^{20} = U^\dagger V^* - V^\dagger U^*$ the upper-left and upper-right corners of the matrix of the particle-number operator in the quasiparticle basis; see equation (8.54) with $\hat{F}$ unity.

The Lipkin–Nogami method

In addition to the specific difficulties of defining MR-EDF kernels for density-dependent pseudo-Hamiltonians, the computational cost of implementing explicit restoration of broken symmetries (especially in VAP) can be very high. It can thus be useful to invoke approximation schemes that mock up the effects of going beyond the mean field.

When aiming only at the lowest state from symmetry restoration, one such scheme has been proposed by Lipkin [97, 98]. It can be applied to any continuous symmetry, see [20] for an application to translational symmetry, but most studies focus on approximate particle-number projection [99]. Making a further approximation proposed by Nogami [100] one arrives at what is now known as the Lipkin–Nogami method, which has been described in great detail in [19, 23, 101–104].

The Lipkin–Nogami method consists in adding terms to the energy functional that cancel the fluctuations of particle number, or any other operator for which the HFB solution is not an eigenstate. The starting point is the generalized constraint operator $\hat{K}$ defined as

$$\hat{K} = \hat{H} - \lambda_1 \hat{N} - \lambda_2 \hat{N}^2.$$

The LN equations are obtained from the conditions

$$\langle \Phi | \hat{K}(\hat{N} - \langle \hat{N} \rangle) | \Phi \rangle = 0$$
$$\langle \Phi | \hat{K}(\hat{N}^2 - \langle \hat{N}^2 \rangle) | \Phi \rangle = 0.$$

By expanding the particle number operator in terms of quasiparticle ladder operators, see (3.35), it is easy to show that

$$\langle \Phi | \hat{K} \hat{N}^{20} | \Phi \rangle = 0$$
$$\langle \Phi | \hat{K}(\hat{N}^{2,20} + \hat{N}^{2,40}) | \Phi \rangle = 0,$$

with

$$\lambda_2 = \frac{\langle 0 | \hat{H}^{40} \hat{N}^{2,40} | 0 \rangle}{\langle 0 | \hat{N}^{2,40} \hat{N}^{2,40} | 0 \rangle}.$$

The explicit calculation of the expectation values in λ_2 is tedious but not especially difficult; the use of the Wick theorem on the HFB vacuum greatly simplifies the calculations. We can show that the final expression for λ_2 reads

$$\lambda_2 = \frac{4 \mathrm{Tr}\, \Gamma' \rho(1-\rho) + 8 \mathrm{Tr}\, \Delta'(1-\rho)\kappa}{8[\mathrm{Tr}\, \rho(1-\rho)]^2 - 16 \mathrm{Tr}\, \rho^2(1-\rho)^2}, \qquad (3.115)$$

with modified mean field and pairing field written as

$$\Gamma'_{ik} = \sum_{jl} \bar{v}_{ijkl} [\rho(1-\rho)]_{lj}$$

$$\Delta'_{kl} = \frac{1}{4} \sum_{ij} \bar{v}_{ijkl} (\kappa^* \rho)_{ij}.$$

This expression can be simplified for pairing forces of the monopole type, $\bar{v}_{ijkl} = -\frac{G}{4}\delta_{j\bar{i}}\delta_{l\bar{k}}$. In this case, we find, in the canonical basis

$$\lambda_2 = \frac{G}{16} \frac{\sum_{i>0} u_i v_i^3 \sum_{i>0} u_i^3 v_i - \sum_{i>0} (u_i v_i)^4}{\left(\sum_{i>0} u_i^2 v_i^2\right)^2 - \sum_{i>0} (u_i v_i)^4}.$$

A procedure to calculate the contributions of any type of energy functional to λ_2 has been proposed in [23, 104] and applied to the Skyrme EDF. The same scheme is used in [105] in the context of covariant density functional theory (cDFT).

Angular-momentum projection

In this paragraph, we will use units where the matrix elements of angular-momentum operators are just numbers, which drops $\hbar$ factors from all equations. The angular-momentum projection operator reads

$$\hat{P}^J_{MK} = \frac{2J+1}{16\pi^2} \int_0^{2\pi} d\alpha \int_0^{\pi} d\beta \sin(\beta) \int_0^{4\pi} d\gamma\, D^{J*}_{MK}(\alpha, \beta, \gamma)\, \hat{R}(\alpha, \beta, \gamma), \quad (3.116)$$

where $D^J_{MK}(\alpha, \beta, \gamma) = e^{-\mathrm{i}\alpha M} d^J_{MK}(\beta) e^{-\mathrm{i}\gamma K}$ is a Wigner rotation matrix and $\hat{R}(\alpha, \beta, \gamma)$ the rotation operator [7]. Although other conventions are possible, both are usually parameterized through the three Euler angles $\Omega = (\alpha, \beta, \gamma) \in [0, 2\pi] \times [0, \pi] \times [0, 4\pi]$

$$\hat{R}(\alpha, \beta, \gamma) = e^{-\mathrm{i}\alpha \hat{J}_z} e^{-\mathrm{i}\beta \hat{J}_y} e^{-\mathrm{i}\gamma \hat{J}_z}, \quad (3.117)$$

with $\hat{J}_\mu$ the μ-component of the total angular momentum of the many-body system. The angular-momentum operator $\hat{J}_\mu = \hat{L}_\mu + \hat{S}_\mu$ has an orbital and a spin part, each of which is the sum over single-particle operators. To construct a rotated q.p. vacuum, each of the states in its q.p. basis has to be rotated. While $e^{-\mathrm{i}\varphi \hat{L}_\mu}$ rotates the q.p. states in space, the rotation in spin $e^{-\mathrm{i}\varphi \hat{S}_\mu}$ mixes their spinor components. The latter ensures that the orientation of spins relative to the principal axes of the nucleus is the same before and after the rotation. The operator $\hat{P}^J_{MM'}$ has the properties

$$\hat{P}^J_{MM'} \hat{P}^{J'}_{KK'} = \delta_{JJ'} \delta_{M'K}\, \hat{P}^J_{MK'} \qquad \left(\hat{P}^J_{MM'}\right)^\dagger = \hat{P}^J_{M'M}. \quad (3.118)$$

Relations (3.118) imply that $\hat{P}^J_{MK}$ is not a projection operator in the sense of being a linear map with $p^2 = p$. Instead, the operator transforms one state out of an irrep into another state of the same irrep. In the literature, such operators are sometimes called *shift operators* [106] or *transfer operators* [107]. The necessity to work with such objects is a consequence of the corresponding group $SU(3)$ being non-Abelian.

As in the case of particle-number projection, it is sufficient to apply the operator $\hat{P}^J_{MM'}$ to only one out of the two states entering a projected matrix element, even if they are different. This is most straightforward for scalar operators such as the norm and Hamiltonian that commute with the rotation operator $\hat{R}(\alpha, \beta, \gamma)$

$$\langle\Phi(\boldsymbol{q})|\hat{P}^J_{KM}\,\hat{H}\,\hat{P}^{J'}_{M'K'}|\Phi(\boldsymbol{q}')\rangle = \langle\Phi(\boldsymbol{q})|\hat{H}\,\hat{P}^J_{KK'}|\Phi(\boldsymbol{q}')\rangle\,\delta_{MM'}\,\delta_{JJ'} \equiv H^{Jqq'}_{KK'}\,\delta_{MM'}\,\delta_{JJ'}$$
$$\langle\Phi(\boldsymbol{q})|\hat{P}^J_{KM}\hat{P}^{J'}_{M'K'}|\Phi(\boldsymbol{q}')\rangle = \langle\Phi(\boldsymbol{q})|\hat{P}^J_{KK'}|\Phi(\boldsymbol{q}')\rangle\,\delta_{MM'}\,\delta_{JJ'} \equiv N^{Jqq'}_{KK'}\,\delta_{MM'}\,\delta_{JJ'}.$$

Except for states projected on $J = 0$, for which $K = K' = 0$, in general the matrices $H^{Jqq'}_{KK'}$ and $N^{Jqq'}_{KK'}$ will contain non-diagonal elements, such that the corresponding states are symmetry-restored, but not yet orthogonal.

The projected states are not completely determined by the application of the operators $\hat{P}^J_{MM'}$ on the states $|\Phi\rangle$. This is a general feature of the restoration of symmetries for non-Abelian groups [106, 107]. To obtain an orthogonal set of symmetry-restored states that can be constructed from a state $|\Phi(\boldsymbol{q})\rangle$, $\hat{H}$ has to be diagonalized within the space spanned by states with different K. To that end, it is necessary to solve the generalized eigenvalue problem

$$\sum_{K'}[H^{Jqq}_{KK'} - E^J_\epsilon N^{Jqq}_{KK'}]f^{Jq}_\epsilon(K') = 0. \tag{3.119}$$

This last step provides the weights f^{JK}_ϵ that can then be used to construct all $(2J + 1)$ states labeled by the z-component of angular momentum in the laboratory frame M in each of the up to $(2J + 1)$ possible different irreps labeled by ϵ of given J that can be projected out from the symmetry-breaking state $|\Phi\rangle$:

$$|\Psi^{JM}_\epsilon(\boldsymbol{q})\rangle = \sum_{K=-J}^{J} f^{Jq}_\epsilon(K)\,\hat{P}^J_{MK}|\Phi(\boldsymbol{q})\rangle, \tag{3.120}$$

with

$$\hat{J}^2\,|\Psi^{JM}_\epsilon(\boldsymbol{q})\rangle = J(J+1)\,|\Psi^{JM}_\epsilon(\boldsymbol{q})\rangle$$
$$\hat{J}_z\,|\Psi^{JM}_\epsilon(\boldsymbol{q})\rangle = M\,|\Psi^{JM}_\epsilon(\boldsymbol{q})\rangle.$$

Note that f^{JK}_ϵ in equation (3.120) contain a normalization factor. In case where the state $|\Phi\rangle$ contains one K-component per angular momentum, which implies that there is one irrep ϵ per J and no need for K-mixing, the corresponding f^{JK}_ϵ in equation (3.120) does not equal one, but $1/\sqrt{\langle\Phi|\hat{P}^J_{KK}|\Phi\rangle}$. In practice, however, it is sufficient to calculate only the matrix elements between states in the irreps with angular momenta J and J' that are connected by a given operator of interest without the need to explicitly construct the projected states $|\Psi^{JM}_\epsilon\rangle$.

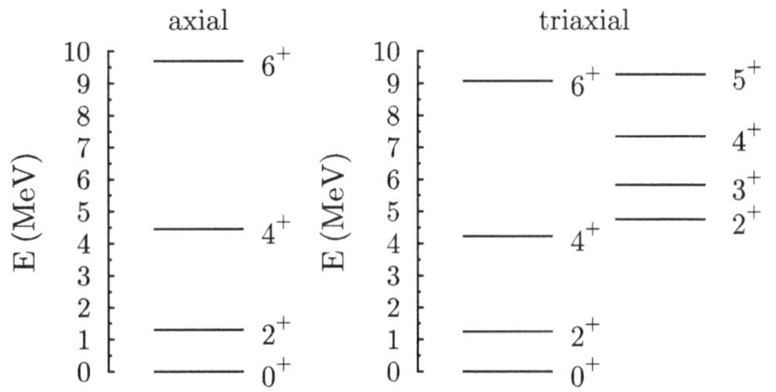

Figure 3.3. Energy spectrum E^{J_μ} of low-lying states of ^{24}Mg, projected from an axially symmetric prolate state (left) and a slightly triaxial state (right) of similar intrinsic quadrupole moment. (Figure created by M Bender.)

The number of irreps with a given J labeled by ϵ that can be projected out from a given state $|\Phi\rangle$ greatly depends on its structure. Because of the symmetries of $|\Phi\rangle$ possibly remaining, it is usually smaller than the maximum possible value of $(2J + 1)$. Two examples of energy spectra obtained from angular-momentum projection are presented in figure 3.3. The panel on the left shows the energy levels of states projected out from the lowest time-reversal-conserving axially symmetric prolate HFB state of ^{24}Mg. The states group into a rotational $\Delta J = 2$ band that roughly follows $E(J) \approx E(J = 0) + J(J + 1)/(2\Theta)$. For symmetry reasons, however, there is only one irrep that can be projected out from an axial state for each value of J, which also has to be even. This is to be contrasted with the panel on the right, which shows the energy spectrum of low-lying states after K-mixing (3.119) that can be projected out from a slightly triaxial HFB state of ^{24}Mg. A low energy, there are now two rotational bands, and additional ones can be found at higher energy. This illustrates that a triaxial state can be decomposed on several irreps of the same angular momentum J, but different energy. The $\Delta J = 2$ *yrast* band in high-spin physics (the *yrast* band denotes the band of lowest energy at a given spin I [17, 108]) built on the $J = 0$ ground state is quite similar to the one projected out from the axial state, whereas the additional excited $\Delta J = 1$ band is built on a $J = 2$ band-head and also contains states with odd angular momentum.

The number of different irreps of the same J that can be projected out from a given symmetry-breaking state has to be distinguished from the degeneracy of the levels within a given irrep. The former depends on the structure of the symmetry-breaking state, whereas the latter is entirely determined by the symmetry group: each of the irreps contains $(2J + 1)$ degenerate states with a different z-component of angular momentum M in the laboratory frame. For further discussion of the decomposition of such states see, for example [42, 44].

Some operators $\hat{T}_{\lambda\mu}$ of interest, such as angular momentum and electromagnetic multipole moments, do not commute with the rotation operator. As irreducible tensor operators of finite rank λ, they transform instead as [7]

$$\hat{R}(\Omega)\, \hat{T}_{\lambda\mu}\, \hat{R}^{\dagger}(\Omega) = \sum_{\mu'=-\lambda}^{\lambda} \hat{T}_{\lambda\mu'}\, \mathcal{D}^{\lambda}_{\mu'\mu}(\Omega). \tag{3.121}$$

Making use of equation (3.118) and various properties of the Wigner rotation matrices [7], their projected matrix elements can be expressed as

$$\langle\Phi|\hat{P}^{J}_{KM}\, \hat{T}_{\lambda\mu}\, \hat{P}^{J'}_{M'K'}|\Phi'\rangle = C^{JM}_{J'M'\lambda\mu} \sum_{\mu'=-\lambda}^{+\lambda} \sum_{\kappa=-J'}^{+J'} C^{JK}_{J'\kappa\lambda\mu'} \langle\Phi|\hat{T}_{\lambda\mu'} \hat{P}^{J'}_{\kappa K'}|\Phi'\rangle, \tag{3.122}$$

where $C^{JM}_{J'M'\lambda\mu}$ are Clebsch–Gordan coefficients. For a scalar operator $\hat{T}_{00}$, equation (3.122) simplifies to $\langle\Phi|\hat{P}^{J}_{KM}\, \hat{T}_{00}\, \hat{P}^{J'}_{M'K'}|\Phi'\rangle = \delta_{JJ'}\delta_{MM'}\langle\Phi|\hat{T}_{00}\, \hat{P}^{J'}_{KK'}|\Phi'\rangle$.

Electromagnetic transitions
Projection on angular momentum restores the selection rules for electromagnetic transitions. The most frequently analyzed electric moments are the $B(E2)$ values for transitions between different irreps

$$B(E2;\, J_{\epsilon'} \to J_{\epsilon}) = \frac{e^2}{2J'+1} \sum_{M=-J}^{+J} \sum_{M'=-J'}^{+J'} \sum_{\mu=-2}^{+2} \left|\langle\Psi^{JM}_{\epsilon}\,\middle|\,\hat{Q}_{2\mu}\,\middle|\,\Psi^{J'M'}_{\epsilon'}\rangle\right|^2.$$

The definition of the $B(E2)$ value contains an average over the $(2J'+1)$ possible initial states with $-J' \leqslant M' \leqslant J'$ in the irrep ϵ' and the sum over all possible final states with $-J \leqslant M \leqslant J$ in the irrep ϵ. We assume throughout this paragraph that the angular-momentum-projected states have good parity, either as a symmetry of the underlying q.p. vacua, or as the result of projection on good parity. Another important electric moment is the spectroscopic quadrupole moment, which is defined as the matrix element between projected states with $M = J$ within a given irrep,

$$Q_s(J_\epsilon) = \sqrt{\frac{16\pi}{5}}\, \langle\Psi^{JJ}_{\epsilon}|\hat{Q}_{20}\, |\Psi^{JJ}_{\epsilon}\rangle.$$

Both are matrix elements of the electric quadrupole operator, which is usually approximated as the quadrupole moments of point protons $Q_{2\mu} = e\, \sum_p \frac{1}{2}\,(1-\hat{\tau}_3)\, r^2\, Y_{2\mu}(\mathbf{r})$. The quadrupole operator connects states with same parity for which $|J'-J| \leqslant 2$.

The most often analyzed matrix elements of magnetic operators are the $B(M1)$ values for transitions between different irreps

$$B(M1;\, J_{\epsilon}' \to J_{\epsilon}) = \frac{3}{4\pi}\frac{1}{2J'+1} \sum_{M=-J}^{+J} \sum_{M'=-J'}^{+J'} \sum_{\kappa=-1}^{+1} \left|\left\langle\Psi^{JM}_{\epsilon}|\hat{\mu}_{\kappa}\, |\Psi^{J'M'}_{\epsilon'}\right\rangle\right|^2$$

and the magnetic dipole moment defined as

$$\mu(J_\nu) = \langle\Psi^{JJ}_{\epsilon}|\hat{\mu}_z\, |\Psi^{JJ}_{\epsilon}\rangle.$$

Both are matrix elements of the magnetic dipole operator $\hat{\boldsymbol{\mu}}$ that is usually assumed to be

$$\hat{\boldsymbol{\mu}} = g_{\ell,p}\hat{\boldsymbol{L}}_p + g_{s,p}\hat{\boldsymbol{S}}_p + g_{s,n}\hat{\boldsymbol{S}}_n,$$

where $\hat{\boldsymbol{L}}_t$ and $\hat{\boldsymbol{S}}_t$ are the total orbital and spin operators for protons and neutrons, $t = p, n$, while $g_{\ell,p} = 1\,\mu_N$, $g_{s,p} = 5.585\,\mu_N$ and $g_{s,n} = -3.826\,\mu_N$ are the orbital and spin g-factors of protons and neutrons in units of the nuclear magneton μ_N. The magnetic dipole operator connects states with same parity for which $|J'-J| \leqslant 1$.

The generalization to other electric and magnetic moments is straightforward, because of parity selection rules. However, odd electric multipoles and even magnetic multipoles only have transition matrix elements between states of opposite parity.

Numerical considerations

From the above discussion it is evident that the restoration of total angular momentum J and corresponding z-component M can be considerably more involved numerically than particle-number projection, in particular when considering projection of non-axial deformed states. There are three main reasons for this. First, the number of parameters in $SU(3)$ is 3, instead of just 1 for $U(1)$. Numerically, this is equivalent to going from one-dimensional to three-dimensional integrals. Second, the numerical application of the spatial rotation operator $\hat{R}(\alpha, \beta, \gamma)$ that mixes all states of same parity in the q.p. basis is much more costly than the application of the gauge-space rotation operator $e^{-i\varphi\hat{N}}$ that only multiplies the U and V matrices with phase factors. Third, the representation of an arbitrary rotation of a deformed state in space implies that all symmetries in the q.p. basis are broken, with the possible exception of parity. We discuss in chapter 8 some of the common symmetries that can be built in an HFB solver in order to accelerate the calculations. Most of these have to be abandoned in the numerical representation of angular-momentum projection.

However, by taking advantage of the symmetries of the q.p. vacuum, the size of the integration interval over Euler angles in equation (3.116) can be reduced, and with that, the computational cost of angular-momentum projection. Quasiparticle vacua are either even or odd under a rotation around 2π about an arbitrary axis, $e^{2i\pi\hat{J}_\mu}|\Phi\rangle = \pm|\Phi\rangle$, which has the consequence that they can only be projected on integer (+) or half-integer (−) angular momenta J, respectively. Note that this symmetry is broken at finite temperature. Knowing which of these two classes of states a given q.p. belongs to allows for reducing the integration interval for γ from $[0, 4\pi]$ to $[0, 2\pi]$. For states that are axial about the z-axis with $\hat{J}_z|\Psi\rangle = K|\Phi\rangle$, such that $e^{-i\gamma\hat{J}_z}|\Phi\rangle = e^{-i\gamma K}|\Phi\rangle$, the integrations over α and γ can be carried out analytically, such that only the integration over β remains to be carried out numerically. Similarly, point-group symmetries of the q.p. vacua [109, 110] such as parity, time-reversal, signature $e^{i\pi\hat{J}_\mu}$, etc, each introduce a relation between pairs of sets of rotation angles Ω, which can be used to reduce the integration interval in equation (3.116), and with that the number of kernels to be explicitly calculated [37]. Point symmetries also reduce the number of possible irreps that the state can be decomposed into.

Because of the numerical cost of angular-momentum projection, many applications, in particular large-scale calculations and applications to very heavy nuclei, are still restricted to the projection of axially symmetric states [48, 111, 112]. The angular-momentum projection of an axial state requires the rotation of this state about an axis perpendicular to its symmetry axis, such that the numerical representation of the kernels with the non-rotated state can no longer assume axial symmetry.

Variation before and after projection
There are several possible strategies to restore symmetries during an MR-EDF calculation. The simplest one is the two-step process where one first solves the HFB equations to construct the symmetry-breaking q.p. vacuum of largest binding energy, which is then projected on the quantum numbers of interest. In the case of the restoration of spatial symmetries, this usually implies projecting out all low-lying states that the q.p. vacuum can be decomposed into. This procedure is usually called projection after variation (PAV) in the literature. While being numerically efficient, it has the disadvantage that there is no guarantee that a reference state that gives the largest binding energy at the SR-EDF level also gives the largest possible energy after projection. This is only achieved when carrying out a VAP instead. In that case, the variational degrees of freedom are still those of the symmetry-breaking q.p. vacuum, but they are optimized to give the largest binding energy *after* projection on specific quantum numbers. This leads to much more complicated equations of motion than the HFB equation discussed in section 3.1.3. In principle the procedure has to be repeated for each set of quantum numbers of interest, since a different optimal q.p. vacuum will be obtained each time. Compared to the q.p. vacuum that minimizes the energy at the SR-EDF level, the q.p. vacuum that minimizes the energy in a VAP calculation might break additional symmetries.

A representation of the VAP equations proposed in [92] has become the standard procedure in modern applications. However, because of the additional cost of carrying out VAP as compared to PAV in an EDF context, full VAP has so far only been implemented for particle-number projection of q.p. vacua [61, 94] and the projection on momentum of spherical HF states [86, 87]. An alternative VAP scheme designed for particle-number-projected HF+BCS calculations has been proposed in [113].

There also is the possibility to apply an intermediate strategy that consists in constructing a set of symmetry-breaking q.p. vacua for which the amplitude of the order parameters of the symmetry to be restored is systematically varied with the help of constraints as explained in section 8.3.1. All of these states are then projected on the quantum number of interest in order to identify the one that gives the largest binding. This strategy is used under the name of restricted variation after projection (RVAP) [114] or minimization after projection (MAP) [48], and can be interpreted as the search for the minimum in the small variational space spanned by different states obtained from PAV calculations.

3.2.4 MR-EDF calculations combining projection and GCM

Projection restores the symmetries of the nuclear Hamiltonian by mixing degenerate reference states that can be transformed into one another by the symmetry transformation. As we have seen above, the weight functions of the corresponding MR-EDF are at least partially determined by the corresponding symmetry group. When combining projection with a configuration mixing of reference states that differ in some order parameters of the broken symmetries, it is of advantage to carry out the calculation in a four-step process that separates the two sources of mixing. We will sketch that for the example of a particle-number and angular-momentum-projected configuration mixing of q.p. vacua, that differ in some order parameters $\tilde{\boldsymbol{q}}$:

$$|\Psi_{\mu}^{JMN_0Z_0}\rangle = \sum_{\tilde{q}} \sum_{K} f_{\mu,\, qK}^{JN_0Z_0} \; \hat{P}_{MK}^{J} \; \hat{P}^{N_0} \; \hat{P}^{Z_0} \; |\Phi(\tilde{\boldsymbol{q}})\rangle.$$

Note that the combination of projection and a GCM on the corresponding order parameter automatically implies a RVAP or MAP. The first step consists in the calculation of the kernels

$$\mathcal{T}_{\alpha\beta\gamma\varphi_n\varphi_p\tilde{\boldsymbol{q}}\tilde{\boldsymbol{q}}'}^{\lambda\mu} = \langle\Phi(\tilde{\boldsymbol{q}})| \; \hat{T}^{\lambda\mu} \; \hat{R}(\alpha,\, \beta,\, \gamma) e^{-\mathrm{i}\varphi_n \hat{N}} e^{-\mathrm{i}\varphi_p \hat{Z}} |\Phi(\tilde{\boldsymbol{q}}')\rangle$$

for all irreducible tensor operators $\hat{T}^{\lambda\mu}$ of interest. Examples of such tensors of interests are: the Hamiltonian $\hat{H}$ and the identity $\hat{I}$ operators, which are scalars hence rank-0 tensors and will generate the energy and norm kernels; the electric or magnetic multipole operators of section 3.2.3, which give access to electromagnetic transitions. In the second step, the integrals over the rotational angles are evaluated

$$\begin{aligned}
\mathcal{T}_{\tilde{\boldsymbol{q}}K\tilde{\boldsymbol{q}}'K'}^{\lambda\mu,JN_0Z_0} = & \frac{1}{2\pi} \int d\varphi_n \, e^{\mathrm{i}\varphi_n N_0} \frac{1}{2\pi} \int d\varphi_p \, e^{\mathrm{i}\varphi_p Z_0} \\
& \times \frac{(2J+1)}{16\pi^2} \int d\alpha \int d\beta \, \sin(\beta) \int d\gamma \, D_{KK'}^{J*}(\alpha,\, \beta,\, \gamma) \, \mathcal{T}_{\alpha\beta\gamma\varphi_n\varphi_p\tilde{\boldsymbol{q}}\tilde{\boldsymbol{q}}'}^{\lambda\mu}.
\end{aligned}$$

The norm and energy kernels thus calculated enter a Hill–Wheeler–Griffin equation of the form

$$\sum_{\tilde{q}'} \sum_{K'} \left[\mathcal{H}_{\boldsymbol{q}K\boldsymbol{q}'K'}^{JN_0Z_0} - E_{\mu}^{JN_0Z_0} \mathcal{N}_{\boldsymbol{q}K\boldsymbol{q}'K'}^{JN_0Z_0}\right] f_{\mu,\, \boldsymbol{q}'K'}^{JN_0Z_0} = 0 \tag{3.123}$$

that has to be solved separately for each combination $(J,\, N_0,\, Z_0)$ of interest with the techniques described in section 3.2.1. While it is customary to project out many angular-momentum components from the reference states $|\Phi(\tilde{\boldsymbol{q}})\rangle$, all applications project only on one combination of proton and neutron number, which in most cases is identical to the one that the reference states have been constrained to when solving the HFB equation. All other observables are then calculated as

$$\langle \Psi_{\mu}^{JMN_0Z_0} | \hat{T}^{\lambda\mu} | \Psi_{\mu'}^{J'M'N_0Z_0} \rangle = C_{J'M'\lambda\mu}^{JM} \sum_{\mu'=-\lambda}^{+\lambda} \sum_{k=-J'}^{+J'} C_{J'k\lambda\mu'}^{JK} f_{\mu,\boldsymbol{q}K}^{JN_0Z_0} \mathcal{T}_{\tilde{\boldsymbol{q}}k\tilde{\boldsymbol{q}}'K'}^{\lambda\mu',JN_0Z_0} f_{\mu',\boldsymbol{q}'K'}^{J'N_0Z_0}.$$

In recent years, GCM calculations of angular-momentum-projected triaxial HFB states along these lines have become a standard application for light and medium-heavy even–even nuclei, using either Skyrme [42], Gogny [44, 68], or covariant [43, 115] functionals. In most of these applications, the authors also perform the projection on good proton and neutron numbers. This corresponds to seven different generator coordinates, $\boldsymbol{q} \equiv (\beta_2, \gamma_2, \alpha, \beta, \gamma, \varphi_p, \varphi_n)$, five of which $(\alpha, \beta, \gamma, \varphi_p, \varphi_n)$ are rotation angles of symmetry restoration. The other two, $\tilde{\boldsymbol{q}} \equiv (\beta_2, \gamma_2)$, are order parameters of rotational symmetry-breaking, namely β_2 is the axial quadrupole deformation and γ_2 the degree of triaxiality.

For reasons of computational cost, most calculations for heavy nuclei are still limited to axial shapes [116–119]. There are also several recent studies of combined (axial) quadrupole and octupole dynamics of angular-momentum- and parity-projected states, either with [51] or without [50] particle-number projection. Combined (axial) quadrupole and pairing dynamics of angular-momentum- and particle-number-projected states has been studied in [54, 55]. For light nuclei, there is recent exploratory work where triaxial states constructed with different constrained intrinsic angular momenta as additional generator coordinates are projected and mixed [56, 120]. Similarly, first calculations for odd-mass nuclei have become available [37, 57], including the mixing of different blocked q.p. states with the same deformation [37].

Combining several modes in MR-EDF calculations not only leads to a richer variational space, it can also remove some ambiguities that arise when mixing symmetry-breaking states. The most important one among these is that the non-diagonal matrix element of the particle-number operator between two different paired q.p. vacua $|\Phi\rangle$ and $|\Phi'\rangle$, that have been constrained to the same particle number N_0 when solving the HFB equation, takes in general a value different from N_0:

$$\frac{\langle \Phi | \hat{N} | \Phi \rangle}{\langle \Phi | \Phi \rangle} = \frac{\langle \Phi' | \hat{N} | \Phi' \rangle}{\langle \Phi' | \Phi' \rangle} = N_0 \neq \frac{\langle \Phi | \hat{N} | \Phi' \rangle}{\sqrt{\langle \Phi | \Phi \rangle \, \langle \Phi' | \Phi' \rangle}}.$$

As the binding energy per nucleon is typically on the order of −8 MeV, already a deviation of the matrix element of $\hat{N}$ as small as 0.1 introduces an error of more than half a MeV into the corresponding normalized energy kernel. Therefore, any mixing of paired Bogoliubov q.p. vacua should be complemented by some correction for particle-number fluctuations.

This problem can be approximately taken care of by adding constraints on the proton and neutron number to the Hill–Wheeler–Griffin equation [35]. Such a procedure has, however, the disadvantage that the corresponding Lagrange multipliers are necessarily different for each solution of the Hill–Wheeler–Griffin equation, which leaves the choice between constructing a set of slightly non-orthogonal collective states that have the same mean particle number or constructing a set of orthogonal collective states that have slightly different mean particle number [35].

Similar corrections are needed when projecting paired q.p. vacua on parity [88] or angular momentum [115, 121] without projecting also on particle number. In this case, it is customary to use the proton and neutron Fermi energies from the HFB equation to define a correction energy. As discussed in [115], the correction energy can in some cases be larger than the typical spacing of energy levels obtained from projection. Both problems are, in principle, rigorously solved by particle-number projection [81], unless one uses an EDF whose extension to the multi-reference case is ill-defined [61–63, 122, 123], in which case one trades one problem for another.

A different kind of ambiguity arises when mixing states with different quadrupole deformations in a GCM without projecting on angular momentum. Consider two different choices for the variational space: a simple one-dimensional GCM calculation that just mixes time-reversal invariant prolate and oblate axial reference states, and a more complete two-dimensional GCM calculation that additionally includes triaxial states. In the first calculation, the axial states could be chosen such that they all have the same symmetry axis, whereas for the second calculation one could choose a sextant of the β–γ plane. These two choices, however, will lead to different kernels between prolate and oblate states; with the former choice, their symmetry axes are aligned, whereas for the latter choice they are perpendicular to each other. The reason for this ambiguity is that the orientation of symmetry-breaking reference states is irrelevant only for their individual properties, but not for the matrix elements between them. A minimal solution of this problem is to consider all possible permutations of the principal axes of the reference states when calculating the kernels, which can be formulated in terms of a discrete group [124]. The rigorous solution is to combine the GCM with angular-momentum projection, as the rotation operator in equation (3.116) automatically covers all relative orientations between the two states.

An illustrative example

To showcase the various correlations obtained in an MR-EDF calculation, figures 3.4, 3.5, 3.6 and 3.7 show results obtained at various stages of a particle-number- and angular-momentum-projected GCM calculation of ^{188}Pb based on axial time-reversal and parity-conserving states [116]. Since the GCM mixing remains one-dimensional, the effect of correlations obtained at each step of the calculation can be easily illustrated. We will focus on angular-momentum projection and dynamical quadrupole correlations from the GCM, such that all states are projected on the same proton ($Z = 82$) and neutron ($N = 106$) number.

This semi-magic nucleus is a prime example of shape coexistence. It has the particular feature that its three lowest states are 0^+ states, each attributed to a state with different intrinsic deformation. The upper panel of figure 3.4 shows the total energy of the particle-number-projected q.p. vacua constructed with the help of a constraint on the dimensionless axial quadrupole moment

$$\beta_2 = \frac{4\pi}{3R^2A}\,\frac{\langle\Phi|r^2 Y_{20}\,|\Phi\rangle}{\langle\Phi|\Phi\rangle} = \sqrt{\frac{5}{16\pi}}\,\frac{4\pi}{3R^2A}\,\frac{\langle\Phi|2z^2-x^2-y^2|\Phi\rangle}{\langle\Phi|\Phi\rangle},$$

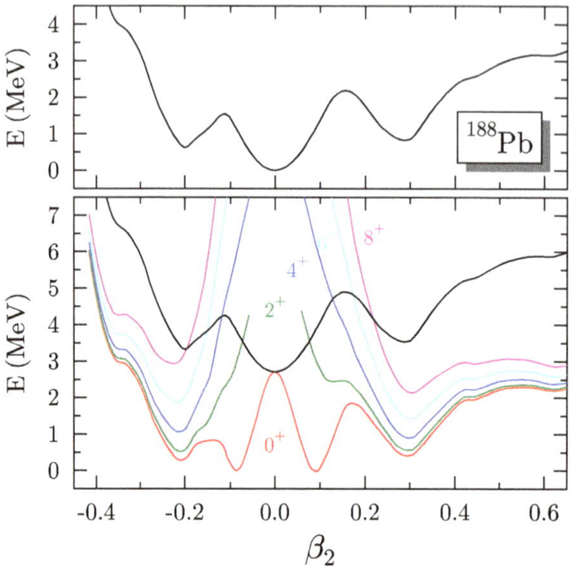

Figure 3.4. Upper panel: total energy of the particle-number-projected q.p. vacua of ^{188}Pb as a function of axial quadrupole deformation β_2. Lower panel: total energy after additional projection on the angular momenta as indicated. The curve of non-projected energies is shown again for comparison. (Figure created by M Bender.)

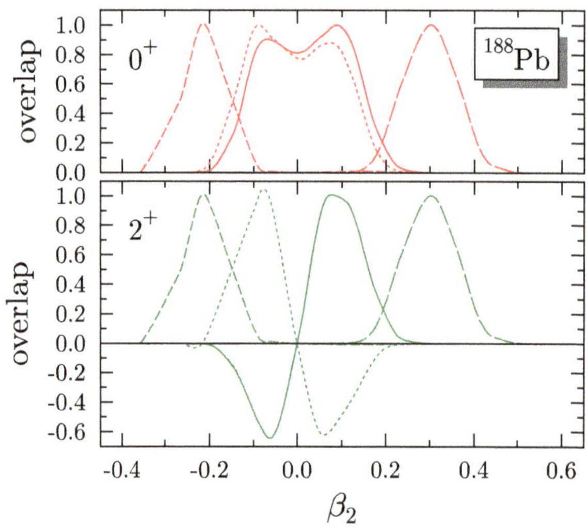

Figure 3.5. Overlap between the projected states $\langle \Phi(\beta_2)|\hat{P}^J_{00}\hat{P}^N\hat{P}^Z|\Phi(\beta_2')\rangle$ entering the GCM calculation for four selected fixed values of β_2' for $J = 0$ (top) and $J = 2$ (bottom). All states have been normalized, such that the overlaps peak at 1 for the diagonal matrix elements. (Figure created by M Bender.)

where $R = 1.2\,A^{1/3}$. Positive β_2 indicate prolate deformation, while configurations with negative β_2 are oblate. The energy surface exhibits three distinct minima at spherical ($\beta_2 = 0$), oblate ($\beta_2 = -0.19$) and prolate ($\beta_2 = 0.3$) shapes. The lower panel shows the energy of the states of angular momentum $J = 0$, 2, 4, 6 and 8 that

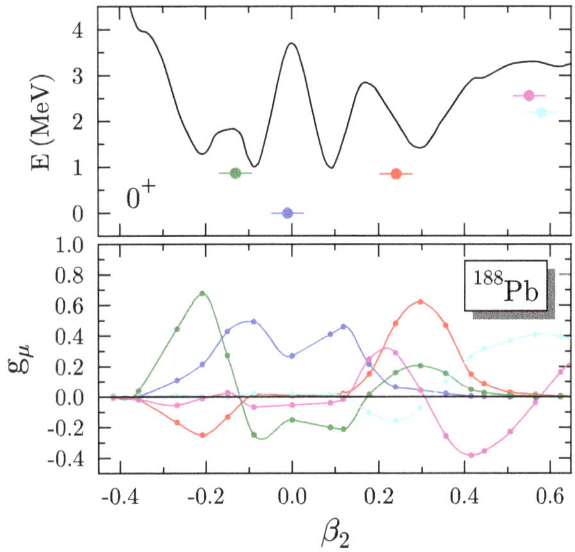

Figure 3.6. Upper panel: energy curve of the $J^\pi = 0^+$ projected states and the energies of the five lowest states obtained from their GCM mixing drawn at the average deformation $\bar{\beta}^{(J)}_{2,\,\mu}$ (see the text). Energies are normalized to the $J^\pi = 0^+$ GCM ground state. Lower panel: collective wave functions $g_\mu(\beta_2)$ of the lowest $J^\pi = 0^+$ states drawn in the same color as the corresponding energies. The dots indicate the discretization points of the GCM. (Figure created by M Bender.)

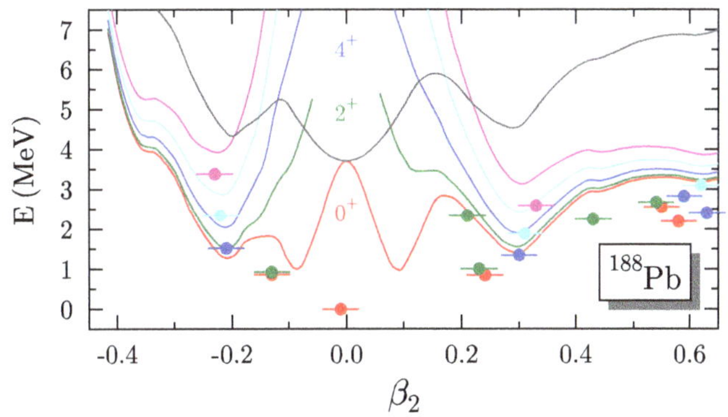

Figure 3.7. Energy curve of the projected states and the energies of the lowest states obtained from their GCM mixing drawn at the average deformation $\bar{\beta}^{(J)}_{2,\,\mu}$ (see the text) for all angular momenta up to $J = 8$ using again the color code of figure 3.4. Energies are normalized to the $J^\pi = 0^+$ GCM ground state. (Figure created by M Bender.)

can be projected out from the q.p. vacua at a given deformation β_2. Because of the axial symmetry of the q.p. vacua, the spectrum of angular-momentum-projected states remains quite simple, see the discussion of figure 3.3. The spherical state ($\beta_2 = 0$) is rotationally invariant; hence, it only contains a $J = 0$ component that necessarily also has the same energy. When deforming the nucleus, the q.p. vacuum is a superposition of angular-momentum eigenstates that are spread over an increasingly wide range of J.

At large deformation $|\beta_2| \gtrsim 0.15$, the topography of the energy curves of the projected states is the same as that of the mean-field energy curve: they all exhibit minima and barriers at near-identical deformations. What slightly changes, however, is the slope of the curves. Compared to the mean-field energy, the energy curves of the projected states are all coming down with increasing deformation, which overall softens the energy surface. The energies of the states with different J projected from the same q.p. vacuum group into a $\Delta J = 2$ rotational band, with excitation energies relative to the $J = 0$ state that approximately follow the $E(J) = J(J+1)/(2\Theta)$ law of a rigid rotor with a moment of inertia Θ that increases with β_2.

For $|\beta_2| \lesssim 0.15$, however, the very rapid increase of correlation energy from projection on $J = 0$ with deformation markedly changes the structure of the energy surface. The spherical state that provides the mean-field minimum delivers a maximum after projection, whereas a new minimum emerges at $|\beta_2| \approx 0.1$, with a total gain in correlation energy of about 3 MeV. There are many indications that the two $J = 0$ states at $\beta_2 = \pm 0.1$ are almost identical, in spite of having been projected from two very different q.p. vacua, and represent a 'correlated spherical state'. This can, for example, be seen from the overlap between projected states plotted for a few exemplary cases in figure 3.5. The two projected states at $\beta_2 = \pm 0.1$ have an overlap larger than 0.9 with each other. In fact, *all* $J = 0$ states projected from q.p. vacua with $|\beta_2| \lesssim 0.15$ have a large overlap with each other. At these small deformations, the representation of the energy as a function of the deformation of the states they have been projected from is highly ambiguous. Note that the overlap between $J = 2$ states projected from prolate and oblate q.p. vacua is negative, indicating that they have a different relative phase than the $J = 0$ components.

Note that the sign of the overlaps between the projected $J = 0$ is subject to a phase convention. Choosing them all to be real and positive is a possibility, but not a necessity. Only this choice, however, leads to collective wave functions g_μ (3.88) whose sign changes as a function of the generator coordinates can easily be interpreted.

For the GCM calculation, states projected out from about 30 different q.p. vacua are then mixed for each set of quantum numbers. Figure 3.6 shows the collective wave functions $g_\mu(\beta_2)$ and the corresponding energy levels E_μ from the mixing of $J =$ 0 states, the latter drawn at the average deformation of the reference states $|\Phi(\beta_2)\rangle$ they are built from:

$$\bar{\beta}_{2,\mu}^{(J)} = \sum_{\beta_2} \beta_2 \left(g_\mu^J(\beta_2)\right)^2.$$

This quantity is not an observable, but helps to visualize the dominant values of generator coordinates in a one-dimensional GCM calculation. Since the GCM corresponds to the diagonalization of the many-body Hamiltonian in the subspace spanned by the projected q.p. vacua, the energy of the GCM ground state is below the minimum of the energy curve of states projected on $J = 0$ that represents the diagonal matrix elements. Note that the discretization points of the GCM ansatz are not equally spaced. The final selection of states has been made according to their overlap with adjacent ones in order to avoid the appearance of redundant states in the GCM, see the discussion in section 3.2.1. The wave function of the $J = 0$ ground state is almost equally spread over states projected from slightly prolate and oblate configurations. From the wave functions $g_\mu(\beta_2)$ of the three lowest 0^+ states it can be seen that the near-spherical ground state is slightly mixed with the dominantly oblate state, which itself is also slightly mixed with the dominantly prolate state.

When combining projected energy curves and GCM energy levels for all angular momenta up to $J = 8$, figure 3.7, two rotational bands of states emerge. These bands have similar $\bar{\beta}_{2,\mu}^{(J)}$ close to the deformations of the oblate and prolate minima of the energy surfaces. That the mean deformations of states within a given band become closer with increasing J indicates that the mixing between the various structures becomes smaller at larger spin, which is also confirmed by the analysis of the collective wave functions and electromagnetic transition moments between these states [48, 119]. There are further rotational bands that are experimentally known for this nucleus. Their description, however, require a larger space for the GCM ansatz, including triaxial deformation (for the γ band) and blocked two-q.p. excitations.

Altogether, the projection on angular momentum and GCM mixing yields an energy gain of almost 4 MeV for this nucleus, which is at the upper end of the range of values typically found in such calculations for the ground-state quadrupole correlation energy [48]. Compared with the total binding energy of almost 1.5 GeV this is a small correction, but its rapid evolution of the correlation energy with N and Z can have a significant impact on separation energies [125]. Other ground-state properties such as charge radii are also affected. The most notable feature of such MR-EDF calculations is that they give access to the mixing of coexisting structures with different shape and the electromagnetic transition moments between them, which is an aspect of nuclear structure that cannot be satisfactorily described at the SR-EDF level.

References

[1] Ring P and Schuck P 2004 *The Nuclear Many-Body Problem Texts and Monographs in Physics* (Berlin: Springer)

[2] Anderson P W 1997 *Basic Notions Of Condensed Matter Physics* (Boulder, CO/Boston, MA: Westview, Addison-Wesley)

[3] Hall B C 2000 *Lie Groups. Lie Algebras and Representations* (Berlin: Springer)

[4] Cornwell J F 1984 *Group Theory in Physics* vol 1 (New York: Academic)

[5] Cornwell J F 1984 *Group Theory in Physics* vol 2 (New York: Academic)

[6] Hamermesh M 1989 *Group Theory and Its Application to Physical Problems* (New York: Dover)

[7] Varshalovich D A, Moskalev A N and Khersonskii V K 1988 *Quantum Theory of Angular Momentum* (Singapore: World Scientific)

[8] Mang H-J 1975 The self-consistent single-particle model in nuclear physics *Phys. Rep.* **18** 325

[9] Bender M, Heenen P-H and Reinhard P-G 2003 Self-consistent mean-field models for nuclear structure *Rev. Mod. Phys.* **75** 121–80

[10] Blaizot J-P and Ripka G 1985 *Quantum Theory of Finite Systems* (Cambridge: MIT Press)

[11] Dobaczewski J *et al* 2016 *TALENT Course: Density Functional Theory and Self-Consistent Methods* url: http://fribtheoryalliance.org/TALENT/courses/course_04.php

[12] Thouless D J 1960 Stability conditions and nuclear rotations in the Hartree–Fock theory *Nucl. Phys.* **21** 225

[13] Klüpfel P *et al* 2009 Variations on a theme by Skyrme: a systematic study of adjustments of model parameters *Phys. Rev.* C **79** 034310

[14] Bloch C and Messiah A 1962 The canonical form of an antisymmetric tensor and its application to the theory of superconductivity *Nucl. Phys.* **39** 95

[15] Zumino B 1962 Normal forms of complex matrices *J. Math. Phys.* **3** 1055

[16] Bertsch G *et al* 2009 Hartree–Fock–Bogoliubov theory of polarized Fermi systems *Phys. Rev.* A **79** 043602

[17] Nilsson S G and Ragnarsson I 1995 *Shapes and Shells in Nuclear Structure* (Cambridge: Cambridge University Press)

[18] Banerjee B, Ring P and Mang H J 1974 On the character of the Hartree–Fock–Bogoliubov solutions in a rotating frame *Nucl. Phys.* A **221** 564–72

[19] Gall B *et al* 1994 Superdeformed rotational bands in the mercury region. A cranked Skyrme–Hartree–Fock–Bogoliubov study *Z. Phys.* A **348** 183

[20] Dobaczewski J *et al* 2009 Solution of the Skyrme–Hartree–Fock–Bogolyubov equations in the Cartesian deformed harmonic-oscillator basis. (VI) HFODD (v.24oh): a new version of the program *Comput. Phys. Commun.* **180** 2361

[21] Schunck N *et al* 2009 Large-scale calculations in odd-mass nuclei *AIP Conf. Proc.* **1128** 40

[22] Duguet T *et al* 2001 Pairing correlations. I. Description of odd nuclei in mean-field theories *Phys. Rev.* C **65** 014310

[23] Bender M *et al* 2000 Pairing gaps from nuclear mean-field models *Eur. Phys. J.* A **8** 59

[24] Duguet T *et al* 2001 Pairing correlations. II. Microscopic analysis of odd–even mass staggering in nuclei *Phys. Rev.* C **65** 014311

[25] Bonneau L, Quentin P and Möller P 2007 Global microscopic calculations of ground-state spins and parities for odd-mass nuclei *Phys. Rev.* C **76** 024320

[26] Pototzky K J *et al* 2010 Properties of odd nuclei and the impact of time-odd mean fields: A systematic Skyrme–Hartree–Fock analysis *Eur. Phys. J.* A **46** 299

[27] Afanasjev A V and Shawaqfeh S 2011 Deformed one-quasiparticle states in covariant density functional theory *Phys. Lett.* B **706** 177–82

[28] Dobaczewski J *et al* 2015 Properties of nuclei in the nobelium region studied within the covariant, Skyrme, and Gogny energy density functionals *Nucl. Phys.* A **944** 388

[29] Tarpanov D *et al* 2014 Polarization corrections to single-particle energies studied within the energy-density-functional and quasiparticle random-phase approximation approaches *Phys. Rev.* C **89** 014307

[30] Perez-Martin S and Robledo L 2008 Microscopic justification of the equal filling approximation *Phys. Rev.* C **78** 014304

[31] Schunck N *et al* 2010 One-quasiparticle states in the nuclear energy density functional theory *Phys. Rev.* C **81** 024316

[32] Eschrig R 1996 *Fundamentals of Density Functional Theory* (Leipzig: Teubner)

[33] Dreizler R M and Gross E K U 1990 *Density Functional Theory: An Approach to the Quantum Many-Body Problem* (Berlin: Springer)

[34] Griffin J J and Wheeler J A 1957 Collective motions in nuclei by the method of generator coordinates *Phys. Rev.* **108** 311

[35] Bonche P *et al* 1990 Analysis of the generator coordinate method in a study of shape isomerism in ^{194}Hg *Nucl. Phys.* A **510** 466

[36] Enami K, Tanabe K and Yoshinaga N 1999 Microscopic description of high-spin states: quantum-number projections of the cranked Hartree–Fock–Bogoliubov self-consistent solution *Phys. Rev.* C **59** 135

[37] Benjamin B 2014 Description des noyaux impairs à l'aide d'une méthode de fonctionnelle énergie de la densité à plusieurs états de référence *PhD thesis* Université de Bordeaux url: http://www.theses.fr/2014BORD0058

[38] Burzyński K and Dobaczewski J 1995 Quadrupole-collective states in a large single-*j* shell *Phys. Rev.* C **51** 1825

[39] Reinhard P-G, Cusson R Y and Goeke K 1983 Time evolution of coherent ground-state correlations and the TDHF approach *Nucl. Phys.* A **398** 141

[40] Verrière M 2017 Description de la dynamique de la fission dans le formalisme de la méthode de la coordonnée génératrice dépendante du temps *PhD thesis* Université Paris-Saclay url: https://tel.archives-ouvertes.fr/tel-01559158

[41] Wong C W 1970 The generator-coordinate theory as a flexible formulation of the many-body Schrödinger equation *Nucl. Phys.* A **147** 545

[42] Bender M and Heenen P-H 2008 Configuration mixing of angular-momentum and particle-number projected triaxial Hartree–Fock–Bogoliubov states using the Skyrme energy density functional *Phys. Rev.* C **78** 024309

[43] Yao J M *et al* 2010 Configuration mixing of angular-momentum-projected triaxial relativistic mean-field wave functions *Phys. Rev.* C **81** 044311

[44] Rodríguez T R and Luis Egido J 2010 Triaxial angular momentum projection and configuration mixing calculations with the Gogny force *Phys. Rev.* C **81** 064323

[45] Rodríguez T R 2014 Structure of krypton isotopes calculated with symmetry-conserving configuration-mixing methods *Phys. Rev.* C **90** 034306

[46] Paul N *et al* 2017 Are there signatures of harmonic oscillator shells far from stability? First spectroscopy of ^{110}Zr *Phys. Rev. Lett.* **118** 032501

[47] Heenen P-H *et al* 1994 Octupole excitations in light xenon and barium nuclei *Phys. Rev.* C **50** 802

[48] Bender M, Bertsch G F and Heenen P-H 2006 Global study of quadrupole correlation effects *Phys. Rev.* C **73** 034322

[49] Zberecki K *et al* 2006 Tetrahedral correlations in ^{80}Zr and ^{98}Zr *Phys. Rev.* C **74** 051302(R)

[50] Yao J M, Zhou E F and Li Z P 2015 Beyond relativistic mean-field approach for nuclear octupole excitations *Phys. Rev.* C **92** 041304(R)

[51] Bernard R N, Robledo L M and Rodríguez T R 2016 Octupole correlations in the ^{144}Ba nucleus described with symmetry-conserving configuration-mixing calculations *Phys. Rev.* C **93** 061302(R)

[52] Meyer J *et al* 1991 Pairing vibrations and stability of superdeformed states *Nucl. Phys.* A **533** 307

[53] Bender M and Duguet T 2007 Pairing correlations beyond the mean field *Int. J. Mod. Phys.* E **16** 222
[54] Vaquero N L, Rodriguez T R and Egido J L 2011 On the impact of large amplitude pairing fluctuations on nuclear spectra *Phys. Lett.* B **704** 520
[55] Vaquero N L, Egido J L and Rodríguez T R 2013 Large-amplitude pairing fluctuations in atomic nuclei *Phys. Rev.* C **88** 064311
[56] Borrajo M, Rodríguez T R and Egido J L 2015 Symmetry conserving configuration mixing method with cranked states *Phys. Lett.* B **746** 341
[57] Borrajo M and Egido J L 2017 Ground-state properties of even and odd magnesium isotopes in a symmetry-conserving approach *Phys. Lett.* B **764** 328
[58] Balian R and Brézin E 1969 Nonunitary Bogoliubov transformations and extension of Wick's theorem *Nuovo Cim.* B **64** 37
[59] Onishi N and Yoshida S 1966 Generator coordinate method applied to nuclei in the transition region *Nucl. Phys.* **80** 367
[60] Flocard H and Vautherin D 1976 Generator coordinate calculations of giant resonances with the Skyrme interaction *Nucl. Phys.* A **264** 197
[61] Anguiano M, Egido J L and Robledo L M 2001 Particle number projection with effective forces *Nucl. Phys.* A **696** 467
[62] Dobaczewski J *et al* 2007 Particle-number projection and the density functional theory *Phys. Rev.* C **76** 054315
[63] Bender M, Duguet T and Lacroix D 2009 Particle-number restoration within the energy density functional formalism *Phys. Rev.* C **79** 044319
[64] Hara K, Hayashi A and Ring P 1982 Exact angular momentum projection of cranked Hartree–Fock–Bogoliubov wave functions *Nucl. Phys.* A **385** 14
[65] Valor A, Heenen P-H and Bonche P 2000 Configuration mixing of mean-field wave functions projected on angular momentum and particle number: application to ^{24}Mg *Nucl. Phys.* A **671** 145
[66] Neergård K and Wüst E 1983 On the calculation of matrix elements of operators between symmetry-projected Bogoliubov states *Nucl. Phys.* A **402** 311
[67] Oi M 2015 Comparison of various HFB overlap formulae *Bulg. J. Phys.* **42** 404
[68] Egido J L, Borrajo M and Rodríguez T R 2016 Collective and single-particle motion in beyond mean field approaches *Phys. Rev. Lett.* **116** 052502
[69] Robledo L M 2009 Sign of the overlap of Hartree–Fock–Bogoliubov wave functions *Phys. Rev.* C **79** 021302
[70] Robledo L M 2011 Technical aspects of the evaluation of the overlap of Hartree–Fock–Bogoliubov wave functions *Phys. Rev.* C **84** 014307
[71] Bertsch G F and Robledo L M 2012 Symmetry restoration in Hartree–Fock–Bogoliubov based theories *Phys. Rev. Lett.* **108** 042505
[72] Mizusaki T, Oi M and Shimizu N 2018 Why does the sign problem occur in evaluating the overlap of HFB wave functions? *Phys. Lett.* B **779** 237
[73] Avez B and Bender M 2012 Evaluation of overlaps between arbitrary fermionic quasi-particle vacua *Phys. Rev.* C **85** 034325
[74] Hagen G *et al* 2014 Coupled-cluster computations of atomic nuclei *Rep. Prog. Phys.* **77** 096302
[75] Hergert H *et al* 2016 The in-medium similarity renormalization group: a novel *ab initio* method for nuclei *Phys. Rep.* **621** 165–222

[76] Duguet T 2015 Symmetry broken and restored coupled-cluster theory: I. Rotational symmetry and angular momentum *J. Phys. G: Nucl. Part. Phys.* **42** 025107
[77] Duguet T and Signoracci A 2017 Symmetry broken and restored coupled-cluster theory: II. Global gauge symmetry and particle number *J. Phys. G: Nucl. Part. Phys.* **44** 015103
[78] Bally B and Duguet T 2018 Norm overlap between many-body states: uncorrelated overlap between arbitrary Bogoliubov product states *Phys. Rev.* C **97** 024304
[79] Robledo L M 1994 Practical formulation of the extended Wick's theorem and the Onishi formula *Phys. Rev.* C **50** 2874
[80] Bonche P *et al* 1990 Quadrupole collective correlations and the depopulation of the superdeformed bands in mercury *Nucl. Phys.* A **519** 509–20
[81] Heenen P-H *et al* 1993 Generator-coordinate method for triaxial quadrupole dynamics in Sr isotopes (II). Results for particle-number-projected states *Nucl. Phys.* A **561** 367
[82] Brink D 1966 The alpha-particle model of light nuclei, *Many-Body Description of Nuclear Structure and Reactions, Proc. International School of Physics 'Enrico Fermi', Course 36* **vol 256**, *Varenna, Italy* ed C Bloch (New York: Academic) p 247
[83] Wong C W 1975 Generator-coordinate methods in nuclear physics *Phys. Rep.* **15** 283
[84] Duguet T 2014 The nuclear energy density functional formalism *The Euroschool on Exotic Beams* ed C Scheidenberger and M Pfützner vol 4 (Berlin: Springer) p 293
[85] Schmid K W and Reinhard P-G 1991 Center-of-mass projection of Skyrme–Hartree–Fock densities *Nucl. Phys.* A **530** 283
[86] Rodríguez-Guzmán R R and Schmid K W 2004 Spherical Hartree–Fock calculations with linear-momentum projection before the variation. Part I: Energies, form factors, charge densities and mathematical sum rules *Eur. Phys. J.* A **19** 45
[87] Rodríguez-Guzmán R R and Schmid K W 2004 Spherical Hartree–Fock calculations with linear-momentum projection before the variation. Part II: Spectral functions and spectroscopic factors *Eur. Phys. J.* A **19** 61
[88] Egido J L and Robledo L M 1991 Parity-projected calculations on octupole deformed nuclei *Nucl. Phys.* A **524** 65
[89] Satuła W *et al* 2010 Isospin-symmetry restoration within the nuclear density functional theory: formalism and applications *Phys. Rev.* C **81** 054310
[90] Satuła W *et al* 2012 Isospin-breaking corrections to superallowed Fermi β S decay in isospin- and angular-momentum-projected nuclear density functional theory *Phys. Rev.* C **86** 054316
[91] Schunck N *et al* 2012 Solution of the Skyrme–Hartree–Fock–Bogolyubov equations in the Cartesian deformed harmonic-oscillator basis. (VII) HFODD (v2.49t): a new version of the program *Comput. Phys. Commun.* **183** 166
[92] Sheikh J A and Ring P 2000 Symmetry-projected Hartree–Fock–Bogoliubov equations *Nucl. Phys.* A **665** 71
[93] Sheikh J A *et al* 2002 Pairing correlations and particle-number projection methods *Phys. Rev.* C **66** 044318
[94] Stoitsov M V *et al* 2007 Variation after particle-number projection for the Hartree–Fock–Bogoliubov method with the Skyrme energy density functional *Phys. Rev.* C **76** 014308
[95] Fomenko V N 1970 Projection in the occupation-number space and the canonical transformation *J. Phys. A: Gen. Phys.* **3** 8
[96] Hupin G, Lacroix D and Bender M 2011 Formulation of functional theory for pairing with particle number restoration *Phys. Rev.* C **84** 014309

[97] Lipkin H J 1960 Collective motion in many-particle systems: Part 1. The violation of conservation laws *Ann. Phys.* **9** 272

[98] Lipkin H J 1961 Collective motion in many-particle systems *Ann. Phys.* **12** 452

[99] Wang X B *et al* 2014 Lipkin method of particle-number restoration to higher orders *Phys. Rev.* C **90** 014312

[100] Nogami Y 1964 Improved superconductivity approximation for the pairing interaction in nuclei *Phys. Rev.* **134** B313

[101] Quentin P *et al* 1990 Approximate energy correction for particle number symmetry breaking in constrained Hartree–Fock plus BCS calculations *Phys. Rev.* C **41** 341

[102] Magierski P *et al* 1993 Approximate particle number projection for rotating nuclei *Phys. Rev.* C **48** 1686

[103] Satuła W, Wyss R and Magierski P 1994 The Lipkin–Nogami formalism for the cranked mean field *Nucl. Phys.* A **578** 45

[104] Reinhard P-G *et al* 1996 Lipkin–Nogami pairing scheme in self-consistent nuclear structure calculations *Phys. Rev.* C **53** 2776

[105] Nikšić T, Vretenar D and Ring P 2006 Beyond the relativistic mean-field approximation. II. Configuration mixing of mean-field wave functions projected on angular momentum and particle number *Phys. Rev.* C **74** 064309

[106] McWeeny R 2002 *An Introduction to Group Theory and its Applications* (New York: Dover)

[107] Tinkham M 1992 *Group Theory and Quantum Mechanics* (New York: Dover)

[108] de Voigt M J A, Dudek J and Szymański Z 1983 High-spin phenomena in atomic nuclei *Rev. Mod. Phys.* **55** 949

[109] Dobaczewski J *et al* 2000 Point symmetries in the Hartree–Fock approach. I. Densities, shapes, and currents *Phys. Rev.* C **62** 014310

[110] Dobaczewski J *et al* 2000 Point symmetries in the Hartree–Fock approach. II. Symmetry-breaking schemes *Phys. Rev.* C **62** 014311

[111] Sabbey B *et al* 2007 Global study of the spectroscopic properties of the first 2^+ state in even–even nuclei *Phys. Rev.* C **75** 044305

[112] Nikšić T, Vretenar D and Ring P 2011 Relativistic nuclear energy density functionals: mean-field and beyond *Prog. Part. Nucl. Phys.* **66** 519

[113] Hupin G and Lacroix D 2012 Number-conserving approach to the pairing problem: application to Kr and Sn isotopic chains *Phys. Rev.* C **86** 024309

[114] Rodríguez T R *et al* 2005 Quality of the restricted variation after projection method with angular momentum projection *Phys. Rev.* C **71** 044313

[115] Yao J M *et al* 2011 Configuration mixing of angular-momentum-projected triaxial relativistic mean-field wave functions. II. Microscopic analysis of low-lying states in magnesium isotopes *Phys. Rev.* C **83** 014308

[116] Bender M, Heenen P-H and Bonche P 2004 Microscopic study of ^{240}Pu: mean field and beyond *Phys. Rev.* C **70** 054304

[117] Bender M *et al* 2004 Configuration mixing of angular momentum projected self-consistent mean-field states for neutron-deficient Pb isotopes *Phys. Rev.* C **69** 064303

[118] Rodríguez-Guzmán R R, Egido J L and Robledo L M 2004 Beyond mean field description of shape coexistence in neutron-deficient Pb isotopes *Phys. Rev.* C **69** 054319

[119] Yao J M, Bender M and Heenen P-H 2013 Systematics of low-lying states of even–even nuclei in the neutron-deficient lead region from a beyond-mean-field calculation *Phys. Rev.* C **87** 034322

[120] Egido J L 2016 State-of-the-art of beyond mean field theories with nuclear density functionals *Phys. Scr.* **91** 073003
[121] Nikšić T, Vretenar D and Ring P 2006 Beyond the relativistic mean-field approximation: configuration mixing of angular-momentum-projected wave functions *Phys. Rev.* C **73** 034308
[122] Lacroix D, Duguet T and Bender M 2009 Configuration mixing within the energy density functional formalism: removing spurious contributions from nondiagonal energy kernels *Phys. Rev.* C **79** 044318
[123] Duguet T *et al* 2009 Particle-number restoration within the energy density functional formalism: nonviability of terms depending on noninteger powers of the density matrices *Phys. Rev.* C **79** 044320
[124] Bonche P *et al* 1991 Generator coordinate method for triaxial quadrupole collective dynamics in strontium isotopes *Nucl Phys.* A **530** 149
[125] Bender M, Bertsch G F and Heenen P-H 2008 Collectivity-induced quenching of signatures for shell closures *Phys. Rev.* C **78** 054312

IOP Publishing

Energy Density Functional Methods for Atomic Nuclei

Nicolas Schunck

Chapter 4

Time-dependent density functional theory

Aurel Bulgac and Michael McNeil Forbes

Phenomena such as nuclear reactions and nuclear fission are intrinsically non-equilibrium time-dependent processes. While the formalism developed in chapters 3, 5 and 6 can be adapted, the extension of energy density functional (EDF) concepts to a time-dependent framework has a wider applicability. In fact, time-dependent density functional theory (TDDFT) encompasses many of the techniques developed to study excited states as we will see, e.g. in the derivation of the linear response in chapter 5 or of collective inertia in chapter 6.

> One might see the term time-dependent energy density functional (TDEDF) used for nuclei instead of TDDFT. This highlights a fundamental difference between DFT for electrons and DFT for nuclei: the microscopic electronic Hamiltonian is two-body and known to high accuracy (in the non-relativistic approximation), while the nuclear Hamiltonian is unknown and contains many-body (possibly up to A-body) terms. As a result, nuclear DFT must encapsulate two 'unknowns', including the nuclear Hamiltonian itself (energy) in addition to the induced many-body effects common to electronic theories. This is one of the reasons that we often use the vocable of EDF for nuclei rather than DFT. Further specializations of the terminology—TDHF, TDHFB and TDSLDA—refer to specific implementations of TDDFT/TDEDF as described in the text.

The starting point for the EDF theory in chapter 3 was the time-independent many-body Schrödinger equation. An obvious generalization is to start from the time-dependent many-body Schrödinger equation which can be recast in the form of a variational (action) principle. Chapter 9 of [1] contains an extended discussion on the derivation of TDDFT from a suitable variational principle. We only recall here the form of the action that will be used to set up the time-dependent extension of the EDF theory,

$$S[\langle\Psi(t)|,\, |\Psi(t)\rangle] = \int_{t_i}^{t_f} \mathrm{d}t\; \langle\Phi(t)|\left(\mathrm{i}\hbar\frac{\partial}{\partial t} - \hat{H}\right)|\Phi(t)\rangle. \tag{4.1}$$

doi:10.1088/2053-2563/aae0edch4

This action is a functional of the many-body (time-dependent) state and its Hermitian conjugate, and the time-dependent many-body Schrödinger equation follows as the Euler–Lagrange equation from the least-action principle $\delta S = 0$. The true meaning of the least-action principle is elucidated within the Feynman path-integral reformulation of quantum mechanics [2]. Here the quantum propagator is represented as a path integral of $\exp(iS/\hbar)$ over all trajectories joining the initial and final configurations, and the condition $\delta S = 0$ merely selects the trajectory with the highest weight or the most probable trajectory along which the quantum system evolves. By selecting only this trajectory the quantum fluctuations are overlooked [2, 3]. However, they are important for restoring broken symmetries; for describing the total kinetic energy distributions, total excitation energy and fragments yields in fission; and for describing the particle transfer distributions during the collision of nuclei. The treatment of quantum fluctuations in a time-dependent framework was discussed in a series of papers by Balian and Veneroni [4, 5]; see also [6] for a recent application to TDDFT.

We can now make additional approximations about the many-body state and use the least-action principle with equation (4.1) to obtain specific time evolution equations. In particular, the time-dependent Hartree–Fock (TDHF) approximation will be obtained by constraining $|\Psi(t)\rangle$ to be a Slater determinant at all times t, and the time-dependent Hartree–Fock–Bogoliubov (TDHFB) equation will similarly be derived after constraining $|\Psi(t)\rangle$ to be a Hartree–Fock–Bogoliubov (HFB) vacuum of the type of equation (3.3) at all times; now with time-dependent U and V blocks of the Bogoliubov matrix or, equivalently, with time-dependent densities. Both approximations should be viewed as specific realizations of a phenomenological TDDFT approach. In each case, the equation of motion for the relevant degree of freedom (d.o.f)—namely, the one-body density matrix $\rho(t)$ for TDHF and the generalized density $\mathcal{R}(t)$ for TDHFB—are obtained from the appropriately constrained variational principle $\delta S = 0$ with the action equation (4.1); see [1].

In this chapter, we will focus on the TDHFB approach since the TDHF approach is well described in the literature, and, as we shall argue below, the self-consistent inclusion of pairing is crucial for nuclear dynamics. For a discussion of TDHF theory, we refer the reader to the standard literature [7] and the recent textbook [8], which describe both the theory, and its applications to nuclear excitations and nuclear reactions. This latter property justifies the use of fully-fledged TDHFB calculations, even though they are considerably more involved numerically than TDHF. To mitigate this cost, TDHFB calculations are almost always performed with a quasi-local EDF of the Skyrme-type; the resulting approach has been dubbed the time-dependent superfluid local density approximation (TDSLDA). The current chapter will briefly review some important aspects of the TDSLDA and highlight the versatility of related time-dependent methods. Complementary discussions can be found in the review articles [9–11].

4.1 Time evolution equations

The theory presented here is the analogue for atomic nuclei (generalized to superfluids) of Kohn and Sham's [12] reformulation for fermions of the original Hohenberg and Kohn [13] DFT theorem.

Derivation of the TDDFT equations

There are multiple ways to derive an equation of motion in TDDFT. Historically, the derivation in the nuclear physics literature (i.e. [7]), begins with the definition of equations (3.1) and (3.17) of the relevant density

$$\rho_{ji}(t) = \frac{\langle \Phi(t)|c_j^\dagger c_i|\Phi(t)\rangle}{\langle \Phi(t)|\Phi(t)\rangle} \qquad \mathcal{R}(t) = \begin{pmatrix} \frac{\langle \Phi(t)|c_j^\dagger c_i|\Phi(t)\rangle}{\langle \Phi(t)|\Phi(t)\rangle} & \frac{\langle \Phi(t)|c_j c_i|\Phi(t)\rangle}{\langle \Phi(t)|\Phi(t)\rangle} \\ \frac{\langle \Phi(t)|c_j^\dagger c_i^\dagger|\Phi(t)\rangle}{\langle \Phi(t)|\Phi(t)\rangle} & \frac{\langle \Phi(t)|c_j c_i^\dagger|\Phi(t)\rangle}{\langle \Phi(t)|\Phi(t)\rangle} \end{pmatrix}$$

where $(c, c^\dagger)$ is an arbitrary basis of the single-particle (s.p.) Hilbert space, and $|\Phi(t)\rangle$ is the exact time-dependent many-body wave function. Taking the time-derivative on each side and using $i\partial_t|\Phi\rangle = \hat{H}|\Phi(t)\rangle$ leads to an equation with the one-body, two-body, ..., up to the N-body density matrix if $\hat{H}$ is a general N-body Hamiltonian. These expressions simplify if one restricts $|\Phi(t)\rangle$ to be either a Slater determinant (TDHF) or an HFB vacuum (TDHFB) at all times. In these cases, the application of the generalized, time-dependent Wick theorem leads to, respectively,

$$i\hbar\dot{\rho} = [h(t), \rho(t)] \qquad \text{TDHF} \tag{4.2a}$$

$$i\hbar\dot{\mathcal{R}} = [\mathcal{H}(t), \mathcal{R}(t)] \qquad \text{TDHFB.} \tag{4.2b}$$

Although the presentation in [1] is based on a more general framework built on the derivation of a classical action, it still follows a similar principle, starting with a many-body Hamiltonian and using the Wick theorem to derive an evolution equation. In electronic DFT such as, e.g. [14], the TDDFT evolution equation is obtained directly by extending the Kohn–Sham (KS) scheme to a time-dependent average potential which is given by the functional derivative of the (time-dependent) energy functional. This approach is summarized in [9].

The time-dependent theory also formally follows from the static theory by simply replacing s.p. energies in the HFB eigenvalue problem equation (3.44) with $i\hbar\partial_t$ as in the time-dependent Schrödinger equation: $H|\psi_n\rangle = E_n|\psi_n\rangle \rightarrow H|\psi\rangle = i\hbar\partial_t|\psi\rangle$. These equations can also be derived from requiring that the functional

$$S = i\hbar \int dt \int d^3\boldsymbol{r} \sum_\mu \left\{ U_\mu^*(\boldsymbol{r}, t)\partial_t U_\mu(\boldsymbol{r}, t) + V_\mu^*(\boldsymbol{r}, t)\partial_t V_\mu(\boldsymbol{r}, t) \right\}$$

$$+ \int dt \int d^3\boldsymbol{r} \{\mathcal{E}[\rho(\boldsymbol{r}, t), \tau(\boldsymbol{r}, t), \boldsymbol{j}(\boldsymbol{r}, t), \kappa(\boldsymbol{r}, t)] + V_{\text{ext}}(\boldsymbol{r}, t)\rho(\boldsymbol{r}, t)\}$$

be stationary.

The coordinate-space representation of the HFB spinors was defined in equation (3.9). The full set of densities derived from the one-body density matrix was introduced in equation (1.22). In the special case where only the local density is considered, they reduce to

$$\rho(\boldsymbol{r},t) = 2\sum_{\mu} |V_\mu(\boldsymbol{r},t)|^2 \qquad \tau(\boldsymbol{r},t) = 2\sum_{\mu} |\nabla V_\mu(\boldsymbol{r},t)|^2$$

$$\boldsymbol{j}(\boldsymbol{r},t) = 2\,\Im\left[i\hbar \sum_{\mu} V_\mu^*(\boldsymbol{r},t)\nabla V_\mu(\boldsymbol{r},t) \right] \qquad \kappa(\boldsymbol{r},t) = \sum_{\mu} V_\mu^*(\boldsymbol{r},t)U_\mu(\boldsymbol{r},t).$$

Here $U_\mu(\boldsymbol{r},t)$ and $V_\mu(\boldsymbol{r},t)$ are the coordinate-space, time-dependent quasiparticle wave functions and $\mathcal{E}$ is the energy density. In the local density approximation (LDA) approximation, the latter depends on the number density $\rho(\boldsymbol{r},t)$, kinetic density $\tau(\boldsymbol{r},t)$, current density $\boldsymbol{j}(\boldsymbol{r},t)$, various spin currents, etc, and anomalous density $\kappa(\boldsymbol{r},t)$ respectively. $V_{\text{ext}}(\boldsymbol{r},t)$ is some external potential in which the system might reside, or a potential exerted by some external probe. Coupling to external gauge potentials, such as electromagnetic fields, can be similarly incorporated through the minimal coupling procedure. In this manner one can also couple the magnetic moments to external magnetic fields.

The main difference between the EDF for stationary phenomena and time-dependent phenomena is in the appearance of time-odd currents. A natural requirement is thus that the description of the intrinsic motion (with the center-of-mass motion excluded) should be Galilean invariant [15]. Since a system can either separate into several fragments or several partners may coalesce during the time evolution, the stronger *local* Galilean invariance needs to be enforced, as the description of the intrinsic motion of any possible separated fragments should satisfy Galilean invariance as well. Another qualitative difference between static and time-dependent formalisms is that spin-densities, and in general time-odd densities, which in even–even static nuclei typically vanish, in time-dependent processes are typically non-vanishing; due to the spin–orbit interaction and also due to the presence of external magnetic fields. For this reason there is no difference between the TDSLDA formalism for even–even, even–odd, odd–even, or odd–odd systems. Another point to mention is that nuclear systems with an odd number of neutrons or protons and with pairing correlations have different chemical potentials for the states with positive and negative total single-particle angular momentum [16, 17].

Let us briefly recall that an equation is invariant under a global gauge transformation of a field $\Psi(\boldsymbol{x})$ if it does not change under a x-independent transformation $\hat{S}$ of the field, $\hat{S}\Psi(\boldsymbol{x}) = (\boldsymbol{x})$. One speaks of local gauge invariance when the transformation $\hat{S}$ is $\boldsymbol{x}$-dependent: $\hat{S}(x)\Psi(\boldsymbol{x}) = (\boldsymbol{x})$. In the context of relativistic quantum field theory, $\boldsymbol{x}$ stands for the four-vector $\boldsymbol{x} \equiv (ct, \boldsymbol{r})$; see section 2.1.1. The use of local gauge invariance has been crucial in the development of field theories. In the nuclear EDF approach, local gauge invariance has also been invoked to impose constraint, e.g. on time-odd fields or higher-order terms of the functional [18–20].

The resulting evolution equations represent an infinite set of coupled non-linear time-dependent 3D partial differential equations for the quasiparticle (q.p.) wave functions

$$\begin{gathered} i\hbar\partial_t\begin{pmatrix} U_{\mu\uparrow}(\boldsymbol{r}, t) \\ U_{\mu\downarrow}(\boldsymbol{r}, t) \\ V_{\mu\uparrow}(\boldsymbol{r}, t) \\ V_{\mu\downarrow}(\boldsymbol{r}, t) \end{pmatrix} = H\begin{pmatrix} U_{\mu\uparrow}(\boldsymbol{r}, t) \\ U_{\mu\downarrow}(\boldsymbol{r}, t) \\ V_{\mu\uparrow}(\boldsymbol{r}, t) \\ V_{\mu\downarrow}(\boldsymbol{r}, t) \end{pmatrix} \\ H = \begin{pmatrix} h_{\uparrow\uparrow} - \lambda_{\mathrm{F}} & h_{\uparrow\downarrow} & 0 & \Delta \\ h_{\downarrow\uparrow} & h_{\downarrow\downarrow} - \lambda_{\mathrm{F}} & -\Delta & 0 \\ 0 & -\Delta^* & -h^*_{\uparrow\uparrow} + \lambda_{\mathrm{F}} & -h^*_{\uparrow\downarrow} \\ \Delta^* & 0 & -h^*_{\downarrow\uparrow} & -h^*_{\downarrow\downarrow} + \lambda_{\mathrm{F}} \end{pmatrix}. \end{gathered} \tag{4.3}$$

For simplicity, we have suppressed the space and time coordinates $(\boldsymbol{r}, t)$ in the s.p. Hamiltonian H. Here both the local mean field $h_{\sigma\sigma'}$ and pairing field Δ depend on the various s.p. densities, equation (1.22), in accordance with the chosen nuclear energy density functional (NEDF). The index μ labels each q.p. wave function, and is both discrete and continuous. The self-consistent potential will generally admit a finite number of bound states (corresponding to discrete μ) but (especially with pairing) will also require an occupation of continuum (i.e. plane-wave) states. This index must also run over isospin so that we have a similar set of equations for both protons and neutrons. We have explicitly included the spin indices $(\sigma, \sigma') \in \{\uparrow, \downarrow\}$ in equation (4.3), allowing for mixing between the spin-up and spin-down states by the spin–orbit interaction. In this manner one captures the effects of proton–proton and neutron–neutron pairing.

By allowing different species to have different chemical potentials, this formalism allows one to study polarized systems, although some care is required in the implementation of the self-consistent strategy (see e.g. [17] for a discussion of these details, and [16] for an example of an interesting super-solid phase predicted to exist in polarized superfluids using these techniques).

As long as these equations remain local and do not involve density matrices, i.e. $H(\boldsymbol{r}, t)$ depends only on the local densities, their derivatives, currents, etc, at the point $\boldsymbol{r}$ and time t, they can be solved using a variety of numerical techniques as described in chapter 8. The formulation of a local DFT for superfluids was a challenge due to the divergence of the local anomalous density $\kappa(\boldsymbol{r}, \boldsymbol{r}', t) \propto 1/\|\boldsymbol{r} - \boldsymbol{r}'\|$. The resolution will be discussed in section 4.3, but first we discuss why pairing is crucial for nuclear dynamics.

In principle, the functional might depend on off-diagonal entries of the density matrix such as $\rho(\boldsymbol{r}, \boldsymbol{r}', t) = \sum_\mu V^*_\mu(\boldsymbol{r}, t)V_\mu(\boldsymbol{r}', t)$. The resulting time-dependent equations from $\delta S = 0$ will no longer be a simple differential equation as in equation (4.3) and will instead be integro-differential equations which are significantly (usually *prohibitively*) more difficult to solve numerically.

4.2 Role of pairing correlations in nuclear dynamics

As discussed in sections 1.3.1 and 9.2, pairing has a relatively small impact on some ground-state observables such as nuclear binding energies. One might thus naïvely expect that a simpler time-dependent theory such as TDHF would have a similar qualitative accuracy as the fully-fledged TDHFB approach characterized by equation (4.3). However, like superconductivity in condensed matter systems, pairing plays a central role in our discussion. Here we discuss why it is essential to qualitatively describe low-energy nuclear dynamics.

As the shape of the nucleus changes during a dynamic evolution, the q.p. (instantaneous) energy levels move up and down, as in the well-known Nilsson diagram scheme, as a function of deformation or rotational frequency [7, 21, 22]. This is schematically illustrated in figure 4.1. Pairing correlations or any other residual interactions lead to 'avoiding' the level crossings. If the occupations of the levels are not able to adjust, then the Fermi surface will become oblate as the nucleus becomes more prolate [23], instead of remaining spherical [24–28].

In even–even nuclei, the s.p. levels display Kramers's degeneracy: time-reversed orbitals are both either occupied or unoccupied. In order to maintain the sphericity of the local Fermi surface at a level crossing (and therefore avoid volume excitations of the system), both fermions occupying the time-reversed states in upward moving states have to be promoted simultaneously to downward moving unoccupied levels. Depending on the speed of the nuclear shape evolution, at each such avoided level crossing, a Landau–Zener transition may occur, with the probability of transition depending exponentially on the rate of evolution and the energy splitting of the crossing.

In the absence of pairing, nucleons in an upward moving level are likely to remain in that state, gradually allowing the Fermi sphere to become deformed. This results in volume excitations with significant collective energy. Imagine a hot quivering blob. The presence of a Bose–Einstein condensate (BEC) of Cooper pairs formed from nucleons in time-reversed orbitals enhances the transitions of paired states, thereby preserving the sphericity of the Fermi surface, suppressing these collective excitations. Since the Pauli exclusion principle strongly suppresses other collision mechanisms in the nucleus, pairing correlations are thus the only effective mechanism for keeping the nucleus thermally cold locally. This results in a low-energy collective motion which follows an almost adiabatic evolution, during which the nucleus remains cold at the deformation reached while its shape evolves without producing entropy. In a sense, pairing thus acts as a 'quantum lubricant,' reducing the tendency of a dynamical nucleus to occupy highly excited q.p. states (hot

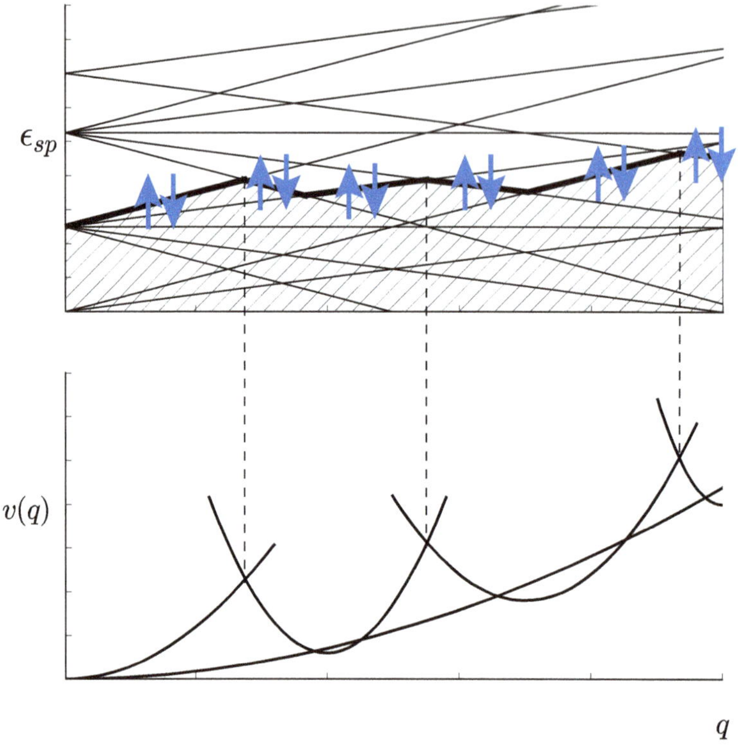

Figure 4.1. Schematic evolution of s.p. energy levels ϵ_{sp} (upper panel), and 'total energy' $v(q)$ (lower panel) is shown as a function of a generic 'deformation' q. The double-occupancy of the Fermi level due to Kramers's degeneracy is indicated with a thick solid line. Redistribution of the double-occupancy of the Fermi level is greatly enhanced in the presence of pairing correlations, which favor pair transitions of nucleons in time-reversed states. (Figure created by A Bulgac.)

quivering blob), and instead enables low-energy, large-amplitude collective motion to proceed as observed, for example, in fission.

> While collisions are strongly suppressed in low-energy large-amplitude collective motion, one-body dissipation (Landau damping) is still very effective [29, 30]. This one-body dissipation allows the transfer of collective energy to internal states via nucleons colliding with moving 'nuclear walls' in dynamical systems (i.e. the time-dependent self-consistent potential), and the exchange of nucleons through an open 'window' when two nuclear sub-systems are in contact.

We can easily obtain from equation (4.3) a continuity equation which guarantees particle number conservation

$$\dot{\rho}(\boldsymbol{r}, t) + \nabla \cdot \boldsymbol{j}(\boldsymbol{r}, t) = 0 \qquad \frac{\partial N}{\partial t} = \frac{\partial}{\partial t} \sum_{\mu,\sigma} \int \mathrm{d}^3\boldsymbol{r}\, |V_{\mu,\sigma}(\boldsymbol{r}, t)| = 0. \tag{4.4}$$

The continuity equation is responsible for one the most fundamental attributes of both classical and quantum theories: the conservation of matter and charge. It also serves as a guiding principle when developing new theoretical frameworks. It ensures the locality of matter transport, thereby preventing non-physical action at distance. It is thus of crucial importance for time-dependent theories to satisfy the continuity equation, which gives a clear advantage to the TDHFB approach over the TDHF +Bardeen–Cooper–Schrieffer (BCS) approach, where the continuity equation is not satisfied [31].

However, note that within the TDHFB, the *individual* occupation probabilities of q.p. levels are not conserved,

$$\dot{\rho}_\mu = \partial_t \sum_\sigma \int \mathrm{d}^3\boldsymbol{r} \; |V_{\mu,\sigma}(\boldsymbol{r}, t)|^2 \neq 0.$$

It is through this mechanism that the TDHFB approach allows an evolving nucleus to maintain a (roughly) spherical Fermi surface in the absence of collisions. Again, this requires a full implementation of pairing as in the TDHFB equation. There have been attempts to mimic this property in simpler approaches based on TDHF, for example, by keeping the occupation probabilities of s.p. levels frozen during the time evolution [31–33]. However, this is not a very good approximation, even for collective excitations of small amplitude such as the giant dipole resonance (GDR). Full TDHFB calculations show that the occupation probabilities change rather dramatically when a GDR is excited in a superfluid nucleus, as shown in figure 4.2.

4.3 Local DFT for superfluids

As described above, computational efficiency demands that the evolution equations (4.3) be local. A local formulation of DFT for superfluids, however, met some early challenges. In condensed matter, the proper description of superconductors within

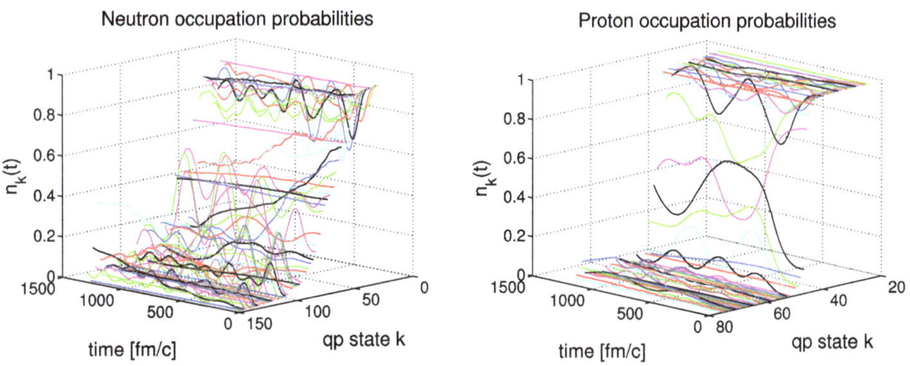

Figure 4.2. Time evolution of the neutron and proton occupation probabilities in ^{238}U near the Fermi level during the excitation of the GDR with a weak time-dependent electromagnetic dipole field. Note that many q.p. states cross during this excitation of modest amplitude. (Figure reproduced with permission from [34], courtesy of Stetcu. Copyright 2011 The American Physical Society.)

the framework of traditional DFT requires the introduction of an anomalous density to track the superfluid component [35]. Unfortunately, the resulting functional was no longer local, which required solving integro-differential equations. While the resulting equations can sometimes be transformed into partial differential equation (PDEs) [36], they are particularly difficult to handle numerically, especially in the case of time-dependent phenomena.

Oliveira *et al* [35] were forced to use integro-differential equations because the anomalous and kinetic energy densities diverge in the case of a local pairing field [37–39]. This divergence can be averted by using an appropriate short-range regularization of the anomalous density and a corresponding renormalization of the pairing coupling constant [38–40]. With this regularization procedure in place, one can follow the same steps as in the original papers of Hohenberg and Kohn [13], and Kohn and Sham [12], to establish the existence of a DFT for superfluid fermions in both the static [41] and time-dependent forms [14, 42].

The generalization of the local Kohn–Sham theory (LDA) for normal systems to a local theory for superfluids leads to the superfluid local density approximation (SLDA) and its time-dependent extension (the TDSLDA) for static and time-dependent phenomena, respectively [43–49]. We now describe the details of this theory, demonstrating how a local form can be obtained with an appropriate pairing regularization.

The origin of this TDSLDA acronym was chosen in analogy with the acronyms for the Kohn–Sham formulation of DFT, the LDA for normal non-polarized systems, since it depends only locally on the densities. As we mentioned above, this is an essential feature, preserved by the TDSLDA, for rendering the approach numerically efficient. The acronym also avoids conflict with the local spin density approximation (LSDA) used for normal spin-polarized ones.

4.3.1 Pairing regularization of the anomalous density

Let us consider the HFB equation (3.44) (corresponding to the middle 2×2 block of equation (4.3)) in homogeneous infinite nuclear matter (INM) where the eigenfunctions are plane waves with quantum numbers $\mu \equiv k$ and the spin–orbit potential is absent,

$$\begin{pmatrix} h_{\uparrow\uparrow} - \lambda_{\mathrm{F}} & \Delta \\ \Delta^* & -h^*_{\downarrow\downarrow} + \lambda_{\mathrm{F}} \end{pmatrix}\begin{pmatrix} U_{\mu\uparrow} \\ V_{\mu\downarrow} \end{pmatrix} \equiv \begin{pmatrix} \varepsilon_k - \lambda_{\mathrm{F}} & \Delta \\ \Delta^* & -\varepsilon_k + \lambda_{\mathrm{F}} \end{pmatrix}\begin{pmatrix} U_{k\uparrow} \\ V_{k\downarrow} \end{pmatrix} = E_k\begin{pmatrix} U_{k\uparrow} \\ V_{k\downarrow} \end{pmatrix}.$$

In this expression, λ_{F} is the chemical potential and the q.p. energies E_k and effective s.p. energies ε_k are given by

$$E_k = \sqrt{\left(\frac{\hbar^2 k^2}{2m} + V_{\mathrm{s.p.}} - \lambda_{\mathrm{F}}\right)^2 + |\Delta|^2} \quad \varepsilon_k = \frac{\hbar^2 k^2}{2m} + V_{\mathrm{s.p.}}.$$

Here $V_{\text{s.p.}} = V_{\text{s.p.}}(\boldsymbol{r}, t)$ is the single-particle potential including any external potential $V_{\text{ext}}(\boldsymbol{r}, t)$ as well as any self-consistent potentials arising from the DFT formulation. The occupation probabilities are given by

$$\| U_{k\uparrow} \|^2 = \frac{1}{2}\left(1 + \frac{\varepsilon_k - \lambda_{\text{F}}}{E_k}\right) \quad \| V_{k\downarrow} \|^2 = \frac{1}{2}\left(1 - \frac{\varepsilon_k - \lambda_{\text{F}}}{E_k}\right).$$

In the absence of currents, the expressions for the number density, anomalous density and kinetic density are

$$\rho = 2\int \frac{\mathrm{d}^3\boldsymbol{k}}{(2\pi)^3} |V_{k\downarrow}|^2 = \int \frac{\mathrm{d}^3\boldsymbol{k}}{(2\pi)^3}\left(1 - \frac{\varepsilon_k - \lambda_{\text{F}}}{E_k}\right) \tag{4.5a}$$

$$\kappa = \int \frac{\mathrm{d}^3\boldsymbol{k}}{(2\pi)^3} V_{k\downarrow}^* U_{k\uparrow} = \int \frac{\mathrm{d}^3\boldsymbol{k}}{(2\pi)^3}\frac{\Delta}{2E_k} \tag{4.5b}$$

$$\frac{\hbar^2}{2m}\tau = 2\frac{\hbar^2}{2m}\int \frac{\mathrm{d}^3\boldsymbol{k}}{(2\pi)^3} k^2 |V_{k\downarrow}|^2 = \frac{\hbar^2}{2m}\int \frac{\mathrm{d}^3\boldsymbol{k}}{(2\pi)^3}\frac{k^2}{2}\left(1 - \frac{\varepsilon_k - \lambda_{\text{F}}}{E_k}\right). \tag{4.5c}$$

In these forms, we follow tradition, expressing without loss of generality only the $E_k > 0$ solutions and choosing Δ to be real. The general form can be obtained by an appropriate inclusion of phases (i.e. Δ^*) and signs. From these expressions, the local divergences, which appear at large momenta $k \equiv |\boldsymbol{k}| \gg 1$, are clear from the integrands,

$$\overbrace{\frac{8\pi}{(2\pi)^3}\frac{\Delta^2 k^2 \mathrm{d}k}{4\left(\frac{\hbar^2 k^2}{2m}\right)^2} \sim \frac{\mathrm{d}k}{k^2}}^{\rho_{\text{integrand}}(k\to\infty)} \quad \overbrace{\frac{4\pi}{(2\pi)^3}\frac{\Delta k^2 \mathrm{d}k}{2\frac{\hbar^2 k^2}{2m}} \sim \mathrm{d}k}^{\kappa_{\text{integrand}}(k\to\infty)} \quad \overbrace{\frac{\hbar^2}{2m}\frac{8\pi}{(2\pi)^3}\frac{\Delta^2 k^4 \mathrm{d}k}{4\left(\frac{\hbar^2 k^2}{2m}\right)^2} \sim \mathrm{d}k}^{\tau_{\text{integrand}}(k\to\infty)}.$$

There are three important points. First, the anomalous density is linearly divergent in k. This can be traced to the behavior of $\kappa(\boldsymbol{r}) = \lim_{\boldsymbol{r}\to\boldsymbol{r}'} \kappa(\boldsymbol{r}, \boldsymbol{r}') \to 1/\|\boldsymbol{r} - \boldsymbol{r}'\|$ in equation (3.1) [37]. This can be solved with an appropriate regulator—e.g. a momentum cutoff $k < k_c$ or $\hbar^2 k^2/2\,m < \varepsilon_{\text{cut}}$—with the corresponding adjustment (renormalization) of the coupling constant such that the pairing gap remains constant. Second, the kinetic density diverges in exactly the same manner. Thus, these two must enter the EDF in the following unique combination to ensure that the functional is cutoff-independent in the limit $\varepsilon_c \to \infty$:

$$\frac{\hbar^2}{2m}\tau - \Delta\kappa_c = \tau + g_{\text{eff}}|\kappa_c|^2. \tag{4.6}$$

Third, the one-body density ρ, although convergent, converges quite slowly as a function of the cutoff $|\rho - \rho_c| \sim \mathrm{O}(1/k_c)$. This latter effect can lead to an appreciable error in the particle number such that, for a cutoff of $\varepsilon_c \approx 50$ MeV, the error in the total binding energy of a heavy nucleus can reach 1 MeV. Similarly affected is the chemical potential $\lambda_F \propto \partial E/\partial N$, whose value should also be independent of the cutoff ε_c. A low cutoff ε_c will thus lead to a higher value of λ_F (for fixed Δ), in order to compensate for the 'missing s.p. states.'

Pairing cutoff and truncation errors

To estimate this error we compute the density ρ of a Fermi gas, and the error induced in this density by cutting off the integral in equation (4.5a) for $k < k_c$,

$$\rho \approx \int_{k<k_F} \frac{d^3\boldsymbol{k}}{(2\pi)^3} = \frac{k_F^3}{6\pi^2} \qquad \delta\rho \approx \frac{|\Delta|^2}{4\varepsilon_F^2}\frac{k_F^4}{2\pi^2 k_c}.$$

Here we have expressed everything in terms of the Fermi momentum k_F and Fermi energy $\varepsilon_F = \hbar^2 k_F^2/2m = \lambda_F$ where m is the nucleon mass and λ_F is the chemical potential appearing in the equations above. From these we can estimate

$$\frac{\delta N}{N} = \frac{\delta\rho}{\rho} \approx \frac{3|\Delta|^2}{4\varepsilon_F^2}\frac{k_F}{k_c} = \frac{3|\Delta|^2}{4\varepsilon_F^2}\sqrt{\frac{\varepsilon_F}{\varepsilon_c}}.$$

Hence the error in the binding energy of a heavy nucleus, say $A \sim 240$, could be as large as δN times the binding energy per nucleon ≈ 8 MeV. In nuclei, the pairing gap $\Delta \sim 1$ MeV and the Fermi energy $\varepsilon_F \sim 35$ MeV. Thus, for a cutoff $\varepsilon_c \approx 50$ MeV, the error in the total binding energy of a heavy nucleus can reach

$$\delta E \approx 1\text{ MeV} \approx \underbrace{\frac{3}{4}\left(\frac{1\text{ MeV}}{35\text{ MeV}}\right)^2\sqrt{\frac{35\text{ MeV}}{50\text{ MeV}}}\cdot \overbrace{240}^{A}}_{\delta N}\cdot\underbrace{8\text{ MeV}}_{E_B/N}.$$

4.3.2 Subtleties with regularization

There are a couple of subtleties associated with the regularization of the locally divergent anomalous density $\kappa(\boldsymbol{r}, \boldsymbol{r}') \to 1/\|\boldsymbol{r} - \boldsymbol{r}'\|$. As mentioned above, rather large errors appear in the particle number that fall off slowly with a spherical cutoff momentum k_c. The convergence can be significantly improved by including the chemical potential in the subtraction to properly position the Fermi surface as described in [37–39], which yields a reasonable approach for dynamical simulations.

Even with this improved regularization scheme, however, dynamical simulations can run into difficulties when level crossings approach the cutoff scale k_c. Level crossings always occur because of dynamical deformation, that is, the fact that the eigenenergies of equation (4.3) depend on time. For sufficiently large amplitude collective motion, s.p. levels with either small or large occupation probabilities will move in and out of the window of energies allowed by the energy cutoff ε_c. As the amplitude of the collective motion increases, more levels will enter and exit the energy window relevant for the pairing correlation, and eventually we find that these crossings adversely affect the numerical stability of the evolution equations. This can be mitigated to some extent by using a smooth cutoff, which works for gentle evolution, but for very large-amplitude excitations such as fission, a new approach is needed.

One solution to obtain stable numerical evolution is to use the natural cutoff provided by the discrete simulation lattice. This 'box' cutoff includes all of the q.p. eigenstates in the simulations, hence, levels never cross past the cutoff—all states are always included. Including all states allows for stable numerical evolution, even with large-amplitude excitations. Reasonable convergence can then be achieved with large enough lattices, although with significant computational cost.

4.4 Validation of the TDSLDA: the unitary Fermi gas

The main limitation of DFT is that the formal existence theorems suggest no procedure for constructing the exact functional. For self-bound systems such as nuclei, where the underlying nuclear Hamiltonian is not well known, the difficulties are even greater. In practice, one uses a physically motivated but approximate EDF which must be carefully validated by comparison with experimental results or truly *ab initio* calculations.

There is a particularly remarkable system, the unitary Fermi gas, which has properties very similar to dilute neutron matter (such as found in the skin of neutron-rich nuclei or in the crust of neutron stars), yet can be studied experimentally and calculated quite accurately using quantum Monte Carlo (QMC) techniques for both homogeneous and inhomogeneous systems. This was the initial motivation for independent microscopic studies of the unitary Fermi gas (UFG) by nuclear theorists [50–52]. A key property that makes the UFG particularly appealing is that for a system of fermions with zero-range interaction and infinite scattering length, the ground-state energy and properties should be determined by the only dimensional scale in the system, namely their density (similarly to a free Fermi gas).

In the case of the UFG, the structure of the functional can thus be fixed with high accuracy by rather general requirements: dimensional arguments require that it can depend only on the fermion mass m, Planck's constant $\hbar$, the density $\rho(\boldsymbol{r})$, the kinetic density $\tau(\boldsymbol{r})$, the anomalous density $\kappa(\boldsymbol{r})$ and the current density $\boldsymbol{j}(\boldsymbol{r})$. Including $\tau(\boldsymbol{r})$ and $\kappa(\boldsymbol{r})$ in the functional should be done in accordance with equation (4.6). The inclusion of the anomalous density is required in order to be able to disentangle the normal and superfluid phases, and the presence of currents is necessary to put in evidence matter flow. These must be combined in an EDF that preserves Galilean invariance and appropriate symmetries (parity, translation, rotation, gauge, etc). As in the implementation by Kohn and Sham [12] of the underlying DFT [12], the kinetic density largely accounts for gradient effects, and additional gradient corrections appear to be rather small [44, 53].

Static properties

Theoretical QMC studies of the Fermi gas near unitarity have achieved percent-level accuracy [52, 54–61], and agree with experiments [62–65] at an accuracy better than 1% for the energy per particle, and a few percent for the magnitude of the pairing gap [58, 66]. From these results, the parameters describing the SLDA functional can be extracted from the properties of homogeneous matter. The resulting functional has been validated against QMC and experimental results in both periodic boxes [61], where it correctly reproduces the shell effects to high precision, and in close to 100 non-homogeneous harmonically trapped clouds containing both normal and superfluid states, as well as unpolarized and polarized systems [47].

To test systems with currents, one can consider rotating superfluids in which quantized superfluid vortices appear. In weakly coupled bosonic superfluids—Bose–Einstein condensate (BECs)—vortices are easily observed since the order parameter, which vanishes in the core of the vortices, corresponds to a density depletion. Thus, vortices may be directly imaged (usually after some expansion to allow the vortex cores to expand, but *in situ* imaging is possible). In weakly coupled fermionic superfluids (such as superconductors), however, quantized vortices are typically much more difficult to observe since pairing is a weak effect on top of the usual Fermi sea. The particles deep in the Fermi sea fill the cores of the vortices, reducing the imaging contrast.

Therefore, it was somewhat unexpected when the SLDA predicted [67, 68] a large depletion in the core of vortices in fermionic superfluids with a large pairing gap. Such a large pairing gap was predicted by *ab initio* theory [52, 69] for dilute neutron matter and the UFG. This large depletion allows the vortices to be imaged in strong-interacting Fermi superfluids, and the imaging of this Abrikosov vortex lattice was used to demonstrate that the low-temperature UFG produced in cold-atom experiments [70] was indeed a superfluid with a large pairing gap. Such results demonstrate that the current form of the SLDA has been validated for static properties in a wide variety of scenarios, and agrees at the percent level with either QMC or experimental results.

Dynamical properties

Validating the accuracy of the TDSLDA for dynamic properties is much more challenging due to due to a lack of high-precision dynamical results from either *ab initio* theoretical techniques or experiments. Indeed, DFT is one of the only techniques for which quantum dynamics can be systematically studied. Nevertheless, we may still proceed by constructing the corresponding time-dependent extension to the SLDA by requiring that it satisfy all of the expected microscopic and symmetry properties (such as Galilean invariance). The resulting TDSLDA functional has not yet been tested at the same level of accuracy as the static SLDA, but appears to correctly capture the relevant dynamics at the current level of experimental accuracy. There exist exact static and time-dependent solutions of a model many-body Hamiltonian with a pairing interaction [71, 72], known as the Richardson model, which are reproduced by the TDSLDA [46]. The TDSLDA has also the expected feature that the superfluidity is correctly lost if a superfluid is stirred with a velocity exceeding Landau's critical velocity [47].

In particular, there have been several experiments colliding fermionic superfluids [73] and exciting solitons [74, 75], which demonstrate several features correctly described by the TDSLDA. In the first case, quantum shock wave were observed in the collisions of two large superfluid UFG clouds [73] and their interpretation as dispersive shock waves was demonstrated using the TDSLDA [47]. In the second case [74], what was thought to be an imprinted domain wall was observed to move almost two orders of magnitude more slowly than expected from theory, suggesting that a new type of quantum excitation dubbed a 'heavy soliton' had been observed [74]. A success of the TDSLDA was the prediction [49, 76] that the imprinted domain wall would rapidly decay into a vortex ring and a vortex soliton which moves at the observed rate. This was later confirmed in more accurate experiments [75], where the evolution from a domain wall to a vortex ring (and finally a single vortex in the presence of gravity) was observed through an improved imaging technique.

All these achievements demonstrate that, at least in the case of the UFG, the extension of DFT to both superfluid and time-dependent phenomena, is in agreement with the Schrödinger many-body description of strongly interacting systems and describes a large range of experimental phenomena. This gives us confidence in using these techniques in more complex systems such as atomic nuclei to predict and describe similar non-equilibrium phenomena.

Memory effects in time-dependent evolution

The TDSLDA which we describe here is known as the adiabatic LDA extension to superfluid dynamics [14]. This approximation contains no memory effects, but should be reasonable when the dynamics are not too violent, as is the case in the experiments described above. Note that the word 'adiabatic' in this context has a very different meaning from that used in chapter 6, where it refers to a slow, quasistatic phenomenon. The same notion of adiabaticity is also discussed in section 5.2.3 in the context of linear response theory.

4.5 Symmetry-breaking

Since the SLDA discussed here is nothing other than a particular case of the single-reference energy density functional (SR-EDF) theory described in section 3.1, the solutions of the static SLDA equations often break various symmetries: translational, rotational, phase (related to particle number) and isospin symmetries. Note that most of the nuclear EDFs currently available neglect so-called beyond mean-field corrections, resulting either from the symmetry-breaking or from inaccurate incorporation of the zero-point fluctuations in the mean field; in other words, these functionals are calibrated at the SR-EDF level, not at the multi-reference energy density functional (MR-EDF) level. Section 9.2 discusses various aspects related to the determination of the parameters of the EDF. Methods have been applied to estimate such corrections in static calculations such as the generator coordinate method (GCM) or projection techniques briefly outlined in section 3.2 and special cases thereof, such as the Gaussian overlap approximation (GOA) approximation of the GCM described in section 6.4. However, their treatment in time-dependent calculations is not settled theoretically. For example, rotational symmetry can be restored in static calculations by performing angular momentum projections. In a similar manner, one can restore the gauge symmetry associated with particle number.

The restoration of various symmetries can be achieved in time-dependent problems by identifying various periodic orbits and imposing their semi-classical quantization [3]

$$\oint p(t)\dot{q}(t)dt \equiv \oint p(q)dq = 2\pi\hbar n, \tag{4.7}$$

where p and q are the corresponding collective momentum and coordinate corresponding to the broken symmetry, and n is an integer (or a half-integer in some cases; see [3] for details). In the case of rotations the generalized momentum p is the angular momentum operator and the coordinate q is the rotation angle. In the case of particle number projection, the generalized momentum is the particle number operator and the coordinate is the phase angle. The angular momentum and particle number projection methods are specific numerical realizations of this kind of semi-classical quantization, which can be used to identify the vibrational states.

Should one attempt to perform similar restoration of expected good quantum numbers in time-dependent calculations? In this section we present arguments that, for most of the dynamical process of interest, the errors incurred by neglecting such projections are small, comparable to other approximations made in the DFT formulation.

Symmetry-breaking for self-bound systems

To understand the issue, consider the problem of modeling a self-bound nucleus. Unlike the electronic systems considered by Hohenberg and Kohn [13] in which the electrons reside in the 'static' Coulomb field created by nuclei, there is no external field for nucleons. This leads to the technical complication in the formulation of the DFT that the free-space density of the ground state is smeared over all space because of the motion of the center-of-mass. While the exact density functional should formally relate this homogeneous density with the binding energy of the nucleus, the result is not of practical interest since all of the interesting information about nuclear structure is now explicitly codified in the functional rather than resulting from the properties of its solution. This issue in particular has resulted in a significant effort by the nuclear physics community to provide a mathematically consistent DFT framework for self-bound systems [77–84], which so far has not been (successfully) implemented in calculations.

We approach the problem of spontaneous symmetry-breaking (SSB) pragmatically. The KS formulation implicit in the formulation of the EDF approach naturally provides self-bound solutions that break translational invariance, displaying very reasonable structural properties. Similarly, these solutions may break rotational invariance, leading to a practical description of deformed nuclei. Part of this limitation is the restriction of the functional to single Slater determinants; the translationally invariant ground state or a state of definite angular momentum can be constructed by taking an appropriate linear superposition of these localized Slater determinant states—this is the essence of performing projection onto states of definite momentum, angular momentum, particle number, etc. Are such projections necessary? If so, then one runs into the formal difficulties alluded to above. Fortunately, we can assess the level of error introduced by neglecting these projections, and we shall show that it is reasonably small for nuclei. This justifies the practical approach of using an orbital-based nuclear EDF which naturally provides a formally incomplete, but extremely practical description of self-bound and deformed nuclei.

4.5.1 Translational motion

To assess the errors introduced by localizing the nucleus, one can solve for the nucleus in an appropriately shaped box (Dirichlet or periodic boundary conditions). The errors introduced by using a localized solution will be proportional to the spacing of the energy levels in the box (or the energy bands if using periodic boundary conditions). As long as the box is spatially large enough that the nucleon density displays the proper asymptotic behavior at the boundaries, the presence of such a potential, which localizes the nucleus and perhaps compresses it a little bit, will be exponentially small [11]. The cost is thus the energy of the center-of-mass motion of a nucleus confined to a large box, which is of order $\hbar^2\pi^2/(2mAL^2)$, where m is the nucleon mass and L is the length of the box. This kind of approach was pioneered by Uhlenbeck and Beth [85], and later in quantum field theory (QFT) by Lüsher [86]. It is now the method of choice in lattice quantum chromodynamics (LQCD) calculations of energy spectra and scattering phase shifts. Thus, the problem of the center-of-mass corrections and the applicability of the DFT formalism to self-bound systems is easily resolved in the case of nuclei as well.

4.5.2 Rotation, pairing and parity symmetries

Let us consider now deformed nuclei and the error incurred by ignoring the projection onto states of definite angular momentum. If a system has a static deformation, then it will have a rotational band characterized by a sequences of energies $E(J)$ with angular momentum J. We refer to the classical textbooks [22, 87] and the review article [21] for a comprehensive discussion of high-spin physics—mostly in the context of phenomenological mean fields. Treating this system as a rigid rotor, the angular frequency of rotation is

$$\hbar\omega = \frac{\mathrm{d}E(J)}{\mathrm{d}J} \approx \frac{E(J+2) - E(J)}{2} \sim 100\ \mathrm{keV} \tag{4.8}$$

or even smaller. The rotational period $T = 2\pi/\omega \sim 12\,000\ \mathrm{fm}\ c^{-1}$ is thus many orders of magnitude larger than other relevant timescales such as the time it takes a nucleon to traverse the diameter of a heavy nucleus ($\sim 50\ \mathrm{fm}\ c^{-1}$ estimated by taking the nucleon to have a kinetic energy of order of the Fermi energy $\varepsilon_F = 35$ MeV) or the $\sim 3000\ \mathrm{fm}\ c^{-1}$ time it takes a fissioning nucleus to descend from the outer saddle to scission [30, 88]. This means that during most of the dynamical processes of interest in TDSLDA simulations (heavy-ion collisions, induced fission, dynamics of quantized vortices in neutron star crust, etc), a nucleus is effectively not in a state of well-defined angular momentum. For pairing correlations, the timescales are about one order of magnitude shorter, but still long enough compared to the evolution time in a dynamical simulation.

> The corresponding Heisenberg uncertainty relation for rotational motion is $\Delta J\ \Delta\phi \approx \Delta J\ \omega\Delta t$, where $\Delta\phi$ is the rotation angle and Δt the characteristic collision or fission time. Based on the estimates given in the text, we find that $\Delta J \gg 1$ (of the order 4 for fission, for instance).

When a nucleus fissions, there is a constant flow of matter on the way from the saddle point to scission over the entire system, and particle number is not conserved in any subsystem of the nucleus. After scission, the system emerges in a highly entangled quantum state [30, 89–92]. A similar situation occurs during the fusion of two heavy ions [92, 93]. The nuclear fragments before scission are deformed and the wave function of the fissioning nucleus is a correlated superposition of many states, with many values of the angular momenta and particle numbers for each fission fragment. Projecting on any of these quantum numbers makes no sense until the fragments are well separated. In this respect the projection on good quantum numbers makes sense only on the final states.

Does one need to perform a projection on good quantum numbers *before* two heavy ions come into contact, or on a fissioning nucleus *before* it starts evolving toward scission configurations? Let us consider two deformed heavy ions slowly approaching each other. The time it takes one of them to undergo a complete rotation is longer than the time it takes them to come into contact and exchange energy, momentum and matter. Thus, unless the two partners simply roll on each other with practically no dissipation, the projection of good quantum numbers of each of them is not physically realized. The projection onto a state with good quantum numbers, particularly for large systems, for which semi-classical estimates are very accurate, is equivalent to the system being able to 'evaluate' the closed loop integral equation (4.7) and thus 'realize' that it is in an eigenstate of the corresponding symmetry operator.

In figure 4.3 we show the potential energy of ^{240}Pu evaluated while constraining the quadrupole moment $q \propto 2z^2 - x^2 - y^2$. This illustrates some important aspects of this discussion pertaining to the specific case of parity projection. The ground state is axially deformed, but the nucleus becomes triaxial in the region of the inner fission barrier. Upon reaching the second fission isomer state, the nucleus becomes axially symmetric again while acquiring an octupole deformation. In the majority of calculations of fission dynamics, only this latter part of the potential energy surface is explored [30, 33, 88, 95]. In order to restore the parity of the nucleus, which is broken by the octupole deformations, one has to consider either the sum or the difference of states with positive and negative octupole moments [96]. These linear combinations involve a pair of states with an exponentially small energy splitting between them, corresponding to an exponentially large time for the nucleus to tunnel between the two different octupole deformations.

Angular momentum projection can be performed using the MR-EDF techniques described in section 3.2.3. More realistic calculations of excited spectra would also involve the combination of projection and configuration mixing via, e.g. the GCM formalism of section 3.2.1. However, Bertsch and Flocard [27] showed that one can obtain similar information about the spectrum of a deformed nucleus by considering a simple hopping model [25, 98] with only discrete states in ^{194}Hg separated by $\Delta q \approx 2$ b in a 60 b range (see figure 4.4). Consequently, in order to restore its good quantum numbers, a deformed ^{194}Hg nucleus would have to probe all quadrupole deformations in the 60 b range. As fully microscopic dynamical calculations of the induced fission of ^{240}Pu demonstrate [30, 88], the change of quadrupole deformation of the nucleus from saddle to scission is in about the same range, and the evolution time is ~ 1000 fm c^{-1}. Only if the physical time required to restore good quantum

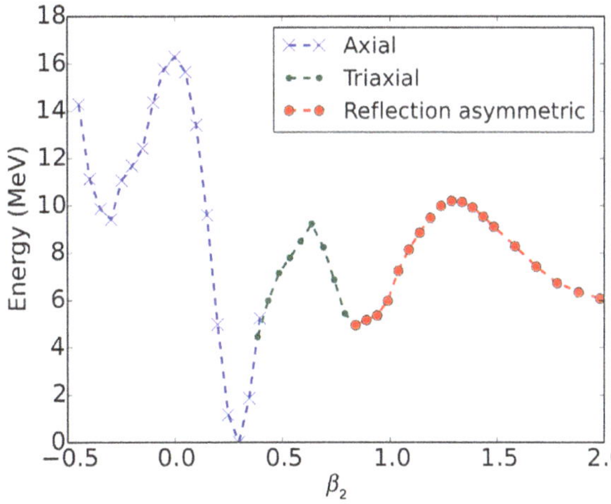

Figure 4.3. Potential energy of ^{240}Pu as a function of the quadrupole deformation evaluated for the SLy4 functional. (Figure reproduced with permission from [94], courtesy of Ryssens. Copyright 2015 by The American Physical Society.)

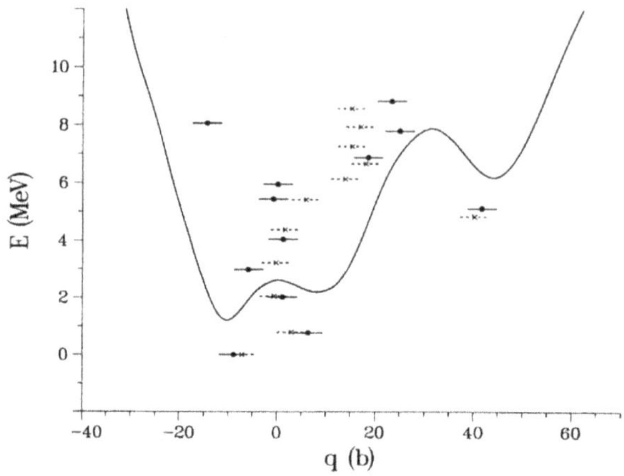

Figure 4.4. The spectrum of ^{194}Hg evaluated in [27] in the hopping model and in [97] within a much more involved GCM calculation. (Figure reproduced with permission from [27], courtesy of Bertsch. Copyright 1991 by The American Physical Society.)

numbers were significantly smaller than the evolution time from saddle to scission, would a nucleus 'know' that it is in a quantum state with good quantum numbers and symmetry restoration would be necessary.

4.6 Time-dependent techniques

Armed with the evolution equations (4.3) and a suitable regularization procedure for the anomalous density, one can use the techniques described in chapter 8 to evolve a nuclear system. Such evolution is still computationally expensive and here we describe several techniques that can be used to help improve the accuracy and efficiency of such calculations.

4.6.1 Accelerating/rotating frames

Within the TDSLDA it is easy to describe the interaction of nuclei with various external probes such as electromagnetic fields that couple to the number density or to currents. In the case of probes that couple to the number or spin density (in particular the time-component of a four-vector) one simply adds an external potential (or in the case of a magnetic field, a coupling to the nucleon spin). In the case of a vector electromagnetic field $\boldsymbol{A}(\boldsymbol{r}, t)$ one follows the minimal coupling procedure. These modifications take the form

$$\begin{aligned} h_{\sigma\sigma'}(\boldsymbol{r}, t) &\to h_{\sigma\sigma'}(\boldsymbol{r}, t) + V_{\text{ext}}(\boldsymbol{r}, t)\delta_{\sigma\sigma'} \\ h_{\sigma\sigma'}(\boldsymbol{r}, t) &\to h_{\sigma\sigma'}(\boldsymbol{r}, t) + g\mu_B \boldsymbol{B}(\boldsymbol{r}, t) \cdot \boldsymbol{\sigma}_{\sigma\sigma'} \\ -\mathrm{i}\hbar\nabla &\to -\mathrm{i}\hbar\nabla - \frac{e}{c}\boldsymbol{A}(\boldsymbol{r}, t), \end{aligned}$$

where g is the gyromagnetic factor of the nucleon and μ_B is the Bohr magneton. The relativistic Coulomb excitation of nuclei [99] is an example where the Coulomb field of a projectile in the reference frame of the target has both electric and magnetic components.

After excitation, the nucleus may start to move, and it can be useful to re-center the box on the nucleus. This can be easily implemented by boosting to an accelerating frame with time-dependent position $\boldsymbol{X}(t)$ so that the wave function depends on the relative coordinate $\boldsymbol{r}'(t) = \boldsymbol{r} - \boldsymbol{X}(t)$. To implement the motion of a nucleus in such an accelerating frame, one needs to update the abscissa, add the collective momentum $\boldsymbol{P} = -i\hbar\nabla$ of the reference frame to spatial derivatives and add an external acceleration term. Specifically, the equations of motion in the transformed frame for the q.p. wave function $\psi'(\boldsymbol{r}', t) = \psi(\boldsymbol{r}, t) = \psi(\boldsymbol{r}' + \boldsymbol{X}(t), t)$ are

$$i\hbar\partial_t\psi(\boldsymbol{r}, t) = (H[\boldsymbol{P}, \boldsymbol{r}]\psi(\boldsymbol{r}, t)) \tag{4.9a}$$

$$i\hbar\partial_t\psi'(\boldsymbol{r}', t) = (H[\boldsymbol{P}, \boldsymbol{r}' + \boldsymbol{X}(t)] - \dot{\boldsymbol{X}}(t)\cdot\boldsymbol{P})\psi'(\boldsymbol{r}', t). \tag{4.9b}$$

The origin of the accelerating frame is given by the instant coordinate of the center-of-mass of the nucleus, which can be determined from the simulation,

$$\boldsymbol{X}(t) = \frac{\int d^3\boldsymbol{r}\; \boldsymbol{r}\rho(\boldsymbol{r})}{\int d^3\boldsymbol{r}\; \rho(\boldsymbol{r})}.$$

A similar transformation can be applied to rotate the frame. This may be of use, for example, if studying an elongated nucleus in an asymmetric box where it may be beneficial to dynamically realign the box with the principle axes of the excited nucleus.

Galilean transformations for quantum systems

Note that the transformation to an accelerating frame discussed here is not the usual Galilean transformation recalled in section 2.1.1, but it provides a simple and coherent implementation of an accelerating frame, equivalent to effecting a canonical transformation in the corresponding classical system with the generating function $G(\boldsymbol{P}', \boldsymbol{r}) = \boldsymbol{P}'\cdot[\boldsymbol{r} - \boldsymbol{X}(t)]$,

$$\boldsymbol{r}' = \frac{\partial G}{\partial \boldsymbol{P}'} = \boldsymbol{r} - \boldsymbol{X}(t) \qquad \boldsymbol{P} = \frac{\partial G}{\partial \boldsymbol{r}} = \boldsymbol{P}' \tag{4.10a}$$

$$H' = H + \frac{\partial G}{\partial t} = H - \dot{\boldsymbol{X}}(t)\cdot\boldsymbol{P}'. \tag{4.10b}$$

Note under this transformation, the momentum does not change $\boldsymbol{P}' = \boldsymbol{P}$.

For reference, the usual Galilean transformation follows from the generating function $G(\boldsymbol{P}', \boldsymbol{r}) = \boldsymbol{P}'\cdot[\boldsymbol{r} - \boldsymbol{X}(t)] + m\boldsymbol{r}\cdot\dot{\boldsymbol{X}}(t)$,

$$\boldsymbol{r}' = \boldsymbol{r} - \boldsymbol{X}(t) \qquad \boldsymbol{P} = \boldsymbol{P}' + m\dot{\boldsymbol{X}}(t) \tag{4.11a}$$

$$H' = H - \dot{X}(t) \cdot \boldsymbol{P}' + m\boldsymbol{r} \cdot \ddot{\boldsymbol{X}}(t). \tag{4.11b}$$

The equivalent quantum transformation includes a phase modification of the wave function. If the functional is manifestly Galilean invariant, then the transformation may be simplified by the conventional Galilean transformation which includes a phase ϕ in the transformed wave function, $\psi'(\boldsymbol{r}', t) = e^{-i\phi}\psi(\boldsymbol{r}, t) = e^{-i\phi}\psi(\boldsymbol{r}' + \boldsymbol{X}(t))$,

$$i\hbar\partial_t\psi'(\boldsymbol{r}', t) = (H[\boldsymbol{P}', \boldsymbol{r}' + \boldsymbol{X}(t)] + m\ddot{\boldsymbol{X}}(t) \cdot \boldsymbol{r}')\psi'(\boldsymbol{r}', t) \tag{4.12a}$$

and the angle ϕ satisfies

$$\hbar\phi = m\dot{\boldsymbol{X}}(t) \cdot \boldsymbol{r} + \int_0^t \frac{m\dot{X}(t)^2}{2}\mathrm{d}t. \tag{4.12b}$$

This is only valid if H is manifestly Galilean invariant, but makes the invariance manifest: the additional phase factor replaces the $-\dot{X} \cdot \boldsymbol{P}$ term in equation (4.9b) with a spatially constant force if the transformation corresponds to an accelerating frame (which vanishes if $\dot{X} = \text{const}$). Such local Galilean invariance is crucial for nuclear EDFs [15, 100, 101], especially when considering dynamics during which a system can partition itself into various pieces, each of which can separate and move with different velocities.

4.6.2 Chemical potentials

The presence of the chemical potential in the evolution equations (4.3) might appear as a nuisance, particularly if colliding two different heavy ions. However, the chemical potential can be easily removed by a trivial gauge transformation performed on each collision partner in their initial state, which will not affect any physical observable,

$$\begin{pmatrix} U_{\mu\uparrow}(\boldsymbol{r}, t) \\ U_{\mu\downarrow}(\boldsymbol{r}, t) \\ V_{\mu\uparrow}(\boldsymbol{r}, t) \\ V_{\mu\downarrow}(\boldsymbol{r}, t) \end{pmatrix} \to \begin{pmatrix} U_{\mu\uparrow}(\boldsymbol{r}, t)e^{+i\frac{\lambda_F t}{\hbar}} \\ U_{\mu\downarrow}(\boldsymbol{r}, t)e^{+i\frac{\lambda_F t}{\hbar}} \\ V_{\mu\uparrow}(\boldsymbol{r}, t)e^{-i\frac{\lambda_F t}{\hbar}} \\ V_{\mu\downarrow}(\boldsymbol{r}, t)e^{-i\frac{\lambda_F t}{\hbar}} \end{pmatrix}, \tag{4.13}$$

where the chemical potential λ_F for the two partners are typically different. Under such a gauge transformation the pairing gap, which for stationary states is time-independent, acquires a phase $\exp(2i\lambda_F t/\hbar)$. In practice, performing such a 'local' gauge transformation on individual partners requires special care. The reason is that while the $V_{\mu\sigma}(\boldsymbol{r}, t)$-components of the q.p. wave functions have a finite norm, the $U_{\mu\sigma}(\boldsymbol{r}, t)$-components belong to the continuum spectrum [37, 102]. Using different chemical potentials for the collision partners introduces undesired discontinuities in these components, which can affect drastically the results of numerical simulations.

There is a relatively simple solution. Before the two collision partners come into contact, they can be subjected to a local constant potential for a period of time T

$$V_{\text{ext}}(\boldsymbol{r}, t) = \lambda_L \theta_L(\boldsymbol{r}, t) + \lambda_R \theta_R(\boldsymbol{r}, t),$$

where $\lambda_{L,\,R}$ are the chemical potentials for the left and right collision partners. The functions $\theta_{L,\,R}(\boldsymbol{r}, t) = T^{-1}$ can then be chosen to smoothly vanish in either the left or right part of the space, respectively, where the two collision partners are initially prepared [49, 76]. With this procedure, one can also control the relative phase of the two colliding partners. This relative phase has been shown to play a highly non-trivial role in the fusion of colliding low-energy superfluid heavy ions [92, 93]. This procedure was also implemented in experiments with colliding cold Fermion atoms, in order to control the relative phase of pairing fields of the two partners [75, 103].

In practice, there is no need to perform the gauge rotation simply to fix the chemical potentials of the two collision partners. One can prepare the nuclei in the same simulation box, keeping them apart either by a suitable constant 'electric field' box [93], or sometimes even by simply placing by hand the densities of the two partners on each side of the box at a relatively large distance. When the two partners are relatively far away from each other, the Coulomb repulsion between them is not strong enough to move them appreciably [92], while determining the self-consistent solution for the entire system [104].

4.6.3 Neutron emission

We close this section by discussing the emission of neutrons from stimulated or fissioning systems. The emission of neutrons in TDSLDA simulations appears as a small non-zero flux of the neutron current $\boldsymbol{j}_n$ through the boundary of the simulation volume. One way of dealing with these emitted neutrons is to implement some sort of absorbing boundary condition. In practice we have found that when this prescription is used together with an initial state corresponding to the ground state of the system, the nucleus starts losing particles rather quickly, even if the simulation box is reasonably large. However, in many cases, it suffices to simply ignore these emitted neutrons. In principle, when implementing periodic boundary conditions, the emitted neutrons will re-enter the simulation box from the opposite side, colliding again with the nucleus. This can introduce errors in the calculation, but such errors can be estimated and reduced by properly considering the infrared (IR) limit when increasing the box size. Increasing the box size has two effects—reducing the density of these emitted neutrons, and increasing the time it takes for the neutrons to exit the box, re-enter on the opposite side and then propagate to the location of the excited nucleus. Obtaining properly IR-converged results as a function of increasing simulation volume will ensure that these effects are minimized. In low-energy dynamics, one emits at most 1–2 neutrons. Strictly speaking one then simulates a nucleus in a very low-density neutron gas and the error incurred by neglecting these effects can be estimated also by estimating the effect on the nucleus of this low-density neutron gas, typically with densities less than 0.0001 fm^{-3}.

4.6.4 Inclusion of pairing correlations within the BCS approximation

The numerical solution of the TDSLDA equations is one of the most challenging computational quantum many-body problems, as it amounts to solving an infinite number of non-linear coupled time-dependent 3D partial differential equations. It is thus natural to seek simpler approaches and the natural candidate is to use some extension of the BCS approximation briefly discussed in section 3.1.3. In condensed matter systems, the BCS approximation is typically good because the pairing correlations are energetically limited to a very narrow energy band around the Fermi surface. The pairing gap in BCS superconductors is thus orders of magnitude smaller than the Fermi energy, and the electron density is hardly affected by the presence or absence of pairing correlations.

The situations in nuclei, neutron stars and cold-atom systems is dramatically different. Even though the pairing gap in nuclei is about an order of magnitude smaller than the Fermi energy, ground-state properties are qualitatively changed in the presence of pairing correlations. In neutron star crusts and cold-atom systems the situation is even more dramatic since in these contexts, the pairing gap becomes comparable to the Fermi energy. Another distinction with electronic superconductivity is the character of the interaction that leads to pairing. In superconductors, the attractive interaction is due to phonons, which weakly couple states close to the Fermi surface.

> Pairing requires an attractive interaction, but electrons naturally repel due to their Coulomb interaction. Superconductivity arises from an induced attractive interaction mediated by phonons propagating through the underlying crystal structure of the substrate and the finite speed of sound must generally be considered.

In contrast, pairing correlations result from direct short-range attractive interactions in nuclear and cold-atom systems, with a range either comparable or much smaller than the average inter-particle separation. In the latter case it often is a good approximation to replace the interaction with a simple zero-range contact interaction which links all occupied and unoccupied single-particle states as described in section 4.3 and requires appropriate regularization.

In the BCS approach, if the corresponding infinite sums over all single-particle states are limited to a finite interval around the Fermi level, it may make sense to replace the corresponding off-diagonal two-body matrix element $\Delta(t)$ with a constant, and approximate the quasiparticle wave functions as

$$\begin{pmatrix} U_\mu(\mathbf{r}, t) \\ V_\mu(\mathbf{r}, t) \end{pmatrix} \rightarrow \begin{pmatrix} U_\mu(t)\phi_\mu(\mathbf{r}, t) \\ V_\mu(t)\phi_{\bar{\mu}}(\mathbf{r}, t) \end{pmatrix}.$$

In the case of stationary states, $U_\mu(t)$ and $V_\mu(t)$ are the usual time-independent BCS quasiparticle amplitudes. In this case, the s.p. wave functions $\phi_\mu(\mathbf{r}, t)$ and $\phi_{\bar{\mu}}(\mathbf{r}, t)$ are formally a couple of time-reversed solutions of the stationary or time-dependent Hartree–Fock (HF) equations [105], which greatly simplifies the numerical calculations.

In the case of time-dependent problems, the BCS amplitudes $U_\mu(t)$ and $V_\mu(t)$ are sometimes chosen as solutions of the stationary BCS equations, while the single-particle wave functions $\phi_\mu(\mathbf{r}, t)$ and $\phi_{\bar{\mu}}(\mathbf{r}, t)$ are solutions of the time-dependent DFT equations with no pairing correlations, except for the time-independent occupation probabilities $\rho_\mu = |V_\mu|^2$ [106, 107]. In this approximation the redistribution of the occupation probabilities discussed in section 4.2 is not allowed.

Another approach is to allow the BCS pairing potential $\Delta(t)$ to vary within the time-dependent BCS approximation [31, 95, 105, 108]. In this case one can derive the following equations for the occupation probabilities $\rho_\mu(t) = |V_\mu(t)|^2$ and the pair probability $\kappa_\mu(t) = U_{\bar{\mu}}^*(t)V_\mu(t)$

$$i\hbar\frac{d\rho_\mu(t)}{dt} = \Delta(t)\kappa_\mu^*(t) - \Delta(t)^*\kappa_\mu(t) \tag{4.14a}$$

$$i\hbar\frac{d\kappa_\mu(t)}{dt} = \Delta(t)(1 - 2\rho_\mu(t)), \tag{4.14b}$$

$$\Delta(t) = g\sum_\mu \kappa_\mu(t) \tag{4.14c}$$

and where g is the pairing coupling constant. Depending on the specific numerical implementation $\Delta(t)$ might show a relatively weak dependence on the single-particle state μ, which for the sake of the simplicity of the argument we neglect here.

It is easy to show that within this approximation the continuity equation (4.4) for the number density $\rho(\mathbf{r}, t) = \sum_\mu n_\mu(t)[|\phi_\mu(\mathbf{r}, t)|^2 + |\phi_{\bar{\mu}}(\mathbf{r}t)|^2]$ is violated and the particle number is conserved only on average [31, 105]. As discussed in section 4.2, the continuity equation is of fundamental importance in dynamical theories, and its violation has observational consequences [31].

Since the sum over the single-particle states μ in the BCS approximation for the number density $\rho(\mathbf{r}, t)$ is limited to bound states only, when approaching either drip-line and when the distance from the Fermi level to the nucleon threshold is less than the pairing gap, the space of available single-particle states to develop pairing correlations is drastically reduced and pairing correlations are artificially suppressed. This is another strong limitation of the applicability of the BCS approximation to either ground states or dynamical phenomena.

One can easily show that equation (4.14) can be solved by

$$\Delta(t) = \Delta(0)\exp\left[-\frac{i}{\hbar}\sum_\mu\left(1 - 2\rho_\mu(t)\right)t\right].$$

If one allows the pairing gap to depend on the state μ as in [31, 105], the actual numerical solution of the time-dependent BCS equations remains qualitatively similar. In both cases, the magnitude of the pairing gap is approximately constant as a function of time, irrespective of the dynamics. This is perhaps the worst deficiency of the time-dependent BCS approximation, as it violates basic properties of superfluids, well established in theory [47] and experiments. In particular, if one applies an external field to a superfluid, superfluidity is lost and the system becomes normal if the speed with which the superfluid is stirred exceeds the Landau critical velocity. The time dependence of the pairing gap in the time-dependent BCS approximation is independent of the presence or absence of an external field and of its time dependence character, at least in the typical approximations used in nuclear physics, and this approximation should be avoided.

4.7 Selected examples

We close this chapter by illustrating some of these considerations with specific examples. The TDSLDA framework is a powerful tool to compute nuclear dynamics, which has been used in its fully unrestricted formulation only recently. We illustrate this versatility by examining three very different physics phenomena: the relativistic Coulomb excitation of a heavy nucleus, fission and the collision of two heavy (superfluid) nuclei. We should note that each of these examples required substantial computation time and the use of leadership class supercomputers.

4.7.1 Relativistic Coulomb excitation of nuclei

The relativistic Coulomb excitation of ^{238}U by an ^{238}U projectile provides a more complex example that is also more difficult to describe microscopically with alternative methods, but emerges naturally from TDSLDA simulations. We first note that because ^{238}U is well-deformed in its ground state, the energy of the first 2^+ is 45 keV, corresponding to a very long rotational period. Therefore, during the simulation time considered here ($\approx 10^{-20}$ s or 300 fm c^{-1}), it can be considered fixed.

The incoming projectile excites various modes in the target nucleus and as a result, the axial symmetry of the initial ground state is lost. The proper identification of these modes requires certain care, since during the collision the system is beyond the linear regime and an analysis using standard linear response theory as presented in section 5.1.3 is not applicable in general. However, the information about the excited nuclear modes is carried through in the subsequent electromagnetic radiation that lead to the nuclear de-excitation. While de-excitation to the ground state via photon emission requires times of about 10^{-16} s, which are four orders of magnitude longer than in the current calculations, it is possible to compute the spectrum of the pre-equilibrium neutrons and gamma radiation, which allows the identification of the excited nuclear modes. We can accurately treat the system as a classical source of electromagnetic radiation and the time dependence of the proton current governs the rate of emission (see [99, 109, 110] for additional discussions):

$$P(t + r/c) = \frac{e^2}{\pi c} \sum_{lm} \left| \int_{-\infty}^{\infty} \boldsymbol{b}_{lm}(k, \omega) e^{-i\omega t} d\omega \right|^2, \tag{4.15}$$

with

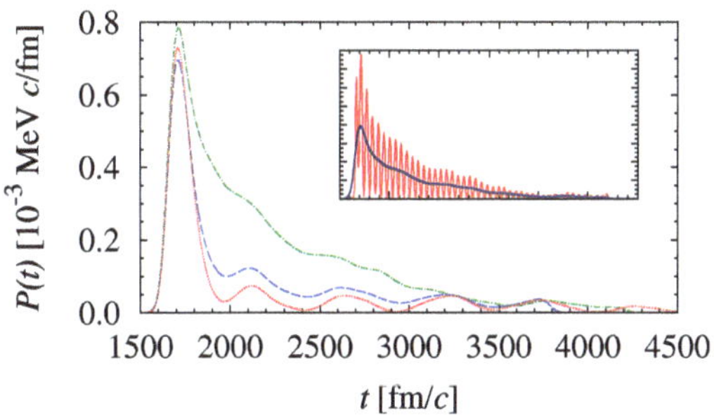

Figure 4.5. Emitted energy rate via electromagnetic radiation for a collision with impact parameter $b = 16.2$ fm, for three relative orientations. In two cases, the nuclear symmetry axis is parallel to the reaction plane and perpendicular (dash-dotted line) or parallel (dashed line) to the trajectory of the incoming projectile, while in the third case, it is both perpendicular to the reaction plane and the incoming projectile (dotted line). The figure shows time-averaged quantities; in the inset frame, we also show the raw, strongly oscillating data for one configuration. The rate at which this quantity changes is directly related to the characteristic damping time, which we estimate at 500 fm c^{-1}, leading to a width $\Gamma_\downarrow \approx 0.4$ MeV. (Figure reproduced with permission from [99], courtesy of Stetcu. Copyright 2014 by The American Physical Society.)

$$\boldsymbol{b}_{lm}(k, \omega) = \int dt \int d^3\boldsymbol{r} e^{-i\omega t}\nabla \times \boldsymbol{j}(\boldsymbol{r}, t) j_l(kr) Y^*_{lm}(\hat{\boldsymbol{r}}).$$

Here $\omega = kc$, $j_l(kr)$ is the spherical Bessel function of order l, and $\boldsymbol{j}(\boldsymbol{r}, t)$ is the proton current. The emission rate P is plotted in figure 4.5. The magnitude of this quantity indicates that the total amount of radiated energy during the evolution time (about 2500 fm c^{-1}) is rather small compared to the total absorbed energy and does not exceed 1 MeV, which is about 2%–3% of the deposited energy reported. This implies that the effect of damping of nuclear motion due to the emitted radiation can be neglected for such short time intervals. Consequently, the decreasing intensity of the radiation, see figure 4.5, is merely related to the rearrangements of the intrinsic structure of the excited nucleus caused by damping of collective modes due to one-body dissipation mechanisms. It has to be emphasized that within the framework of the current approach, one is able to extract only a small fraction of the spreading width $\Gamma_\downarrow$, the one that is caused by one-body dissipation mechanisms. The two-body effects require, for example, stochastic extension of TDSLDA, which would allow for dynamic hopping between various mean fields, and thus could account for collisional damping as well. The extracted averaged power photon energy spectrum is shown in figure 4.6.

4.7.2 Induced fission

Nuclear fission is one of the most complex problems in science. A theoretical description of fission requires an accurate understanding of in-medium nuclear forces; a full account of all types of many-body correlations, including intrinsic excitations (odd–even effects, fission isomers), large-amplitude collective modes and statistical excitations (induced fission with 'fast' neutrons); a proper description of particle emission and the coupling to

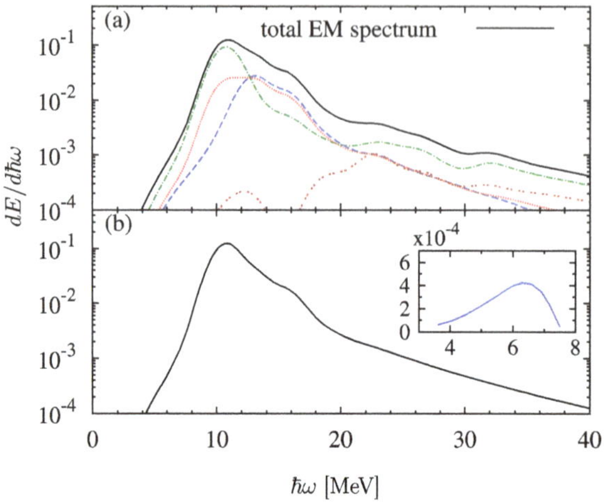

Figure 4.6. (a) The total energy spectrum (solid line) of emitted electromagnetic radiation, averaged over the target-projectile configurations, at the impact parameter $b = 12.2$ fm. We show the total quadrupole contribution (double-dotted line), as well as the contributions from the three target-projectile orientations using the same symbols as in figure 4.5. (b) The radiation emitted from the target nucleus when only the dipole component of the projectile electromagnetic field is used. The insert shows the pygmy resonance contribution to the emitted spectrum, visible in the main figure as the slope change at low energies. (Figure reproduced with permission from [99], courtesy of Stetcu. Copyright 2014 by The American Physical Society.)

the continuum (pre- and post-scission neutrons), etc. Currently, EDF methods are in fact the most promising approach to achieve a quantitative prediction of basic fission observables such as fission half-lives, fission fragment distributions, etc, in a microscopic approach based only on nuclear forces and many-body methods [111].

Fission lends itself naturally to a TDSLDA description, but the computational requirements are still significant. Early simulations were based on the TDHF approach [106–108]. Since pairing correlations were not included (by construction), the mechanisms described in section 4.2 to go over level crossings were not available, and as a result, it was impossible for the nucleus to reach fission unless some non-physical excitation boosts were provided to the system. The inclusion of pairing correlations in the BCS approach, with occupation frozen during the dynamics, represented a real progress since fission could then be achieved [33, 95]. The TDSLDA simulations of fission to fully include pairing correlations obtained surprisingly good agreement with experimental data [30, 88].

Figure 4.7 illustrates some of these aspects. It shows the evolution of the nuclear shape, represented by the matter density, as the nucleus moves from a compact deformed shape to two well-separated fragments. Most importantly, the figure also illustrates the crucial role of pairing, not only through its amplitude (what was described as the order parameter $|\kappa|$ in section 3.1.1), but also its phase (the gauge angle φ). As discussed earlier in section 4.2, time-dependent extensions of the EDF approach provide a unique opportunity to probe the phase of the pairing field. This phase is particularly important in nuclear reactions, as we will show below.

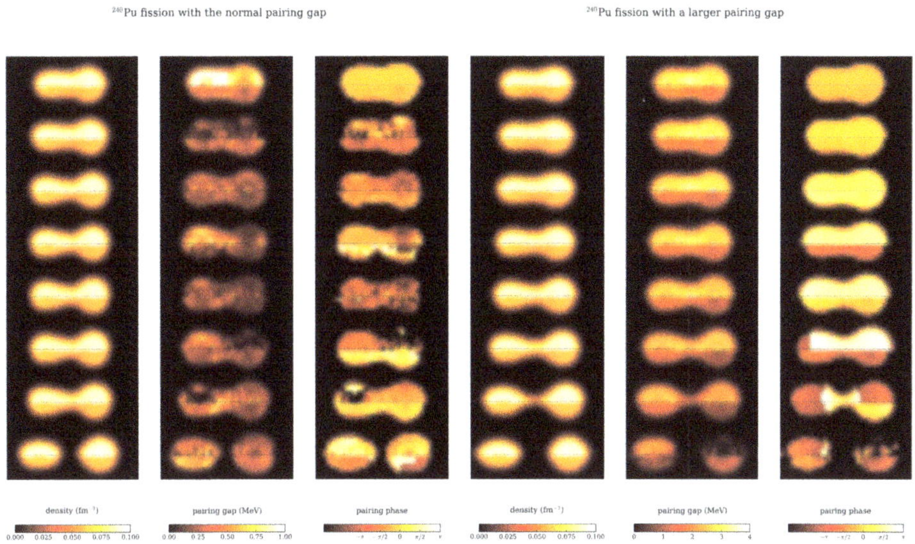

Figure 4.7. Snapshots of the real time evolution of ^{240} during fission as simulated within the TDSLDA approach described in this chapter. Nuclear fission with normal pairing (left) and enhanced pairing (right). For each frame, the left-most panel shows the evolution of the matter density, the middle panel the pairing gap $\Delta(t) = \text{tr}\ \Delta(t)\rho(t)$ and the right-most panel the pairing phase. (Figure reproduced with permission from [88], courtesy of Bulgac. Copyright 2016 by The American Physical Society.)

4.7.3 Collisions of superfluid nuclei

The pairing field is characterized not only by its magnitude, but also by its phase. In the ground state, where all (time-odd) currents vanish, the phase is uniform, and typically the pairing field is assumed to be positive throughout the entire system. The phase of the pairing field is canonically conjugated with the particle number, and when performing particle projection, the wave function becomes a superposition of states with all values of the pairing phase; see section 3.2.3. As a consequence, the phase of the pairing condensate and the particle number cannot both have well-defined values, and a generalized Heisenberg uncertainty relation exists [112, 113]. However, in macroscopic systems, the phase and the particle number acquire a classical character and remarkable phenomena such as Josephson currents, when two superconductors are put in contact, can emerge [114, 115]. Similar phenomena have been observed as well in the case of two colliding BEC [116]. The role played by the relative phase of two condensates can be rather subtle and hard to guess [92, 117].

The role of the pairing phase in nuclear collisions, as schematically sketched in figure 4.8, was investigated within the TDSLDA approach in [93, 118]. It was observed that while the pairing energy in nuclei can hardly ever exceed a few MeV, the relative phase between two condensates can lead to changes of the order of 20 MeV to 30 MeV in the total kinetic energy (TKE) of the two fragments after the collisions or in their apparent fusion barrier; see figure 4.9(b). Even though the

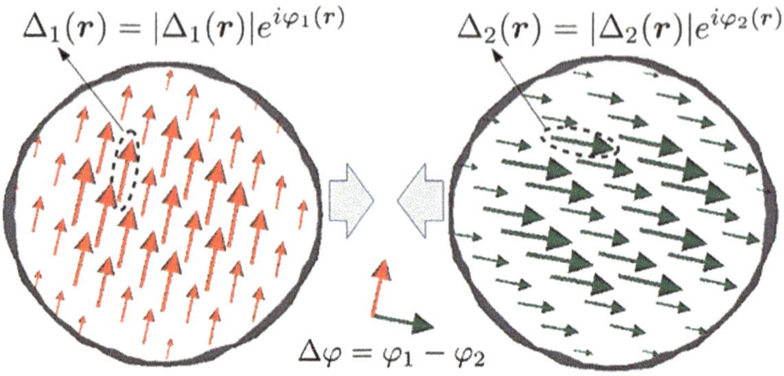

Figure 4.8. Schematic picture of the collision of two superfluid nuclei with different phases of the pairing fields. Each disc represents a cross section of a nucleus. The length of each arrow indicate the local amplitude of the pairing field $\Delta(\boldsymbol{r})$, while the direction indicates its phase $\varphi(\boldsymbol{r})$. In the ground state, the phase is uniform across each nucleus and the phase difference is well-defined. (Figure reproduced with permission from [93], courtesy of Magierski. Copyright 2017 The American Physical Society.)

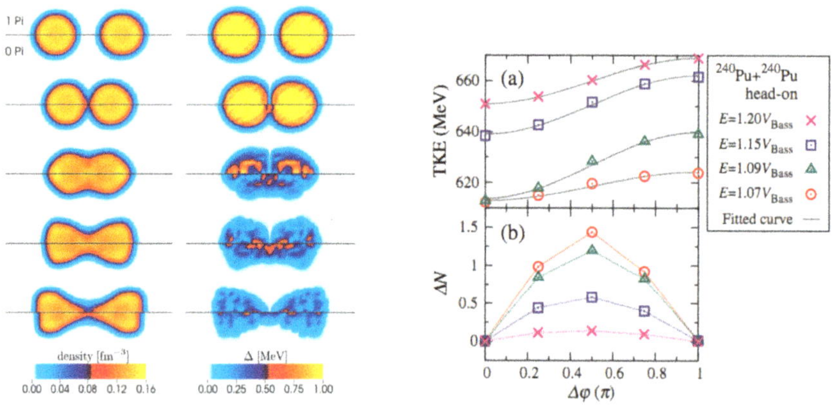

(a) Snapshot of the collision ^{240}Pu+^{240}Pu for two extreme values of the relative phase difference, $\Delta\varphi = 0$ and π at about 1.1 times the phenomenological fusion barrier. The left panel shows the total density distribution while the right panel shows the pairing field. For each panel, the top half corresponds to $\varphi = 0$ and the bottom half to $\varphi = \pi$.

(b) Total kinetic energy (top) and average number of transferred nucleons (bottom) as a function of the relative phase difference in the collision ^{240}Pu+^{240}Pu.

Figure 4.9. TDDFT simulation of the collision ^{240}Pu + ^{240}Pu. (Figure reproduced with permission from [93], courtesy of Magierski. Copyright 2017 The American Physical Society.)

relative phase of the two condensates in nuclei cannot be controlled individually, they lead to a noticeable widening of experimentally observable distributions with respect to the collisions of normal nuclei.

References

[1] Blaizot J-P and Ripka G 1985 *Quantum Theory of Finite Systems* (Cambridge, MA: MIT Press)
[2] Negele J W and Orland H 1988 *Quantum Many-Particle Systems* (Cambridge, MA: Perseus)
[3] Bulgac A 2010 The long journey from *ab initio* calculations to density functional theory for nuclear large amplitude collective motion *J. Phys. G: Nucl. Part. Phys.* **37** 064006
[4] Balian R and Vénéroni M 1984 Fluctuations in a time-dependent mean-field approach *Phys. Lett.* B **136** 301
[5] Balian R and Vénéroni M 1992 Correlations and fluctuations in static and dynamic mean-field approaches *Ann. Phys.* **216** 351
[6] Bulgac A, Jin S and Stetcu I 2018 Unitary evolution with fluctuations and dissipation, arxiv: 1805.08908
[7] Ring P and Schuck P 2004 *The Nuclear Many-Body ProblemTexts and Monographs in Physics* (Berlin: Springer)
[8] Lacroix D, Avez B and Simenel C 2010 *Quantum Many-Body Dynamics: Applications to Nuclear Reactions* (Riga: VDM)
[9] Nakatsukasa T 2012 Density functional approaches to collective phenomena in nuclei: time-dependent density functional theory for perturbative and non-perturbative nuclear dynamics *Prog. Theor. Exp. Phys.* **2012** 01A207
[10] Nakatsukasa T *et al* 2012 Density functional approaches to nuclear dynamics *J. Phys.: Conf. Ser.* **387** 012015
[11] Bulgac A and Forbes M M N 2013 *Time-Dependent Superfluid Local Density Approximation* ed N Proukakis *et al* (Singapore: World Scientific) ch 32 pp397–406
[12] Kohn W and Sham L J 1965 Self-consistent equations including exchange and correlation effects *Phys. Rev.* **140** A1133
[13] Hohenberg P and Kohn W 1964 Inhomogeneous electron gas *Phys. Rev.* **136** B864–71
[14] Marques M A L *et al* (ed) 2012 *Fundamentals of Time-Dependent Density Functional TheoryLecture Notes in Physics* vol 837 (Heidelberg: Springer)
[15] Bender M, Heenen P-H and Reinhard P-G 2003 Self-consistent mean-field models for nuclear structure *Rev. Mod. Phys.* **75** 121–80
[16] Bulgac A and Forbes M M 2008 Unitary Fermi supersolid: the Larkin–Ovchinnikov phase *Phys. Rev. Lett.* **101** 215301
[17] Bertsch G *et al* 2009 Hartree–Fock–Bogoliubov theory of polarized fermi systems *Phys. Rev.* A **79** 043602
[18] Dobaczewski J and Dudek J 1996 Time-odd components in the rotating mean field and identical bands *Acta Phys. Pol.* B **27** 45
[19] Carlsson B G, Dobaczewski J and Kortelainen M 2008 Local nuclear energy density functional at next-to-next-to-next-to-leading order *Phys. Rev.* C **78** 044326
[20] Raimondi F, Carlsson B G and Dobaczewski J 2011 Effective pseudopotential for energy density functionals with higher-order derivatives *Phys. Rev.* C **83** 054311
[21] de Voigt M J A, Dudek J and Szymański Z 1983 High-spin phenomena in atomic nuclei *Rev. Mod. Phys.* **55** 949
[22] Nilsson S G and Ragnarsson I 1995 *Shapes and Shells in Nuclear Structure* (Cambridge: Cambridge University Press)

[23] Hill D L and Wheeler J A 1953 Nuclear constitution and the interpretation of fission phenomena *Phys. Rev.* **89** 1102–45
[24] Arve P *et al* 1988 Static path approximation for the nuclear partition function *Ann. Phys.* **183** 309
[25] Barranco F, Broglia R A and Bertsch G F 1988 Exotic radioactivity as a superfluid tunneling phenomenon *Phys. Rev. Lett.* **60** 507–10
[26] Barranco F *et al* 1990 Large-amplitude motion in superfluid Fermi droplets *Phys. Rev.* A **512** 253–74
[27] Bertsch G and Flocard H 1991 Pairing effects in nuclear collective motion: generator coordinate method *Phys. Rev.* C **43** 2200–4
[28] Bertsch G F and Bulgac A 1997 Comment on 'Spontaneous fission: a kinetic approach' *Phys. Rev. Lett.* **79** 3539
[29] Blocki J *et al* 1978 One-body dissipation and the super-viscosity of nuclei *Ann. Phys.* **113** 330–86
[30] Bulgac A *et al* 2018 Fission dynamics, arXiv: 1806.00694
[31] Scamps G *et al* 2012 Pairing dynamics in particle transport *Phys. Rev.* C **85** 034328
[32] Scamps G and Lacroix D 2013 Effect of pairing on one- and two-nucleon transfer below the Coulomb barrier: a time-dependent microscopic description *Phys. Rev.* C **87** 014605
[33] Scamps G, Simenel C and Lacroix D 2015 Superfluid dynamics of ^{258}Fm fission *Phys. Rev.* C **92** 011602
[34] Stetcu I *et al* 2011 Isovector giant dipole resonance from the 3D time-dependent density functional theory for superfluid nuclei *Phys. Rev.* C **84** 051309(R)
[35] Oliveira L N, Gross E K U and Kohn W 1988 Density-functional theory for superconductors *Phys. Rev. Lett.* **60** 2430–3
[36] Bulgac A 1988 Semilocal approach to nonlocal equations *Nucl. Phys.* A **487** 251–68
[37] Bulgac A 1980 Hartree–Fock–Bogoliubov approximation for finite systems, arXiv: nucl-th/9907088
[38] Bulgac A and Yu Y 2002 Renormalization of the Hartree–Fock–Bogoliubov equations in the case of a zero range pairing interaction *Phys. Rev. Lett.* **88** 042504
[39] Bulgac A 2002 Local density approximation for systems with pairing correlations *Phys. Rev.* C **65** 051305(R)
[40] Borycki P J *et al* 2006 Pairing renormalization and regularization within the local density approximation *Phys. Rev.* C **73** 044319
[41] Gross E K U *et al* 1995 Density functional theory of the superconducting state *Density Functional Theory NATO ASI Series B: Physics* ed E K U Gross and R M Dreizler vol 337 (New York: Plenum) p 676
[42] Runge E and Gross E K U 1984 Density-functional theory for time-dependent systems *Phys. Rev. Lett.* **52** 997
[43] Bulgac A and Yu Y 2004 Superfluid LDA (SLDA): local density approximation for systems with superfluid correlations *Int. J. Mod. Phys.* E **13** 147–56
[44] Bulgac A 2007 Local density functional theory for superfluid fermionic systems: the unitary gas *Phys. Rev.* A **76** 040502
[45] Bulgac A and Roche K J 2008 Time-dependent density functional theory applied to superfluid nuclei *J. Phys.: Conf. Ser.* **125** 012064
[46] Bulgac A and Yoon S 2009 Large amplitude dynamics of the pairing correlations in a unitary Fermi gas *Phys. Rev. Lett.* **102** 085302

[47] Bulgac A *et al* 2011 Real-time dynamics of quantized vortices in a unitary fermi superfluid *Science* **332** 1288–91

[48] Bulgac A 2013 Time-dependent density functional theory and the real-time dynamics of Fermi superfluids *Annu. Rev. Nucl. Part. Sci.* **63** 97–121

[49] Bulgac A, Forbes M M N and Sharma R 2013 Strength of the vortex-pinning interaction from real-time dynamics *Phys. Rev. Lett.* **110** 241102

[50] Baker G A Jr 2001 The MBX challenge competition: a neutron matter model *Int. J. Mod. Phys.* B **15** 1314–20

[51] Baker G A Jr 1999 Neutron matter model *Phys. Rev.* C **60** 054311

[52] Carlson J *et al* 2003 Superfluid Fermi gases with large scattering length *Phys. Rev. Lett.* **91** 050401

[53] Forbes M M N, Gandolfi S and Gezerlis A 2012 Effective-range dependence of resonantly interacting fermions *Phys. Rev.* A **86** 053603

[54] Astrakharchik G E *et al* 2004 Equation of state of a Fermi gas in the BEC–BCS crossover: a quantum monte carlo study *Phys. Rev. Lett.* **93** 200404

[55] Carlson J and Reddy S 2005 Asymmetric two-component fermion systems in strong coupling *Phys. Rev. Lett.* **95** 060401

[56] Blume D, von Stecher J and Greene C H 2007 Universal properties of a trapped two-component Fermi gas at unitarity *Phys. Rev. Lett.* **99** 233201

[57] Blume D 2008 The trapped polarized Fermi gas at unitarity *Phys. Rev.* A **78** 013635

[58] Carlson J and Reddy S 2008 Superfluid pairing gap in strong coupling *Phys. Rev. Lett.* **100** 150403

[59] Pilati S and Giorgini S 2008 Phase separation in a polarized Fermi gas at zero temperature *Phys. Rev. Lett.* **100** 030401

[60] Carlson J *et al* 2011 Auxiliary field quantum Monte Carlo for strongly paired fermions *Phys. Rev.* A **84** 061602

[61] Forbes M M N, Gandolfi S and Gezerlis A 2011 Resonantly interacting fermions in a box *Phys. Rev. Lett.* **106** 235303

[62] Lobo C *et al* 2006 Normal state of a polarized Fermi gas at unitarity *Phys. Rev. Lett.* **97** 200403

[63] Shin Y *et al* 2008 Realization of a strongly interacting Bose–Fermi mixture from a two-component Fermi gas, arXiv: 0805.0623

[64] Shin Y *et al* 2008 Phase diagram of a two-component Fermi gas with resonant interactions, arXiv: 0709.3027

[65] Ku M J H *et al* 2012 Revealing the superfluid lambda transition in the universal thermodynamics of a unitary Fermi gas *Science* **335** 563–7

[66] Schirotzek A *et al* 2008 Determination of the superfluid gap in atomic Fermi gases by quasiparticle spectroscopy *Phys. Rev. Lett.* **101** 140403

[67] Bulgac A and Yu Y 2003 Vortex state in a strongly coupled dilute atomic fermionic superfluid *Phys. Rev. Lett.* **91** 190404

[68] Yu Y and Bulgac A 2003 Spatial structure of a vortex in low density neutron matter *Phys. Rev. Lett.* **90** 161101

[69] Gezerlis A and Carlson J 2008 Strongly paired fermions: cold atoms and neutron matter *Phys. Rev.* C **77** 032801

[70] Zwierlein M W *et al* 2005 Vortices and superfluidity in a strongly interacting Fermi gas *Nature* **435** 1047–51

[71] Yuzbashyan E A, Tsyplyatyev O and Altshuler B L 2006 Relaxation and persistent oscillations of the order parameter in fermionic condensates *Phys. Rev. Lett.* **96** 097005
[72] Yuzbashyan E A *et al* 2005 Nonequilibrium Cooper pairing in the nonadiabatic regime *Phys. Rev.* B **72** 220503
[73] Joseph J A *et al* 2011 Observation of shock waves in a strongly interacting Fermi gas *Phys. Rev. Lett.* **106** 150401
[74] Yefsah T *et al* 2013 Heavy solitons in a fermionic superfluid *Nature* **499** 426–30
[75] Ku M J H *et al* 2014 Motion of a solitonic vortex in the BEC–BCS crossover *Phys. Rev. Lett.* **113** 065301
[76] Wlazłowski G *et al* 2015 Life cycle of superfluid vortices and quantum turbulence in the unitary Fermi gas *Phys. Rev.* A **91** 031602
[77] Engel J 2007 Intrinsic-density functionals *Phys. Rev.* C **75** 014306
[78] Barnea N 2007 Density functional theory for self-bound systems *Phys. Rev.* C **76** 067302
[79] Giraud B G 2008 Scalar nature of the nuclear density functional *Phys. Rev.* C **78** 014307
[80] Giraud B G, Jennings B K and Barrett B R 2008 Existence of a density functional for an intrinsic state *Phys. Rev.* A **78** 032507
[81] Giraud B G 2008 Density functionals in the laboratory frame *Phys. Rev.* C **77** 014311
[82] Messud J, Bender M and Suraud E 2009 Density functional theory and Kohn–Sham scheme for self-bound systems *Phys. Rev.* C **80** 054314
[83] Messud J 2011 Generalization of internal density-functional theory and Kohn–Sham scheme to multicomponent self-bound systems, and link with traditional density-functional theory *Phys. Rev.* A **84** 052113
[84] Messud J 2013 Alternate, well-founded way to treat center-of-mass correlations: proposal of a local center-of-mass correlations potential *Phys. Rev.* C **87** 024302
[85] Uhlenbeck G E and Beth E 1937 The quantum theory of the non-ideal gas. II. Behaviour at low temperatures *Physica* **4** 0915–24
[86] Lüsher M 1986 Volume dependence of the energy spectrum in massive quantum field theories *Commun. Math. Phys.* **105** 153–88
[87] Bohr A and Mottelson B R 1998 *Single-Particle Motion Nuclear Structure* vol 1 (Singapore: World Scientific)
[88] Bulgac A *et al* 2016 Induced fission of ^{240}Pu within a real-time microscopic framework *Phys. Rev. Lett.* **116** 122504
[89] Younes W and Gogny D 2011 Nuclear scission and quantum localization *Phys. Rev. Lett.* **107** 132501
[90] Schunck N *et al* 2014 Description of induced nuclear fission with Skyrme energy functionals: static potential energy surfaces and fission fragment properties *Phys. Rev.* C **90** 054305
[91] Schunck N *et al* 2015 Quantification of uncertainties in nuclear density functional theory *Nucl. Data Sheets* **123** 115
[92] Bulgac A and Jin S 2017 Dynamics of fragmented condensates and macroscopic entanglement *Phys. Rev. Lett.* **119** 052501
[93] Magierski P, Sekizawa K and Wlazłowski G 2017 Novel role of superfluidity in low-energy nuclear reactions *Phys. Rev. Lett.* **119** 042501
[94] Ryssens W, Heenen P-H and Bender M 2015 Numerical accuracy of mean-field calculations in coordinate space *Phys. Rev.* C **92** 064318
[95] Tanimura Y, Lacroix D and Ayik S 2017 Microscopic phase-space exploration modeling of Fm 258 spontaneous fission *Phys. Rev. Lett.* **118** 152501

[96] Egido J L and Robledo L M 1991 Parity-projected calculations on octupole deformed nuclei *Nucl. Phys.* A **524** 65

[97] Bonche P *et al* 1990 Quadrupole collective correlations and the depopulation of the superdeformed bands in mercury *Nucl. Phys.* A **519** 509–20

[98] Bertsch G 1980 The nuclear density of states in the space of nuclear shapes *Phys. Lett.* B **95** 157–9

[99] Stetcu I *et al* 2014 Relativistic Coulomb excitation within time dependent superfluid local density approximation *Phys. Rev. Lett.* **114** 012701

[100] Engel Y M *et al* 1975 Time-dependent Hartree–Fock theory with Skyrme's interaction *Nucl. Phys.* A **249** 215–38

[101] Dobaczewski J, Nazarewicz W and Werner T R 1995 Closed shells at drip-line nuclei *Phys. Scr.* **1995** 15

[102] Dobaczewski J, Flocard H and Treiner J 1984 Hartree–Fock–Bogolyubov description of nuclei near the neutron-drip line *Nucl. Phys.* A **422** 103–39

[103] Mark J H *et al* 2015 From planar solitons to vortex rings and lines: cascade of solitonic excitations in a superfluid Fermi gas *Phys. Rev. Lett.* **116** 045304

[104] Jin S *et al* 2017 Coordinate-space solver for superfluid many-fermion systems with the shifted conjugate-orthogonal conjugate-gradient method *Phys. Rev.* C **95** 044302

[105] Ebata S *et al* 2010 Canonical-basis time-dependent Hartree–Fock–Bogoliubov theory and linear-response calculations *Phys. Rev.* C **82** 034306

[106] Goddard P, Stevenson P and Rios A 2015 Fission dynamics within time-dependent Hartree–Fock: deformation-induced fission *Phys. Rev.* C **92** 054610

[107] Goddard P, Stevenson P and Rios A 2016 Fission dynamics within time-dependent Hartree–Fock. II. Boost-induced fission *Phys. Rev.* C **93** 014620

[108] Simenel C and Umar A S 2014 Formation and dynamics of fission fragments *Phys. Rev.* C **89**

[109] Baran V *et al* 1996 Giant dipole emission as a probe of the entrance channel dynamics *Nucl. Phys.* A **600** 111–30

[110] Oberacker V E *et al* 2012 Dynamic microscopic study of pre-equilibrium giant resonance excitation and fusion in the reactions ^{132}Sn + ^{48}Ca and ^{124}Sn + ^{40}Ca *Phys. Rev.* C **85** 034609

[111] Schunck N and Robledo L M 2016 Microscopic theory of nuclear fission: a review *Rep. Prog. Phys.* **79** 116301

[112] Anderson P W 1966 Considerations on the flow of superfluid helium *Rev. Mod. Phys.* **38** 298–310

[113] Carruthers P and Nieto M M 1968 Phase and angle variables in quantum mechanics *Rev. Mod. Phys.* **40** 411–40

[114] Josephson B D 1964 Coupled superconductors *Rev. Mod. Phys.* **36** 216–20

[115] Josephson B D 1974 The discovery of tunnelling supercurrents *Rev. Mod. Phys.* **46** 251–4

[116] Shin Y *et al* 2004 Atom interferometry with Bose–Einstein condensates in a double-well potential *Phys. Rev. Lett.* **92** 050405

[117] Philip W and Anderson A 1986 *Career in Theoretical Physics* (Singapore: World Scientific) pp 23–34

[118] Hashimoto Y and Scamps G 2016 Gauge angle dependence in time-dependent Hartree–Fock–Bogoliubov calculations of ^{20}O + ^{20}O head-on collisions with the Gogny interaction *Phys. Rev.* C **94** 014610

IOP Publishing

Energy Density Functional Methods for Atomic Nuclei

Nicolas Schunck

Chapter 5

Small-amplitude collective motion

Jonathan Engel

Excited states in nuclei vary considerably in their properties. Some involve collective motion, others just the excitation of a single particle–hole state. Some are easily excited by the operators responsible for photon absorption or beta decay, others are hardly touched. It is difficult to treat all these states within a single framework, but as long as the excitation bears a good resemblance to the ground state, the quasiparticle random-phase approximation (QRPA) works well. The method does not produce an excited-state wave function, but it is the most important tool the field has for computing the rates of transitions to or from a ground state. It can provide a complete spectrum of excited-state energies and transition densities $\langle\nu|a_a^\dagger c_b|0\rangle$ between the ground state $|0\rangle$ and each excited state $|\nu\rangle$. The distribution of 'strength' (the square of the transition matrix element of a one-body operator) to excited states satisfies certain sum rules that ensure reasonable results for inclusive processes, i.e. those in which we do not care about the final nuclear state. Collective motion is particularly easy to describe. No other approach shares all these features.

> *A word on notation.* Here and thereafter, we will often refer to excited states with the generic label $|\nu\rangle$. As usual throughout this book $(c,\ c^\dagger)$ are creation and annihilation operators for arbitrary single-particle (s.p.) states, and we use the $(a,\ a^\dagger)$ notation to refer to specific s.p. operators (usually the ones corresponding to the Hartree–Fock (HF) basis).

With access to a genuine many-body Hamiltonian, one can derive the QRPA in at least four very different ways. Not all of these generalize to a pure energy density functional (EDF) approach, in which the energy functional does not necessarily derive from a pseudo-potential. Perhaps the easiest EDF-based derivation involves the theory of linear response, the small-amplitude motion induced by a weak external potential. Thus, after briefly presenting the equations-of-motion method in which the random phase approximation (RPA) naturally emerges from a Hamiltonian, we will focus on the linear-response function as we move to a general EDF. We will also not touch upon many-body diagrammatic theory (see, e.g. the textbooks [1–3]) or on the generator coordinate method [4], both of which allow derivations of the RPA and QRPA from a Hamiltonian. Much of the material in this chapter has been presented elsewhere in greater depth, in particular in [1, 5, 6]. The connection between small-amplitude motion as treated in nuclear physics and the adiabatic approximation to linear response in density functional theory (DFT) is an exception; it is difficult to find a description anywhere else.

The QRPA in all its aspects is similar to, but formally more complicated than, the ordinary RPA. For pedagogical purposes, we use the RPA to develop many ideas, and postpone the extension to the QRPA until a little later. We will become familiar with linear response in the particle–hole context before including pairing and moving to quasiparticles.

Strength functions and giant resonances

Photo-absorption provides a good example of a transition-strength function (or strength function for short). Fermi's golden rule implies that for photons with energy $E \ll \hbar c/R$, where R is the nuclear radius, the rate of absorption into state $|\nu\rangle$ is

$$\sigma_\nu(E) = \frac{4\pi^2 e^2}{\hbar c}(E_\nu - E_0)|\langle\nu|\hat{D}_z|0\rangle|^2 \delta(E - E_\nu + E_0).$$

Here $\hat{D}_z$ is the z-component of the electric-dipole operator

$$\hat{\boldsymbol{D}} \propto \sum_{i=1}^{A} e_i \boldsymbol{r}_i,$$

with $e_i = 1$ if particle i is a proton and $e_i = 0$ if it is a neutron. $\hat{D}$ is just the first in a series of 'multipole' operators, the others of which become more important as E_ν grows. A relatively complete treatment can be found in appendix B of [6].

Figure 5.1 shows the experimental cross sections for several light nuclei to absorb a photon, as a function of photon energy. The cross section is small except in a broad region centered at about 22 MeV (carbon and oxygen) and 17 MeV (copper). The peak in that region, which shifts to lower energy as a function of mass, corresponds to the excitation of a resonant oscillation of the protons in the nucleus against the neutrons. This 'giant resonance' is spread out over 5 or 6 MeV because the Hamiltonian mixes the simple state corresponding to a coherent oscillation with more complicated 'background states.' We will see that the RPA and QRPA are particularly good at reproducing the collective states, if not the background states that give the resonance its width.

In what follows we will work with general one-body transition operators. As long as these operators do not change the charge of the nucleus, one can imagine them representing the electric-dipole operator, the electric quadrupole operator that is relevant to Coloumb excitation and electron scattering, or other multipole operators.

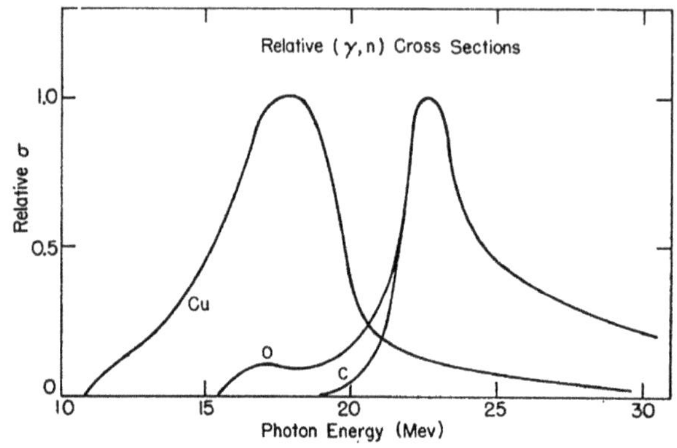

Figure 5.1. Average experimental cross section to absorb a photon for ^{12}C, ^{16}O and ^{63}Cu. (Figure reproduced with permission from [7], courtesy of Penfold. Copyright 1959 The American Physical Society.)

5.1 RPA with a Hamiltonian

5.1.1 Equations of motion method

Following [6] closely, we begin with the full many-body Schrödinger equation

$$\hat{H}|\nu\rangle = E_\nu|\nu\rangle, \tag{5.1}$$

with $\hat{H}$ the nuclear Hamiltonian, which we take to have the form $\hat{H} = \hat{T} + \hat{V}$. Here $\hat{T}$ is the kinetic energy operator and $\hat{V}$ is a two-body potential. One can define operators that turn the ground state $|0\rangle$ into excited states,

$$\hat{Q}_\nu^\dagger = |\nu\rangle\langle 0| \qquad \nu \neq 0,$$

with the restriction of ν ensuring that $\hat{Q}_\nu|0\rangle = 0$. Then,

$$\left[\hat{H}, \hat{Q}_\nu^\dagger\right]|0\rangle = (E_\nu - E_0)\hat{Q}_\nu^\dagger|0\rangle \equiv \Omega^\nu \hat{Q}_\nu^\dagger|0\rangle$$

and taking the commutator of both sides with any other operator $\hat{G}$—think of the electric-dipole as a canonical example—and acting with $\langle 0|$ gives

$$\langle 0|\left[\hat{G}, \left[\hat{H}, Q_\nu^\dagger\right]|0\rangle = \Omega^\nu\langle 0|\left[\hat{G}, Q_\nu^\dagger\right]|0\rangle.$$

The operator $\hat{G}$ could be, for instance, one of the operators involved in the multipole expansion of the radiation field. In the long-wavelength limit often used in nuclear physics, these moments read

$$\hat{Q}_{IM} \propto r^I Y_{IM}(\theta, \varphi)$$
$$\hat{M}_{IM} \propto \left(g_s\hat{\boldsymbol{S}} + \frac{2}{I+1}g_l\hat{\boldsymbol{L}}\right)\nabla[r^I Y_{IM}(\theta, \varphi)],$$

where g_l and g_s are the usual orbital and spin gyromagnetic factors; see appendix B of [6] for a comprehensive discussion. The sign $\propto$ indicates that the normalization constant of these operators is arbitrary (several conventions coexist in the literature).

Now let $\hat{G} = a_m^\dagger a_i$ or $\hat{G} = a_i^\dagger a_m$, where m and i label states in the HF basis, $\epsilon_m > \lambda_F$ and $\epsilon_i < \lambda_F$, with λ_F the Fermi energy in the HF version of the ground state. The operators $a^\dagger$, a are the usual creation/annihilation operators for the s.p. states corresponding to the HF ground state, and ϵ are their energies. Then,

$$\langle 0|[a_m^\dagger a_i, [\hat{H}, \hat{Q}_\nu^\dagger]]|0\rangle = \Omega^\nu \langle 0|[a_m^\dagger a_i, \hat{Q}_\nu^\dagger]|0\rangle \tag{5.2a}$$

$$\langle 0|[a_i^\dagger a_m, [\hat{H}, \hat{Q}_\nu^\dagger]]|0\rangle = \Omega^\nu \langle 0|[a_i^\dagger a_m, \hat{Q}_\nu^\dagger]|0\rangle. \tag{5.2b}$$

Finally, let us choose the form of $\hat{Q}_\nu^\dagger$ so that it creates particle–hole (p.h.) excitations of the ground state of a particular form

$$\hat{Q}_\nu^\dagger = \sum_{mi} X_{mi}^\nu a_m^\dagger a_i - \sum_{mi} Y_{mi}^\nu a_i^\dagger a_m.$$

So far all this is exact.

The crucial step in developing an approximation, the one that motivates all the commutators above, is to assume that for one- and two-body operators $\hat{O}$ that *result from commutations*,

$$\langle 0|\hat{O}|0\rangle \approx \langle \mathrm{HF}|\hat{O}|\mathrm{HF}\rangle.$$

In this expression, $|\mathrm{HF}\rangle$ is the many-body state formed by the anti-symmetrized s.p. states that are solutions to the HF equation,

$$|\mathrm{HF}\rangle = \prod_{i=1}^{A} a_i^\dagger |-\rangle,$$

with $|-\rangle$ the state with no particles; see below for a quick review of HF theory. In other words, the expectation value of the operator $\hat{O}$ in the true ground state, when the operator results from commutations of s.p. creation and annihilation operators, is approximately equal to its value in the HF ground state. Some algebra then gives

$$\begin{pmatrix} A & B \\ -B^* & -A^* \end{pmatrix}\begin{pmatrix} X^\nu \\ Y^\nu \end{pmatrix} \equiv \mathcal{M}\begin{pmatrix} X^\nu \\ Y^\nu \end{pmatrix} = \Omega_{\mathrm{RPA}}^\nu \begin{pmatrix} X^\nu \\ Y^\nu \end{pmatrix}, \tag{5.3}$$

where $\mathcal{M}$ is the RPA matrix (the reasons for the odd name are historical) with

$$A_{mi,nj} = (\epsilon_m - \epsilon_i)\,\delta_{mn}\delta_{ij} + \bar{v}_{mjin}, \tag{5.4a}$$

$$B_{mi,nj} = \bar{v}_{mnij} \tag{5.4b}$$

and the anti-symmetrized matrix elements of the potential are $\bar{v}_{abcd} = v_{abcd} - v_{abdc}$. Recall that the indices $i, j, k, \ldots$ refer to occupied s.p. levels below the Fermi level while indices $m, n, p, \ldots$ refer to empty levels above the Fermi level. The amplitudes X and Y, which must be determined by solving equation (5.3), are elements of the 'transition-density' matrix,

$$X_{mi}^{\nu *} = \langle \nu | a_m^\dagger a_i | 0 \rangle = \langle \mathrm{HF} | [\hat{Q}_\nu, a_m^\dagger a_i] | \mathrm{HF} \rangle$$
$$Y_{mi}^{\nu *} = \langle \nu | a_i^\dagger a_m | 0 \rangle = \langle \mathrm{HF} | [\hat{Q}_\nu, a_i^\dagger a_m] | \mathrm{HF} \rangle.$$

Recall that $|0\rangle$ is the true ground state, which is why $a_m|0\rangle \neq 0$. In contrast, $a_m|\mathrm{HF}\rangle = 0$ since, by definition, all levels m above the Fermi level are empty in the HF ground state. As their definition implies, the X and Y amplitudes can be interpreted as the overlaps between the excited state $|\nu\rangle$ and one-particle one-hole excitations of the true ground state.

Thus, even without an explicit expression for the true ground state $|0\rangle$, which must be different from $|\mathrm{HF}\rangle$, we can calculate the transition matrix elements of any operator $\hat{G} \equiv \sum_{mi}(G_{mi}a_m^\dagger a_i + G_{im}a_i^\dagger a_m)$

$$\langle \nu | \hat{G} | 0 \rangle_{\mathrm{RPA}} = \sum_{mi} (X_{mi}^{\nu *} G_{mi} + Y_{mi}^{\nu *} G_{im}). \tag{5.5}$$

The RPA approximation is based on the evaluation of the commutators in equation (5.2) before taking the expectation values. The commutators reduce the number of creation and annihilation operators by one each from the number in the products. By writing the equations in terms of commutators, therefore, one takes the expectation value of a one-body operator instead of a two-body operator, or a two-body operator instead of a three-body operator. The Slater determinant that one uses for the ground state in those equations contains no correlations but has close to the exact (one-body) density, so the lower the number of creation and annihilation operators in the expectation value, the better we can expect the result to be. The omission of the commutator would yield the Tamm–Dancoff approximation, which is usually less accurate than the RPA. In the Tam–Dancoff approximation, the second term on the right side of equation (5.5) vanishes because the Slater determinant contains no configurations in which there are particles above the Fermi surface or holes below it.

The dimension of the matrix in equation (5.3) is actually twice the number of particle–hole excitations. It is not hard to show [6] that for every energy eigenvalue $\Omega_{\mathrm{RPA}}^{\nu}$, there is another equal to $-\Omega_{\mathrm{RPA}}^{\nu}$. Only the positive-energy solutions are physically significant, of course.

Summary of Hartree–Fock theory

Let us briefly recall here the main features of HF theory. The starting point is the definition of the one-body density matrix in an arbitrary many-body reference state $|0\rangle$

$$\rho_{ab} = \langle 0 | a_b^\dagger a_a | 0 \rangle,$$

defined so that the expectation value of a one-body operator $\hat{G}$ is

$$\langle 0|\hat{G}|0\rangle = \sum_{ab} G_{ab}\rho_{ba} = \mathrm{tr}(G\rho),$$

where tr is the trace over s.p. states. Here $\hat{\rho}$ without subscripts is an operator defined only in the space of s.p. states and characterized by the matrix ρ with elements ρ_{ab}. For a ground-state Slater determinant $|\mathrm{HF}\rangle$, $\rho^2 = \rho$, and in the basis of s.p. eigenstates

$$\rho_{ab} = \begin{cases} \delta_{ab} & a, b < \lambda_F \\ 0 & a > \lambda_F \text{or } b > \lambda_F \end{cases},$$

with λ_F the Fermi energy. If the EDF is constructed out of an effective pseudo-Hamiltonian $\hat{H}$, we can write the energy as

$$\mathcal{E}[\rho] = \langle \mathrm{HF}|\hat{H}|\mathrm{HF}\rangle = \sum_{ab} t_{ab}\rho_{ba} + \frac{1}{2}\sum_{abcd} \bar{v}_{abcd}\rho_{ca}\rho_{db}.$$

Unlike in DFT, $\mathcal{E}$ is a functional of the entire density matrix, not just local densities. Setting $\delta[\mathcal{E} - \Lambda(\rho^2 - \rho)] = 0$ under small variations $\delta\rho$ (where Λ in any basis is a matrix of Lagrange multipliers) leads to the requirement that the mean-field Hamiltonian, defined in the space of s.p. states as the operator with matrix elements

$$h_{ab}[\rho] = \frac{\partial\mathcal{E}}{\partial\rho_{ba}} = t_{ab} + \sum_{bd} \bar{v}_{acbd}\rho_{dc}, \tag{5.6}$$

obeys

$$[h, \rho] = 0.$$

In other words, h and ρ can be made simultaneously diagonal. Diagonalizing both leads to a non-linear eigenvalue equation,

$$h_{ab}[\rho] = \epsilon_a\delta_{ab},$$

where, in the basis of eigenvectors (the HF basis),

$$h_{ab} = t_{ab} + \sum_{j<\lambda_F} \bar{v}_{ajbj}.$$

Also, the constraint $\rho^2 = \rho$ implies that in this basis the small variations $\eta\,\delta\rho$ (where η is a small constant so that $\delta\rho$ is of order ρ) around ρ satisfy

$$\delta\rho_{ij} = \delta\rho_{mn} = 0,$$

where, again, m labels states above the Fermi surface and i labels states below it.

5.1.2 Linear response

One way to determine properties of excited states is through the *linear response* of the underlying ground state to a weak, time-dependent, one-body operator. We will briefly describe general linear-response theory before looking at the RPA approximation. Our presentation follows that of [1, 5, 8].

Consider the system described by the Hamiltonian $\hat{H}$, with eigenvalues and eigenvectors determined by equation (5.1). The wave function of this unperturbed system satisfies the usual time-dependent (many-body) Schrödinger equation

$$i\hbar\frac{\partial|\Psi(t)\rangle}{\partial t} = \hat{H}|\Psi(t)\rangle,$$

the solution of which is formally given by

$$|\Psi(t)\rangle = e^{-i\hat{H}t/\hbar}|\Psi(0)\rangle = \hat{U}(t)|\Psi(0)\rangle.$$

We now subject the system to a weak time-dependent perturbation $\hat{G}(t)$, starting at time $t = 0$ so that $\hat{G}(t)$ contains the Heaviside step function $\Theta(t)$. The perturbed wave function satisfies the equation

$$i\hbar\frac{\partial|\bar{\Psi}(t)\rangle}{\partial t} = \left(\hat{H} + \hat{G}(t)\right)|\bar{\Psi}(t)\rangle.$$

If we write the solution in the form

$$|\bar{\Psi}(t)\rangle = e^{-i\hat{H}t/\hbar}\hat{A}(t)|\Psi(0)\rangle$$

it is straightforward to show that the operator $\hat{A}(t)$ satisfies the differential equation

$$i\hbar\frac{\partial\hat{A}(t)}{\partial t} = e^{+i\hat{H}t/\hbar}\hat{G}(t)e^{-i\hat{H}t/\hbar}\hat{A}(t) = \hat{G}_H(t)\hat{A}(t),$$

where the subscript H refers to the Heisenberg representation of operators. With the assumption that $A(t)$ is invertible, we have

$$\frac{1}{\hat{A}(t)}\frac{\partial\hat{A}(t)}{\partial t} = -\frac{i}{\hbar}\hat{G}_H(t).$$

The left side is the derivative of the logarithm of $\hat{A}(t)$; exponentiating and expanding in the small perturbation $\hat{G}(t)$, we have to first order in $\hat{G}$:

$$\hat{A}(t) = 1 - \frac{i}{\hbar}\int_0^t d\tau\ \hat{G}_H(\tau).$$

With this result, one can easily compute the difference between the expectation value of an operator $\hat{O}$ in $|\bar{\Psi}(t)\rangle$ and in $|\Psi(t)\rangle$; the difference is the effect of the perturbation on the expectation value at time t,

$$\delta O(t) = \langle \bar{\Psi}(t)|\hat{O}|\bar{\Psi}(t)\rangle - \langle \Psi(t)|\hat{O}|\Psi(t)\rangle \\ = -\frac{i}{\hbar}\int_0^t d\tau \, \langle \Psi(0)|[\hat{O}_H(t), \hat{G}_H(\tau)]|\Psi(0)\rangle. \tag{5.7}$$

Note that apart from the linear approximation to the exponential, everything above is exact. Thus, if $\hat{H}$ is the true Hamiltonian of the system and $|\Psi(t)\rangle$ the true state vector, then equation (5.7) gives the exact linear response to the perturbation.

Now let us consider the evolution of the one-body density matrix $\rho_{ab}(t) = \langle \bar{\Psi}(t)|c_b^\dagger c_a|\bar{\Psi}(t)\rangle$, we work in an arbitrary s.p. basis and use the associated creation and annihilation operators $(c, c^\dagger)$. Our one-body operators can thus be written as $\hat{G} = \sum_{ab} G_{ab} c_a^\dagger c_b$.

$$\delta\rho_{ab}(t) = -\frac{i}{\hbar}\sum_{cd}\int_0^t d\tau \, G_{cd}(\tau)\langle 0|\left[e^{-i\hat{H}\tau}c_b^\dagger c_a e^{i\hat{H}\tau}, e^{-i\hat{H}t}c_c^\dagger c_d e^{i\hat{H}t}\right]|0\rangle,$$

where $G_{cd}(t) \equiv \langle c|\hat{G}(t)|d\rangle$. Inserting a complete set of unperturbed energy eigenstates and using equation (5.1) to replace the action of the Hamiltonian leads to

$$\delta\rho_{ab}(t) = -\frac{i}{\hbar}\sum_{cd}\int_0^t d\tau \, G_{cd}(\tau)\sum_\nu \langle 0|\Big(e^{-iE_0\tau}c_b^\dagger c_a e^{iE_\nu\tau}|\nu\rangle\langle\nu|e^{-iE_\nu t}c_c^\dagger c_d e^{iE_0 t} \\ - e^{-iE_0 t}c_c^\dagger c_d e^{iE_\nu t}|\nu\rangle\langle\nu|e^{-iE_\nu\tau}c_b^\dagger c_a e^{iE_0\tau}\Big)|0\rangle,$$

which reduces to

$$\delta\rho_{ab}(t) = -\frac{i}{\hbar}\sum_{cd}\int_0^t d\tau \, G_{cd}(\tau)\sum_\nu \Big(e^{i(E_0-E_\nu)(t-\tau)}\langle 0|c_b^\dagger c_a|\nu\rangle\langle\nu|c_c^\dagger c_d|0\rangle \\ - e^{-i(E_0-E_\nu)(t-\tau)}\langle 0|c_c^\dagger c_d|\nu\rangle\langle\nu|c_b^\dagger c_a|0\rangle\Big).$$

Note that $\Omega_e^\nu = E_\nu - E_0$ is the exact energy of the excited state $|\nu\rangle$. We define the response function of the system,

$$R_{ab,\,cd}(t) \equiv -\frac{i}{\hbar}\sum_\nu\Big(e^{i\Omega_e^\nu t}\langle 0|c_b^\dagger c_a|\nu\rangle\langle\nu|c_c^\dagger c_d|0\rangle - e^{-i\Omega_e^\nu t}\langle 0|c_c^\dagger c_d|\nu\rangle\langle\nu|c_b^\dagger c_a|0\rangle\Big)\Theta(t).$$

As before $\Theta(t)$ represents the Heaviside distribution, or step function

$$\Theta(t) = \begin{cases} 0 & t < 0 \\ 1 & t > 0. \end{cases}$$

It follows that

$$\delta\rho_{ab}(t) = \sum_{cd}\int_0^t d\tau R_{ab,cd}(t-\tau)G_{cd}(\tau).$$

It is customary to work with the Fourier transform of this convolution integral, which is a simple product. Except for a factor of $1/\sqrt{2\pi}$, the Fourier transform of the response function is

$$R_{ab,cd}(\omega) = -\frac{i}{\hbar}\sum_{\nu}\left(\frac{\langle 0|a_b^{\dagger}a_a|\nu\rangle\langle\nu|a_c^{\dagger}a_d|0\rangle}{\omega - \Omega_e^{\nu} + i\epsilon} - \frac{\langle 0|a_c^{\dagger}a_d|\nu\rangle\langle\nu|a_b^{\dagger}a_a|0\rangle}{\omega + \Omega_e^{\nu} + i\epsilon}\right) \tag{5.8}$$

and we have, in matrix-vector form with pairs of single-particle indices labeling columns and rows,

$$\delta\rho(\omega) = R(\omega)G(\omega). \tag{5.9}$$

Again, everything in these equations is exact, including the ground and excited states. The poles of R are thus at the exact excited-state energies Ω_e^{ν} and the residues are the squares of the corresponding transition densities. Note also that the rate for exciting a state of energy $\hbar\omega$ through the action of $\hat{G}(t)$ is

$$\text{Rate}_{\,0\to\omega} = \frac{2\pi}{\hbar}\sum_{\nu>0}|\langle 0|\hat{G}(t)|\nu\rangle|^2\delta(\omega - \Omega_e^{\nu}) = -\frac{2}{\hbar}\,\text{Im Tr}[G^{\dagger}(\omega)\delta\rho(\omega)]$$

even for states in the continuum where the sum becomes an integral. The restriction of the sum is to positive-energy states. The square of the matrix element of $\hat{G}$ between the ground state and the excited state $|\nu\rangle$ is called the *transition strength* (or just the strength) of $\hat{G}$ to that state. When considered as a function of excitation energy, the set of squared matrix elements is, as we have noted, the strength function or distribution.

5.1.3 RPA response

The simplest approximation to R is the HF, or s.p., response R^{HF}, in which the true ground state $|0\rangle$ is approximated by the HF Slater determinant $|\text{HF}\rangle$, so that excited states are simple p.h. excitations. A better approximation comes from assuming that the *time-dependent* density matrix in the presence of a time-dependent one-body operator $\hat{G}$ satisfies the time-dependent Hartree–Fock (TDHF) equation (4.2*a*)

$$i\dot{\rho} = [h[\rho] + G(t), \rho],$$

where the equation relates matrices (or vectors in the notation introduced at the end of the last subsection) and we have set $\hbar$ to 1. The TDHF ansatz implies that the ground state is $|\text{HF}\rangle$, but that excursions from the ground state in the presence of $\hat{G}(t)$ involve more than the simple uncorrelated p.h. excitations in the single-particle response. If $\hat{G}$ is small and harmonic with frequency ω, i.e.

$$\hat{G}(t) = \eta\left(G(\omega)e^{-i\omega t} + G^{\dagger}(\omega)e^{i\omega t}\right),$$

with $\eta \ll 1$, so that we can write

$$\rho(t) = \rho^{(0)} + \eta\left(\delta\rho(\omega)e^{-i\omega t} + \delta\rho^{\dagger}(\omega)e^{i\omega t}\right),$$

with $\rho^{(0)}$ the static HF density, then the Fourier transform of the TDHF equation becomes, to first order in $\delta\rho$,

$$\omega\,\delta\rho(\omega) = [h^{(0)}, \delta\rho] + [\delta h(\omega), \rho^{(0)}] + [G(\omega), \rho^{(0)}] \tag{5.10}$$

$$= [h^{(0)}, \delta\rho] + \sum_{mi}\left[\frac{\partial h}{\partial\rho_{mi}}\delta\rho_{mi} + \frac{\partial h}{\partial\rho_{im}}\delta\rho_{im}, \rho^{(0)}\right] + [G, \rho^{(0)}], \tag{5.11}$$

where $h^{(0)} = h[\rho^{(0)}]$ and

$$\delta h(\omega) = \lim_{\eta\to 0}\frac{h[\rho + \eta\delta\rho(\omega)] - h^{(0)}}{\eta}. \tag{5.12}$$

To obtain an equation relating the elements of $\delta\rho$ to those of the perturbing operator $\hat{G}$, we work in the HF basis, in which both the density matrix and the HF potential are diagonal

$$\rho_{ij}^{(0)} = \delta_{ij} \qquad \rho_{im}^{(0)} = \rho_{mn}^{(0)} = 0 \qquad h_{ab}^{(0)} = \epsilon_a\delta_{ab} \qquad \frac{\partial h_{ab}}{\partial\rho_{cd}} = \bar{v}_{adbc}, \tag{5.13}$$

where i, j label occupied orbitals, m, n label empty ones and a, b, c, d label orbitals that can either be occupied or empty. As already noted before, we work in this section under the assumption that the system is described by a given Hamiltonian $\hat{H}$, which we also assume contains only two-body potentials $\hat{v}$. With these relations, equation (5.11) becomes

$$\left\{\begin{pmatrix}\omega & 0\\ 0 & -\omega\end{pmatrix} - \begin{pmatrix}A & B\\ B^* & A^*\end{pmatrix}\right\}\begin{pmatrix}\delta\rho^{ph}\\ \delta\rho^{hp}\end{pmatrix} = \begin{pmatrix}G^{ph}\\ G^{hp}\end{pmatrix}, \tag{5.14}$$

with the same A, B matrices as before! Here, the indices ph and hp refer to particle–hole (with indices of the form mi, $m > \lambda_F$, $i < \lambda_F$) and hole–particle (im) parts of the corresponding one-body matrices.

The response function $R_{ab,cd}(\omega)$ in this approximation is just the inverse of the matrix $\{\cdots\}$ on the left side of equation (5.14). Compare equation (5.14) with (5.9), remembering that G^{ph}, and $\delta\rho^{ph}$ are functions of ω. It vanishes unless one of the indices in pairs a, b and c, d labels a particle m and the other index labels a hole i. The response function has poles at $\omega = \pm\Omega^{\nu}_{\mathrm{RPA}}$, where the $\Omega^{\nu}_{\mathrm{RPA}}$ are the eigenvalues of the RPA matrix (which come in pairs); see equation (5.3). Residues are RPA eigenvectors X^{ν}, Y^{ν}, which represent transition densities.

If the poles and residue are unfamiliar, it might be helpful to look at a similar example from single-particle quantum mechanics: a Hamiltonian $\hat{h}$ with eigenvalues e_n and eigenvectors $|n\rangle$. One finds

$$\langle a|(\omega - \hat{h})^{-1}|b\rangle = \sum_n \frac{\langle a|n\rangle\langle n|b\rangle}{\omega - e_n}$$

simply by inserting the resolution of identity $\sum_n |n\rangle\langle n| = 1$ and using the fact that

$$(\omega - \hat{h})^{-1}|n\rangle = \frac{1}{\omega - e_n}|n\rangle.$$

This development shows that small-amplitude motion around the ground state in TDHF leads to the RPA response function [6]

$$R^{\rm RPA}_{ab,\,cd}(\omega) = \sum_{\nu}\left(\frac{\langle 0|a_b^\dagger a_a|\nu\rangle_{\rm RPA}\langle\nu|a_c^\dagger a_d|0\rangle_{\rm RPA}}{\omega - \Omega^{\nu}_{\rm RPA} + i\epsilon} - \frac{\langle 0|a_c^\dagger a_d|\nu\rangle_{\rm RPA}\langle\nu|a_b^\dagger a_a|0\rangle_{\rm RPA}}{\omega + \Omega^{\nu}_{\rm RPA} + i\epsilon}\right),$$

where now the energies and transition matrix elements are given by the RPA, as in equations (5.3) and (5.5).

5.1.4 Bethe–Salpeter equation for RPA response

As, we did in the last side note, let us briefly consider one-particle quantum mechanics, with a Hamiltonian given by $\hat{h} = \hat{h}_0 + \hat{v}$, where $\hat{h}_0$ is, for example, the kinetic energy. We can obtain an equation for the quantity $(\omega - \hat{h})^{-1}$, for any ω for which the inverse exists, as follows,

$$\begin{aligned}\frac{1}{\omega - \hat{h}} &= \frac{1}{\omega - \hat{h}_0}[\omega - \hat{h}_0]\frac{1}{\omega - \hat{h}} = \frac{1}{\omega - \hat{h}_0}[\omega - \hat{h} + \hat{v}]\frac{1}{\omega - \hat{h}} \\ &= \frac{1}{\omega - \hat{h}_0}\left[1 + \hat{v}\frac{1}{\omega - \hat{h}}\right],\end{aligned}$$

that is,

$$\frac{1}{\omega - \hat{h}} = \frac{1}{\omega - \hat{h}_0} + \frac{1}{\omega - \hat{h}_0}\hat{v}\frac{1}{\omega - \hat{h}}. \tag{5.15}$$

Equations like this one, which arise frequently in scattering theory, are called Bethe–Salpeter equations in the Green's function theory of linear response. For an introduction to many-body Green's functions in the context of nuclear physics, we refer the reader to [1, 2, 5, 6]. A much more expansive presentation of the theory can be found in [9]. To see how such an equation applies here, recall that the RPA response function is just the inverse of the term in braces in equation (5.14), and that, from equation (5.13), the elements of the second term in braces (the matrix built from A and B) can be represented as a sum of matrix elements of the HF Hamiltonian $h^{(0)}$ and matrix elements of a potential given by $\partial h/\partial\rho$. After multiplying equation (5.14) by the diagonal matrix with 1 in the upper left and (− 1) in the lower right, the inverse of the term in braces has the same form as the left-hand side of equation (5.15). The result, after manipulations similar to those above, is a linear equation for the RPA response function,

$$R^{\text{RPA}}_{ab,cd}(\omega) = R^{\text{HF}}_{ab,cd}(\omega) + \sum_{ef,pq} R^{\text{HF}}_{ab,ef}(\omega)\frac{\partial h_{ef}}{\partial \rho_{pq}} R^{\text{RPA}}_{pq,cd}(\omega) \tag{5.16}$$

or

$$R^{\text{RPA}}(\omega) = R^{\text{HF}}(\omega) + R^{\text{HF}}(\omega)\frac{\partial h}{\partial \rho} R^{\text{RPA}}(\omega) \tag{5.17}$$

for short. In this equation R^{HF}, which corresponds to $(\omega - \hat{h}_0)^{-1}$ in equation (5.15), is the response when the residual interaction is neglected entirely, i.e. the response produced by the time-independent HF approximation.

Equation (5.16) is particularly useful for Skyrme functionals when written in coordinate space, i.e. when *ab*, *cd* label spatial coordinates (plus spin and isospin coordinates). Skyrme functionals resemble zero-range interactions, and when written in coordinate space the four indices collapse to two, so that equation (5.8) becomes an integral equation for a two-coordinate response function $R^{\text{RPA}}(\boldsymbol{r}, \boldsymbol{r}', \omega)$. The non-interacting response $R^{\text{HF}}(\boldsymbol{r}, \boldsymbol{r}', \omega)$ can be arranged to include continuum boundary conditions so that the RPA response includes the 'escape width' (a slight broadening of peaks from continuum states) that eludes calculations that are formulated in a discrete basis. A pedagogical introduction to this approach appears in [6] and a comprehensive report in [10]. We will return to the Bethe–Salpeter equation when we discuss numerical implementations of the RPA.

5.2 RPA in density functional theory

5.2.1 Brief introduction to DFT for atoms and molecules

For electronic systems, time-independent DFT is based on the important Hohenberg–Kohn theorem [11] and a procedure for taking advantage of the theorem due to Kohn and Sham [12]. Recall that the Hamiltonian describing electrons in an atom is

$$\hat{H} = \hat{T} + \sum_{i=1}^{N} \hat{v}_{\text{ext}}(\boldsymbol{r}_i) + \sum_{ij} \frac{e^2}{4\pi\epsilon_0} \frac{1}{|\boldsymbol{r}_i - \boldsymbol{r}_j|},$$

where $\hat{v}_{\text{ext}}(\boldsymbol{r})$ is a one-body external potential, typically the confining potential for the electrons created by the atomic nuclei. As already mentioned in the introduction to chapter 4, one of the key differences between DFT for electrons and for nuclei is that the nuclear Hamiltonian is not known. Although the ideas to follow were developed for use with this system, they apply equally to any many-body system, with any interaction. The framework is based on the following theorem, paraphrased crudely here.

Theorem 2 (Hohenberg–Kohn (HK)). *There exists a functional of the density $E[\rho]$ that is minimized by the ground-state one-body density, producing the ground-state energy at that density. The functional can be written as*

$$E[\rho] = \int d^3 r \hat{v}_{\text{ext}}(\boldsymbol{r})\rho(\boldsymbol{r}) + F_{\text{HK}}[\rho].$$

The second piece, F_{HK}, is universal, meaning that it is the same no matter what the external potential $\hat{v}_{\text{ext}}$.

Although the theorem seems mysterious at first sight, it can be understood in terms familiar to physicists if one realizes that the energy functional $E[\rho]$ is the Legendre transform of the ground-state energy when written as a functional of the external potential $\hat{v}_{\text{ext}}$. Legendre transforms, used frequently in thermodynamics and classical mechanics, relate the dependence of a quantity (e.g. the Lagrangian in classical mechanics) on a particular variable (the velocity in classical mechanics) to the dependence of a similar quantity (the Hamiltonian) on a complementary variable (the canonical momentum). In DFT, $\hat{v}_{\text{ext}}$ and ρ are a set of complementary variables, one for each point in space; they are complementary functions, making $E[\rho]$ a functional. The ground-state energy is clearly a functional of the external potential, and $E[\rho]$ is very nearly the Legendre transform of that functional. (See [13] for pedagogical details.) $E[\rho]$ can also be defined as the smallest possible average energy (expectation value for the Hamiltonian $\hat{H}$) that can be produced by a state with density constrained to be ρ by an external potential that is not included in $\hat{H}$ [14]. This definition makes it clear that the functional has its minimum at the exact ground-state density; any state that yields a density different from the ground-state density can obviously not be the ground state, and therefore must have a higher average energy than does the ground state.

Although the Legendre transform offers a means for constructing $E[\rho]$ in principle, in practice the construction relies on some kind of approximation. The Kohn–Sham procedure [15] makes constructing good approximations easier. The idea is that the kinetic energy of a correlated system depends much more on the system's overall density distribution than on the precise nature of the correlations. Thus, one can approximate a system's kinetic energy quite well by the kinetic energy of a non-interacting system with the same density. Much of the rest of the energy is given by the mean-field (Hartree) energy associated with the same non-interacting system. Kohn and Sham showed how to represent the rest of the energy, called the exchange-correlation energy, as a correction to the density functional for the non-interacting system. The result is a Hartree-like equation, called the Kohn–Sham (KS) equation, for single-particle states associated with non-interacting particles that nevertheless produce the exact correlated ground-state energy and density

$$\left(-\frac{\hbar^2}{2m}\hat{\nabla}^2 + \hat{V}_{\text{KS}}[\rho(\boldsymbol{r})]\right)\phi_a(\boldsymbol{r}) = \epsilon_a \phi_a(\boldsymbol{r}),$$

with the KS self-consistent potential

$$\hat{V}_{\text{KS}}[\rho(\boldsymbol{r})] = \hat{v}_{\text{ext}}(\boldsymbol{r}) + \int d^3 r' \frac{\rho(\boldsymbol{r}')}{|\boldsymbol{r} - \boldsymbol{r}'|} + \frac{\delta E_{\text{XC}}}{\delta \rho}[\rho(\boldsymbol{r})]. \tag{5.18}$$

Here the second term is the Hartree potential, which would have a different form if the underlying interaction were not of Coulomb form and, E_{XC} is the exchange-correlation energy. Because much of the energy is in the kinetic piece, which is approximated well by the Laplacian term, and in the Hartree potential (the second term in $\hat{V}_{KS}$), even crude approximations to the 'exchange-correlation potential'—the last term in $\hat{V}_{KS}$, written there as the functional derivative of the exchange-correlation energy functional—can give reasonable results.

Note that, if the original HK theorem was formulated in terms of the *local* density $\rho(\boldsymbol{r})$, it can just as well be expressed in terms of the non-local density $\rho(\boldsymbol{r}, \boldsymbol{r}')$. The functional is different, and the KS scheme contains explicitly the HF contribution to the energy, that is, also the exchange term (1.37) (adapted for electrons).

One important difference between electronic systems and nuclei is that the latter are self-bound, and so the density one is really interested in is the intrinsic density, defined with respect to the center of mass, rather than the 'laboratory density,' which is spread out over all space. Recent work has shown that it is straightforward to extend the HK theorem and the KS procedure to the intrinsic density in nuclei [16–20]. Some of the virtues of DFT are lost because nuclei are not contained inside confining potentials, so the universality of the functional F_{HK} offers no simplification. In addition, extensions of the important theorems to independent-particle states that break rotational and particle-number symmetries, while they very probably exist, have not been presented. (One must define these densities in a model-independent way to apply the theorems as they stand. Doing so is complicated.) In practice, nuclear theorists operate as if there existed an equivalent of the HK theorem for the symmetry-breaking local densities that appear in the Skyrme functional (DFT can be generalized to more than one kind of density), and one may view the Hartree–Fock–Bogoliubov (HFB) equation as the analog of the KS equation for number-nonconserving (superfluid) systems. And the EDF approach with Skyrme functionals turns out to closely resemble DFT as typically practiced in atomic systems, with gradient corrections to the local-density approximation [21–23]. We will exploit the DFT interpretation of Skyrme functionals in the following sections.

5.2.2 Time-dependent density functional theory

The extension of DFT to time-dependent phenomena, time-dependent density functional theory (TDDFT), is based on the time-dependent version of the HK theorem, known as the Runge–Gross (RG) theorem, and of the KS procedure [24]. There are two important differences between TDDFT and the time-independent version. First, the time-dependent Kohn–Sham Hamiltonian, which we denote by $h[\rho](t)$, depends on the full time-dependent density, that is on the density at all times prior to t. (It also depends implicitly on the initial state of the system.) Thus, one cannot write h as the functional derivative of an energy-density functional, as we do in time-independent DFT, and we cannot write it as $h[\rho(t)]$, a functional of the density at the present time t only. See chapter 1 of [24] for more on these statements.

We can apply the RG theorem to the case of a time-dependent perturbation by a one-body operator $\hat{G}(t)$ that is added to the *static* Hamiltonian. We assume that the system starts in its ground state. The combination of the RG theorem and KS procedure just discussed tells us that there is a unique mean-field Hamiltonian $h[\rho](t)$ giving the exact expectation value at each time for the local density. It can be written as

$$h[\rho](t) = h^{\mathrm{KS}}[\rho_0] + \tilde{G}(t), \tag{5.19}$$

where ρ_0 is the initial ground-state density, $h^{\mathrm{KS}}[\rho_0]$ is the familiar time-independent mean-field Hamiltonian associated with that density, and it is important to note that

$$\tilde{G}(t) \neq G(t).$$

In other words, when the full non-local mean-field Hamiltonian h is replaced by its static counterpart, the field that represents the perturbation is more complicated than G, and absorbs the dependence on the full density history that was in h. Again, ρ is just the local density, not the full one-body density matrix, but can be generalized to include more than one kind of local density.

Consider the linear response for small $G(t)$. We have seen that in matrix form, the exact linear response function R obeys

$$\delta\rho(\omega) = R(\omega)G(\omega). \tag{5.9}$$

For the system of independent particles described by the full time-dependent KS Hamiltonian h, the RG theorem says that the full time-dependent density is exactly the same as for the real system, hence we should also have

$$\delta\rho(\omega) = R^{\mathrm{KS}}(\omega)\tilde{G}(\omega).$$

Now let us assume we can find a nice function f, again implicitly dependent on the density history ρ, such that

$$\tilde{G}(\omega) \equiv G(\omega) + f(\omega)\delta\rho(\omega).$$

Such a function should exist because $\tilde{G}$ should be similar in size to G. Then we have

$$\delta\rho(\omega) = R^{\mathrm{KS}}(\omega)[G(\omega) + f(\omega)\,\delta\rho(\omega)]$$

or, using $\delta\rho = RG$,

$$R(\omega)G(\omega) = R^{\mathrm{KS}}(\omega)[G(\omega) + f(\omega)R(\omega)G(\omega)].$$

So, since G is arbitrary, we have

$$R(\omega) = R^{\mathrm{KS}}(\omega) + R^{\mathrm{KS}}(\omega)f(\omega)R(\omega).$$

This looks like the equation for the RPA response, but the response in the static Kohn–Sham potential replaces that in the HF potential, and the function $R(\omega)$ on the left is the *exact* response. To obtain it, we need only the 'kernel' $f(\omega)$. The problem, of course, is that we have really only reformulated the original Schrödinger equation. But a reformulation can suggest approximations. The simplest for f (beyond $f = 0$) is the 'adiabatic approximation', which is the density-functional version of the RPA.

5.2.3 Adiabatic approximation

The quantum-mechanical adiabatic theorem (see chapter 17, section 10 of [25]) tells us that if a system's Hamiltonian changes very slowly and the system is initially in its ground state, then that state evolves to be very nearly the instantaneous ground state of the time-dependent Hamiltonian at each subsequent time. The simplest example: a spin initially aligned with a slowly rotating magnetic field will rotate so that it stays aligned. The implication for linear response is that if the external perturbation and thus the density change slowly enough, then the density at each time should very nearly be the instantaneous ground-state density, and the static Kohn–Sham Hamiltonian, evaluated at the instantaneous ground-state density, should very nearly represent the system's dynamics. In other words, we should have,

$$h[\rho](t) \approx h^{\mathrm{KS}}[\rho(t)] + G(t). \tag{5.20}$$

Recall that the exact time-dependent single-particle Hamiltonian is given by equation (5.19) so in the adiabatic approximation,

$$\tilde{G}_{ab}(t) - G_{ab}(t) \approx h_{ab}^{\mathrm{KS}}[\rho(t)] - h_{ab}^{\mathrm{KS}}[\rho_0] \approx \sum_{cd} \left.\frac{\partial h_{ab}^{\mathrm{KS}}[\rho]}{\partial \rho_{cd}}\right|_{\rho_0} \delta\rho_{cd}(t). \tag{5.21}$$

It follows that

$$f(\omega) \approx \left.\frac{\partial h^{\mathrm{KS}}}{\partial \rho}\right|_{\rho_0}$$

and the equation for the response function becomes

$$R(\omega) = R^{KS}(\omega) + R^{KS}(\omega)\left.\frac{\partial h^{KS}}{\partial \rho}\right|_{\rho_0} R(\omega). \tag{5.22}$$

In the adiabatic approximation, therefore, the kernel is independent of frequency because it is the functional derivative of the EDF *taken in the KS ground state*—which is by definition frequency-independent. This means that the approximate density functional depends only on the current time. And though in configuration space f is a four-index quantity, in real space it depends only on two spatial coordinates (because $\rho(t)$ in equations (5.20) and (5.21) is the local density).

5.2.4 Significance of adiabatic approximation

Compare equation (5.22) with (5.17): the function R^{KS} is a mean-field response function like R^{HF}. The adiabatic approximation thus produces a response function in RPA form, with $f_{ab,cd}(\omega) = \partial h_{ab}^{\mathrm{KS}}/\partial \rho_{cd}$ in place of the matrix element $\bar{v}_{adbc}$. In nuclei, Skyrme RPA can therefore be considered an attempt to approximate the adiabatic limit of the exact response function.

Going beyond the adiabatic limit would require a frequency-dependent f. Once frequencies reach typical single-particle energy spacings, frequency dependence in f is probably necessary. One example of a theory with a frequency-dependent kernel f is second RPA (see, e.g. [26]). Unfortunately, its response does not become the RPA response in the adiabatic ($\omega = 0$) limit. We will see shortly, when we discuss the inverse-energy-weighted sum rule, that the RPA response is exact at zero frequency.

The second RPA response can be modified in an ad hoc way so that it does become the RPA response at zero frequency, however. One procedure for doing so is the 'subtraction method', proposed in [27] and applied with Skyrme functionals in [28] and [29]. The result is a better description that the RPA can offer of the spreading of strong excitations such as the giant dipole resonance, without sacrificing the RPA's good predictions for the position of the resonance.

5.3 Sum rules

Sum rules typically involve expressions for the sum of all transition strengths weighted by powers of the excitation energy. The two most important such expressions apply to the energy-weighted and inverse-energy-weighted strength. The energy-weighted sum is given byRight-hand side of Eq. 5.23 is wong, it should read:

$$\sum_{\nu}(E_{\nu} - E_0)|\langle\nu|\hat{G}|0\rangle|^2 = \frac{1}{2}\langle 0|[[\hat{G}, \hat{H}], \hat{G}]|0\rangle. \tag{5.23}$$

One can verify this rule by inserting a complete set of states. The inverse-energy-weighted sum is

$$\sum_{\nu}\frac{1}{(E_{\nu} - E_0)}|\langle\nu|\hat{G}|0\rangle|^2 = -\frac{1}{2}\frac{d}{d\lambda}\langle 0_{\lambda}|\hat{G}|0_{\lambda}\rangle\bigg|_{\lambda=0}, \tag{5.24}$$

where $|0_{\lambda}\rangle$ is the ground state of $\hat{H} + \lambda\hat{G}$. This one follows from first-order perturbation theory. The equation-of-motion method shows that the energy-weighted sum rule holds for the RPA transition strength if the double commutators are evaluated in mean-field ground states. When one uses an energy functional rather than a Hamiltonian, the sum rule must be modified as described, e.g. in [30].

Regarding the inverse-energy-weighted sum rule, the TDHF derivation of the RPA response function implies that for the HF state $|HF_{\lambda}\rangle$ of the perturbed Hamiltonian $\hat{H} + \lambda\hat{G}$,

$$\delta\rho = \lambda R^{\mathrm{RPA}}(\omega = 0)G \tag{5.25}$$

and

$$\langle HF_\lambda|\hat{G}|HF_\lambda\rangle = \mathrm{tr}(G\rho_0) + \mathrm{tr}(G\delta\rho) + \mathcal{O}(\lambda^2).$$

Thus,

$$-\frac{1}{2}\frac{d}{d\lambda}\langle HF_\lambda|\hat{G}|HF_\lambda\rangle|_{\lambda=0} = -\frac{1}{2}\sum_{abcd} G_{ba}R^{\mathrm{RPA}}_{ab,cd}(0)G_{cd}$$
$$= \sum_{abcd} G_{ba}\sum_{\nu}\frac{\langle 0|a_b^\dagger a_a|\nu\rangle_{\mathrm{RPA}}\langle\nu|a_c^\dagger a_d|0\rangle_{\mathrm{RPA}}}{\Omega^\nu_{\mathrm{RPA}}}G_{cd},$$

that is,

$$-\frac{1}{2}\frac{d}{d\lambda}\langle HF_\lambda|\hat{G}|HF_\lambda\rangle|_{\lambda=0} = \sum_{\nu}\frac{\langle 0|\hat{G}|\nu\rangle_{\mathrm{RPA}}\langle\nu|\hat{G}|0\rangle_{\mathrm{RPA}}}{\Omega^\nu_{\mathrm{RPA}}}. \tag{5.26}$$

Thus, both the energy-weighted and inverse-energy-weighted sums hold in the RPA *if* the ground-state expectation values in each rule are replaced by their mean-field counterparts. Since mean-field expectation values of one-body operators are typically reasonably accurate, both sums must be reasonably close to their exact counterparts, and the two rules together put tight and realistic constraints on the distribution of strength. The RPA (and QRPA) are thus extremely useful in inclusive processes, when one must integrate over a large part of the strength function.

In case we have not been clear enough about this: the HF ground state in Skyrme EDF theory would essentially be the KS ground state if the energy functional were exact. The Coulomb interaction is treated differently in EDF theory and symmetries are broken, but otherwise the two theories are essentially equivalent. This means that a Skyrme-like functional that reproduces the experimental ground state energy and density will also reproduce the zero-frequency response.

Exact linear response

Going back to electronic DFT, we note that if we use a KS ground state $|0^{\mathrm{KS}}_\lambda\rangle$ in place of the HF ground state $|HF_\lambda\rangle$, then equation (5.26) becomes

$$-\frac{1}{2}\frac{d}{d\lambda}\left\langle 0^{\mathrm{KS}}_\lambda|\hat{G}|0^{\mathrm{KS}}_\lambda\right\rangle\Big|_{\lambda=0} = \sum_{\nu}\frac{\langle 0|\hat{G}|\nu\rangle\langle\nu|\hat{G}|0\rangle_{\text{“RPA”}}}{\Omega^\nu_{\text{“RPA”}}},$$

where ‘RPA’ means the adiabatic approximation discussed earlier. If the KS vacuum on the left-hand side were exact, then the right-hand side would be as well. Thus, even though the energies and matrix elements come from adiabatic approximations to TDDFT, the sum on the left-hand side above is exact because (from equation 5.25) it reflects the response function at $\omega = 0$, i.e. at the adiabatic limit.

5.3.1 Symmetries

As discussed extensively in sections 1.1.3 and 3.1.1, the EDF approach incorporates the concept of spontaneous symmetry-breaking (SSB), which produces a set of ground states related to one another by symmetry transformations. Mean-field ground states are localized in space, for example, causing at least two problems: (1) the true ground state is not localized and (2) when one localized state is picked to represent the ground state, other equivalent states with centers-of-mass at different positions can mix with real excited states unless one is very careful. The electric-dipole operator discussed earlier can directly excite the 'spurious' mode corresponding to simple motion of the center-of-mass. Although the RPA does not eliminate the localization induced by mean-field theory it handles the problem of spurious excitation in an automatic way.

To see why, consider a symmetry operation $\hat{U} = e^{-i\lambda\hat{S}}$, with $[\hat{H}, \hat{S}] = 0$. If the ground state $|0\rangle$ is not an eigenstate of $\hat{S}$, then the operation $\hat{U}|0\rangle = |0_\lambda\rangle$ produces another state that has the same energy as $|0\rangle$ and differs from it only by a phase. See figure 1.1 in section 1.1.3 for a visual explanation of the action of $\hat{U}$: all states $|0_\lambda\rangle$ are lying at the bottom of the Mexican hat. In EDF theory, the new density ρ_λ must obey the same equation as the original one,

$$\left[h[\rho_\lambda], \rho_\lambda\right] = 0.$$

For small λ, symmetry transformation rules of the density (3.74) lead to

$$\rho_{\lambda,\, ab} = \sum_{cd} U^*_{ac}\rho_{cd} U_{db} = \rho_{ab} + \lambda[S, \rho]_{ab} + \dots.$$

Now if one repeats the expansion of the TDHF Hamiltonian and density, as in the derivation of the RPA in equations (5.10) and (5.11) (with $\omega = 0$), one obtains, instead of equation (5.14),

$$\begin{pmatrix} A & B \\ -B^* & -A^* \end{pmatrix}\begin{pmatrix} S \\ -S^* \end{pmatrix} = 0.$$

This means that there is an RPA eigenstate with

$$\Omega_{\rm sp} = 0 \qquad X^{\rm sp}_{mi} = S_{mi} \qquad Y^{\rm sp}_{mi} = -S_{im},$$

where 'sp' means 'spurious'. Knowing this ahead of time allows us to remove that eigenstate from consideration.

The spurious state is a 'zero mode' like the Goldstone bosons associated with spontaneous symmetry-breaking in field theories. It is spurious in our context because translational and rotational symmetries are never really broken in finite systems. Although the multi-reference version of EDF theory is faithful to the symmetries, the single-reference energy density functional (SR-EDF) approximation, around which EDF oscillations take place, breaks them (see section 1.1). If the symmetry is translational, then the RPA produces three spurious states, each corresponding to uniform translational motion in some direction. The states have excitation energy zero, and can be removed from the spectrum ahead of time by restricting the space of excitations to their complement. The same is true if the reference Slater determinant is deformed, breaking rotational symmetry.

5.3.2 Numerical implementations of RPA

Outside of the finite-amplitude method (FAM), which we will discuss in the next section, there are two basic routes to obtaining the RPA excitation energies and/or response. Here, we only consider the codes that solve the RPA equation, not the QRPA equation that we will discuss in section 5.4.

1. Solve (5.3) in some convenient basis. This requires discretizing the continuum, usually by putting the system in a harmonic oscillator or 'spherical box'. Examples of such implementations have been developed for Skyrme functionals [31] and covariant functionals [32, 33]. Early applications of the theory were performed mostly with Skyrme functionals in spherical symmetry because of the low computational cost; see for instance [34–38].
2. Solve the Bethe–Salpeter equation in coordinate space, a task that is relatively easy with zero-range effective pseudo-potentials or in Kohn–Sham energy density functional (KS-EDF) theory; one needs only $R(\boldsymbol{r}_1, \boldsymbol{r}_1'; \boldsymbol{r}_2, \boldsymbol{r}_2')$ at $\boldsymbol{r}_1 = \boldsymbol{r}_1'$, $\boldsymbol{r}_2 = \boldsymbol{r}_2'$ [39–42]. With that simplification, equation (5.16) takes the form,

$$R^{\mathrm{RPA}}(\boldsymbol{r}_1, \boldsymbol{r}_2; \omega) = R^{\mathrm{HF}}(\boldsymbol{r}_1, \boldsymbol{r}_2; \omega) + \int d\boldsymbol{r}' R^{\mathrm{HF}}(\boldsymbol{r}_1, \boldsymbol{r}'; \omega) \tilde{V}(\boldsymbol{r}') R^{\mathrm{RPA}}(\boldsymbol{r}', \boldsymbol{r}_2; \omega),$$

where $\tilde{V}$ is just a constant when the interaction is a pure delta function. One can solve this equation by discretizing the integral and treating the response function as a matrix. One is helped by a nice expression for R^{HF}

$$\begin{aligned} R^{\mathrm{HF}}(\boldsymbol{r}_1, \boldsymbol{r}_2; \omega) &= \sum_{mi} \left(\frac{\phi_m(\boldsymbol{r}_1)\phi_i^*(\boldsymbol{r}_1)\phi_m^*(\boldsymbol{r}_2)\phi_i(\boldsymbol{r}_2)}{\omega - \epsilon_m + \epsilon_i + i\eta} + \frac{\phi_i^*(\boldsymbol{r}_2)\phi_m(\boldsymbol{r}_2)\phi_m^*(\boldsymbol{s})\phi_i(\boldsymbol{r}_1)}{-\omega - \epsilon_m + \epsilon_i + i\eta} \right) \\ &= \sum_i \left(\phi^*(\boldsymbol{r}_1) \langle \boldsymbol{r}_1 | \frac{1}{\omega + \epsilon_i - h + i\eta} | \boldsymbol{r}_2 \rangle \phi_i(\boldsymbol{r}_2) \right. \\ &\quad \left. + \phi^*(\boldsymbol{r}_2) \langle \boldsymbol{r}_2 | \frac{1}{-\omega + \epsilon_i - h + i\eta} | \boldsymbol{r}_1 \rangle \phi_i(\boldsymbol{r}_1) \right), \end{aligned}$$

where ϕ are the HF s.p. wave functions. R^{HF} is just the ordinary Green's function for a particle scattering from the one-body mean-field potential in h (see [6]). As we mentioned previously, the Bethe–Salpeter equation has the advantage of treating outgoing boundary conditions of continuum particles correctly, and leads to a continuum of excited states with a non-zero 'escape width' associated with each resonance [43, 44].

5.3.3 Finite-amplitude method

The idea of the finite-amplitude method (FAM) [45] is to solve equation (5.10) directly for the linear response, without expanding δh as in equation (5.11). In the HF basis, $\rho^{(0)}$ is diagonal, with matrix elements equal to 1 for hole states and 0 for particle states, i.e.

$$\rho^{(0)} = \begin{pmatrix} 1 & 0 \\ 0 & 0 \end{pmatrix}.$$

The mean-field Hamiltonian $h^{(0)}$ is also diagonal, with matrix elements ε_i for hole states and ε_m for particle states. In order that $\rho^{(0)} + \delta\rho(\omega)$ be a projector, as we have seen, $\delta\rho$ can have non-zero matrix elements only in the particle–hole and hole–particle subspaces, that is,

$$\delta\rho = \begin{pmatrix} 0 & Y \\ X & 0 \end{pmatrix}.$$

For each p.h. excitation, equation (5.10) then takes the form

$$(\varepsilon_m - \varepsilon_i - \omega)X_{mi}(\omega) + \delta h_{mi}(\omega) = -G_{mi}(\omega) \tag{5.27a}$$

$$(\varepsilon_m - \varepsilon_i + \omega)Y_{mi}(\omega) + \delta h_{mi}(\omega) = -G^{*}_{mi}(\omega), \tag{5.27b}$$

where, from equation (5.6), h_{mi} is given by

$$h_{mi}(\omega) = \frac{\delta\mathcal{E}}{\delta\rho_{im}(\omega)}. \tag{5.28}$$

Thus, δh depends on X and Y implicitly through $\delta\rho$.

Because δh depends on $\delta\rho$, the equations in (5.27) are nonlinear. Linearizing them in the usual way, by expanding δh to first order in $\eta\delta\rho$ as in equation (5.11), would yield the usual matrix RPA of (5.14) (which reduces to equation (5.3) in the absence of a driving perturbation). In the FAM, one instead takes advantage of the ease with which the two-index quantities h_{mi} can be computed compared to the many more four-index A and B matrix elements in equation (5.4) or (5.13). Instead of linearizing analytically, one (usually) does so numerically (and approximately), evaluating equation (5.12) for small but finite η.

The FAM procedure in its crudest form then boils down, for each ω, to starting with a guess for the Xs and Ys and thus for $\delta\rho$, computing δh through equations (5.28) and (5.12), solving equation (5.27) for new Xs and Ys, and repeating until convergence. Avoiding the calculation of the A and B matrices can shorten computation time by an order of magnitude or two [46–49].

It is possible to analytically linearize explicit functionals of the local densities without actually constructing the A and B matrices. Kortelainen *et al* [50] takes this tack to avoid subtle problems associated with numerical differentiation. The FAM method is a tool of choice for large-scale calculations of multipole response and β-decay, which involve deformed nuclei for which the direct, matrix-based approach is too expensive [51, 52].

5.4 Pairing correlations and QRPA formalism

The RPA formalism can be extended to describe the linear response of nuclei in the presence of pairing correlations. One can think about the resulting approximation—the quasiparticle RPA (QRPA)—either as an equation of motion for excited states built from two-quasiparticle and two quasi-hole excitations of a correlated ground state, or as the linear response of the HFB vacuum. The extension to quasiparticles can be technically complicated, but introduces no fundamentally new ideas.

The HFB theory was introduced in chapter 3. Sections 3.1.2, 3.1.3 and 3.1.4 contain a detailed presentation of the concept of the quasiparticle (q.p.), of the Bogoliubov transformation and the densities relevant to describe a q.p. vacuum, and of an expansion of the HFB energy up to second order in perturbations. Some of the formal manipulations performed in section 6.4.3 and in section 8.3.1 could also be relevant.

5.4.1 QRPA through the equations-of-motion method

The idea here is the same as in the RPA, except that we have more general commutator relations,

$$\left\langle 0\middle|\beta_a^\dagger\beta_b^\dagger, \left[\hat{H}, Q_\nu^\dagger\right]\middle|0\right\rangle = \Omega^\nu\left\langle 0\middle|\left[\beta_a^\dagger\beta_b^\dagger, Q_\nu^\dagger\right]\middle|0\right\rangle,$$
$$\left\langle 0\middle|\left[\beta_a\beta_b, \left[\hat{H}, Q_\nu^\dagger\right]\right]\middle|0\right\rangle = \Omega^\nu\left\langle 0\middle|\left[\beta_a\beta_b, Q_\nu^\dagger\right]\middle|0\right\rangle,$$

and that

$$Q_\nu^\dagger = \sum_{a>b} X_{ab}^\nu\beta_a^\dagger\beta_b^\dagger - \sum_{a>b} Y_{ab}^\nu\beta_b\beta_a.$$

Roughly speaking, one goes from HF/RPA to HFB/QRPA by replacing a product of the type $a_m^\dagger a_i$ with $\beta_\mu^\dagger\beta_\nu^\dagger$ and $a_i^\dagger a_m$ (Hermitian conjugate) with $\beta_\nu\beta_\mu$. Although the indices i, m have a specific meaning, implying that the corresponding states are below or above the Fermi level, q.p. indices are more general.

Now assume that we can substitute $|\text{HFB}\rangle$ for $|0\rangle$ after carrying out the commutators. When calculating expectation values in the HFB vacuum, we take advantage of the fact that $\beta_\nu|\text{HFB}\rangle = 0$ and thus that $\langle\text{HFB}|\beta_\mu^\dagger = 0$. Again, we obtain

$$\begin{pmatrix} A & B \\ -B^* & -A^* \end{pmatrix}\begin{pmatrix} X^\nu \\ Y^\nu \end{pmatrix} = \Omega^\nu\begin{pmatrix} X^\nu \\ Y^\nu \end{pmatrix}, \tag{5.29}$$

but now with more complicated expressions for A and B. Here, we will give the expression for the case in which the total energy is determined by an EDF, rather than by a Hamiltonian, i.e. in which it is given by (and only by) $\mathcal{E} \equiv \mathcal{E}[\rho, \kappa, \kappa^*]$. In this case, one must use the linear response to determine the elements of these matrices; see section 5.4.3. In the canonical basis, where the one-body density matrix ρ is diagonal and κ has the canonical form (3.48) (see section 3.1.3), they are

$$\begin{aligned} A_{ab,cd} &= E_{ac}\delta_{bd} - E_{bc}\delta_{ad} - E_{ad}\delta_{bc} + E_{bd}\delta_{ac} \\ &- V^{\mathrm{ph}}_{a\bar{c}\bar{b}d} u_d v_c u_a v_b + V^{\mathrm{ph}}_{b\bar{c}\bar{a}d} u_d v_c u_b v_a + V^{\mathrm{ph}}_{a\bar{d}\bar{b}c} u_c v_d u_a v_b \\ &- V^{\mathrm{ph}}_{b\bar{d}\bar{a}c} u_c v_d u_b v_a - V^{\mathrm{pp}}_{\bar{c}\bar{d}\bar{b}\bar{a}} v_c v_d v_b v_a - V^{\mathrm{pp}}_{abdc} u_a u_b u_c u_d \\ &- V^{\mathrm{3p1h}}_{\bar{c}\bar{d}a\bar{b}} v_c v_d u_a v_b + V^{\mathrm{3p1h}}_{\bar{c}\bar{d}b\bar{a}} v_c v_d u_b v_a - V^{\mathrm{3p1h}}_{ab\bar{c}d} u_a u_b u_d v_c \\ &+ V^{\mathrm{3p1h}}_{ab\bar{d}c} u_a u_b u_c v_d - V^{\mathrm{1p3h}}_{\bar{c}d\bar{b}\bar{a}} u_d v_c v_b v_a + V^{\mathrm{1p3h}}_{\bar{d}c\bar{b}\bar{a}} u_c v_d v_b v_a \\ &- V^{\mathrm{1p3h}}_{a\bar{b}dc} u_a v_b u_c u_d + V^{\mathrm{1p3h}}_{b\bar{a}dc} u_b v_a u_c u_d \end{aligned} \tag{5.30a}$$

$$\begin{aligned} B_{ab,cd} &= V^{\mathrm{ph}}_{bd\bar{a}\bar{c}} u_d v_c u_b v_a - V^{\mathrm{ph}}_{ad\bar{b}\bar{c}} u_d v_c u_a v_b - V^{\mathrm{ph}}_{bc\bar{a}\bar{d}} u_c v_d u_b v_a \\ &+ V^{\mathrm{ph}}_{ac\bar{b}\bar{d}} u_c v_d u_a v_b + V^{\mathrm{pp}}_{ba\bar{c}\bar{d}} v_c v_d u_a u_b + V^{\mathrm{pp}}_{dc\bar{a}\bar{b}} v_a v_b u_c u_d \\ &+ V^{\mathrm{3p1h}}_{bad\bar{c}} u_d v_c u_a u_b - V^{\mathrm{3p1h}}_{bac\bar{d}} u_c v_d u_a u_b + V^{\mathrm{3p1h}}_{dcb\bar{a}} u_b v_a u_c u_d \\ &- V^{\mathrm{3p1h}}_{dca\bar{b}} u_a v_b u_c u_d + V^{\mathrm{1p3h}}_{b\bar{a}\bar{c}\bar{d}} v_c v_d u_b v_a - V^{\mathrm{1p3h}}_{a\bar{b}\bar{c}\bar{d}} v_c v_d u_a v_b \\ &+ V^{\mathrm{1p3h}}_{d\bar{c}\bar{a}\bar{b}} v_a v_b u_d v_c - V^{\mathrm{1p3h}}_{c\bar{d}\bar{a}\bar{b}} v_a v_b u_c v_d, \end{aligned} \tag{5.30b}$$

where

$$V^{\mathrm{ph}}_{acbd} = \frac{\delta^2 \mathcal{E}[\rho, \kappa, \kappa^*]}{\delta \rho_{ba} \delta \rho_{dc}} \quad V^{\mathrm{pp}}_{badc} = \frac{\delta^2 \mathcal{E}[\rho, \kappa, \kappa^*]}{\delta \kappa^*_{ba} \delta \kappa_{dc}}$$

$$V^{\mathrm{3p1h}}_{badc} = \frac{\delta^2 \mathcal{E}[\rho, \kappa, \kappa^*]}{\delta \kappa^*_{ba} \delta \rho_{cd}} = V^{\mathrm{1p3h}\,*}_{cdba}.$$

If the energy functional can be strictly derived from the expectation value of a true Hamiltonian, then the 'cross-terms' V^{3p1h}_{badc} and V^{1p3h}_{cdba} vanish.

5.4.2 Implementation

These A and B matrices are much larger than in the regular RPA because the indices run over *all* quasiparticles, rather than being confined to particles or holes. The set of states must be truncated at some point, with energy or occupation the criterion. Note that zero-range pairing must be renormalized at the HFB level, usually by putting an upper limit on the single-q.p. spectrum. A more robust method to renormalize pairing is based on the procedure outlined in section 4.3.1. For a given energy cutoff, the renormalization procedure is independent of the size of the basis or the size of the simulation box. In addition to these truncations, and this remark applies to the RPA as well as the QRPA, the discretization of the continuum introduces uncertainty in energy. It takes about $t = 2R_{\mathrm{box}}/c$ for an emitted particle to bounce off a wall of a box and come back (rather than escaping). This introduces uncertainty [53] in the energy of a state of about

$$\Delta E \approx \frac{\hbar}{t} \approx \frac{100\text{fm}}{R_{\text{box}}}.$$

For a confining box of dimension $R_{\text{box}} = 20$ fm, the uncertainty is about 5 MeV. One needs to smooth out the strength distribution by hand to avoid spurious peaks.

5.4.3 HFB linear response and FAM generalization

Let us now consider a small generalized s.p. operator $\hat{G}(t)$, defined in the HFB q.p. basis by

$$\hat{G} = \frac{1}{2}\sum_{ab}\left(G^{11}_{ab}\beta^{\dagger}_{a}\beta_{b} + G^{20}_{ab}\beta^{\dagger}_{a}\beta^{\dagger}_{b} + G^{02}_{ab}\beta_{a}\beta_{b} - G^{11}_{ba}\beta_{a}\beta^{\dagger}_{b}\right). \tag{5.31}$$

The time-dependent Hartree–Fock–Bogoliubov (TDHFB) equation for the generalized density $\mathcal{R}$ in the presence of $\hat{G}$ is

$$i\dot{\mathcal{R}} = [\mathcal{H}(\mathcal{R}) + G(t), \mathcal{R}].$$

As before, we consider harmonic perturbations $\hat{G}$

$$\hat{G}(t) = \eta\left(\hat{G}(\omega)e^{-i\omega t} + \hat{G}^{\dagger}(\omega)e^{i\omega t}\right)$$

so that

$$\mathcal{R}(t) = \mathcal{R}^{(0)} + \eta\left(\delta\mathcal{R}(\omega)e^{-i\omega t} + \delta R(\omega)^{\dagger}e^{i\omega t}\right).$$

In this case, the TDHFB equation yields to first order in η

$$\omega\,\delta\mathcal{R}(\omega) = [\mathcal{H}^{(0)}, \delta\mathcal{R}] + [\delta\mathcal{H}(\omega), \mathcal{R}^{(0)}] + [G(\omega), \mathcal{R}^{(0)}]. \tag{5.32}$$

> We draw the reader's attention to the central role of this equation—for small amplitude, time-dependent oscillation around the HFB minimum. It will appear later in adjustments of Lagrange multipliers in constrained calculations (see equation 8.51) and in estimates of the derivative $\partial\mathcal{R}/\partial\boldsymbol{q}$ in theories of large amplitude collective motion (see equation 6.65).

Instead of working with a diagonal $\rho^{(0)}$, we can now work in the q.p. basis that diagonalizes the generalized density. Following the notation of chapter 3 we use a tilde to denote matrices in the q.p. basis. Recall that we have

$$\tilde{\mathcal{R}}^{(0)} = \begin{pmatrix} 0 & 0 \\ 0 & 1 \end{pmatrix} \quad \tilde{\mathcal{H}}^{(0)} = \begin{pmatrix} E & 0 \\ 0 & -E \end{pmatrix}.$$

To insure that $\tilde{\mathcal{R}} = \tilde{\mathcal{R}}^{(0)} + \eta\delta\tilde{\mathcal{R}}(\omega)$ is a projector, $\delta\tilde{\mathcal{R}}$ must be zero except in the two-quasiparticle and two-quasihole subspaces, that is,

$$\delta\tilde{\mathcal{R}} = \begin{pmatrix} 0 & X \\ Y & 0 \end{pmatrix}. \tag{5.33}$$

The perturbed mean field, in the same basis, can be written in the generic form

$$\delta\tilde{\mathcal{H}} = \begin{pmatrix} H^{11} & H^{20} \\ H^{02} & -(H^{11})^T \end{pmatrix}.$$

These forms imply that the q.p.-basis representation of equation (5.32) is

$$(E_a + E_b - \omega)X_{ab} + H_{ab}^{20} = -G_{ab}^{20} \tag{5.34a}$$

$$(E_a + E_b + \omega)Y_{ab} + H_{ab}^{02} = -G_{ab}^{02}. \tag{5.34b}$$

The derivation of the energy up to second order in $\mathcal{R}$ given in section 3.1.4 also shows how one can express the perturbation $\delta\tilde{\mathcal{H}}$ (specifically, its matrix H^{20}) in terms of $\delta\tilde{\mathcal{R}}$. With an expression for H^{20} in terms of X and Y, one can solve the resulting non-linear equations for X and Y. This is the QRPA FAM [47, 54–56].

The expression one needs for $\delta\tilde{\mathcal{H}}$ can be obtained by working in the original s.p. basis, in which $\mathcal{R}$ and $\mathcal{H}$ are given by matrices of the forms in equations (3.17) and (3.42). Transforming $\mathcal{R}$ in equation (5.33) back to that basis by multiplying with W on the left and $W^\dagger$ on the right, one obtains,

$$\delta\mathcal{R}(\omega) \equiv W\begin{pmatrix} 0 & X \\ Y & 0 \end{pmatrix}W^\dagger \equiv \begin{pmatrix} \delta\rho & \delta\kappa \\ -\delta\bar{\kappa}^* & -\delta\rho^* \end{pmatrix},$$

with

$$\begin{aligned} \delta\rho &= +UXV^T + V^*YU^\dagger \\ \delta\kappa &= +UXU^T + V^*YV^\dagger \\ \delta\bar{\kappa}^* &= -U^*YU^\dagger - VXV^T, \end{aligned}$$

Parameterizing the perturbed mean field in this original basis as

$$\delta\mathcal{H}(\omega) \equiv \begin{pmatrix} \delta h & \delta\Delta \\ -\delta\bar{\Delta}^* & -\delta h^T \end{pmatrix}$$

one obtains δh, $\delta\Delta$ and $\delta\Delta$ by taking numerical derivatives of $\mathcal{H}$, as in the HF case in equation (5.12). Finally, transforming back to the quasiparticle basis, one finds

$$\begin{aligned} H^{20}(\omega) &= U^\dagger\delta hV^* - V^\dagger\delta h^TU^* + U^\dagger\delta\Delta U^* - V^\dagger\delta\bar{\Delta}^*V^* \\ H^{02}(\omega) &= V^T\delta hU - U^T\delta h^TU + V^T\delta\Delta V - U^T\delta\bar{\Delta}^*U. \end{aligned}$$

The FAM procedure then consists of starting with a trial set of Xs and Ys, computing $\delta\rho$, $\delta\kappa$ and $\delta\bar{\kappa}$, then δh, $\delta\Delta$ and $\delta\bar{\Delta}$, and finally H^{20} and H^{02}, solving equation (5.34) for a new set of Xs and Ys, and repeating until convergence.

Alternatively, one can use the development above to obtain explicit expressions for the QRPA matrix from the density functional by explicitly constructing δh, $\delta\Delta$ and $\delta\bar{\Delta}$ as linear quantities in $\delta\rho$, $\delta\Delta$ and $\delta\bar{\Delta}$ through the relation

$$\delta\mathcal{H}_{ab} = \sum_{c\leq d} \frac{\partial\mathcal{H}_{ab}}{\partial\mathcal{R}_{cd}} \delta\mathcal{R}_{cd}.$$

In the quasiparticle basis, equation (5.34) can be written in the standard form

$$\left\{\begin{pmatrix} \omega & 0 \\ 0 & -\omega \end{pmatrix} - \begin{pmatrix} A & B \\ B^* & A^* \end{pmatrix}\right\}\begin{pmatrix} X \\ Y \end{pmatrix} = \begin{pmatrix} G^{20} \\ G^{02} \end{pmatrix}. \tag{5.35}$$

If the functional is built from a real Hamiltonian, the derivatives that define $\delta\mathcal{H}$ lead to

$$\begin{aligned} A_{ab,cd} &= (E_a + E_b)\delta_{ac}\delta_{bd} \\ &+ U^\dagger_{ar} V^*_{sb} \bar{v}_{rusv} U_{vc} V^T_{du} - V^\dagger_{ar} V^*_{sb} \bar{v}_{rsuv} V_{ud} U^T_{cv} \\ &+ U^\dagger_{ar} U^*_{sb} \bar{v}_{rsuv} U_{uc} U^T_{dv} - V^\dagger_{ar} U^*_{sb} \bar{v}_{rusv} V_{vd} U^T_{cu} \end{aligned} \tag{5.36a}$$

$$\begin{aligned} B_{ab,cd} = &- U^\dagger_{ar} V^*_{sb} \bar{v}_{rusv} V_{vd} U^\dagger_{cu} + V^\dagger_{ar} V^*_{sb} \bar{v}_{rusv} U^*_{uc} U^\dagger_{dv} \\ &- U^\dagger_{ar} U^*_{sb} \bar{v}_{rsuv} V^*_{ud} V^\dagger_{cv} + V^\dagger_{ar} U^*_{sb} \bar{v}_{rusv} U^*_{vc} V^\dagger_{du}. \end{aligned} \tag{5.36b}$$

If the functional is more general, the expressions contain additional 'rearrangement terms', such as those in equation (5.30).

The direct solution of equation (5.35), a process called matrix-QRPA, is time-consuming in heavy, deformed nuclei, but nevertheless possible [57–62]. Calculations with covariant functionals were also performed in spherical nuclei [63]. Figure 5.2 shows the results of a matrix-QRPA calculation of the isoscalar and isovector strength functions in a large number of tin isotopes. The isovector functions are directly related to the cross section for absorbing photons; when the coordinates of the dipole moment operator are in the center-of-mass frame (as they should be), the isoscalar piece of the operator **D** disappears.

5.5 Charge-changing QRPA

Sometimes one is interested in processes that change neutrons into protons or vice versa. Our canonical example here, one that is entirely analogous to the photon absorption discussed earlier, is β-decay. A related process—charged current neutrino scattering—is important for nuclear astrophysics as well as neutrino physics. In the long-wavelength approximation, in which the energy release is much smaller than the inverse size of the nucleus, and in heavy nuclei for which the isobar analog resonance is above the decay threshold, β-decay rates are determined by the matrix elements between initial and final states of the Gamow–Teller operator,

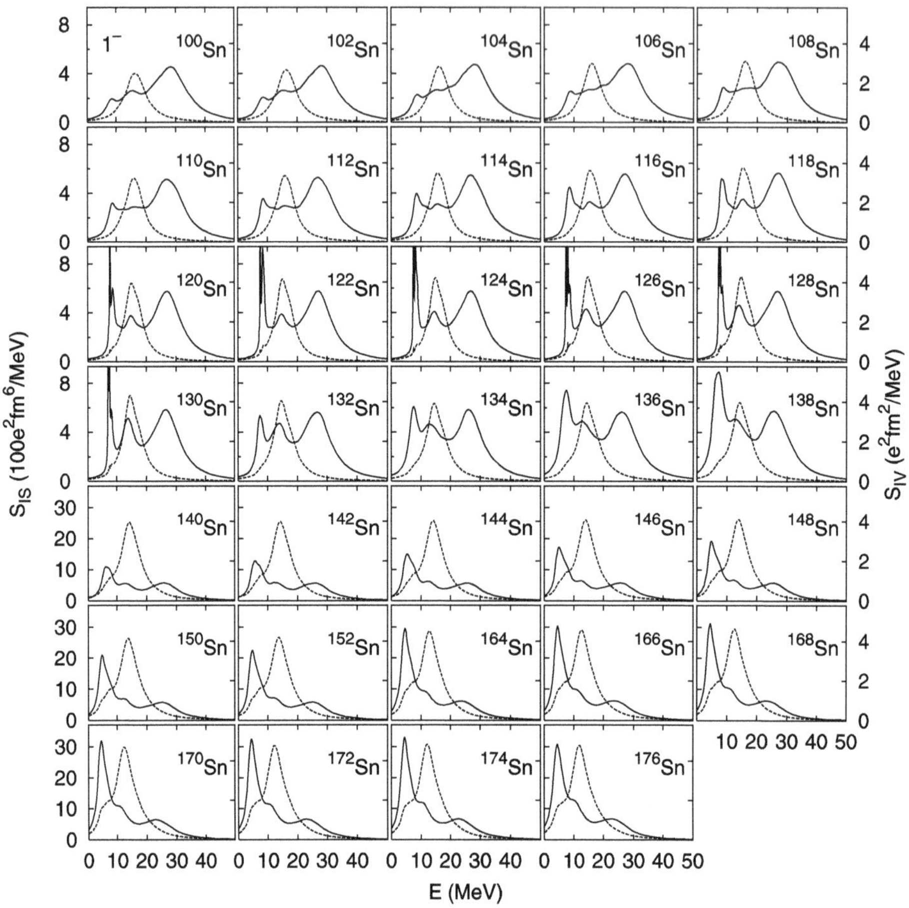

Figure 5.2. The isoscalar (solid lines) and isovector (dotted lines) strength functions for a larger number of tin isotopes, computed in the matrix-QRPA with Skyrme functionals. (Figure reproduced with permission from [58], courtesy of Terasaki. Copyright 2010 The American Physical Society.)

$$\mathbf{O}_{GT}^{\pm} \propto \boldsymbol{\sigma}\tau^{\pm}, \tag{5.37}$$

where the operator τ^+ turns a proton into a neutron and the operator τ^- does the opposite. We omit here the constants that precede $\boldsymbol{\sigma}\tau^{\pm}$ in the weak interaction. The strength produced by $\mathbf{O}_{GT}^{-}$ has a peak that corresponds to the resonant oscillation of neutrons into protons: the so-called Gamow–Teller giant resonance. Figure 5.3 shows cross sections for neutrons emerging at 0° from the charge-changing (p, n) reaction of 198 MeV protons for two representative examples of medium-mass and heavy nuclei; the cross sections show strong peaks at excitation energies between 10 and 20 MeV. (The reaction replaces a nuclear neutron with a proton, and in $N \geqslant Z$ nuclei, the replacement is Pauli blocked.) The cross sections are proportional to the Gamow–Teller strength. The RPA and QRPA have the same advantages in reproducing this kind of strong peak in the charge-changing strength as they do in reproducing the ordinary resonances we considered before.

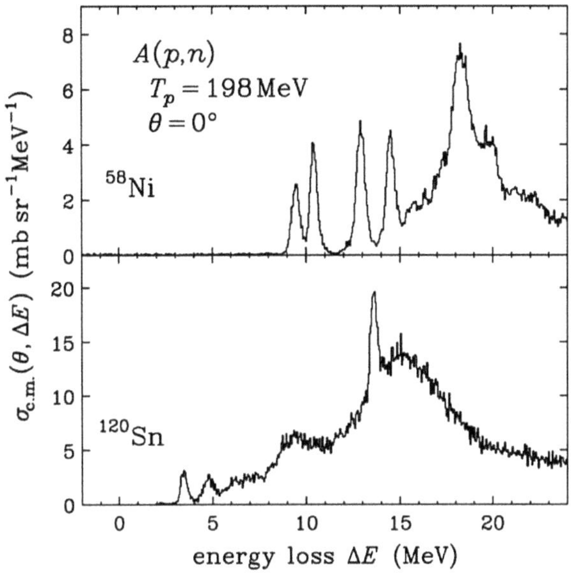

Figure 5.3. Double differential cross sections for the (p, n) reaction at 198 MeV on ^{58}Ni and ^{120}Sn as a function of energy loss of the reaction. (Figure reproduced with permission from [64], courtesy of Sasano. Copyright 2009 The American Physical Society.)

5.5.1 The equations-of-motion method

If we want to calculate transition strength induced by charge-changing operators, we must construct final states that have one more proton and one fewer neutrons (or vice versa) than the nucleus that we are initially considering. To do that, we require the phonon creation operators to contain one proton q.p. operator and one neutron q.p. operator. Assuming that the HFB vacuum is a product of proton and neutron q. p. vacua, we have

$$Q_\nu^\dagger = \sum_{pn}\left(X_{pn}^\nu \beta_p^\dagger \beta_n^\dagger - Y_{pn}^\nu \beta_p \beta_n\right), \tag{5.38}$$

where the subscripts p and n label proton and neutron single-quasiparticle states. In the canonical basis, we obtain,

$$\begin{aligned} A_{pn,p'n'} &= E_{pp'}\,\delta_{nn'} + E_{nn'}\,\delta_{pp'} \\ &\quad + V^{\mathrm{ph}}_{pn,p'n'}(u_p v_n u_{p'} v_{n'} + v_p u_n v_{p'} u_{n'}) \\ &\quad + V^{\mathrm{pp}}_{pn,p'n'}(u_p u_n u_{p'} u_{n'} + v_p v_n v_{p'} v_{n'}) \end{aligned} \tag{5.39}$$

and

$$\begin{aligned} B_{pn,p'n'} &= V^{\mathrm{ph}}_{pn,p'n'}(v_p u_n u_{p'} v_{n'} + u_p v_n v_{p'} u_{n'}) \\ &\quad - V^{\mathrm{pp}}_{pn,p'n'}(u_p u_n v_{p'} v_{n'} + v_p v_n u_{p'} u_{n'}) \end{aligned} \tag{5.40}$$

if the energy functional contains no terms containing the product $\rho\kappa$. In part because of the charge-changing piece of the density operator has a vanishing expectation value in the HFB vacuum, these equations are simpler than those of the like-particle QRPA. In addition, the Coulomb potential, which is typically more time-consuming to treat than the zero-range Skyrme effective pseudo-potential, does not act between neutrons and protons and so is absent from equation (5.40). The charge-changing QRPA has been applied with Skyrme [65], Gogny [66, 67] and covariant functionals [68].

5.5.2 Finite-amplitude method

The FAM was adapted to the charge-changing QRPA in [49], where it was called the pnFAM (with 'pn' standing for 'proton–neutron'). The driving operator is now charge-changing and thus non-Hermitian, and the nonlinear equations for $X_{pn}(\omega)$ and $Y_{pn}(\omega)$ take the form

$$\begin{aligned}\left(E_p + E_n - \omega\right)X_{pn} + \delta\tilde{\mathcal{H}}^{20}_{pn} &= -\, G^{20}_{pn}\\ \left(E_p + E_n + \omega\right)Y_{pn} + \delta\tilde{\mathcal{H}}^{02}_{pn} &= -\, G^{02}_{pn}.\end{aligned} \tag{5.41}$$

The functional $\delta\mathcal{H}$ now depends on the difference between mean-field Hamiltonians at slightly different *charge-changing* densities ρ_{pn} and κ_{pn}.

To use the method to compute a β-decay rate, one must weight the strength function by phase space factors and integrate up to the decay threshold. For Gamow–Teller decay (usually the dominant mode in nuclei with $N \neq Z$), the decay rate is

$$\begin{aligned}\lambda = &\frac{G^2 g_A^2}{2\pi^3\hbar(m_p c^2)^4}\int_{m_e} dE_e\, E_e\sqrt{E_e^2 - m_e^2}\,(E_I - E_F - E_e)^2\\ &\times \sum_F |\langle F|\sum_n \sigma_n \tau_n^+ |I\rangle|^2 F(Z, E_e),\end{aligned} \tag{5.42}$$

where $G \approx 10^{-5}$, $g_A \approx 1.27$, m_p is the proton mass, I and F label initial and final states, n labels particles and $F(Z, E_e)$ is a 'Fermi function' that reflects the effects of the Coulomb interaction between the nucleus and the outgoing electron. The upper limits of the sum over final states F and the integral over electron energies are determined by the requirement that the factor in parentheses that is squared is positive (we have neglected the very small neutrino mass).

The matrix elements in equation (5.42) are the residues associated with poles in the response function, and so one weights the response function by the other factors in the equation and integrates in the complex plane along a contour that encloses the piece of the real axis that contains the poles below threshold, i.e. for which $E_I > E_F$. Details can be found in [49, 69, 52].

References

[1] Fetter A L and Walecka J D 1971 *Quantum Theory of Many-Particle Systems* (Boston, MA: McGraw-Hill)

[2] Negele J W and Orland H 1988 *Quantum Many-Particle Systems* (Cambridge, MA: Perseus)
[3] Shalit A and Feshbach H 1974 *Theoretical Nuclear Physics: Nuclear Structure* (New York: Wiley)
[4] Jancovici B and Schiff D H 1964 The collective vibrations of a many-fermion system *Nucl. Phys.* **58** 678
[5] Blaizot J-P and Ripka G 1985 *Quantum Theory of Finite Systems* (Cambridge, MA: MIT Press)
[6] Ring P and Schuck P 2004 *The Nuclear Many-Body ProblemTexts and Monographs in Physics* (Berlin: Springer)
[7] Penfold A S and Garwin E L 1959 Nuclear photon absorption in carbon and oxygen *Phys. Rev.* **114** 1324
[8] Nozières P 1964 *Theory of Interacting Fermi Systems* (Boca Raton, FL: CRC Press)
[9] Dickhoff W H and Van Neck D 2005 *Many-Body Theory Exposed! Propagator Description of Quantum Mechanics in Many-Body Systems* (Singapore: World Scientific)
[10] Bertsch G F and Tsai S F 1975 A study of the nuclear response function *Phys. Rep.* **18** 125
[11] Hohenberg P and Kohn W 1964 Inhomogeneous electron gas *Phys. Rev* B **136** 864–71
[12] Kohn W and Sham L J 1965 Self-consistent equations including exchange and correlation effects *Phys. Rev.* A **140** 1133
[13] Argaman N and Makov G 2000 Density functional theory: an introduction *Amer. J. Phys.* **68** 69
[14] Fiolhais C *et al* 2003 *A Primer in Density Functional Theory* (Berlin: Springer) p 56
[15] Kohn W and Sham L J 1965 Self-consistent equations including exchange and correlation effects *Phys. Rev.* A **140** 1133
[16] Barnea N 2007 Density functional theory for self-bound systems *Phys. Rev.* C **76** 067302
[17] Engel J 2007 Intrinsic-density functionals *Phys. Rev.* C **75** 14306
[18] Giraud B G 2008 Scalar nature of the nuclear density functional *Phys. Rev.* C **78** 14307
[19] Messud J, Bender M and Suraud E 2009 Density functional theory and Kohn–Sham scheme for self-bound systems *Phys. Rev.* C **80** 054314
[20] Messud J 2011 Generalization of internal density-functional theory and Kohn–Sham scheme to multicomponent self-bound systems, and link with traditional density-functional theory *Phys. Rev.* A **84** 052113
[21] Parr R G and Yang W 1989 *Density Functional Theory of Atoms and Molecules International Series of Monographs on Chemistry* (New York: Oxford University Press)
[22] Dreizler R M and Gross E K U 1990 *Density Functional Theory: An Approach to the Quantum Many-Body Problem* (Berlin: Springer)
[23] Eschrig R 1996 *Fundamentals of Density Functional Theory* (Leipzig: Teubner)
[24] Marques M A L and Gross E K U 2004 Time-dependent density functional theory *Annu. Rev. Phys. Chem.* **55** 427
[25] Messiah A 1962 *Quantum Mechanics* vol 2 (Amsterdam: North-Holland)
[26] Da Providencia J 1965 Variational approach to the many-body problem *Nucl. Phys.* **61** 87
[27] Tselyaev V I 2013 Subtraction method and stability condition in extended random-phase approximation theories *Phys. Rev.* C **88** 054301
[28] Gambacurta D, Grasso M and Engel J 2015 Subtraction method in the second random-phase approximation: first applications with a Skyrme energy functional *Phys. Rev.* C **92** 034303

[29] Gambacurta D, Grasso M and Vasseur O 2017 Electric dipole strength and dipole polarizability in ^{48}Ca within a fully self-consistent second random-phase approximation *Phys. Lett.* B **777** 163
[30] Lipparini E and Stringari S 1989 Sum rules and giant resonances in nuclei *Phys. Rep* **175** 103
[31] Colò G *et al* 2013 Self-consistent RPA calculations with Skyrme-type interactions: the skyrme_rpa program *Comput. Phys. Commun.* **184** 142
[32] Nikšić T, Vretenar D and Ring P 2002 Relativistic random-phase approximation with density-dependent meson–nucleon couplings *Phys. Rev.* C **66** 064302
[33] Nikšić T, Vretenar D and Ring P 2005 Random-phase approximation based on relativistic point-coupling models *Phys. Rev.* C **72** 014312
[34] Blaizot J P 1976 The Skyrme energy functional and RPA calculations *Phys. Lett.* B **60** 435
[35] Abbas A and Zamick L 1980 Nuclear vibrations with a zero-range interaction and the multipole condition *Phys. Rev.* C **22** 1755
[36] Dumitrescu T S and Serr F E 1983 Self-consistent calculations of dipole and quadrupole compression modes *Phys. Rev.* C **27** 811
[37] Serr F E *et al* 1983 Microscopic description of current distributions for collective excitations in spherical nuclei *Nucl Phys.* A **404** 359
[38] Navarro J and Barranco M 1989 The dipole isovector M3 sum rule in the random phase approximation *Nucl. Phys.* A **505** 173
[39] Shlomo S and Bertsch G 1975 Nuclear response in the continuum *Nucl. Phys.* A **243** 507–18
[40] Liu K F and Van Giai N 1976 A self-consistent microscopic description of the giant resonances including the particle continuum *Phys. Lett.* B **65** 23
[41] Krewald S *et al* 1977 On the use of Skyrme forces in self-consistent RPA calculations *Nucl. Phys.* A **281** 166
[42] Shigehara T, Shimizu K and Arima A 1989 The continuum RPA with the exchange term explicitly included and its applications to the spin–isospin and longitudinal response functions *Nucl. Phys.* A **492** 388
[43] Van Giai N *et al* 1987 Underlying structure of continuum response functions in random phase approximation *Phys. Lett.* B **199** 155
[44] Vertse T *et al* 1991 Continuum RPA calculation of escape widths *Phys. Lett.* B **264** 1
[45] Nakatsukasa T, Inakura T and Yabana K 2007 Finite amplitude method for the solution of the random-phase approximation *Phys. Rev.* C **76** 024318
[46] Inakura T, Nakatsukasa T and Yabana K 2009 Self-consistent calculation of nuclear photoabsorption cross sections: finite amplitude method with Skyrme functionals in the three-dimensional real space *Phys. Rev.* C **80** 044301
[47] Avogadro P and Nakatsukasa T 2011 Finite amplitude method for the quasiparticle random-phase approximation *Phys. Rev.* C **84** 014314
[48] Stoitsov M *et al* 2011 Monopole strength function of deformed superfluid nuclei *Phys. Rev.* C **84** 041305
[49] Mustonen M T *et al* 2014 Finite-amplitude method for charge-changing transitions in axially deformed nuclei *Phys. Rev.* C **90** 024308
[50] Kortelainen M, Hinohara N and Nazarewicz W 2015 Multipole modes in deformed nuclei within the finite amplitude method *Phys. Rev.* C **92** 051302
[51] Mustonen M T and Engel J 2016 Global description of β-decay in even–even nuclei with the axially-deformed Skyrme finite-amplitude method *Phys. Rev.* C **93** 014304

[52] Shafer T *et al* 2016 β decay of deformed *r*-process nuclei near $A = 80$ and $A = 160$, including odd-*A* and odd–odd nuclei, with the Skyrme finite-amplitude method *Phys. Rev.* C **94** 055802
[53] Nakatsukasa T and Yabana K 2003 Giant resonances in the deformed continuum *Eur. Phys. J.* A **20** 163
[54] Hinohara N, Kortelainen M and Nazarewicz W 2013 Low-energy collective modes of deformed superfluid nuclei within the finite-amplitude method *Phys. Rev.* C **87** 064309
[55] Nikšić T *et al* 2013 Implementation of the finite amplitude method for the relativistic quasiparticle random-phase approximation *Phys. Rev.* C **88** 044327
[56] Hinohara N *et al* 2015 Complex-energy approach to sum rules within nuclear density functional theory *Phys. Rev.* C **91** 044323
[57] Terasaki J and Engel J 2006 Self-consistent description of multipole strength: systematic calculations *Phys. Rev.* C **74** 044301
[58] Terasaki J and Engel J 2010 Self-consistent Skyrme quasiparticle random-phase approximation for use in axially symmetric nuclei of arbitrary mass *Phys. Rev.* C **82** 034326
[59] Péru S and Goutte H 2008 Role of deformation on giant resonances within the quasiparticle random-phase approximation and the Gogny force *Phys. Rev.* C **77** 044313
[60] Péru S *et al* 2011 Giant resonances in ^{238}U within the quasiparticle random-phase approximation with the Gogny force *Phys. Rev.* C **83** 014314
[61] Lechaftois F, Deloncle I and Péru S 2015 Introduction of a valence space in quasiparticle random-phase approximation: impact on vibrational mass parameters and spectroscopic properties *Phys. Rev.* C **92** 034315
[62] Martini M *et al* 2016 Large-scale deformed quasiparticle random-phase approximation calculations of the γ-ray strength function using the Gogny force *Phys. Rev.* C **94** 014304
[63] Paar N *et al* 2003 Quasiparticle random phase approximation based on the relativistic Hartree–Bogoliubov model *Phys. Rev.* C **67** 034312
[64] Sasano M *et al* 2009 Gamow–Teller unit cross sections of the *(p, n)* reaction at 198 and 297 MeV on medium-heavy nuclei *Phys. Rev.* C **79** 024602
[65] Engel J *et al* 1999 Beta decay rates of *r*-process waiting-point nuclei in a self-consistent approach *Phys. Rev.* C **60** 014302
[66] Martini M, Goriely S and Péru S 2014 Charge-exchange QRPA with the Gogny force for axially-symmetric deformed nuclei *Nucl. Data Sheets* **120** 133
[67] Goriely S *et al* 2016 Gogny–Hartree–Fock–Bogolyubov plus quasiparticle random-phase approximation predictions of the M1 strength function and its impact on radiative neutron capture cross section *Phys. Rev.* C **94** 044306
[68] Paar N *et al* 2004 Quasiparticle random phase approximation based on the relativistic Hartree–Bogoliubov model. II. Nuclear spin and isospin excitations *Phys. Rev.* C **69** 054303
[69] Mustonen M T and Engel J 2016 Global description of β decay in even–even nuclei with the axially-deformed Skyrme finite-amplitude method *Phys. Rev.* C **93** 014304

Chapter 6

Large-amplitude collective motion

Takashi Nakatsukasa and Nicolas Schunck

In order to describe large-amplitude collective phenomena in nuclear physics, such as nuclear fusion/fission processes in low-energy heavy-ion reactions or shape coexistence/fluctuations in low-lying levels, there has always been a strong demand to go beyond the linear regime of time-dependent density functional theory (TDDFT) described in chapter 5. The real-time TDDFT presented in chapter 4, which solves the time-dependent Kohn–Sham (TDKS) equations in real time given an initial state, is a way to investigate non-linear regimes. Thanks to increasing computational resources in recent years, it is now feasible to perform realistic numerical simulations of heavy-ion collisions using modern energy density functionals. As far as time-dependent mean values of one-body observables are concerned, the TDDFT simulations achieved a great success. Recently, there have been significant efforts to improve the description of the distribution width of these one-body observables. Despite these achievements, a number of low-energy nuclear collective phenomena, for which quantum fluctuations play an essential role, are still beyond the range of TDDFT real-time simulations.

A well-known failure of real-time TDDFT is a description of collective quantum tunneling processes. If the energy of the collision is lower than the height of its Coulomb barrier, the TDDFT only simulates the (elastic) Coulomb scattering and cannot describe the sub-barrier fusion process. Similarly, it is impossible to describe spontaneous fission, which is a many-body quantum tunneling phenomenon. The restriction to the common mean-field (Kohn–Sham) potential prevents these quantum tunneling processes from taking place. A possible solution to this problem is the imaginary-time time-dependent Hartree–Fock (TDHF) with re-quantization as outlined in, e.g. [1]. However, the re-quantized imaginary-time TDHF has scarcely been applied to nuclear many-body problems. The method requires us to find periodic orbits in both classically allowed and forbidden regions which join together at the turning points. This is never trivial in multi-dimensional (many-body) systems.

doi:10.1088/2053-2563/aae0edch6

There is an alternative approach to address this problem. Instead of searching for periodic orbits, we may aim at the extraction of a small-dimensional collective submanifold embedded in the infinite-dimensional TDDFT phase space. It will then be assumed that the dynamics of the system is restricted to this collective subspace. In this chapter, we will present some of the techniques that have been invented over the years, both to extract a collective subspace and to define equations of motion in that space. We will begin in section 6.1 by showing how a collective subspace might be defined rigorously by recasting the TDDFT equation into classical Hamilton equations. In section 6.2, we will give a brief description of the adiabatic time-dependent Hartree–Fock (ATDHF) theory, which is often used, albeit in an approximate version, for calculation of large amplitude collective motion. The adiabatic self-consistent collective coordinate (ASCC) theory, which has been developed in recent years to solve some of the limitations of the ATDHF approach, will be presented in section 6.3. Finally, we will give in section 6.4 an extended description of the generator coordinate method (GCM) in the Gaussian overlap approximation (GOA), when Hartree–Fock–Bogoliubov (HFB) states are used as generators. One of the authors (T N) is mainly responsible for sections 6.1, 6.2 and 6.3, while N S is mainly responsible for section 6.4.

6.1 Collective subspace

In the class of theories for large amplitude collective motion discussed in the present chapter, the main issue is to identify a subspace that is, at least approximately, decoupled from the other (non-collective) degrees of freedom as well as the canonical variables which span this subspace. Once this is done, re-quantization becomes feasible, even if it is not uniquely defined. For instance, if we obtain a decoupled subspace for the quadrupole shape degrees of freedom, we may construct the collective Hamiltonian of the Bohr model with the deformation parameters of $\alpha_{2\mu}$ ($\mu = -2, \ldots, 2$). The Bohr model usually adopts the Pauli prescription for quantization [2].

Why theories of large-amplitude collective motion?

At least in electronic systems, the fundamental theorems of TDDFT [3, 4] ensure that TDDFT should be able to probe all the excited states. As emphasized in chapter 5, linear response theory should not be an approximation but the exact linear response of the system. However, there is no equivalent of the Kohn–Sham (KS) procedure to handle self-bound systems such as nuclei, where the relevant densities are the symmetry-breaking densities in the body-fixed frame. In any case, existence theorems do not provide recipes to identify the form of the energy density functional (EDF). As a result, the nuclear EDF currently used in the TDDFT framework of chapter 4 cannot reproduce low-lying excited states associated with strong anharmonicity and quantum fluctuations. Therefore, theoretical tools capable of describing discrete quantum levels corresponding to the stationary collective eigenstates remain especially relevant. The extraction of the collective subspace and its re-quantization may serve this purpose.

6.1.1 Classical Hamilton form of TDDFT

In order to understand the concept of the decoupled collective subspace, it may be useful to rewrite the TDKS equations in the form of the classical Hamilton equations [5]. In this section, we show how this can be done and results in classical equations of motion for canonical variables. For non-superfluid systems, the number of such variables is twice that of the particle–hole (p.h.) degree of freedom (d.o.f.). There are many different ways to define the classical canonical variables, among which we show here the simplest one, the instantaneous (local) canonical variables. For the definition of other variables, see [5, 6], for example.

At each instantaneous time t, we define occupied (hole) natural orbitals $\psi_h(t)$ as the solution of the TDKS equation,

$$i\frac{\partial|\psi_h(t)\rangle}{\partial t} = h[\rho(t)]|\psi_h(t)\rangle \quad h = 1, \ldots, N.$$

The TDKS equation is formally identical to the TDHF equation, which can be obtained as the limit of the time-dependent Hartree–Fock–Bogoliubov (TDHFB) equation described in chapter 4 when pairing correlations vanish at all times. In contrast to superfluid dynamics, the TDHF method involves only as many (coupled) differential equations as the number of particles. Recall that $|\psi_h(t)\rangle$ refers to a time-dependent vector in the single-particle (s.p.) space. The corresponding wave function would be $\langle \boldsymbol{r}\sigma|\psi_h(t)\rangle = \psi_h(\boldsymbol{r}\sigma, t)$. The index $h(p)$ refers to the occupied (unoccupied) nature of the s.p. state.

The unoccupied (particle) orbitals $|\psi_p(t)\rangle$ are arbitrary as long as they are orthogonal to all the occupied orbitals and among themselves. In this instantaneous basis $\{|\psi_k(t)\rangle\}$, the density $\rho(t)$ is diagonal,

$$\rho_{kl} = \begin{cases} \delta_{kl} & k \leqslant N \\ 0 & k > N \end{cases},$$

where $N = N_h$ is the particle number, which is equal to the number of hole orbitals. Starting from the TDKS equation

$$i\frac{\partial \rho(t)}{\partial t} = [h[\rho(t)], \rho(t)] \tag{6.1}$$

and calculating the matrix elements of the density matrix among unoccupied (particle) and occupied (hole) states, we can easily obtain

$$i\dot{\rho}_{hh'} = i\dot{\rho}_{pp'} = 0 \qquad i\dot{\rho}_{ph} = +h_{ph} \qquad i\dot{\rho}_{hp} = -h_{hp}. \tag{6.2}$$

Here, the KS s.p. Hamiltonian is given by the derivative of the total energy with respect to the density

$$h_{kl} = \frac{\partial \mathcal{E}[\rho(t)]}{\partial \rho_{lk}}.$$

The latter two equations in (6.2) are identical to

$$i\frac{\partial \rho_{ph}}{\partial t} = +\frac{\partial \mathcal{E}[\rho]}{\partial \rho^*_{ph}} \qquad i\frac{\partial \rho^*_{ph}}{\partial t} = -\frac{\partial \mathcal{E}[\rho]}{\partial \rho_{ph}}.$$

This particular form suggests that we can identify the real and imaginary parts of the p.h. matrix elements of the density matrix as the coordinates $\boldsymbol{\xi}$ and momenta $\boldsymbol{\pi}$, respectively, with

$$\rho_{ph} = (\xi^{ph} + i\pi_{ph})\big/\sqrt{2}. \tag{6.3}$$

This leads to the Hamiltonian equation of motion for the canonical variables $(\boldsymbol{\xi}, \boldsymbol{\pi})$, with the classical Hamiltonian $\mathcal{H}(\boldsymbol{\xi}, \boldsymbol{\pi}) = \mathcal{E}[\rho]$,

$$\frac{d\xi^\alpha(t)}{dt} = \frac{\partial \mathcal{H}(\boldsymbol{\xi}, \boldsymbol{\pi})}{\partial \pi_\alpha} \qquad \frac{d\pi_\alpha(t)}{dt} = -\frac{\partial \mathcal{H}(\boldsymbol{\xi}, \boldsymbol{\pi})}{\partial \xi^\alpha}, \tag{6.4}$$

where the index α here denotes a particle and hole pair, $\alpha \equiv ph$. We also adopt a vector notation, with $\boldsymbol{\xi} \equiv (\xi^1, \ldots, \xi^\alpha, \ldots)$. The TDKS equation is thus identical to the classical Hamiltonian equation of motion with the (infinite set of) canonical variables $(\boldsymbol{\xi}, \boldsymbol{\pi})$. In practice, one would typically truncate this phase space of classical dynamics to some finite size, $2M = 2N_{ph}$, where $N_{ph} = N_h N_p$ is the total number of possible p.h. excitations allowed.

When pairing correlations are present, we can use very similar arguments and end up with analogous equations, this time for the generalized density $\mathcal{R}(t)$:

$$i\frac{\partial \mathcal{R}_{\mu\nu}}{\partial t} = \frac{\partial \mathcal{E}[\mathcal{R}]}{\partial \mathcal{R}^*_{\mu\nu}} \qquad i\frac{\partial \mathcal{R}^*_{\mu\nu}}{\partial t} = -\frac{\partial \mathcal{E}[\mathcal{R}]}{\partial \mathcal{R}_{\mu\nu}}.$$

Here, $\mathcal{R}_{\mu\nu}$ are defined as $\mathcal{R}_{\mu\nu}(t) \equiv \Phi^\dagger_\mu \mathcal{R} \bar{\Phi}_\nu$ where Φ_μ and $\bar{\Phi}_\nu$ correspond to instantaneous eigenvectors of the time-dependent $\mathcal{R}(t)$ of equation (3.17). The form of $\mathcal{R}(t)$ is the same as equation (3.68) in this instantaneous quasiparticle (q.p.) basis. At time t, the generalized density $\mathcal{R}(t)$ is expressed by the time-dependent q.p. orbits as $\mathcal{R}(t) = \sum_\nu \bar{\Phi}_\nu(t)\bar{\Phi}^\dagger_\nu(t) = 1 - \sum_\mu \Phi_\mu(t)\Phi^\dagger_\mu(t)$. Φ_μ and $\bar{\Phi}_\nu$ are 'unoccupied' ($\mathcal{R}\Phi_\mu = 0$) and 'occupied' ($\mathcal{R}\bar{\Phi}_\mu = \bar{\Phi}_\mu$) q.p. states. Identifying the real and imaginary parts of $\mathcal{R}_{\mu\nu}$ with the coordinates ξ^α and the momenta π_α, respectively, the time-dependent Kohn–Sham–Bogoliubov–de-Gennes equations are expressed by the Hamiltonian equation (6.4). In this case, the index α of the canonical variables (ξ^α, π_α) corresponds to the two-q.p. index $\mu\nu$. The dimension of the phase space $(\boldsymbol{\xi}, \boldsymbol{\pi})$ is, in this case, $2 \times N_m(N_m - 1)$, where N_m is the dimension of the s.p. model space.

The whole TDDFT space is spanned by the variables $(\boldsymbol{\xi}, \boldsymbol{\pi})$. This means that there is one-to-one correspondence between the time-dependent one-body density $\rho(t)$ $(R(t))$ and the time-dependent variables $(\boldsymbol{\xi}(t), \boldsymbol{\pi}(t))$. In other words, any time-dependent (generalized) Slater determinant $|\phi(t)\rangle$ corresponds to a single trajectory $(\boldsymbol{\xi}(t), \boldsymbol{\pi}(t))$ in the phase space. In this section, we have shown the simplest definition of the variables, namely, the local canonical variables. Although the local definition is enough in the following arguments, if we want, we can extend it to the global variables using the Thouless form for Slater determinants; see [5, 6] for more details.

Notation

In the following sections, we follow mathematical notations and conventions that are often used in the theory of general relativity. Although each notation is explained in the text, we summarize them here.

ξ^α, q^μ	Contravariant vectors (coordinates)
π_α, p_μ	Covariant vectors (momenta)
$\mathcal{A}^\alpha\mathcal{B}_\alpha$	Einstein summation convention; $\mathcal{A}^\alpha\mathcal{B}_\alpha = \sum_\alpha \mathcal{A}^\alpha\mathcal{B}_\alpha$
$G_{\alpha\beta}$, $G^{\alpha\beta}$	Metric tensor and its reciprocal tensor ($G^{\alpha\gamma}G_{\gamma\beta} = \delta^\alpha_\gamma$)
$\Gamma^\gamma_{\alpha\beta}$	Affine connection defined with the metric as (6.24)
$\mathcal{A}_{,\alpha}$	Partial derivatives of a scalar function $\mathcal{A}(\boldsymbol{\xi})$; $\mathcal{A}_{,\alpha} = \partial\mathcal{A}/\partial\xi^\alpha$
$\mathcal{V}_{\beta,\alpha}$	Derivatives of a covariant vector, $\mathcal{V}_\beta(\boldsymbol{\xi})$; $\mathcal{V}_{\beta,\alpha} = \partial\mathcal{V}_\beta/\partial\xi^\alpha$
$\mathcal{T}_{\beta\gamma,\alpha}$	Derivatives of a tensor, $\mathcal{T}_{\beta\gamma}(\boldsymbol{\xi})$; $\mathcal{T}_{\beta\gamma,\alpha} = \partial\mathcal{T}_{\beta\gamma}/\partial\xi^\alpha$
$\mathcal{A}_{,\alpha\beta}$	Second partial derivatives; $\mathcal{A}_{,\alpha\beta} = \partial^2\mathcal{A}/\partial\xi^\alpha\partial\xi^\beta$
$\mathcal{A}_{;\alpha\beta}$	Covariant second derivative of $\mathcal{A}$; $\mathcal{A}_{;\alpha\beta} = \partial^2\mathcal{A}/\partial\xi^\alpha\partial\xi^\beta - \Gamma^\gamma_{\alpha\beta}\mathcal{A}_{,\gamma}$

6.1.2 Basic concepts of a decoupled subspace

We use the fact that the TDKS equations of motion can be written in the classical form of equation (6.4), with a large number of classical variables $(\boldsymbol{\xi}, \boldsymbol{\pi})$. This facilitates understanding the concept of the collective submanifold, in terms of a suitable coordinate transformation of a classical system. Let us start with a $2M$-dimensional phase space with canonical variables (ξ^α, π_α), $\alpha = 1, \dots, M$ and the equations of motion (6.4). We then consider a set of point transformations $\boldsymbol{f}$ from the original generalized coordinates $\boldsymbol{\xi}$ to new coordinates $\boldsymbol{q}$ and their inverse transformation $\boldsymbol{g}$:

$$q^\mu = f^\mu(\boldsymbol{\xi}) \qquad \xi^\alpha = g^\alpha(\boldsymbol{q}). \tag{6.5}$$

The conjugate momenta $\boldsymbol{p}$ associated with $\boldsymbol{q}$ are given by

$$p_\mu = \sum_\alpha \frac{\partial\xi^\alpha}{\partial q^\mu}\pi_\alpha \equiv g^\alpha_{,\mu}\pi_\alpha \qquad \pi_\alpha = \sum_\mu \frac{\partial q^\mu}{\partial\xi^\alpha}p_\mu \equiv f^\mu_{,\alpha}p_\mu, \tag{6.6}$$

where we introduced the comma index for partial derivatives and the Einstein summation convention for repeated upper and lower indices.

The canonicity of the new variables $(\boldsymbol{q}, \boldsymbol{p})$ is guaranteed by the conservation of the Poisson brackets, which is easily proven by using the chain-rule relations, $g^\alpha_{,\mu}f^\mu_{,\beta} = \delta^\alpha_\beta$ and $f^\mu_{,\alpha}g^\alpha_{,\nu} = \delta^\mu_\nu$. For example,

$$\{q^\mu, p_\nu\}_{\rm PB} = \frac{\partial q^\mu}{\partial\xi^\alpha}\frac{\partial p_\nu}{\partial\pi_\alpha} - \frac{\partial p_\nu}{\partial\xi^\alpha}\frac{\partial q^\mu}{\partial\pi_\alpha} = f^\mu_{,\alpha}g^\alpha_{,\nu} = \delta^\mu_\nu.$$

Recall that the Poisson brackets for two functions f and g of the canonical coordinates $\boldsymbol{\xi}$ and $\boldsymbol{\pi}$ is

$$\{f, g\}_{\mathrm{PB}} = \sum_{\mu}\left(\frac{\partial f}{\partial \xi^{\alpha}}\frac{\partial g}{\partial \pi_{\alpha}} - \frac{\partial f}{\partial \pi_{\alpha}}\frac{\partial g}{\partial \xi^{\alpha}}\right).$$

The canonical coordinates must verify the following identities

$$\{q^{\mu}, q^{\nu}\}_{\mathrm{PB}} = \{p_{\mu}, p_{\nu}\}_{\mathrm{PB}} = 0$$
$$\{q^{\mu}, p_{\nu}\}_{\mathrm{PB}} = \delta_{\nu}^{\mu}.$$

The main purpose of this coordinate transformation is to identify a $2K$-dimensional decoupled sub-manifold in the $2M$-dimensional phase space ($K \ll M$). In the new coordinate system of equation (6.5), the $\{q^{\mu}\}$ are divided into two subsets, $q^{i}, i = 1, \ldots, K$ (collective) and the rest $q^{a}, a = K + 1, \ldots, M$ (intrinsic/non-collective). Hereafter, we use indices $i, j, \ldots$ for collective variables ($1 \leqslant i, j \leqslant K$), $a, b, \ldots$ for non-collective variables ($K + 1 \leqslant a, b \leqslant M$), and $\mu, \nu, \ldots$ for both ($1 \leqslant \mu, \nu \leqslant M$). For the original coordinate system of $\boldsymbol{\xi}$, we use indices $\alpha, \beta, \cdots$. We impose the condition that the $2K$-dimensional collective subspace Σ_K, which is defined by $q^{a} = p_{a} = 0$ and spanned by the canonical variables (q^{i}, p_{i}), is decoupled from the rest of the space. In classical mechanics, decoupling means that if the system is initially located on the subspace Σ_K, it will stay inside Σ_K at all times. In other words, if the system is located at $q^{a} = p_{a} = 0$ at time $t = 0$, we have $q^{a}(t) = p_{a}(t) = 0$ at any later times t. If this is strictly satisfied for all $t > 0$, the decoupling is exact. We know a trivial example of this exact decoupled collective motion, namely, the center-of-mass motion of an isolated nucleus. However, we are more interested in non-trivial collective motions that correspond to an *approximately decoupled sub-manifold*. We want to determine the decoupled subspace Σ_K, shown by the meshed pattern in figure 6.1, to which the motion in the phase space is approximately confined.

The present concept of a decoupled collective sub-manifold can be generalized by removing the restriction of a point transformation. It can be extended to a general canonical transformation between $(\boldsymbol{\xi}, \boldsymbol{\pi})$ and $(\boldsymbol{q}, \boldsymbol{p})$. In this case, the linear relation between momenta $\boldsymbol{\pi}$ and $\boldsymbol{p}$ in equation (6.6) no longer holds. Thus, the expansion with respect to momenta is not uniquely defined, because it depends on the choice of the coordinate system. In theories for slow motion in section 6.2.1, we adopt the expansion with respect to momenta to find a decoupled subspace. Thus, we will focus on the point transformation hereafter. An extension to more general transformation will be discussed in section 6.3, based on the expansion with respect to the momenta.

6.1.3 Decoupling conditions

After the transformation, $\boldsymbol{\xi} \to \boldsymbol{q}$, Hamilton equations of motion read

$$\frac{dq^{\mu}}{dt} = \frac{\partial H(\boldsymbol{q}, \boldsymbol{p})}{\partial p_{\mu}} \qquad \frac{dp_{\mu}}{dt} = -\frac{\partial H(\boldsymbol{q}, \boldsymbol{p})}{\partial q^{\mu}}. \tag{6.7}$$

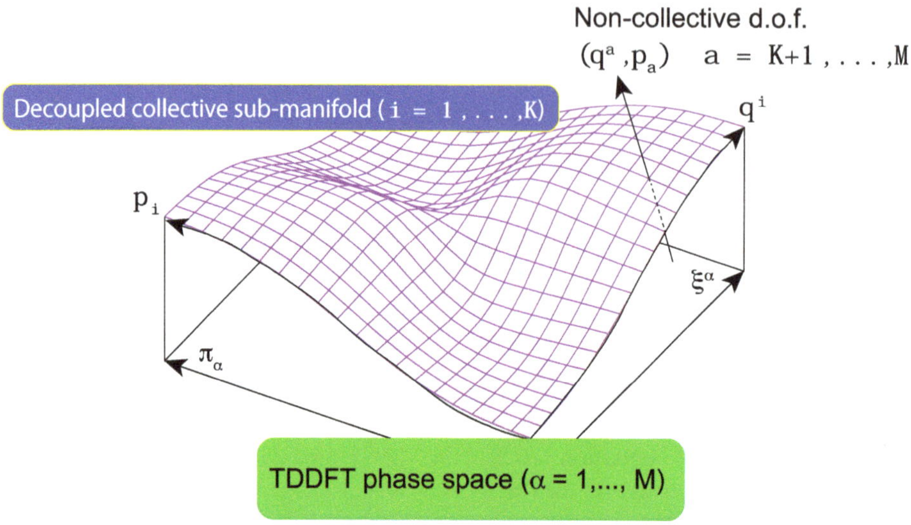

Figure 6.1. Schematic illustration of a decoupled collective subspace. In the space of the original coordinates $\boldsymbol{\xi}$ and momenta $\boldsymbol{\pi}$, the decoupled subspace is a 'curved' subspace of dimension $2K$, which are spanned by the collective variables (q^i, p_i) with $i = 1, \cdots, K$. In the case of $K = 1$, it is a two-dimensional curved plane. (Figure created by T Nakatsukasa.)

Since the functional form of $H(\boldsymbol{q}, \boldsymbol{p})$ may be different from $\mathcal{H}(\boldsymbol{\xi}, \boldsymbol{\pi})$, we use the symbol H for the collective Hamiltonian with variables $(\boldsymbol{q}, \boldsymbol{p})$ and $\mathcal{H}$ for that with $(\boldsymbol{\xi}, \boldsymbol{\pi})$. The point transformation implies that $\mathcal{H}(\xi^\alpha, \pi_\alpha) = H(f^\mu(\boldsymbol{\xi}), g^\alpha_{,\mu}\pi_\alpha)$. In classical mechanics, the decoupling condition of the $2K$-dimensional subspace Σ_K is written as $\dot{q}^a = \dot{p}_a = 0$, on Σ_K, where the dot refers to the time-derivative. Therefore, Hamilton equations (6.7) become on Σ_K

$$\left.\frac{\partial H(\boldsymbol{q}, \boldsymbol{p})}{\partial p_a}\right|_{\Sigma_K} = 0 \qquad \left.\frac{\partial H(\boldsymbol{q}, \boldsymbol{p})}{\partial q^a}\right|_{\Sigma_K} = 0, \tag{6.8}$$

where Σ_K is defined as $q^a = p_a = 0$. It is easy to understand that if the system is initially located on Σ_K, it will stay on Σ_K at all times. These decoupling conditions restrict the tangent vectors, $f^a_{,\beta}$ and $g^\beta_{,a}$, of the point transformation on Σ_K.

6.2 Adiabatic time-dependent Hartree–Fock theory

In this section, we show how the ATDHF theory [7–21] naturally fits within the class of large amplitude collective motion based on the decoupling of a collective subspace and classical equations of motion. As we will discuss in section 6.2.2, there are different types of ATDHF theories in the literature. We will focus here on a self-contained implementation of ATDHF, where the collective variables are not arbitrarily chosen but determined from the decoupling conditions in a collective space of dimension K.

6.2.1 Slow collective motion

Let us now assume that the collective motion of interest is characterized by 'slow' degrees of freedom compared to the ones describing intrinsic motion. In this case, we may expand the decoupling conditions (6.8) in powers of the collective momentum. This approximation is often referred to as the 'adiabatic' approximation. However, one should emphasize that it is somewhat different from the famous adiabatic theory in molecular physics known as the Born–Oppenheimer approximation. In the Born–Oppenheimer approximation, we fix the nuclear configurations and determine the ground state for the electrons. In nuclear physics, the slow collective degrees of freedom are not as trivially determined, and one of the main goals of the theory is, in fact, to identify them.

In the ATDHF theory, the word 'adiabatic' is used to refer to slow collective degree of freedom compared to the ones describing intrinsic motion. In [16], this is quantified in the case of a single collective variable by the condition

$$E_{\text{coll}} \ll N\Delta,$$

where Δ is the average p.h. energy, $\Delta = \overline{e_p - e_h}$ and N the number of s.p. 'single-particle degrees of freedom contributing to the collective motion'. For very collective motion such as, e.g. nuclear fission, there is a large number of such s.p. degrees of freedom so that this condition is in practice easily verified.

6.2.2 Overview

The most well-known theory for the adiabatic regime is the ATDHF theory. However, in the literature, the same name 'ATDHF' is often used to refer to somewhat different concepts. For instance, in their pioneering work, Baranger and Kumar [7–11] derive the five-dimensional collective model (Bohr model) from the pairing-plus-quadrupole (P+Q) Hamiltonian, assuming the quadrupole shapes (β, γ) of the mean-field potential are the collective variables. In the P+Q model of Baranger and Kumar, pairing correlations are treated at the Bardeen–Cooper–Schrieffer (BCS) approximation, and exchange (i.e. Fock) terms are neglected. Although the collective coordinates are determined by hand, taking an advantage of the P+Q model, this is often referred to as the ATDHF.

The work of Baranger and Veneroni in [16] is often cited as the reference for the ATDHF theory. In contrast to the Baranger and Kumar model, it is not restricted to separable interactions. The one-body density matrix in TDHF is expressed in the form $\rho(t) = e^{i\chi(t)}\rho_0(t)e^{-i\chi(t)}$, where $\rho_0(t)$ and $\chi(t)$ are both time-dependent and time-even Hermitian matrices. The TDHF equation is then expanded with respect to powers of χ yielding a pair of coupled equations called the ATDHF equations. After introducing the collective coordinates $q^i(t)$, $i = 1, \ldots, K$ to parameterize the density $\rho_0(\boldsymbol{q}(t))$, we may derive a collective Hamiltonian $H(\boldsymbol{q}, \boldsymbol{p})$ with conjugate momenta p_i. As we will see later, a similar collective Hamiltonian can be derived from the GCM–GOA approach. The main advantage of the Baranger–Veneroni ATDHF approach over GCM–GOA is a proper account of the time-odd components in the density. As a result, the theory gives the exact collective mass for the particular case of center-of-mass motion [16]. However, the theory still relies on our physical intuition to choose the collective coordinates $\boldsymbol{q}$.

The two ATDHF equations obtained after expanding $\rho(t) = \rho^{(0)}(t) + \rho^{(1)}(t) + \rho^{(2)}(t)$ are

$$i\hbar\dot{\hat{\rho}}^{(0)} = [\hat{h}^{(0)}, \hat{\rho}^{(1)}] + [\hat{\Gamma}^{(1)}, \hat{\rho}^{(0)}]$$
$$i\hbar\dot{\hat{\rho}}^{(1)} = [\hat{h}^{(0)}, \hat{\rho}^{(0)}] + [\hat{h}^{(0)}, \hat{\rho}^{(2)}] + [\hat{\Gamma}^{(1)}, \hat{\rho}^{(1)}] + [\hat{\Gamma}^{(2)}, \hat{\rho}^{(0)}].$$

With rare exceptions [24], only the first ATDHF equation is used, often on a predetermined collective path and even dropping any residual interaction (the term $\Gamma^{(1)}$). From there, one introduces collective variables through $\dot{\rho}^{(0)} = \sum_i \frac{\partial \rho}{\partial q^i}\dot{q}^i$. Properties of the Hartree–Fock (HF) equation allow the expression of $\frac{\partial \rho}{\partial q^i}$ in simple terms; see [25] for a recent summary.

In the third class of ATDHF approaches, one tries to determine the optimum collective coordinates $\boldsymbol{q}$ self-consistently without using arbitrary physics arguments [15, 17]. The basic equations are obtained from the time-dependent variational principle by taking into account TDHF equations up to the first order in momenta. The method determines a constraint operator for the constrained mean-field calculation to define the mean-field state $|\phi(\boldsymbol{q})\rangle$. Compared to the two previous approaches, it is the most advanced, in the sense that it does not rely on any assumption about the choice of collective coordinate. However, the theory is plagued by the non-uniqueness of the resulting collective path. We will discuss this problem later.

All of these theories are called 'ATDHF' in the literature. Since the levels of achievement and main purposes of the theories are different to each other, we had better distinguish them. In sections 6.2.3 and 6.2.4, we focus our discussion on the third category of the ATDHF theory. A good summary of the second version of ATDHF approaches, generalized to the superfluid regime, can be found in [25].

The concept of the collective path

One of the most often used concepts in adiabatic theories of large amplitude collective motion is that of the 'collective path' (collective space in the multi-dimensional case). The collective path is defined by a decoupled one-dimensional collective coordinate q^1 (the case of $K = 1$ in figure 6.1). The dependence of the Hamiltonian on the conjugate momentum p_1 determines its rate of change as equation (6.7). Various techniques to compute the collective path will be discussed in the following sections. In many practical applications, so far, the collective path (space) is often *preset* to, e.g. a set of constrained HF or HFB calculations with constraints on the expectation values of given one-body operators as the collective variables, $q^i \equiv \langle\Psi(\boldsymbol{q})|\hat{Q}^i|\Psi(\boldsymbol{q})\rangle$. This allows setting up simple and computationally manageable formulas to determine the collective potential energy and the collective inertial mass (i.e. the 'resistance' to the collective motion). However, obtaining the proper inertial mass is not a simple task, even if the collective coordinate is trivial, such as the center-of-mass for the translational motion, since it should involve, in principle, solutions of full quasiparticle random-phase approximation (QRPA) equations [25].

6.2.3 Equation for the collective path

Villars presented the basic ATDHF equations for the case of one collective coordinate ($K = 1$) [15]. We will first present the basic ATDHF equations to determine the optimum collective path within the framework of classical dynamics. They are identical to the original equations proposed by Villars, even though they may look somewhat different at first sight.

We start by expanding the (classical) Hamiltonian in powers of the momenta,

$$\mathcal{H}(\boldsymbol{\xi}, \boldsymbol{\pi}) = \mathcal{H}^{(0)}(\boldsymbol{\xi}) + \frac{1}{2}\mathcal{H}^{(2)\alpha\beta}(\boldsymbol{\xi})\pi_\alpha\pi_\beta + \mathcal{O}(\boldsymbol{\pi}^4) \tag{6.9a}$$

$$H(\boldsymbol{q}, \boldsymbol{p}) = H^{(0)}(\boldsymbol{q}) + \frac{1}{2}H^{(2)\mu\nu}(\boldsymbol{q})p_\mu p_\nu + \mathcal{O}(\boldsymbol{p}^4), \tag{6.9b}$$

where $\mathcal{H}^{(0)}(\boldsymbol{\xi}) = \mathcal{H}(\boldsymbol{\xi}, \boldsymbol{\pi} = 0)$, $H^{(0)}(\boldsymbol{q}) = H(\boldsymbol{q}, \boldsymbol{p} = 0)$, $\mathcal{H}^{(2)\alpha\beta}(\boldsymbol{\xi}) \equiv \partial^2\mathcal{H}/\partial\pi_\alpha\partial\pi_\beta|_{\boldsymbol{\pi}=0}$ and $H^{(2)\mu\nu}(\boldsymbol{q}) \equiv \partial^2 H/\partial p_\mu \partial p_\nu|_{\boldsymbol{p}=0}$. $\mathcal{H}(\boldsymbol{\xi}, \boldsymbol{\pi})$ and $H(\boldsymbol{q}, \boldsymbol{p})$ are nothing other than the energy density functional; see equation (6.4). The transformation of the momenta, equation (6.6), leads to $H^{(2)\mu\nu} = f^{\mu}_{,\alpha}\mathcal{H}^{(2)\alpha\beta}f^{\nu}_{,\beta}$ and $\mathcal{H}^{(2)\alpha\beta} = g^{\alpha}_{,\mu}H^{(2)\mu\nu}g^{\beta}_{,\nu}$. Here, we assume that the Hamiltonian is invariant with respect to time reversal and that the momenta are time-odd quantities. As a result, the first-order term in the expansion is automatically zero. Remember that, at least in principle, we know the actual functions of $\boldsymbol{\xi}$ such as $\mathcal{H}^{(0)}(\boldsymbol{\xi})$ and $\mathcal{H}^{(2)\alpha\beta}(\boldsymbol{\xi})$. In contrast, all functions of $\boldsymbol{q}$ are unknown, because we do not know the new coordinates $\boldsymbol{q}$ themselves.

The minimization of $\mathcal{H}^{(0)}(\boldsymbol{\xi})$ leads to the HF (HFB) ground-state density $\boldsymbol{\xi} = \boldsymbol{\xi}_0$ and its energy $\mathcal{H}^{(0)}(\boldsymbol{\xi}_0)$. There, all the time-odd densities vanish ($\boldsymbol{\pi} = 0$). $\mathcal{H}^{(2)\alpha\beta}(\boldsymbol{\xi})$ are the second derivatives of the energy with respect to the time-odd densities. Once the functional form of the EDF is specified, such derivatives can be explicitly calculated.

Since the motion in the decoupled collective subspace Σ_K is confined in Σ_K, we may define the collective Hamiltonian as

$$H_c(q^i, p_i) = H(\boldsymbol{q}, \boldsymbol{p})|_{\Sigma_K} = H^{(0)}(q^i) + \frac{1}{2}H^{(2)ij}(q^i)p_i p_j + \mathcal{O}(\boldsymbol{p}^4) \tag{6.10}$$

in terms of the collective variables only, $i = 1, \cdots, K$.

Let us apply the decoupling conditions (6.8) to the expression (6.9b), order by order in momenta. For the zeroth-order term in momenta, only derivatives with respect to q^a give non-trivial conditions. The first-order terms only come from derivatives to the momenta:

$$\mathcal{O}(\boldsymbol{p}^0) = \mathcal{O}(\boldsymbol{\pi}^0): \quad \frac{\partial H^{(0)}}{\partial q^a} \mathrel{\circeq} 0 \Rightarrow H^{(0)}_{,a} = \mathcal{H}^{(0)}_{,\beta} g^{\beta}_{,a} \mathrel{\circeq} 0 \tag{6.11a}$$

$$\mathcal{O}(\boldsymbol{p}^1) = \mathcal{O}(\boldsymbol{\pi}^1): \quad \frac{\partial}{\partial p_a}\left(\frac{1}{2} H^{(2)\mu\nu} p_\mu p_\nu\right) \mathrel{\circeq} 0 \Rightarrow H^{(2)ia} = H^{(2)ai} = f^{i}_{,\alpha} \mathcal{H}^{(2)\alpha\beta} f^{a}_{,\beta} \mathrel{\circeq} 0. \tag{6.11b}$$

These conditions are valid on the decoupled sub-manifold Σ_K. Hereafter, we will use the equation symbol $\circeq$ for equality on Σ_K and will omit the symbol $|_{\Sigma_K}$.

For practical convenience, we rewrite the conditions (6.11a) and (6.11b) as follows. The summation over the index $\mu = 1, \ldots, M$ in the following identities,

$$\mathcal{H}^{(0)}_{,\beta} - H^{(0)}_{,\mu} f^{\mu}_{,\beta} = 0 \quad H^{(2)i\mu} g^{\beta}_{,\mu} = f^{i}_{,\alpha} \mathcal{H}^{(2)\alpha\beta},$$

can be restricted to the collective variables $i, j = 1, \ldots, K$ on Σ_K. We can use these to rewrite the decoupling conditions (6.11a) and (6.11b) as

$$\mathcal{O}(\boldsymbol{p}^0) = \mathcal{O}(\boldsymbol{\pi}^0): \quad \mathcal{H}^{(0)}_{,\beta} - H^{(0)}_{,i} f^{i}_{,\beta} \mathrel{\circeq} 0 \tag{6.12a}$$

$$\mathcal{O}(\boldsymbol{p}^1) = \mathcal{O}(\boldsymbol{\pi}^1): \quad H^{(2)ij} g^{\beta}_{,j} \mathrel{\circeq} f^{i}_{,\alpha} \mathcal{H}^{(2)\alpha\beta}. \tag{6.12b}$$

These constitute the equations defining the decoupled collective sub-manifold in the ATDHF theory.

In the case $K = 1$, we have only one collective coordinate q^1, which is a function of the M original variables ξ^α. Conversely, there exist M functions giving ξ^α in terms of q^1,

$$\xi^\alpha(q^1) = g^\alpha(q^1, q^a = 0)$$

on the decoupled subspace Σ_1. The collective Hamiltonian (6.10) thus becomes

$$H_c(q^1, p_1) = H^{(0)}(q^1) + \frac{1}{2} H^{(2)11}(q^1)(p_1)^2.$$

In the case $K = 1$ of a single collective variable, (6.12) can be solved easily for $f^{1}_{,\beta}$ since

$$f^{1}_{,\alpha} \mathrel{\circeq} \mathcal{H}^{(0)}_{,\alpha} \big/ H^{(0)}_{,1}. \tag{6.13}$$

Since the scale of the collective coordinate q^1 is arbitrary as far as it is non-zero, we can choose $H_{,1}{}^{(0)}$ as we like. Let us choose it so as to have

$$H^{(2)11} = f^1_{,\alpha}\mathcal{H}^{(2)\alpha\beta}f^1_{,\beta} \overset{\circ}{=} \frac{\mathcal{H}^{(0)}_{,\alpha}\mathcal{H}^{(2)\alpha\beta}\mathcal{H}^{(0)}_{,\beta}}{(H^{(0)}_{,1})^2} = 1. \tag{6.14}$$

Then, equation (6.12b) leads to the 'equation for the collective path' [19, 26],

$$g^\beta_{,1} \overset{\circ}{=} \frac{d\xi^\beta}{dq^1} \overset{\circ}{=} f^1_{,\alpha}\mathcal{H}^{(2)\alpha\beta} \overset{\circ}{=} \frac{\mathcal{H}^{(0)}_{,\alpha}\mathcal{H}^{(2)\alpha\beta}}{H^{(0)}_{,1}}. \tag{6.15}$$

Solving equation (6.15) means that we determine the M functions $\xi^\alpha = g^\alpha(q^1)$ on Σ_1, which determine the time-even densities. We expect that the solution provides an approximate decoupling between the collective motion, characterized by the variable q^1, and the rest (intrinsic excitations) q^a. The direction of the tangent vectors on the collective path Σ_1 is given by $g^\beta_{,1} \propto \mathcal{H}^{(0)}_{,\alpha}\mathcal{H}^{(2)\alpha\beta}$. The trivial identity $f^1_{,\alpha}g^\alpha_{,1} = 1$, determines the magnitude of $H^{(0)}_{,1}$ as

$$\left(H^{(0)}_{,1}\right)^2 \overset{\circ}{=} \mathcal{H}^{(0)}_{,\alpha}\mathcal{H}^{(2)\alpha\beta}\mathcal{H}^{(0)}_{,\beta}.$$

We can use a simple Euler method to solve equation (6.15). Starting from the initial state $\boldsymbol{\xi}_0$ at $q^1 = 0$, a series of states $\boldsymbol{\xi}_n$ at $q^1 = n \times \Delta q$ are obtained by integrating equation (6.15),

$$\xi^\beta_n \approx \xi^\beta_{n-1} + \frac{\Delta q}{H^{(0)}_{,1}}\mathcal{H}^{(0)}_{,\alpha}\mathcal{H}^{(2)\alpha\beta},$$

with a small increment Δq of the coordinate q^1. The obtained series of states $\{\boldsymbol{\xi}_n\} \equiv \{\boldsymbol{\xi}(q^1 = n\Delta q)\}$ determines the collective path Σ_1. In other words, it determines the values of the original variables $\boldsymbol{\xi}$ that the system will take during its collective motion. Recall that $K = 1$ means that there is a single pair of decoupled, canonical variables (q^1, p_1), which are able to describe the collective motion originally written in terms of the $2M$ variables $(\boldsymbol{\xi}, \boldsymbol{\pi})$.

In addition to determining the collective path Σ_1, we can also extract the (decoupled) collective coordinate q^1, its conjugate momentum $p_1 = g^\alpha_{,1}\pi_\alpha$ and the classical collective Hamiltonian (6.10). Once we obtain the collective path Σ_1, we may always redefine the scale of q^1, according to a certain operator $\hat{Q}$ intuitively chosen by setting the expectation value $Q(q^1, q^a = 0) = \mathcal{Q}(\boldsymbol{\xi}, \boldsymbol{\pi} = 0)$ with $\xi^\alpha = g^\alpha(q^1, q^a = 0)$. As long as the one-to-one correspondence between q^1 and Q is guaranteed ($Q(q^1)$ and $q^1(Q)$), the operator $\hat{Q}$ can be arbitrary. The transformation of the functions in the collective Hamiltonian (6.10) are as follows,

$$H^{(0)}(q^1) \to \bar{H}^{(0)}(Q) = H^{(0)}(q^1(Q)) \quad H^{(2)11} \overset{\circ}{=} 1 \to \bar{H}^{(2)QQ} \overset{\circ}{=} \left(\frac{dQ}{dq^1}\right)^2.$$

6.2.4 Problems with the equation of path

It is a nice feature of the ATDHF theory that it is capable of determining the optimum collective sub-manifold, at least for the case of $K = 1$. The collective inertia $1/H^{(2)11}(q^1)$ takes into account the time-odd effects, thus it is able to produce the exact total mass Am for the translational center-of-mass motion. However, there are some problems with the equation of path (6.15).

Let us comment on the inertial mass and its notation. The contravariant tensors $\mathcal{H}^{(2)\alpha\beta}$ correspond to the inverse mass, while the covariant ones $\mathcal{H}^{(2)}_{\alpha\beta}$, defined by $\mathcal{H}^{(2)\alpha\gamma}\mathcal{H}^{(2)}_{\gamma\beta} = \delta^{\alpha}_{\beta}$, correspond to the mass. It is also customary to use the symbol B, $\mathcal{B}^{\alpha\beta} = \mathcal{H}^{(2)\alpha\beta}$ and $\mathcal{B}_{\alpha\beta} = \mathcal{H}^{(2)}_{\alpha\beta}$. For $K = 1$, the inertial mass is given by $B_{11}(\boldsymbol{q}^1) = H^{(2)}_{11}(q^1) = 1/H^{(2)11}$.

Multi-dimensional cases

First, extending the theory to a multi-dimensional collective submanifold $K > 1$ is not trivial. The total number of ATDHF equations in (6.11a) and (6.11b) is $M - K + K(M - K)$. In addition, we have K^2 chain-rule relations $f^i_{,\alpha} g^{\alpha}_{,j} = \delta^i_j$. On the other hand, the number of unknown quantities, $(f^i_{,\alpha}, g^{\alpha}_{,i})$, is $2KM$. Therefore, in the general case $K > 1$, we simply do not have enough equations to determine all the unknown quantities. The case of $K = 1$ is exceptional: although we have then $2M - 1$ equations for $2M$ unknown, the scale of the coordinate q is arbitrary, yielding just enough equations. This is one of the reasons why the ATDHF theory is most often implemented in what we called its second variant, where the collective variables are preset and the collective path is pre-computed from a set of, e.g. constrained HF or HFB calculations.

Extension to superfluid phase

Another problem involves the extension of the ATDHF to adiabatic time-dependent Hartree–Fock–Bogoliubov (ATDHFB) theory. Such an extension is much needed because pairing correlations are important for the description of low-frequency collective motion in open-shell nuclei. Moreover, large amplitude collective motion involves a number of s.p. level crossings; in order to follow adiabatic configurations, pairing correlations are indispensable.

The problem is somewhat related to the previous one, the extension to the multi-dimensional problem, because the breaking of gauge symmetry in the superfluid phase produces spurious excitations in the form of pair rotations known as the Anderson–Nambu–Goldstone (ANG) modes [27]. The concept of spontaneous symmetry-breaking was introduced in section 1.1.3 and discussed in more detail in section 3.1.1. In ATDHFB, dealing with such modes requires involving a 'quasi-multi-dimensional' sub-manifold that can account simultaneously for two kinds of collective motion, (q^1, p_1) and the pair rotational motion (q^r, p_r). For example, the ground state is no longer the energy minimum at equilibrium, but corresponds to a non-equilibrium state—since there is a constraint on the expectation value of particle number. In other words, instead of $\mathcal{H}^{(0)}_{,\alpha} = 0$, we have

$$\mathcal{H}^{(0)}_{,\alpha} - H^{(0)}_{,r} f^{r}_{,\alpha} \circeq 0$$

for the ground state. Here, $H^{(0)}_{,r}$ plays the role of the chemical potential needed to fix the average particle number to N.

A straightforward extension of the basic equations defining the collective sub-manifold to $K = 2$ leads to equations (6.12a) and (6.12b) with $i, j = 1$ and r. If we adopt q^r as the average particle number of the system, the zeroth-order equation (6.12a) becomes

$$\mathcal{H}^{(0)}_{,\beta} - H^{(0)}_{,r} f^{r}_{,\beta} - H^{(0)}_{,1} f^{1}_{,\beta} \circeq 0. \tag{6.16}$$

For the first-order equation (6.12b), assuming vanishing off-diagonal elements, $H^{(2)1r} = 0$, we have

$$g^{\beta}_{,i} \circeq \frac{1}{H^{(2)ii}} f^{i}_{,\alpha} \mathcal{H}^{(2)\alpha\beta} \quad i = 1, r, \tag{6.17}$$

where we take no summation with respect to i on the right-hand side. From equations (6.16) and (6.17), the equation of collective path becomes

$$g^{\beta}_{,1} = \frac{d\xi^{\beta}}{dq^1} \circeq \frac{1}{H^{(2)11}} f^{1}_{,\alpha} \mathcal{H}^{(2)\alpha\beta} \circeq \frac{1}{H^{(2)11} H^{(0)}_{,1}} \left(\mathcal{H}^{(0)}_{,\alpha} - H^{(0)}_{,r} f^{r}_{,\alpha} \right) \mathcal{H}^{(2)\alpha\beta}. \tag{6.18}$$

Again, the pre-factor of $1/(H^{(2)11} H^{(0)}_{,1})$ can be arbitrarily chosen to fix the scale of q^1. Calculating the value of $H^{(0)}_{,r} = \mathcal{H}^{(0)}_{,\beta} g^{\beta}_{,r}$ at each point, we may solve equation (6.18) to obtain the collective path.

There still remain some questions/problems to solve. First, can we justify the assumption of vanishing off-diagonal elements, $H^{(2)1r} = f^{1}_{,\alpha} \mathcal{H}^{(2)\alpha\beta} f^{r}_{,\beta} = 0$? Another related question is whether the collective variables (q^1, p_1) are orthogonal to those of the pair rotation (ANG mode) (q^r, p_r)? There is no clear answer to these questions. The third question is about the ANG coordinate q^r. Can we achieve the transformation $\xi \to q^r$ by a point transformation? The average particle number q^r may, in general, contain even orders of momenta,

$$q^r(\boldsymbol{\xi}, \boldsymbol{\pi}) = f^r(\boldsymbol{\xi}) + \frac{1}{2} f^{(1)r\alpha\beta} \pi_\alpha \pi_\beta + \mathcal{O}(\boldsymbol{\pi}^4). \tag{6.19}$$

In other words, can we safely neglect the second-order terms in this expansion? The ATDHFB theory does not really provide answers to these questions. The more complete theory presented section 6.3 will solve these problems.

Non-uniqueness problem

Among several problems of the ATDHF equation of path, perhaps the most well-known and serious one is the non-uniqueness problem. This can be easily understood. The ATDHF equation of path (6.15) is a differential equation. A full characterization of the path requires an initial state $\boldsymbol{\xi}_0$; the equation of path is a kind of initial-value problem, even though there is no 'time' t. Unfortunately, we cannot choose this initial state as the minimum-energy state—at the minimum point, defined by $\mathcal{H}_{,\alpha} = 0$, the tangent vector of Σ_1 (6.13) vanishes, and the equation of path is identically zero then, $g_{,1}^{\beta} = 0$. Therefore, we must start from a non-equilibrium point $\mathcal{H}_{,\alpha} \neq 0$. For each value of this point that we choose, we obtain a different collective path.

In numerical applications, the equation of path (6.15) is solved starting from many initial states chosen by physical intuition. For example, in studies of nucleus–nucleus collisions [26], the two nuclei in the ground state are placed at different distances R, and they are adopted as the initial states $\boldsymbol{\xi}_0$ for the equation of path. Computing many trajectories in this way, the energy $H^{(0)}$ is plotted as a function of the relative distance $R(q^1)$. These result in many fall lines which converge to a valley of $H^{(0)}$. In the end, an envelope curve of many fall lines is chosen as the collective path Σ_1. Numerically, it is difficult to reach the saddle point.

A solution to this non-uniqueness problem was later proposed by adding equations in the second order in momenta [28]. The method in section 6.3 is free from this non-uniqueness problem.

6.2.5 Conventional representation of ATDHF

In the preceding sections, the ATDHF theory was presented in the context of the adiabatic approximation to the TDHF theory using a formalism based on classical mechanics developed by Klein and collaborators [29]. In this section, we rewrite the ATDHF (ATDHFB) equation for the collective path in a more conventional manner, which is closer to the historical work of Villars. We assume that the system is described by a state-independent quantum many-body Hamiltonian $\hat{H}$ and time-dependent Slater determinants (HF states) $|\phi(t)\rangle$.

> As discussed in the introduction to this chapter, different variants of the ATDHF theory have been developed over the years, from the early work of Baranger and Kumar [7–11], to the approach pioneered by Baranger, Veneroni and collaborators [16, 18, 20, 21, 30], the works of Villars [15], and subsequent work by Goeke and Reinhard [17, 19, 31]. Here, we will connect the previous section to the formulation of Villars, and of Goeke and Reinhard.

As we have shown in section 6.1.1, the time dependence of the (generalized) density is through $\boldsymbol{\xi}(t)$ and $\boldsymbol{\pi}(t)$. This means that the time-dependent (generalized) Slater determinants can be written as $|\phi(\boldsymbol{\xi}, \boldsymbol{\pi})\rangle$, and the time dependence of $|\phi(t)\rangle$ is given by the time-dependent variables $(\boldsymbol{\xi}, \boldsymbol{\pi})$. The adiabatic limit is given by $\boldsymbol{\pi} \to 0$. However, it should be noted that, for a proper description of nuclear dynamics, we cannot neglect $\boldsymbol{\pi}$.

First, we define generators of canonical variables, which are locally given by one-body operators, $\hat{\Xi}^{\alpha}(\boldsymbol{\xi}, \boldsymbol{\pi})$ and $\hat{\Pi}_{\alpha}(\boldsymbol{\xi}, \boldsymbol{\pi})$, at each state $|\phi(\boldsymbol{\xi}, \boldsymbol{\pi})\rangle$. At the limit of $\boldsymbol{\pi} \to 0$, they are defined as

$$\hat{\Xi}^{\alpha}(\boldsymbol{\xi})|\phi(\boldsymbol{\xi})\rangle \equiv \lim_{\pi \to 0}\left(-i\frac{\partial}{\partial \pi_{\alpha}}|\phi(\boldsymbol{\xi}, \boldsymbol{\pi})\rangle \right)$$

$$\hat{\Pi}_{\alpha}(\boldsymbol{\xi})|\phi(\boldsymbol{\xi})\rangle = \lim_{\pi \to 0}\left(i\frac{\partial}{\partial \xi^{\alpha}}|\phi(\boldsymbol{\xi}, \boldsymbol{\pi})\rangle \right) = i\frac{\partial}{\partial \xi^{\alpha}}|\phi(\boldsymbol{\xi})\rangle,$$

where the time-even state $|\phi(\boldsymbol{\xi})\rangle$ is defined by $|\phi(\boldsymbol{\xi})\rangle = |\phi(\boldsymbol{\xi}, \boldsymbol{\pi} = 0)\rangle$.

For some readers, these signs may look opposite to our common understanding. This is related to the fact that $(\boldsymbol{\xi}, \boldsymbol{\pi})$ (and $(\boldsymbol{q}, \boldsymbol{p})$) are 'parameters' to express the Slater determinants $|\phi(\boldsymbol{\xi}, \boldsymbol{\pi}\rangle$, not the dynamical variables. Let us take a simple Gaussian whose center is located at ξ and the momentum atπ, $\psi(x|\xi, \pi) = e^{i\pi(x-\xi)}e^{-\alpha(x-\xi)^2}$. The motion of q corresponds to the translational motion. It is easy to find

$$\hat{p}\psi(x|\xi, \pi) = -i\partial_x\psi(x|\xi, \pi) = i\partial_{\xi}\psi(x|\xi, \pi)$$
$$\hat{x}\psi(x|\xi, \pi) = \{\xi + (x - \xi)\}\phi(x|\xi, \pi) = (\xi - i\partial_{\pi})\psi(x|\xi, \pi).$$

Do not assume $\hat{\Xi}^{\alpha}(\boldsymbol{\xi})|\phi(\boldsymbol{\xi})\rangle = \partial/\partial\pi_{\alpha}|\phi(\boldsymbol{\xi})\rangle = 0$. They are the generators to boost the momenta from $\boldsymbol{\pi} = 0$ to $\boldsymbol{\pi} \neq 0$. In the same manner, the generators for the new variables $(\boldsymbol{q}, \boldsymbol{p})$ at the adiabatic limit are defined as

$$\hat{Q}^{\mu}(\boldsymbol{q})|\phi(\boldsymbol{q})\rangle = \lim_{p \to 0}\left(-i\frac{\partial}{\partial p_{\mu}}|\phi(\boldsymbol{q}, \boldsymbol{p})\rangle \right)$$

$$\hat{P}_{\mu}(\boldsymbol{q})|\phi(\boldsymbol{q})\rangle = \lim_{p \to 0}\left(i\frac{\partial}{\partial q^{\mu}}|\phi(\boldsymbol{q}, \boldsymbol{p})\rangle \right) = i\frac{\partial}{\partial q^{\mu}}|\phi(\boldsymbol{q})\rangle.$$

In the point transformation (6.5), the adiabatic limit of $\boldsymbol{\pi} \to 0$ is identical to that of $\boldsymbol{p} \to 0$. Certain trivial generators, such as those for the center-of-mass coordinate and the total momentum, can be globally defined. However, the infinitesimal generators for the collective variables associated with the most general type of collective motion change as functions of $\boldsymbol{\xi}$ or $\boldsymbol{q}$.

The condition that variables $(\boldsymbol{\xi}, \boldsymbol{\pi})$ and $(\boldsymbol{q}, \boldsymbol{p})$ are canonical leads to the 'weakly' canonical commutation relations [27, 32–33]

$$\langle\phi(\boldsymbol{\xi})|\left[\hat{\Xi}^{\alpha}(\boldsymbol{\xi}), \hat{\Pi}_{\beta}(\boldsymbol{\xi})\right]|\phi(\boldsymbol{\xi})\rangle = i\delta^{\alpha}_{\beta} \tag{6.20a}$$

$$\langle\phi(\boldsymbol{q})|[\hat{Q}^{\mu}(\boldsymbol{q}), \hat{P}_{\nu}(\boldsymbol{q})]|\phi(\boldsymbol{q})\rangle = i\delta^{\mu}_{\nu} \tag{6.20b}$$

$$\langle[\hat{\Xi}^\alpha, \hat{\Xi}^\beta]\rangle = \langle[\hat{\Pi}_\alpha, \hat{\Pi}_\beta]\rangle = \langle[\hat{Q}^\mu, \hat{Q}^\nu]\rangle = \langle[\hat{P}_\mu, \hat{P}_\nu]\rangle = 0. \tag{6.20c}$$

Trivial arguments are omitted in equation (6.20c).

After this introduction, we rewrite the ATDHF equation of path (6.15) using the generators $(\hat{Q}, \hat{P})$ and the adiabatic states $|\phi\rangle$. We will assume here the case of a one dimensional ($K = 1$) collective sub-manifold. Let us first consider the variation of $\langle\phi|\hat{Q}^1|\phi\rangle$. The variation for the state $|\phi\rangle$ can be taken as

$$|\delta\phi\rangle = \delta\xi^\alpha|\frac{\partial\phi}{\partial\xi^\alpha}\rangle + \delta\pi_\alpha|\frac{\partial\phi}{\partial\pi_\alpha}\rangle = -i\delta\xi^\alpha\hat{\Pi}_\alpha|\phi\rangle + i\delta\pi_\alpha\hat{\Xi}^\alpha|\phi\rangle.$$

The variation $\delta\langle\phi(\boldsymbol{q})|\hat{Q}^1(\boldsymbol{q})|\phi(\boldsymbol{q})\rangle$ involves first replacing $\hat{Q}^1(\boldsymbol{q})$ by the generator in the old variables, $\hat{\Xi}^\alpha$, then taking variations with respect to the bra and the ket. Using the chain-rule relations, it is easy to find $\hat{Q}^1 = f^1_{,\alpha}\hat{\Xi}^\alpha$.

Thus, using the transformation (6.6), we have

$$\begin{aligned}\delta\langle\phi(\boldsymbol{q})|\hat{Q}^1(\boldsymbol{q})|\phi(\boldsymbol{q})\rangle &= f^1_{,\alpha}(\langle\delta\phi|\hat{\Xi}^\alpha|\phi\rangle + \langle\phi|\hat{\Xi}^\alpha|\delta\phi\rangle)\\ &= if^1_{,\alpha}\delta\xi^\beta\langle\phi|\left[\hat{\Pi}_\beta, \hat{\Xi}^\alpha\right]|\phi\rangle - if^1_{,\alpha}\delta\pi_\beta\langle\phi|[\hat{\Xi}^\beta, \hat{\Xi}^\alpha]|\phi\rangle\\ &= f^1_{,\alpha}\delta\xi^\alpha = \delta q^1.\end{aligned}$$

Similarly, we can easily find

$$\delta\langle\phi(\boldsymbol{q})|\hat{P}_1(\boldsymbol{q})|\phi(\boldsymbol{q})\rangle = g^\alpha_{,1}\delta\pi_\alpha.$$

The adiabatic limit ($\boldsymbol{\pi} \to 0$) leads to $\delta p_1 = g^\alpha_{,1}\delta\pi_\alpha$. Thus, it may read $\delta\langle\phi(\boldsymbol{q})|\hat{P}_1(\boldsymbol{q}) \to \delta p_1$.

Next, let us consider the quantity $\delta\langle\phi|[\hat{H}, i\hat{Q}^1]|\phi\rangle$. This contains terms proportional to $\delta\xi^\alpha$ and to $\delta\pi_\alpha$. Since the expectation value of the Hamiltonian is given as equation (6.9a) or (6.9b), $\langle\phi|[\hat{H}, i\hat{Q}^1]|\phi\rangle = \mathcal{O}(\boldsymbol{p}) = \mathcal{O}(\boldsymbol{\pi})$. The terms proportional to $\delta\xi^\alpha$ in $\delta\langle\phi|[\hat{H}, i\hat{Q}^1]|\phi\rangle$ are also $\mathcal{O}(\boldsymbol{\pi})$ and vanish at the adiabatic limit $\boldsymbol{\pi} \to 0$. Therefore, we have

$$\begin{aligned}\delta\langle\phi|[\hat{H}, i\hat{Q}]|\phi\rangle &= f^1_{,\alpha}\left(\langle\delta\phi|[\hat{H}, i\hat{\Xi}^\alpha]|\phi\rangle + \langle\phi|[\hat{H}, i\hat{\Xi}^\alpha]|\delta\phi\rangle\right)\\ &= f^1_{,\alpha}\langle\phi|[[\hat{H}, i\hat{\Xi}^\alpha], i\hat{\Xi}^\beta]|\phi\rangle\delta\pi_\beta\\ &= f^1_{,\alpha}\left.\frac{\partial^2\mathcal{H}}{\partial\pi_\alpha\partial\pi_\beta}\right|_{\boldsymbol{\pi}=0}\delta\pi_\beta = f^1_{,\alpha}\mathcal{H}^{(2)\alpha\beta}\delta\pi_\beta.\end{aligned}$$

The decoupling conditions (6.12a) and (6.12b) are now rewritten as

$$\delta\langle\phi(q^1)|\hat{H} - H^{(0)}_{,1}(q^1)\hat{Q}^1(q^1)|\phi(q^1)\rangle \hat{=} 0 \qquad (6.21a)$$

$$\delta\langle\phi(q^1)|[\hat{H}, i\hat{Q}(q^1)] - H^{(2)11}(q^1)\hat{P}(q^1)|\phi(q^1)\rangle \hat{=} 0. \qquad (6.21b)$$

The operators and states originally depend on $\boldsymbol{\xi}$ or $\boldsymbol{q}$. However, the states satisfying equations (6.21a) and (6.21b) are on the collective path Σ_1, which is spanned only by q^1 ($q^a = 0$, $a = 2, \ldots, M$). Thus, in equations (6.21a) and (6.21b), we have replaced the arguments $\boldsymbol{q}$ ($\boldsymbol{\xi}$) simply by q^1. These are the basic equations of the ATDHF theory by Villars [15].

The state $|\phi\rangle$ describes a Slater determinant state. Its variations are expressed in terms of p.h. excitations, $|\delta\phi\rangle = c_p^\dagger c_h|\phi\rangle$. Therefore, the decoupling condition (6.21) can be understood as

$$\hat{H}_{\rm ph}(q^1) \hat{=} H^{(0)}_{,1}(q^1)\hat{Q}^1(q^1), \qquad (6.22)$$

where $\hat{H}_{\rm ph}(q^1)$ is defined as the p.h. ($c_p^\dagger c_h$ and $c_h^\dagger c_p$) parts of the Hamiltonian, which correspond to the fourth term in the right-hand side of the following:

$$\hat{H} = \langle\phi(q^1)|\hat{H}|\phi(q^1)\rangle + \sum_{p,p'} H^{pp}_{{\rm pp}'} c_p^\dagger c_{p'} + \sum_{h,h'} H^{\rm hh}_{hh'} c_h c_{h'}^\dagger$$
$$+ \sum_{p,h}\left(H^{\rm ph}_{ph} c_p^\dagger c_h + \text{h.c.}\right) + \text{terms with } c^\dagger c^\dagger cc,$$

where, for the sake of simplicity, we have assumed a two-body Hamiltonian only. In this expression, the creation and annihilation (c, $c^\dagger$) operators and the coefficients such as $H^{\rm ph}_{ph}$ all depend on the collective coordinate q^1, even though we omit the argument for simplicity. Combining (6.22) with (6.21b), we obtain

$$\frac{\partial}{\partial q^1}|\phi(q^1)\rangle \hat{=} \left(H^{(2)11}\right)^{-1}\left[\hat{H}, \hat{H}_{\rm ph}\right]_{\rm ph}|\phi(q^1)\rangle.$$

This is nothing other than the equation of the collective path obtained by Goeke, Reinhard and collaborators [34, 35], and is equivalent to equation (6.15).

We have derived the equation of the collective path for generalized Slater determinants, that is, the ATDHFB theory. As already mentioned in section 6.2.4, the extension from ATDHF to ATDHFB is not as straightforward as it looks. Therefore, so far, the applications of the equation for the collective path to the superfluid phase have been very limited.

6.3 Adiabatic self-consistent collective coordinate method

There have been a number of attempts to improve the ATDHF theory and to solve some of the problems discussed in the previous sections, which led to further developments in theories of large amplitude collective motion. The survey of these historical and recent developments can be found in [5, 27]. In this section, we focus only on one of those developments, the ASCC theory of [27, 36]. Here, the theory is presented in a different way to the aforementioned references. We instead follow the general framework of [5] by keeping a classical mechanics point of view and following a similar logic as in sections 6.1 and 6.2.

Covariant derivatives

Let us consider a vector field $\mathcal{A}(\boldsymbol{x})$ in Cartesian coordinates. We introduce the covariant Cartesian basis vectors $\boldsymbol{e}_x$, $\boldsymbol{e}_y$ and $\boldsymbol{e}_z$ of $\mathbb{R}^3$. These vectors form an orthonormal basis $\mathcal{B}$ of the space. The reciprocal basis $\bar{\mathcal{B}}$ of $\mathcal{B}$ is the set of vectors $\boldsymbol{e}^x$, $\boldsymbol{e}^y$, $\boldsymbol{e}^z$ that verify

$$\boldsymbol{e}_i \cdot \boldsymbol{e}^k = \delta_{ik}.$$

In the particular case of the Cartesian basis of $\mathbb{R}^3$, the covariant and contravariant bases coincide. We denote by A_i the covariant coordinates of $\mathcal{A}$ in $\mathcal{B}$ and by A^i its contravariant coordinates in $\bar{\mathcal{B}}$. We now consider an arbitrary local basis $\mathcal{B}(\boldsymbol{x}) = (\boldsymbol{e}_1(\boldsymbol{x}), \boldsymbol{e}_2(\boldsymbol{x}), \boldsymbol{e}_3(\boldsymbol{x}))$ of space. Since the basis is position-dependent, the partial derivatives of the vector fields become

$$\frac{\partial \mathcal{A}}{\partial x^k} = \frac{\partial A^j}{\partial x^k}\boldsymbol{e}_j + A^j\frac{\partial \boldsymbol{e}_j}{\partial x^k} = \frac{\partial A_j}{\partial x^k}\boldsymbol{e}^j + A_j\frac{\partial \boldsymbol{e}^j}{\partial x^k}. \tag{6.23}$$

The covariant derivative of the vector field are then defined as

$$A^i_{;k} \equiv \frac{\partial \mathcal{A}}{\partial x^k} \cdot \boldsymbol{e}^i \qquad A_{i;k} \equiv \frac{\partial \mathcal{A}}{\partial x^k} \cdot \boldsymbol{e}_i.$$

In tensor analysis, covariant derivatives are often denoted with the semicolon index. Using equation (6.23), we may easily find

$$A^i_{;k} = \frac{\partial A^i}{\partial x^k} + \Gamma^i_{jk}A^j \qquad A_{i;k} = \frac{\partial A_i}{\partial x^k} - \Gamma^j_{ik}A_j,$$

where $\partial \boldsymbol{e}_j/\partial x^k \equiv \Gamma^i_{jk}\boldsymbol{e}_i$.

In the usual case of a flat, constant-metric Cartesian space, the second term in equation (6.23) vanishes and the covariant derivatives of a vector field are simply the set of partial derivatives of its components. In this case, it is customary to use the comma symbol,

$$A^i_{,k} = \frac{\partial A^i}{\partial x^k} A_{i,\,k} = \frac{\partial A_i}{\partial x^k}.$$

If $\mathcal{A}$ is in fact the scalar field A, then we can use similar notations to refer to partial derivatives, $A_{,k} = \partial A/\partial x^k$. A general introduction to tensor analysis on manifolds can be found in chapter 5 of [37].

6.3.1 Covariant derivatives with Riemannian connection

For preparation, let us first introduce the concept of a (covariant) metric tensor $G_{\alpha\beta}$; see [37] for details. The contravariant components $G^{\alpha\beta}$ are given by the identity $G^{\alpha\beta}G_{\beta\gamma} = \delta_\gamma^\alpha$. In the following we will choose $G^{\alpha\beta} \equiv \mathcal{H}^{(2)\alpha\beta}$, thus our metric tensor $G_{\alpha\beta}$ is nothing other than the collective inertia tensor. Using this metric, we define the affine connection $\Gamma^\alpha_{\beta\gamma}$ (also commonly called the Christoffel symbol, see equation (6.61)) as

$$\Gamma^\alpha_{\beta\gamma} = \frac{1}{2}G^{\alpha\delta}\left(G_{\beta\delta,\,\gamma} + G_{\gamma\delta,\,\beta} - G_{\beta\gamma,\,\delta}\right), \tag{6.24}$$

where, as before, the symbol α is a short-hand notation for the derivative with respect to ξ^α and repeated indices imply a summation (Einstein convention). The definition (6.24) is called Riemannian connection in [5]. Since we are now working in a space with a non-constant metric, defined by the tensor $G_{\alpha\beta}$, the covariant derivative is different from the regular, partial derivative. The energy $\mathcal{H}^{(0)}(\boldsymbol{\xi}, \boldsymbol{\pi})$ is a scalar field. We can form a vector field by considering its derivatives with respect to the coordinates $\boldsymbol{\xi}$. The covariant derivative of this vector with respect to coordinate ξ^γ is, by definition,

$$\mathcal{H}^{(0)}_{;\beta\gamma} \equiv \mathcal{H}^{(0)}_{,\beta\gamma} - \Gamma^\alpha_{\beta\gamma}\mathcal{H}^{(0)}_{,\alpha}. \tag{6.25}$$

According to the definition of the covariant derivative recalled earlier, the derivatives of $\mathcal{H}^{(0)}$ with respect to $\boldsymbol{\xi}$ give a vector field, $\mathcal{H}^{(0)'} = \mathcal{H}^{(0)}_{,\alpha}\boldsymbol{e}^\alpha$. Then, its derivatives with respect to ξ^γ provide

$$\mathcal{H}^{(0)}_{;\beta\gamma} \equiv \frac{\partial \mathcal{H}^{(0)'}}{\partial \xi^\gamma} \cdot \boldsymbol{e}_\beta$$

which is equal to equation (6.25).

In a flat space, we would have

$$\mathcal{H}^{(0)}_{;\beta\gamma} = \mathcal{H}^{(0)}_{,\beta\gamma} = \frac{\partial^2 \mathcal{H}^{(0)}}{\partial \xi^\beta \partial \xi^\gamma}$$

and the quantity $\mathcal{H}^{(0)}_{;\beta\gamma}$ thus plays the role of a generalized Hessian matrix (in a curved space defined by the metric $G_{\alpha\beta}$). Under the point transformation $\boldsymbol{\xi} \to \boldsymbol{q}$, the affine connection is transformed as [38]

$$\Gamma^\mu_{\nu\rho} = f^{\mu}_{,\alpha} g^{\beta}_{,\nu} g^{\gamma}_{,\rho} \Gamma^\alpha_{\beta\gamma} + f^{\mu}_{,\alpha} g^{\alpha}_{,\nu\rho} = f^{\mu}_{,\alpha} g^{\beta}_{,\nu} g^{\gamma}_{,\rho} \Gamma^\alpha_{\beta\gamma} - f^{\mu}_{,\alpha\beta} g^{\alpha}_{,\nu} g^{\beta}_{,\rho}, \tag{6.26}$$

where we used in the last equation, $f^{\mu}_{,\alpha} g^{\alpha}_{,\nu\rho} = -f^{\mu}_{,\alpha\beta} g^{\alpha}_{,\nu} g^{\beta}_{,\rho}$, which comes from the chain-rule relations. Recall that indices $a, b, \ldots$ refer to non-collective d.o.f.s while indices $i, j, \ldots$ refer to collective ones—and $\mu, \nu, \ldots$ refer to all d.o.f.s. The indices $\alpha, \beta, \cdots$ refer to the original coordinate system $\boldsymbol{\xi}$. For simplicity, we used the same symbol

Γ for both the left- and right-hand sides. However, keep in mind that $\Gamma^{\mu}_{\nu\rho}$ is defined in the coordinate system $\boldsymbol{q}$ while $\Gamma^{\alpha}_{\beta\gamma}$ is defined in the coordinate system $\boldsymbol{\xi}$. In the new coordinate system, $\Gamma^{\mu}_{\nu\rho}$ is defined in exactly the same way as in equation (6.24), only with $G^{\mu\nu} = f^{\mu}_{,\alpha} G^{\alpha\beta} f^{\nu}_{,\beta} = H^{(2)\mu\nu}$ and $G_{\mu\nu} = g^{\alpha}_{,\mu} G_{\alpha\beta} g^{\beta}_{,\nu} = (H^{(2)-1})_{\mu\nu}$. The first-order decoupling condition of (6.11b) tells us that the affine connection on the collective subspace Σ_K is closed inside Σ_K,

$$\Gamma^{i}_{jk} \overset{\circ}{=} \frac{1}{2} G^{il}\left(G_{lj,\,k} + G_{lk,\,j} - G_{jk,\,l}\right).$$

6.3.2 Equations of collective subspace

The zeroth-order decoupling condition (6.12a) must be valid everywhere in the collective submanifold Σ_K. Therefore, the derivatives of the left-hand side of equation (6.12a) along the collective coordinates q^i should vanish, which leads to the condition

$$\left(\mathcal{H}^{(0)}_{,\beta\gamma} - H^{(0)}_{,j} f^{j}_{,\beta\gamma}\right) g^{\gamma}_{,i} \overset{\circ}{=} H^{(0)}_{,ji} f^{j}_{,\beta}. \tag{6.27}$$

In fact, the quantity inside the bracket of the left-hand side of (6.27) can be regarded as another kind of covariant derivative with the metric given by $G_{\alpha\beta} \equiv \sum_{\mu} f^{\mu}_{,\alpha} f^{\mu}_{,\beta}$. The connection with this metric is called 'symplectic connection' in [5]. Multiplying equation (6.27) by $\mathcal{H}^{(2)\alpha\beta}$, we obtain the following equations,

$$\mathcal{H}^{(2)\alpha\beta}\left(\mathcal{H}^{(0)}_{,\beta\gamma} - H^{(0)}_{,j} f^{j}_{,\beta\gamma}\right) g^{\gamma}_{,i} \overset{\circ}{=} H^{(2)jk} H^{(0)}_{,ij} g^{\alpha}_{,k}. \tag{6.28}$$

At equilibrium, where $\mathcal{H}^{(0)}_{,\alpha} = H^{(0)}_{,\mu} = 0$, equation (6.28) reduces to the usual harmonic approximation, without the term $-H^{(0)}_{,j} f^{j}_{,\beta\gamma}$ inside the bracket on the left-hand side. Equation (6.28) can thus be regarded as a generalized harmonic approach for non-equilibrium states.

By definition, the HF solution corresponds to the minimum energy state, namely, an equilibrium with respect to variations of the density matrix ρ. If we denote this equilibrium point by $\boldsymbol{\xi}_0$, we must thus have

$$\frac{\partial \mathcal{H}}{\partial \xi^{\alpha}}(\boldsymbol{\xi}_0, \boldsymbol{\pi} = 0) = \mathcal{H}^{(0)}_{,\alpha} = 0 \qquad \forall\, \alpha.$$

The harmonic approximation at equilibrium is given by expanding the Hamiltonian up to the second order both in coordinates and momenta,

$$\mathcal{H} - \mathcal{H}^{(0)}(\boldsymbol{\xi}_0) \approx \frac{1}{2}\mathcal{H}^{(0)}_{,\alpha\beta}\xi^{\alpha}\xi^{\beta} + \frac{1}{2}\mathcal{H}^{(2)\alpha\beta}\pi_{\alpha}\pi_{\beta}.$$

The normal modes of excitation are defined by eigenvalues and eigenvectors of the matrix $\mathcal{H}^{(2)\alpha\beta}\mathcal{H}^{(0)}_{,\beta\gamma}$.

A difficulty of this generalized harmonic approach is that the equation contains 'curvature terms' which are associated with $f^i_{,\beta\gamma}$. The solutions of equation (6.27) or (6.28) determine the tangent vectors $f^i_{,\alpha}$ and $g^{\alpha}_{,i}$, but do not provide the curvature terms, $f^i_{,\beta\gamma}$. Therefore, the equation is not self-contained and even an iterative approach will not work. In order to eliminate the curvature terms, we need to take one more step beyond the standard ATDHF approach by including terms up to second order in momenta. The second equation in equation (6.8) gives

$$H^{(2)ij}_{,a} \circeq 0 \tag{6.29}$$

in the second order in momenta $\boldsymbol{p}$. This second-order condition, together with the first-order condition of equation (6.11b), leads to

$$H^{(2)ij}_{,a} \circeq H^{(2)ia}_{,j} \circeq H^{(2)}_{ij,\,a} \circeq H^{(2)}_{ia,\,j} \circeq 0.$$

From this, we immediately obtain $\Gamma^i_{ja} \circeq 0$ (recall that the metric term in equation (6.24) is precisely $\mathcal{H}^{(2)\alpha\beta}$). Therefore, equation (6.26) in the collective submanifold can be modified into

$$\Gamma^i_{jk} f^k_{,\gamma} \circeq \Gamma^i_{j\mu} f^{\mu}_{,\gamma} \circeq f^i_{,\alpha} g^{\beta}_{,j} \Gamma^{\alpha}_{\beta\gamma} - f^i_{,\alpha\gamma} g^{\alpha}_{,j}. \tag{6.30}$$

This relation between $\Gamma^{\alpha}_{\beta\gamma}$ and Γ^i_{jk} on the collective submanifold Σ_K can be utilized to eliminate the curvature terms. Multiplying equation (6.30) with $H^{(0)}_{,i}$, we find

$$H^{(0)}_{,i} f^i_{,\alpha\gamma} g^{\alpha}_{,j} \circeq -\Gamma^i_{jk} H^{(0)}_{,i} f^k_{,\gamma} + \Gamma^{\alpha}_{\beta\gamma} H^{(0)}_{,i} f^i_{,\alpha} g^{\beta}_{,j} \circeq -\Gamma^i_{jk} H^{(0)}_{,i} f^k_{,\gamma} + \Gamma^{\alpha}_{\beta\gamma} \mathcal{H}^{(0)}_{,\alpha} g^{\beta}_{,j},$$

where in the last equality we used equation (6.11). We eliminate the second term in the left-hand side of equation (6.27) by substituting this relation and immediately obtain

$$\left(\mathcal{H}^{(0)}_{,\beta\gamma} - \Gamma^{\alpha}_{\beta\gamma} \mathcal{H}^{(0)}_{,\alpha}\right) g^{\gamma}_{,i} \circeq \left(H^{(0)}_{,ij} - \Gamma^k_{ij} H^{(0)}_{,k}\right) f^j_{,\beta}.$$

Each of the brackets above is nothing other than the covariant derivative defined in equation (6.25), therefore equation (6.27) reduces to

$$\mathcal{H}^{(0)}_{;\beta\gamma} g^{\gamma}_{,i} \circeq H^{(0)}_{;ij} f^j_{,\beta}. \tag{6.31}$$

Multiplying equation (6.31) by $\mathcal{H}^{(2)\alpha\beta}$, we find

$$\mathcal{M}^{\alpha}_{\gamma} g^{\gamma}_{,i} \circeq M^k_i g^{\alpha}_{,k}, \tag{6.32}$$

where

$$\mathcal{M}^{\alpha}_{\gamma} \equiv \mathcal{H}^{(2)\alpha\beta} \mathcal{H}^{(0)}_{;\beta\gamma} \quad M^k_i \equiv H^{(2)kj} H^{(0)}_{;ji}.$$

Alternatively, multiplying equation (6.31) by $H^{(2)ki}$, and using equation (6.12b), we obtain

$$\mathcal{M}^{\alpha}_{\gamma} f^i_{,\alpha} \circeq M^i_k f^k_{,\gamma}. \tag{6.33}$$

Equations (6.32) and (6.33) constitute a generalized version of the harmonic approximation in non-equilibrium states. Again, they become identical to those of the harmonic approximation for equilibrium states. In contrast to equation (6.28), the matrix $\mathcal{M}_\gamma^\alpha$ on the left-hand side of equations (6.32) and (6.33) is calculable. In the ASCC theory, the collective subspace is determined by searching for self-consistent solutions which simultaneously satisfy equation (6.33) and the decoupling conditions (6.12a).

Equations (6.32) and (6.33) represent a set of coupled linear equations to determine the derivatives of unknown functions g^α and f^i, respectively, which define the point transformation $\boldsymbol{\xi} \leftrightarrow \boldsymbol{q}$. The matrix $\mathcal{M}_\gamma^\alpha$ reads

$$\mathcal{M}_\gamma^\alpha = \mathcal{H}^{(2)\alpha\beta}\left(\mathcal{H}_{,\beta\gamma}^{(0)} - \Gamma_{\beta\gamma}^\alpha \mathcal{H}_{,\alpha}^{(0)}\right)$$

and we recall that the Christoffel symbol, (6.24), also involves $\mathcal{H}^{(2)\alpha\beta}$, that is, the second derivative of the energy with respect to the momenta.

In the $K = 1$ case, there is a single collective variable q^1, hence the tensor M_k^i reduces to a single number M_1^1 (within the collective submanifold Σ_K) and the equations for the collective path reduce to

$$\mathcal{M}_\beta^\alpha f_{,\alpha}^1 \overset{\circ}{=} \omega^2 f_{,\beta}^1 \qquad \mathcal{M}_{\beta}^\alpha g_{,1}^\beta \overset{\circ}{=} \omega^2 g_{,1}^\alpha, \tag{6.34}$$

with $\omega^2 \equiv M_1^1 = H^{(2)11} H_{,11}^{(0)}$ and

$$\mathcal{H}_{,\beta}^{(0)} - H_{,1}^{(0)} f_{,\beta}^1 \overset{\circ}{=} 0. \tag{6.35}$$

Thus, the gradient of the energy surface $\mathcal{H}^{(0)}(\xi)$ is parallel to a normal mode $f_{,\alpha}^1$ of equation (6.34).

6.3.3 Beyond point transformations

We discuss here a possibility to extend the point transformation to a more general canonical transformation, $(\boldsymbol{q}(\boldsymbol{\xi}, \boldsymbol{\pi}), \boldsymbol{p}(\boldsymbol{\xi}, \boldsymbol{\pi}))$. Since we expand the decoupling conditions with respect to momenta, we do the same for $\boldsymbol{q} = \boldsymbol{q}(\boldsymbol{\xi}, \boldsymbol{\pi})$ and its inverse $\boldsymbol{\xi} = \boldsymbol{\xi}(\boldsymbol{q}, \boldsymbol{p})$, up to the second order in momenta. Equations (6.5) are thus generalized by

$$q^\mu = f^\mu(\xi) + \frac{1}{2} f^{(1)\mu\alpha\beta}(\xi)\pi_\alpha \pi_\beta + \mathcal{O}(\pi^4) \tag{6.36a}$$

$$\xi^\alpha = g^\alpha(q) + \frac{1}{2} g^{(1)\alpha\mu\nu}(q) p_\mu p_\nu + \mathcal{O}(p^4). \tag{6.36b}$$

The transformation of the momenta is the same as equation (6.6), because we neglect the terms that are cubic in momenta. Using (6.6), the independence of canonical variables, $\partial\xi^\alpha / \partial\pi_\beta = 0$, leads to the condition

$$g^{(1)\alpha\mu\nu} g^{\beta}_{,\mu} g^{\gamma}_{,\nu} = -f^{(1)\lambda\beta\gamma} g^{\alpha}_{,\lambda}. \tag{6.37}$$

From the canonicity condition $\{q^\mu, q^\nu\}_{\text{PB}} = 0$, we also find

$$f^{\mu}_{,\alpha} f^{(1)\nu\alpha\beta} = f^{\nu}_{,\alpha} f^{(1)\mu\alpha\beta}. \tag{6.38}$$

The Hamiltonian in the new coordinate system (6.9b) is given in the same form as before, but with an extended collective inertia tensor,

$$H^{(2)\mu\nu} = f^{\mu}_{,\alpha} \mathcal{H}^{(2)\alpha\beta} f^{\nu}_{,\beta} + \mathcal{H}^{(0)}_{,\gamma} g^{(1)\gamma\mu\nu}.$$

The major difference between the use of the extended adiabatic transformation and the point transformation is the modification of the inertial tensor,

$$\begin{aligned} \tilde{\mathcal{H}}^{(2)\alpha\beta} &\equiv g^{\alpha}_{,\mu} H^{(2)\mu\nu} g^{\beta}_{,\nu} \\ &= \mathcal{H}^{(2)\alpha\beta} - H^{(0)}_{,\mu} f^{(1)\mu\alpha\beta} \end{aligned} \tag{6.39a}$$

$$\mathrel{\circeq} \mathcal{H}^{(2)\alpha\beta} - H^{(0)}_{,i} f^{(1)i\alpha\beta}. \tag{6.39b}$$

Here we have used the relation (6.37) to obtain equations (6.39a) and (6.11a) for the last equality. Equations (6.32) and (6.33) retain the same form with the extended point transformation, but replacing $\mathcal{H}^{(2)\alpha\beta}$ with $\tilde{\mathcal{H}}^{(2)\alpha\beta}$.

The extension of the theory can be summarized as

$$\mathcal{M}^{\alpha}_{\gamma} g^{\gamma}_{,i} \mathrel{\circeq} M^{k}_{i} g^{\alpha}_{,k} \qquad \mathcal{M}^{\alpha}_{\gamma} f^{i}_{,\alpha} \mathrel{\circeq} M^{i}_{k} f^{k}_{,\gamma}, \tag{6.40}$$

where

$$\mathcal{M}^{\alpha}_{\gamma} \equiv \tilde{\mathcal{H}}^{(2)\alpha\beta} \mathcal{H}^{(0)}_{;\beta\gamma} \quad M^{k}_{i} \equiv H^{(2)kj} H^{(0)}_{;ji}.$$

This final form looks very similar to equations (6.32) and (6.33). However, these equations are not self-contained. The solution of equation (6.40) provides the tangent vectors $f^{i}_{,\alpha}$ and $g^{\alpha}_{,i}$, while the calculation of $\tilde{\mathcal{H}}^{(2)\alpha\beta}$ also requires the quantities $f^{(1)i\alpha\beta}$. Currently, the ASCC theory does not provide a well-founded solution to determine $f^{(1)i\alpha\beta}$.

So far, numerical applications of the ASCC approach have adopted one of the following 'prescriptions' to determine $f^{(1)i\alpha\beta}$. The simplest prescription is to neglect the term, $f^{(1)1\alpha\beta} = 0$. A better prescription is to assume a diagonal form, $f^{(1)1\alpha\beta} = f^{(1)1\alpha}\delta^{\alpha\beta}$. In this case, $f^{(1)1\alpha}$ can be determined from equation (6.38). A third prescription is to assume that q^1 is locally given by the expectation value of a Hermitian one-body operator $q^1 \approx \langle \hat{Q}^1 \rangle$. When the symmetry represented by an operator $\hat{S}$ is broken, requesting the 'strong canonicity condition', $[\hat{Q}^1, \hat{S}] = 0$, can determine $f^{(1)1\alpha\beta}$. It should be noted that the development of a method founded by the theory itself is desired in the future. The extended version of the theory has not been completed yet.

6.3.4 Anderson–Nambu–Goldstone modes: constants of motion

As discussed in section 3.1.1, the single-reference energy density functional (SR-EDF) approach often provides a ground state which violates the symmetry of the true nuclear Hamiltonian, such as translational and rotational symmetries. This spontaneous symmetry-breaking (SSB) produces spurious excitation, or ANG modes, which correspond to well-decoupled collective motion. Since the generators of the ANG modes can be globally defined, we are more interested in the extraction of the collective submanifold perpendicular to the ANG modes. Thus, it is important for the theory to guarantee that the ANG modes are decoupled from other collective degrees of freedom. In this section, we show that the ASCC theory indeed has this desired property. In fact, the ASCC theory was originally proposed for large amplitude collective motion in superfluid nuclei [36], aiming at constructing a proper theory in the presence of the pair rotations.

Cyclic variables

Let us consider a one-body symmetry operator $\hat{S}$, globally defined, which commutes with the original nuclear Hamiltonian,

$$[\hat{S}, \hat{H}] = 0. \tag{6.41}$$

For instance, translational symmetry is represented by the total momentum $\hat{P}$, rotational symmetry by the total angular momentum $\hat{J}_k$ ($k = x, y, z$), isospin symmetry by the total isospin $\hat{T}_k$ ($k = 1, 2, 3$) and gauge symmetry by the particle number operator $\hat{N}$. When the symmetry is spontaneously violated, the ground state $|\Phi_0\rangle$ is no longer an eigenstate of this operator $\hat{S}$,

$$\hat{S}|\Phi_0\rangle \neq c|\Phi_0\rangle,$$

with any $\mathbb{C}$-number c. The operator $\hat{S}$ becomes a generator for the ANG mode.

In theories of large amplitude collective motion, the ANG modes associated with symmetry-breaking are relevant not only in the ground state, but also during the collective motion. Even if the ground state is invariant under the symmetry, departures from equilibrium may break the symmetry. Then, the ANG modes are *dynamically* activated. A typical example is the shape oscillations of a spherical nucleus: the ground state is spherically symmetric, but the shape oscillations correspond to dynamical deformations.

After the symmetry transformation, each ANG mode is described by a pair of canonical variables (q^r, p_r). Now, we divide the set $\{q^\mu, p_\mu\}$ ($\mu = 1, \ldots, M$) into three subsets, the collective coordinates $\{q^i, p_i\}$, $i = 1, \ldots, K$, the cyclic coordinates $\{q^r, p_r\}$, $r = K + 1, \ldots, K + K'$ and the non-collective coordinates $\{q^a, p_a\}$, $a = K + K' + 1, \ldots, M$.

For the quantum mechanical system satisfying equation (6.41), the expectation value of the operator $\hat{S}$ is conserved. In TDDFT, this is translated by the fact that a classical variable $\mathcal{S}(\boldsymbol{\xi}, \boldsymbol{\pi})$, which corresponds to the expectation value of $\hat{S}$, is a constant of motion, $d\mathcal{S}/dt = 0$. This means that the Poisson bracket between $\mathcal{S}$ and $\mathcal{H}$ should vanish,

$$\{\mathcal{S}, \mathcal{H}\} = \frac{\partial \mathcal{S}}{\partial \xi^\alpha}\frac{\partial \mathcal{H}}{\partial \pi_\alpha} - \frac{\partial \mathcal{S}}{\partial \pi_\alpha}\frac{\partial \mathcal{H}}{\partial \xi^\alpha} = 0. \tag{6.42}$$

This is the classical counterpart of the quantum relation (6.41). $\mathcal{S}$ can be expanded as

$$\mathcal{S}(\boldsymbol{\xi}, \boldsymbol{\pi}) = \mathcal{S}^{(0)}(\boldsymbol{\xi}) + \mathcal{S}^{(1)\alpha}\pi_\alpha + \frac{1}{2}\mathcal{S}^{(2)\alpha\beta}\pi_\alpha\pi_\beta + \mathcal{O}(\boldsymbol{\pi}^3).$$

In fact, depending on the properties of the symmetry operator, $\mathcal{S}$ can fall into one of two categories. If $\mathcal{S}$ is even with respect to time reversal symmetry, we have $\mathcal{S}^{(1)} = 0$ and $q^r = \mathcal{S}$ as in (6.19). In contrast, if $\mathcal{S}$ is time-odd, we have $\mathcal{S}^{(0)} = \mathcal{S}^{(2)} = 0$ and $p_r = \mathcal{S}$. For instance, the total momenta $\hat{P}_k$ and the angular momenta $\hat{J}_k$ $(k = x, y, z)$ are time-odd operators, and the particle number $\hat{N}$ is time-even. Note that we know the explicit form of the q^r (p_r) corresponding to $\mathcal{S}$. However, their conjugate variables are not known from the beginning. Thus, it is not a trivial task to impose the orthogonality conditions, $\{q^r, p_i\} = \{p_r, q^i\} = 0$, on Σ_K.

From equation (6.42), the terms of the zeroth and first orders in $\boldsymbol{\pi}$ give

$$\mathcal{S}^{(1)\alpha}\mathcal{H}^{(0)}_{,\alpha} = 0 \tag{6.43a}$$

$$\mathcal{S}^{(0)}_{,\alpha}\mathcal{H}^{(2)\alpha\beta} - \mathcal{S}^{(2)\alpha\beta}\mathcal{H}^{(0)}_{,\alpha} = 0. \tag{6.43b}$$

Equations (6.43a) and (6.43b) hold at any point $\boldsymbol{\xi}$ in the space, not only on the collective subspace Σ_K.

Remember section 6.1.1: a point in the space $(\boldsymbol{\xi}, \boldsymbol{\pi})$ corresponds to a single Slater determinant (or a density). The conservation of $\mathcal{S}(t)$ $(d\mathcal{S}/dt = 0)$ is a consequence of the symmetry of the Hamiltonian and the energy density functional, thus, it should not depend on the state.

The decoupling condition (6.12a) and the collective inertia mass of the ASCC theory, (6.39), should be modified with additional terms with respect to q^r as

$$\mathcal{H}^{(0)}_{,\alpha} - H^{(0)}_{,i}f^{i}_{,\alpha} - H^{(0)}_{,r}f^{r}_{,\alpha} \doteq 0 \tag{6.44a}$$

$$\tilde{\mathcal{H}}^{(2)\alpha\beta} \doteq \mathcal{H}^{(2)\alpha\beta} - H^{(0)}_{,i}f^{(1)i\alpha\beta} - H^{(0)}_{,r}f^{(1)r\alpha\beta}. \tag{6.44b}$$

When applying the ASCC to superfluid nuclei, we are interested in the large amplitude collective motion (q^i, p_i) at the given number of particles, $q^r = \langle N \rangle = N_0$. In this case, since we explicitly know the form of $f^r(\xi)$, it is no problem to calculate the various terms, $H^{(0)}_{,r}f^{r}_{,\alpha}$ and $H^{(0)}_{,r}f^{(1)r\alpha\beta}$. In cases where the constants of motion correspond to the momentum p_r, we do not know the explicit form of $f^r(\xi)$. However, $H^{(0)}_{,r}$ corresponds to the derivative with respect to the angle variable, thus $H^{(0)}_{,r} = 0$. This comes from (6.7), $dp_r/dt = -H^{(0)}_{,r} + \mathcal{O}(p^2) = 0$. Therefore, the terms, $H^{(0)}_{,r} f^r_{,\alpha}$ and $H^{(0)}_{,r} f^{(1)r\alpha\beta}$, vanish in this case.

Solutions of equation (6.40) provide $f^i_{,\alpha}$, which is enough to solve equation (6.44). The self-consistent solution of equations (6.44) and (6.40) defines the decoupled collective subspace Σ_K. The question is whether the theory guarantees the orthogonality property between q^r and q^i.

Orthogonality of cyclic variables

Let us prove that $f^r_{,\alpha}$ or $g^\alpha_{,r}$ correspond to the special solution of equation (6.40), and thus are orthogonal to $f^i_{,\alpha}$ and $g^\alpha_{,i}$. We start from the case where the coordinates q^r are conserved, with $f^r(\xi) = \mathcal{S}^{(0)}(\xi)$ and $f^{(1)r\alpha\beta}(\xi) = \mathcal{S}^{(2)\alpha\beta}(\xi)$. In this case, we can show that they correspond to the exact zero modes:

$$
\begin{aligned}
\mathcal{M}^\alpha_\beta f^r_{,\alpha} &= \tilde{\mathcal{H}}^{(2)\alpha\gamma}\mathcal{H}^{(0)}_{;\gamma\beta} f^r_{,\alpha} \\
&\doteq (\mathcal{H}^{(2)\alpha\gamma} f^r_{,\alpha} - H^{(0)}_{,\mu} f^{(1)\mu\alpha\gamma} f^r_{,\alpha})\mathcal{H}^{(0)}_{;\gamma\beta} \\
&\doteq (\mathcal{H}^{(2)\alpha\gamma} f^r_{,\alpha} - H^{(0)}_{,\mu} f^{(1)r\alpha\gamma} f^\mu_{,\alpha})\mathcal{H}^{(0)}_{;\gamma\beta} \\
&\doteq (\mathcal{H}^{(2)\alpha\gamma} f^r_{,\alpha} - \mathcal{H}^{(0)}_{,\alpha} f^{(1)r\alpha\gamma})\mathcal{H}^{(0)}_{;\gamma\beta} \doteq 0.
\end{aligned}
$$

Here, equation (6.38) was used for the third equality and equation (6.43b) for the last equality. Thus, $f^r_{,\alpha}$ is a special solution of (6.40) with $M^r_k = M^k_r = 0$ on the collective subspace Σ_K. As far as the solutions $f^i_{,\alpha}$ ($g^\alpha_{,i}$) correspond to non-zero values of M^k_i (M^i_k), they are orthogonal to the ANG modes.

Next, we discuss the case that the momenta p_r are conserved, with $g^\alpha_{,r}(\xi) = \mathcal{S}^{(1)\alpha}(\xi)$. Differentiating the chain relation $g^\alpha_{,\mu} f^\mu_{,\beta} = \delta^\alpha_\beta$ with respect to q^ν, we obtain

$$
g^\alpha_{,\mu\nu} f^\mu_{,\beta} = -g^\alpha_{,\mu} g^\gamma_{,\nu} f^\mu_{,\beta\gamma}.
$$

Differentiating equation (6.43) with respect to ξ^β, we have

$$
\mathcal{H}^{(0)}_{,\alpha\beta} g^\alpha_{,r} + \mathcal{H}^{(0)}_{,\alpha} g^\alpha_{,r\mu} f^\mu_{,\beta} = 0.
$$

Utilizing these equations, we may prove

$$
\begin{aligned}
\left(\mathcal{H}^{(0)}_{,\gamma\beta} - f^i_{,\gamma\beta} H^{(0)}_{,i}\right) g^\beta_{,r} &\doteq \left(\mathcal{H}^{(0)}_{,\gamma\beta} - f^\mu_{,\gamma\beta} H^{(0)}_{,\mu}\right) g^\beta_{,r} \\
&\doteq \mathcal{H}^{(0)}_{,\gamma\beta} g^\beta_{,r} - f^\mu_{,\gamma\beta}\mathcal{H}^{(0)}_{,\delta} g^\delta_{,\mu} g^\beta_{,r} \\
&\doteq \mathcal{H}^{(0)}_{,\gamma\beta} g^\beta_{,r} + f^\mu_{,\gamma} g^\delta_{,r\mu}\mathcal{H}^{(0)}_{,\delta} \doteq 0.
\end{aligned}
$$

In this sense, g^α_r correspond to the zero-mode solutions of equation (6.28). However, they are not necessarily the exact eigenmodes of equation (6.40), since the decoupling is not exact in general. When the inertial tensor is decoupled from the collective subspace, $H^{(2)ri} = 0$, the affine connection is also decoupled and, in this case, $g_{,r}^{\alpha}$ become the eigenmodes of equation (6.40). This is exactly true for the center-of-mass motion. For the case of rotational motion, there may be some contamination in the solution of equation (6.40). Nevertheless, the decoupling is guaranteed, at least approximately.

6.3.5 Arbitrariness of coordinate systems

One-dimensional case (K = 1)

The basic formulation to determine the collective submanifold is given by equations (6.34) and (6.35). These equations provide a unique solution, except for the scale of the collective coordinate q^1. Indeed, there is an arbitrariness of the scale transformation, $q^1 \to cq^1$ and $p_1 \to c^{-1}p_1$, namely,

$$f_{,\alpha}^{1} \to cf_{,\alpha}^{1} \quad g_{,1}^{\alpha} \to c^{-1}g_{,1}^{\alpha}, \tag{6.45}$$

where the scale parameter c is arbitrary. All the equations in section 6.3 are invariant with respect to this scale transformation. During the numerical calculation, the scale should be fixed with an additional constraint, such as fixing the inertial mass equal to a constant value.

Multi-dimensional case (K > 1)

For the multi-dimensional collective manifold ($K > 1$), the solution of equations (6.12) and (6.32), or equivalently (6.33), is not only invariant under an arbitrary scale transformation such as equation (6.45), it is also invariant under a rearrangement of coordinates in the collective subspace. To see this arbitrariness, let us consider a point transformation among a pair of collective variables (q^k, p_k) and (q^l, p_l), $k \neq l$,

$$q^k \to q^k + cq^l \qquad p_l \to p_l - cp_k, \tag{6.46}$$

with an arbitrary parameter c, keeping the other variables unchanged. If (q^k, p_k) and (q^l, p_l) are a solution of equations (6.12) and (6.32), $(q^k + cq^l, p_k)$ and $(q^l, p_l - cp_k)$ also provide another solution. This can be easily shown as follows.

The transformation (6.46) gives

$$f_{,\alpha}^{k} \to f_{,\alpha}^{k} + cf_{,\alpha}^{l} \qquad g_{,l}^{\alpha} \to g_{,l}^{\alpha} - cg_{,k}^{\alpha}. \tag{6.47}$$

The derivative of $H^{(0)}$, $H_{,l}^{(0)} = \mathcal{H}_{,\alpha}^{(0)} g_{,l}^{\alpha}$, is transformed as

$$H_{,l}^{(0)} \to H_{,l}^{(0)} - cH_{,k}^{(0)}.$$

Therefore, the combination of $H_{,i}^{(0)} f_{,\alpha}^{i}$ is invariant with respect to this transformation, which means that the transformed vectors of equation (6.47) are the solutions of equation (6.12).

The matrix M^i_j on the right-hand side of equations (6.32) and (6.33) transforms as

$$M^k_i \to M^k_i + cM^l_i$$
$$M^j_l \to M^j_l - cM^j_k$$
$$M^k_l \to M^k_l + cM^l_l - cM^k_k - c^2M^l_k$$

for $i \neq l$ and $j \neq k$. The other components M^i_j are all invariant. This can be easily obtained from the relation $M^i_j = f^i_{,\alpha}\mathcal{M}^\alpha_\beta g^\beta_{,j}$. From these properties, it is easy to see that $\{f^i_{,\alpha}\}_{i=1,\cdots,K}$ with $f^k_{,\alpha}$ replaced by $f^k_{,\alpha} + cf^l_{,\alpha}$ also provides a solution of equation (6.33). In the same manner, we can show that $\{g^\alpha_{,i}\}_{i=1,\cdots,K}$ with $g^\alpha_{,l}$ replaced by $g^\alpha_{,l} - cg^\alpha_{,k}$ is a solution of equation (6.32) as well. This proves that the new coordinates and momenta of equation (6.46) are also self-consistent solutions of the ASCC equations.

When the cyclic variables (q^r, p_r) exist, this rearrangement of coordinates is present even for $K = 1$. Suppose q^1 is a collective coordinate, which is a self-consistent solution of equations (6.12) and (6.40). Then, the following transformation provides another solution

$$q^1 \to q^1 + cq^r \qquad p_r \to p_r - cp_1. \tag{6.48}$$

The proof is exactly the same as for equation (6.46).

> As discussed in section 6.3.4, spontaneous symmetry-breaking is characterized by the emergence of ANG modes and the cyclic variables (q^r, p_r) that describe them. In this case, we need to practically work in a collective manifold of dimension $K + K_r$, where K_r is the number of cyclic variables.

This property of arbitrariness tells us that we need to fix the parameter c when we solve the equations. For instance, a possible choice could be requiring $H^{(0)}_{,r1} = 0$ which was adopted in [39]. One can make other choices if they are more convenient [39–41], and the physical quantities should not depend on this choice.

6.3.6 Summary and practical implementation

Let us summarize the ASCC formulation we have obtained so far. We may regard the present formulation as the local harmonic formulation at equilibrium with the *moving* Hamiltonian

$$\mathcal{H}_{\mathrm{mv}}(\boldsymbol{\xi}, \boldsymbol{\pi}) \equiv \mathcal{H}(\boldsymbol{\xi}, \boldsymbol{\pi}) - \lambda_r q^r - \lambda_i q^i.$$

In the end, $\lambda_r = H^{(0)}_{,r} = \mathcal{H}^{(0)}_{,\alpha} g^\alpha_{,r}$ and $\lambda_i = H^{(0)}_{,i} = \mathcal{H}^{(0)}_{,\alpha} g^\alpha_{,i}$ must satisfy the following equilibrium condition (6.49). Equations (6.12), (6.33) and (6.40) can be rewritten as

$$\delta\{\mathcal{H}_{\mathrm{mv}}(\boldsymbol{\xi}, \boldsymbol{\pi} = 0)\} = (\mathcal{H}_{\mathrm{mv}}(\boldsymbol{\xi}, \boldsymbol{\pi} = 0))_{,\alpha} \stackrel{\circ}{=} 0 \tag{6.49a}$$

$$\mathcal{M}^\alpha_\beta f^i_{,\alpha} \stackrel{\circ}{=} M^i_j f^j_{,\beta} \quad \text{or} \quad \mathcal{M}^\alpha_\beta g^\beta_{,j} \stackrel{\circ}{=} M^i_j g^\alpha_{,i}. \tag{6.49b}$$

Here, the matrix $\mathcal{M}^{\alpha}_{\beta}$ is given by a product of the second (covariant) derivatives of $\mathcal{H}_{\rm mv}$ with respect to coordinates and momenta, $\mathcal{M}^{\alpha}_{\beta} = (\tilde{\mathcal{H}}^{(2)}_{\rm mv})^{\alpha\gamma}(\mathcal{H}^{(0)}_{\rm mv})_{,\gamma\beta}$, with

$$(\tilde{\mathcal{H}}^{(2)}_{\rm mv})^{\alpha\beta} \equiv \left.\frac{\partial^2 \mathcal{H}_{\rm mv}}{\partial\pi_\alpha\partial\pi_\beta}\right|_{\pi=0} \qquad (\mathcal{H}^{(0)}_{\rm mv})_{,\alpha\beta} \equiv \left.\frac{\partial^2 \mathcal{H}_{\rm mv}}{\partial\xi^\alpha\partial\xi^\beta}\right|_{\pi=0}.$$

The latter quantity, $(\mathcal{H}^{(0)}_{\rm mv})_{,\alpha\beta}$, can be replaced by the covariant derivative, $\mathcal{H}^{(0)}_{;\alpha\beta}$, according to the argument in section 6.3. Therefore, we can avoid the calculation of the curvature terms associated with $f^{i}_{,\alpha\beta}$. It should be noted that the terms $-\lambda_r q^r - \lambda_i q^i$ are not merely the constraints in the ASCC. These terms change the inertial mass $(\tilde{\mathcal{H}}^{(2)}_{\rm mv})^{\alpha\beta}$ and the covariant derivatives of the potential $(\mathcal{H}^{(0)}_{\rm mv})_{,\alpha\beta}$ in equation (6.49b).

> Collective rotational motion can be treated microscopically within the 'cranking' model [42, 43]. The Hamiltonian
>
> $$\hat{H} \to \hat{H}' = \hat{H} - \boldsymbol{\omega}\cdot\hat{\boldsymbol{J}}$$
>
> can be regarded as one in a rotating frame of the nucleus. In this case, a rotating nucleus corresponds to a stationary state in the rotating frame. A similar modification of the Hamiltonian is found here in the ASCC theory. However, it should be noted that we do not work within the 'rotating' frame, thus, the state is not stationary. During a course of the collective motion, $(\boldsymbol{\xi}(t), \boldsymbol{\pi}(t))$ move in time within the collective subspace Σ_K; see [27] for discussion on this point.

Practical solution of the ASCC

The theory to define a decoupled submanifold consists of equations (6.49a) and (6.49b). The first equation (6.49a) is the potential minimization with constraints on q^i and q^r, which defines the position $\boldsymbol{\xi}$ on Σ_K. The second equation (6.49b) defines the normal modes $f^{i}_{,\alpha}$ $(g^{\alpha}_{,i})$ at the same position $\boldsymbol{\xi}$, which should provide $f^{i}_{,\alpha}$ used in equation (6.49a). Therefore, these equations should be solved self-consistently.

Let us take the $K = 1$ case as an example ($i = j = 1$), without the ANG modes. In this case, equation (6.49b) is an eigenvalue problem with $f^{1}_{,\alpha}$ $(g^{\alpha}_{,1})$ as an eigenvector and M^1_1 as an eigenvalue. First, we start from the equilibrium ground state $\xi^\alpha = \xi^\alpha_{(0)}$ $(q^1 = q^1_{(0)})$, where equation (6.49) is trivially satisfied because of $\mathcal{H}_{,\alpha}(\xi_{(0)}) = 0$ and $\mathcal{H}_{\rm mv} = \mathcal{H}$. Thus, $\xi^\alpha_{(0)}$ is on the collective subspace Σ_1. We can calculate the matrix $\mathcal{M}^{\alpha}_{\beta} = \mathcal{H}^{(2)\alpha\gamma}\mathcal{H}^{(0)}_{,\gamma\beta}$, then, the solution of equation (6.49b) provides many eigenmodes. Among them, we choose one of the lowest excitations, $f^{1}_{,\alpha}$ $(g^{\alpha}_{,1})$. This determines the direction of the collective coordinate q^1. As we mentioned in section 6.3.5, we may choose the scale of q^1 as we like. The arbitrary parameter c in equation (6.45) should be fixed, for instance, by the constant inertial mass condition, $M_0^{-1} = H^{(2)11} = f^{1}_{,\alpha}\mathcal{H}^{(2)\alpha\beta}f^{1}_{,\beta}$.

Then, we move to the next point $\xi^{\alpha}_{(1)}$ ($q^1_{(1)} = q^1_{(0)} + \Delta q$) on the collective path Σ_1, $\xi^{\alpha}_{(1)} = \xi^{\alpha}_{(0)} + g^{\alpha}_{,1}\Delta q$. The calculation of the matrix involves $\Gamma^{\alpha}_{\beta\gamma}$ and $f^{(1)1\alpha\beta}$. In addition, away from equilibrium it is not trivial whether the point $\xi_{(1)}$ simultaneously satisfies both equations (6.49a) and (6.49b). Therefore, an iterative approach to the self-consistent solution is necessary. After finding the self-consistent point $\xi_{(1)}$ ($q^1_{(1)}$), the same procedure leads to the next point $\xi_{(2)}$ ($q^1_{(2)} = q^1_{(1)} + \Delta q$). We repeat this to construct the collective subspace Σ_1, as a series of the self-consistent points $\{\xi_{(0)}, \xi_{(1)}, \cdots\}$. Of course, there is a one-to-one correspondence between $\xi_{(n)}$ and $q^1_{(n)}$.

In this manner, we obtain the collective subspace Σ_1. At each point on Σ_1, the solutions of equations (6.49a) and (6.49b) provide us with the eigenmodes, ($f^1_{,\alpha}$, $g^{\alpha}_{,1}$) and the eigenvalues M^1_1. From these, we may construct the collective Hamiltonian.

ASCC and QRPA

The HF ground state is an equilibrium state, $\mathcal{H}^{(0)}_{,\alpha} = 0$ by definition. In this case, the covariant derivative in equation (6.25) is identical to the second derivative, $\mathcal{H}^{(0)}_{;\alpha\beta} = \mathcal{H}^{(0)}_{,\alpha\beta}$. Then, equations (6.33) and (6.40) are identical to the harmonic approximation around the ground state for the Hamiltonian (6.9), which is equivalent to the random phase approximation (RPA) in section 5.1. The QRPA of section 5.4 is also regarded as a harmonic approximation for the Hamiltonian, $\hat{H}' \equiv \hat{H} - \lambda\hat{N}$. It is natural to use $\hat{H}'$ for constructing the HFB ground state, because the constraint term $-\lambda\hat{N}$ is necessary to obtain a state with the proper expectation value $\mathcal{N} = \langle\hat{N}\rangle = N_0$. However, even if this is widely done in practice, it is not obvious why we should also use $\hat{H}'$ to set up the QRPA matrix and extract the QRPA normal modes of excitation. The ASCC theory provides a physical justification for this choice. In contrast to the HF ground state, the HFB ground state in the superfluid phase ($\Delta \neq 0$) is not in equilibrium. There is a constant of motion, $q^r = \mathcal{N}$, associated with the ANG mode. In general, the gradient $\mathcal{H}^{(0)}_{,\alpha}$ at the ground state is non-zero, but parallel to q^r ($\partial E/\partial N = H^{(0)}_{,r} \neq 0$). Therefore, the QRPA should be formulated as a generalized harmonic approximation at a non-equilibrium state. According to equation (6.28), we need to modify the normal second derivative into $\mathcal{H}^{(0)}_{,\alpha\beta} - H^{(0)}_{,r}f^{r}_{,\alpha\beta}$. In addition, the extension of the point transformation leads to the modified inertial tensor $\tilde{\mathcal{H}}^{(2)\alpha\beta}$. This is simply done by replacing $\hat{H}$ by $\hat{H}'$, in the case of QRPA.

6.3.7 Collective Hamiltonian and requantization

Collective Hamiltonian

The collective subspace Σ_K is determined by solving equations (6.49a) and (6.49b). The collective Hamiltonian $H(\boldsymbol{q}, \boldsymbol{p})$ is defined as equation (6.10) on Σ_K. Since Σ_K is spanned by the collective coordinates q^i ($i = 1, \cdots, K$), ξ^{α} on Σ_K is parameterized by q^i. The collective potential $H(\boldsymbol{q}, \boldsymbol{p} = 0)$ on Σ_K is simply given by

$$V(q^i) \equiv H^{(0)}(q^i)|_{\Sigma_K} \triangleq \mathcal{H}^{(0)}(\xi(q^i)). \tag{6.50}$$

At each point on Σ_K, the tangent vectors $(f^i_{,\alpha}, g^\alpha_{,i})$, $i = 1, \cdots, K$, are solutions of equation (6.49b). From these, we may calculate the collective inertial tensors,

$$B^{ij}(q^k) \equiv H^{(2)ij}(q^k)|_{\Sigma_K} \triangleq f^i_{,\alpha} \tilde{\mathcal{H}}^{(2)\alpha\beta} f^j_{,\beta}, \tag{6.51}$$

which is in fact the reciprocal one. Again, we note here the difference between contravariant and covariant inertial tensors. B^{ij} in equation (6.51) are contravariant tensors and correspond to the inverse mass. The inertial tensor B_{ij} is defined by $B_{ik}B^{kj} = \delta_i^{\,j}$. To second order in momenta, the collective Hamiltonian on the decoupled subspace Σ_K is given by

$$H_c(\boldsymbol{q}, \boldsymbol{p}) \triangleq \frac{1}{2} B^{ij}(\boldsymbol{q}) p_i p_j + V(\boldsymbol{q}), \tag{6.52}$$

where $q = (q^1, \cdots, q^K)$ and $p = (p_1, \cdots, p_K)$.

Requantization

The collective Hamiltonian (6.52) is subsequently quantized. Quantization always involves some ambiguity. Since (q^i, p_i) are guaranteed to be canonical variables, the canonical quantization with the Pauli prescription is a standard choice and leads to the collective Hamiltonian operator

$$\hat{H}_c = -\frac{1}{2} \frac{1}{\sqrt{\gamma}} \frac{\partial}{\partial q^i} \sqrt{g}\, B^{ij} \frac{\partial}{\partial q^j} + V(\boldsymbol{q}), \tag{6.53}$$

where $\boldsymbol{p} = -i\partial_{\boldsymbol{q}}$ and $\gamma(\boldsymbol{q}) \equiv \det\{B_{ij}(\boldsymbol{q})\}$. B_{ij} play the role of a metric tensor and the volume element is given by $dv_K = \sqrt{\gamma}\, d\boldsymbol{q} = \sqrt{\gamma}\, dq^1 \cdots dq^K$.

> The quantized collective Hamiltonian obtained here from either the ATDHF or ASCC theory is formally identical to the one obtained from the GCM with the GOA approximation, see equation (6.59) below. The differences are in how the collective inertia tensors B^{ij} and potential $V(\boldsymbol{q})$ are calculated. In addition, the ATDHF/ASCC theories provide practical methods to non-empirically determine the collective submanifold according to dynamics of the system.

The requantization is a practical way to take into account quantum fluctuations associated with the collective motions described by q^i $(i = 1, \cdots, K)$. For instance, spontaneous fission never takes place in TDDFT if the initial condition is taken from a meta-stable initial state corresponding to the ground state of a heavy nucleus. This is due to the fact that the density functional lacks quantum shape fluctuation. The requantization on the collective subspace Σ_K may recover the effect of quantum fluctuations. Here, the requantization is performed only with respect to the collective degrees of freedom associated with slow motion. Fast degrees of freedom, such as the s.p. motion and giant resonances, are well accounted for by the standard density functional. Thus, the requantization should not be performed for all the degrees of freedom, but only for Σ_K.

Another possible way to take into account missing correlations in TDDFT is to improve the EDF itself. This is a standard way in TDDFT for electronic systems [44]. The generalization of TDDFT to the multi-reference energy density functional (MR-EDF) framework is another possible solution, that would involve solving the time-dependent Hill–Wheeler equation [45], which poses considerable computational difficulties [46]. In practice, the GOA approximation presented in the next section is a way to alleviate these. Path integral methods are another approach to incorporate large-amplitude collective fluctuations [1, 47–50].

The solution of the collective Schrödinger equation

$$\hat{H}_c \Psi_n(\boldsymbol{q}) = E_n \Psi_n(\boldsymbol{q})$$

provides collective wave functions $\Psi_n(\boldsymbol{q})$ and eigen-energies E_n. Observables of one-body operators $\hat{O}$ are treated in the following manner. Each point $(\boldsymbol{\xi}, \boldsymbol{\pi})$ corresponds to a (generalized) Slater determinant $|\boldsymbol{\xi}, \boldsymbol{\pi}\rangle$, from which the classical image of $\hat{O}$, namely, the expectation value at the state corresponding to $(\boldsymbol{\xi}, \boldsymbol{\pi})$, $\mathcal{O}(\boldsymbol{\xi}, \boldsymbol{\pi})= \langle \boldsymbol{\xi}, \boldsymbol{\pi}|\hat{O}|\boldsymbol{\xi}, \boldsymbol{\pi}\rangle = \mathrm{tr}[\hat{O}\rho(\boldsymbol{\xi}, \boldsymbol{\pi})]$, is constructed. This collective representation on Σ_K is given by the transformation from $(\boldsymbol{\xi}, \boldsymbol{\pi})$ to $(\boldsymbol{q}, \boldsymbol{p})$, and expanded in terms of $\boldsymbol{p}$.

$$\mathcal{O}(\boldsymbol{q}, \boldsymbol{p})|_{\Sigma_K} = \mathcal{O}^{(0)}(\boldsymbol{q}) + p_i \mathcal{O}^{(1)i}(\boldsymbol{q}) + \frac{1}{2}\mathcal{O}^{(1)ij}(\boldsymbol{q})p_i p_j + \ldots.$$

The quantization, $\mathcal{O}(\boldsymbol{q}, \boldsymbol{p}) \to O(\boldsymbol{q}, -i\partial_q)$, again involves some ambiguity with respect to the ordering between $\boldsymbol{q}$ and $\boldsymbol{p}$. However, this is not so serious in practice, since the present collective approach may not be applicable, in any case, to 'non-collective' quantities which strongly depend on this ordering. For time-even quantities which represent well the collectivity of the system, the zeroth-order term of $O^{(0)}(\boldsymbol{q})$ is dominant:

$$\langle \Psi_m | O | \Psi_n \rangle \approx \int d\boldsymbol{q} \Psi_m^*(\boldsymbol{q}) O^{(0)}(\boldsymbol{q}) \Psi_n(\boldsymbol{q}),$$

We note again the formal analogy between the quantized version of semi-classical approaches to large amplitude collective motion and the GCM+GOA formalism. See equation (3.102), for example. However, it is important to bear in mind their conceptual difference. The GCM+GOA uses 'off-diagonal' Hamiltonian kernels $\langle \Phi(\boldsymbol{q}|\hat{H}|\Phi(\boldsymbol{q}')\rangle$ to take into account correlation beyond the mean field. On the other hand, the requantized TDDFT utilizes only the 'diagonal' ones.

Application to the $\alpha + \alpha$ reaction

Let us show an application in one of the simplest cases, namely the reaction of two α particles. In figure 6.2, we show the calculated collective inertial mass and collective potential for this reaction. In this case, we derive the collective path with $K = 1$ and determine a single pair of canonical variables (q, p). In order to obtain an intuitive

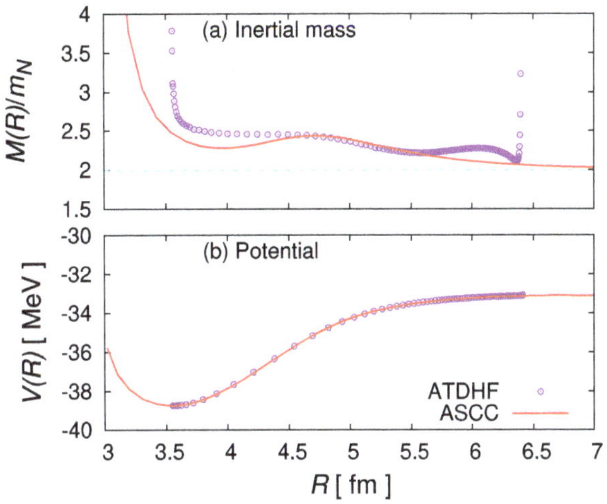

Figure 6.2. Collective mass in units of the nucleon mass m_N and potential energy curves in units of MeV, as functions of the distance between two α particles, calculated with the BKN energy functional [51]. The results of the ASCC method are shown by solid curves and those of the ATDHF by open circles. The dashed line in (a) indicates the reduced mass for the $\alpha + \alpha$ system, $\mu = 2m_N$. (Figure reproduced with permission from [52], courtesy of Nakatsukasa. Copyright 2016 by The American Physical Society.)

understanding, we map the collective coordinate q to R, where R is the distance between the two α particles. In fact, the distance R is not well-defined after the two α touch each other and merge to form the nucleus of ^{8}Be. However, this ambiguity is not a problem in the present case, because R is merely a parameter to change the scale of the collective coordinate q. Therefore, observables calculated with the Hamiltonians, $H_c(q, p)$ and $H_c(R, P_R)$, are identical to each other. $H_c(q, p)$ in equation (6.52) is transformed into

$$H_c(R, P_R) \overset{\circ}{=} \frac{1}{2}B(R)P_R^2 + V(R).$$

As long as there exists a one-to-one correspondence between q and R, i.e. the functions $q(R)$ and $R(q)$ exist, the potential $V(R)$ is given by equation (6.50),

$$V(R) \equiv H^{(0)}(q(R))|_{\Sigma_1} = \mathcal{H}^{(0)}(\xi(R)). \tag{6.54}$$

Although the choice of R does not affect the final result, the magnitude of the inertial mass depends on it. The collective inertia with respect to the coordinate q is given by equation (6.51). If we change the scale of the coordinate as $q \to R(q)$, it becomes

$$B(R) = \frac{dR}{dq}B(q)\frac{dR}{dq}. \tag{6.55}$$

The quantity plotted in figure 6.2(a) is the collective mass, which is the inverse of the collective inertia, $M(R) = 1/B(R)$.

The potential has a minimum at $R \approx 3.5$ fm, which corresponds to the (meta-stable) ground state of ^{8}Be. The ASCC collective mass increases rapidly for $R < 3.5$ fm. This comes from the factor $(dR/dq)^2$ in equation (6.55) with $dR/dq \to 0$. This indicates that the obtained collective path becomes nearly orthogonal to R at small R. This is quite natural because the notion of a distance between two α particles loses its meaning in this region. In the opposite limit of $R \to \infty$, R is a well-defined quantity and should be a good collective coordinate. This is true in the present calculation. The collective mass at $R \to \infty$ is identical to the reduced mass between two α particles, $\mu = 2m_N$ where m_N is the nucleon mass.

For comparison, we also report results obtained with the ATDHF method. As emphasized in section 6.2, the ATDHF method is an initial value problem. We start from the two α particles at a distance of $R = 6.4$ fm. Then, following the equation of path (6.15), we generate the ATDHF collective path, which is shown by circles in figure 6.2. The potential energy curve nearly matches that of the ASCC. However, it becomes numerically unstable to calculate the collective path at short relative distances.

The collective mass in the ATDHF method approximately coincides with that of the ASCC theory in the region $4.5 < R < 5.5$ fm. However, outside this region, they deviate from one other. In particular, the ATDHF collective mass shows a dramatic increase near the end-points of the calculation at $R \approx 3.5$ fm and $R \approx 6.5$ fm. This increase is not physical. The ATDHF method requires many fall-line trajectories from different initial states. What is shown in figure 6.2 merely represents one of these trajectories, and choosing the physically correct trajectory at each value of R is not trivial [26]; see also the discussion of the 'non-uniqueness problem' in section 6.2.4.

Once the collective Hamiltonian is constructed, we may calculate observables in the $\alpha + \alpha$ reaction by requantizing H_c. For example, we may obtain scattering phase shifts, the life time of ^{8}Be, etc. As long as the Pauli prescription is adopted for quantization, the choice of $H_c(q, p)$ or $H_c(R, P_R)$ does not matter. The final results should be identical to each other; see [52] for calculation of the scattering phase shifts.

6.3.8 Derivation of the Bohr Hamiltonian

The ASCC theory and its requantization procedure described in section 6.3.7 can be used to derive the five-dimensional Bohr Hamiltonian [2]. The Bohr model assumes two quadrupole shape degrees of freedom, typically parameterized by the deformations (β, γ) in the intrinsic frame, in addition to three rotational degrees of freedom. Historically, the developments of the ATDHF theory in particular were largely motivated by the need to anchor the phenomenological Bohr collective Hamiltonian to a microscopic description of nuclear structure; see the introduction in [16]. The derivation of the Bohr Hamiltonian requires the following conditions:

1. A two-dimensional ($K = 2$) collective space spanned by the coordinates (q^1, q^2).
2. A one-to-one correspondence between (q^1, q^2) and the quadrupole shape degrees of freedom (β, γ).
3. The applicability of symmetry properties of the Bohr model.

Since the ASCC guarantees the separation of the ANG modes induced by symmetry-breaking, here the rotational symmetry of the nuclear Hamiltonian, the collective subspace Σ_2 of condition 1 is orthogonal to the subspace Σ_r associated with the rotational motion of deformed systems. Therefore, it is most likely that the collective coordinates (q^1, q^2) are associated only with the shape degrees of freedom of the nucleus. Condition 2 assumes the existence of functions $q^i = \bar{g}^i(\beta, \gamma)$ with $i = 1,2$, and their inverse transformation, $\beta = \bar{f}^{\beta}(q^1, q^2)$ and $\gamma = \bar{f}^{\gamma}(q^1, q^2)$. It should be emphasized that this does not mean that the decoupled collective subspace Σ_2 corresponds to pure quadrupole shape oscillations only. Condition 2 does not forbid other quantities changing as functions of (q^1, q^2). For instance, the pairing gap may depend on these, $\Delta(q^1, q^2)$, which means that coupling to pairing vibrations can be taken into account, at least in part. Condition 3 is necessary, because the symmetry is used in the quantization of the Bohr Hamiltonian, which restricts the spin and the parity. Thus, the collective subspace Σ_2 cannot contain octupole deformations, for instance.

The collective Hamiltonian for the shape oscillation is now given by equation (6.52) with $K = 2$. Because of the one-to-one correspondence (condition 2), we may rewrite it in terms of β and γ. The potential $\bar{V}(\beta, \gamma)$ is simply given by

$$\bar{V}(\beta, \gamma) = V\left(q^1(\beta, \gamma), q^2(\beta, \gamma)\right).$$

The components of the collective inertia tensor $\bar{B}^{kl}$ $(k, l = \beta, \gamma)$ transform according to

$$\bar{B}^{kl} = \bar{f}^{k}_{,i} B^{ij} \bar{f}^{l}_{,j} \quad i, j = 1, 2 \quad \text{and} \quad k, l = \beta, \gamma.$$

This transformation from (q^1, q^2) to (β, γ) does not change the classical dynamics of the collective Hamiltonian (6.52).

The rotational motion of triaxially deformed systems corresponds to ANG modes (q^r, p_r), $r = 1, 2, 3$. The Hamiltonian should not depend on the orientation of the nucleus q^r. Thus, only the kinetic energy terms contribute to the Hamiltonian. Normalizing the collective momenta p_r so as to coincide with the total angular momenta, the collective inertia tensor is given by the moments of inertia along each of the principal axes of the nucleus and is thus diagonal in the principal-axis frame.

Eventually, we obtain a classical collective Hamiltonian as

$$H_c(\beta, \gamma; p_\beta, p_\gamma, \boldsymbol{I}) \doteq \frac{1}{2}\bar{B}^{\beta\beta}p_\beta^2 + \frac{1}{2}\bar{B}^{\gamma\gamma}p_\gamma^2 + \bar{B}^{\beta\gamma}p_\beta p_\gamma + V(\beta, \gamma) + \sum_{k=1,2,3} \frac{I_k^2}{2\mathcal{J}_k},$$

where the inertia parameters, $\bar{B}^{kl}$ (quadrupole shape vibrations) and $\mathcal{J}_k$ (collective rotational motion) are functions of (β, γ). In order to obtain the collective Hamiltonian operator, we need to perform the requantization. Adopting the Pauli prescription for the quantization as in equation (6.53), we obtain the five-dimensional Bohr Hamiltonian,

$$\hat{H}_c = \hat{T}_{\text{rot}} + \hat{T}_{\text{vib}} + \hat{V}_{\text{vib}}(\beta, \gamma),$$

where

$$\hat{T}_{\rm rot} = \sum_{k=1,2,3} \frac{\hat{I}_k^2}{2\mathcal{J}_k}$$

$$\hat{T}_{\rm vib} = \frac{1}{2\sqrt{WR}}\left\{\frac{1}{\beta^4}\left[\frac{\partial}{\partial\beta}\left(\beta^2\sqrt{\frac{R}{W}}\bar{B}_{\gamma\gamma}\frac{\partial}{\partial\beta}\right) - \frac{\partial}{\partial\beta}\left(\beta^2\sqrt{\frac{R}{W}}\bar{B}_{\beta\gamma}\frac{\partial}{\partial\gamma}\right)\right]\right.$$
$$+ \frac{1}{\beta^2\sin 3\gamma}\left[-\frac{\partial}{\partial\gamma}\left(\sqrt{\frac{R}{W}}\sin 3\gamma\bar{B}_{\beta\gamma}\frac{\partial}{\partial\beta}\right)\right.$$
$$\left.\left.+ \frac{\partial}{\partial\gamma}\left(\sqrt{\frac{R}{W}}\sin 3\gamma\bar{B}_{\beta\beta}\frac{\partial}{\partial\gamma}\right)\right]\right\},$$

with the following notation: $\beta^2 W(\beta, \gamma) \equiv \bar{B}_{\beta\beta}\bar{B}_{\gamma\gamma} - \bar{B}_{\beta\gamma}$, the moments of inertia are $\mathcal{J}_k = 4\beta^2 D_k \sin^2(\gamma - 2\pi k/3)$, and $R(\beta, \gamma) \equiv D_1 D_2 D_3$. Here, the covariant inertia tensor $\bar{B}_{kl}$ is defined by $\bar{B}^{ll'}\bar{B}_{l'k} = \delta^l_k$.

If the collective subspace Σ_2 is obtained according to the ASCC theory, it is interesting to check whether it fulfills (or violates) conditions 1, 2, and 3. However, it has not been done so far, because the self-consistent determination of the two-dimensional collective coordinates (q^1, q^2) is a difficult task. In this sense, the derivation of the Bohr Hamiltonian strictly following the ASCC theory has never been achieved in practice. Instead of solving the ASCC equations, one typically adopts approximations, for example assuming that the collective coordinates are globally defined as the expectation value of the quadrupole operators ($\hat{Q}^1 \equiv r^2 Y_{20}$, $\hat{Q}^2 \equiv (r^2 Y_{22} + r^2 Y_{2-2})/\sqrt{2}$); $q^i = \langle\phi(\boldsymbol{q})|\hat{Q}^i|\phi(\boldsymbol{q})\rangle$. Using this approximation, one can relate (q^1, q^2) to (β, γ) and equation (6.49) is nothing other than the minimization of the energy at given (β, γ). The solutions of equation (6.49b) provide the inertia parameters. However, the choice of the eigenvectors $f^i_{,\alpha}$ ($g^\alpha_{,i}$) is somewhat arbitrary, because we neglect the self-consistency between equations (6.49a) and (6.49b) by the adopted approximation. Thus, a possibly better prescription is to choose the two eigen-solutions of equation (6.49b) which have the largest overlap with the quadrupole shape fluctuations (β, γ). This approximation and prescription were proposed and the numerical calculation were performed by Hinohara and collaborators [53].

In most applications reported so far in the literature, further approximations are made to derive the Bohr Hamiltonian. Instead of solving equation (6.49b), one calculates the collective mass using the cranking formula with the adiabatic perturbation [54], assuming that the collective coordinates are (β, γ). This approximation significantly simplifies the numerical calculations. However, one loses one of the great advantages of the time-dependent approaches over the GCM, namely, the effect of the time-odd fields. Because of this deficiency, the calculated inertia masses are often too small to reproduce experimental data.

As noted early by Goeke and Reinhard in [55], the effects of time-odd fields can be properly taken into account in the GCM, but at the price of adding a collective variable—for each collective variable q, one also needs to add the conjugate variable p in the GCM ansatz (3.79).

6.4 Gaussian overlap approximation of the GCM

In section 3.2, the MR-EDF framework was introduced to take into account correlations missing at the SR-EDF level. The implementation of the MR-EDF is currently very challenging. On the one hand, there are formal difficulties in defining GCM kernels that are free from spuriosities for density-dependent energy functionals such as the popular Skyrme and Gogny functionals [56–59]. While there exist recipes to solve this problem, solving the full Hill–Wheeler–Griffin equation (3.80) directly is both computationally demanding and requires addressing the problem of zero-norm states—pairs of states with vanishingly small overlap $\langle \boldsymbol{q}|\boldsymbol{q}'\rangle$; see the discussion in section 3.2.1. The GOA provides an interesting solution to many of these problems, since it recasts the Hill–Wheeler–Griffin equation into a much simpler collective Schrödinger equation.

6.4.1 The Gaussian overlap approximation

The local GOA consists in assuming that the overlap between generator states, the norm kernel (3.81a), has a Gaussian shape [60, 61],

$$\mathcal{N}(\boldsymbol{a}, \boldsymbol{a}') \propto \exp\left[-\frac{1}{2}(\boldsymbol{a} - \boldsymbol{a}')\boldsymbol{G}(\boldsymbol{a}, \boldsymbol{a}')(\boldsymbol{a} - \boldsymbol{a}')\right]. \tag{6.56}$$

If the tensor $G(\boldsymbol{a}, \boldsymbol{a}')$ is not a constant but a true function of the collective variables, it can be used to define a metric in the collective space. A common choice is to take $\boldsymbol{G} \equiv \boldsymbol{G}(\bar{\boldsymbol{a}})$ with $\bar{\boldsymbol{a}} = (\boldsymbol{a} + \boldsymbol{a}')/2$ [60, 62], which defines the metric

$$\gamma(\boldsymbol{a}) = \det(\boldsymbol{G}(\boldsymbol{a})).$$

More general options are discussed, e.g. in [63]. Derivations with such a non-constant metric are a little more involved, but it is always possible to go back to the constant metric case through a change of variable of the type [64, 65]

$$\boldsymbol{\alpha}(\boldsymbol{q}) = \int^{\boldsymbol{q}} \sqrt{\gamma(\boldsymbol{q}')}\, d\boldsymbol{q}'.$$

An additional hypothesis in the GOA is to assume that the energy kernel is closely related to the norm kernel by the relation

$$\mathcal{H}(\boldsymbol{a}, \boldsymbol{a}') = \mathcal{N}(\boldsymbol{a}, \boldsymbol{a}')h(\boldsymbol{a}, \boldsymbol{a}'), \tag{6.57}$$

where $h(\boldsymbol{a}, \boldsymbol{a}')$ is a polynomial of degree 2 for the collective variables $\boldsymbol{a}$ and $\boldsymbol{a}'$. The polynomial $h(\boldsymbol{a}, \boldsymbol{a}')$ is often called the reduced Hamiltonian. This second hypothesis comes from the second-order Taylor expansion of the reduced Hamiltonian around a point $\boldsymbol{q}$,

$$\begin{aligned} h(\boldsymbol{a}, \boldsymbol{a}') &= h_0 + h_{\boldsymbol{a}}(\boldsymbol{a} - \boldsymbol{q}) + h_{\boldsymbol{a}'}(\boldsymbol{a}' - \boldsymbol{q}) \\ &+ \frac{1}{2}\left[h_{\boldsymbol{aa}}(\boldsymbol{a} - \boldsymbol{q})^2 + 2h_{\boldsymbol{aa}'}(\boldsymbol{a} - \boldsymbol{q})(\boldsymbol{a}' - \boldsymbol{q}) + h_{\boldsymbol{a}'\boldsymbol{a}'}(\boldsymbol{a}' - \boldsymbol{q})^2\right], \end{aligned} \tag{6.58}$$

where $h_{\boldsymbol{a}'}$ is a shorthand notation for

$$h_{\boldsymbol{a}} \equiv \left.\frac{\partial h(\boldsymbol{a}, \boldsymbol{a}')}{\partial \boldsymbol{a}}\right|_{a=a'=q} \equiv \left(\left.\frac{\partial h(\boldsymbol{a}, \boldsymbol{a}')}{\partial a_1}\right|_{a=a'=q}, \ldots, \left.\frac{\partial h(\boldsymbol{a}, \boldsymbol{a}')}{\partial a_n}\right|_{a=a'=q}\right)$$

and similarly for the tensor of second derivatives $h_{\boldsymbol{a}\,\boldsymbol{a}'}$, etc. Note that the derivatives are taken at point $\boldsymbol{q}$, hence are independent of the variables of integration $\boldsymbol{a}$ and $\boldsymbol{a}'$.

Derivation of a collective Schrödinger equation

We give here some basic pointers to complete the demonstration of the form (6.59) below of the collective Hamiltonian in the easier case of a constant metric, $G(\boldsymbol{a}) = G$. The starting point is the property (3.102) for the Hamiltonian, which we write explicitly

$$\langle\Psi|\hat{H}|\Psi\rangle = \int_{\boldsymbol{a}} g^*(\boldsymbol{a})\hat{\mathcal{H}}_{\text{coll}}(\boldsymbol{a})g(\boldsymbol{a})d\boldsymbol{a}.$$

The goal is to compute the left-hand side of the equation and put it in the form of the right-hand side. Introducing the definition (3.79) of the GCM ansatz and of the reduced kernel, we can write the left-hand side as

$$\langle\Psi|\hat{H}|\Psi\rangle = \int d\boldsymbol{a} \int d\boldsymbol{a}' \int d\boldsymbol{q}\, f^*(\boldsymbol{a})\mathcal{N}^{1/2}(\boldsymbol{a}, \boldsymbol{q})h(\boldsymbol{a}, \boldsymbol{a}')f(\boldsymbol{a}')\mathcal{N}^{1/2}(\boldsymbol{q}, \boldsymbol{a}').$$

We then introduce equation (6.58) into the integral above and use the fact that the collective wave function reads

$$g(\boldsymbol{q}) = \int d\boldsymbol{a}\, \mathcal{N}^{1/2}(\boldsymbol{q}, \boldsymbol{a})f(\boldsymbol{a}) = \int d\boldsymbol{a}\, \mathcal{N}^{1/2}(\boldsymbol{a}, \boldsymbol{q})f(\boldsymbol{a}).$$

The next step is to take the norm kernel (6.56), compute partial derivatives with respect to $\boldsymbol{a}$ and $\boldsymbol{q}$, and use these relations to express terms such as $(\boldsymbol{a} - \boldsymbol{q})$ as a function of derivatives of the norm kernel and metric tensor. With just a little bit of effort, one can express $\langle\Psi|\hat{H}|\Psi\rangle$ as an integral over $\boldsymbol{q}$ of a sum of terms involving not $f(\boldsymbol{a})$ but the collective function $g(\boldsymbol{q})$. Using integration by parts, one can express everything in the form of a scalar product in the collective space of $g(\boldsymbol{q})$ and $\hat{\mathcal{H}}_{\text{coll}}(\boldsymbol{q})g(\boldsymbol{q})$, as requested.

6.4.2 The collective Hamiltonian

In the most general case of a non-constant metric, one can still follow the overall procedure given above to derive the form of the collective Hamiltonian. It reads

$$\hat{\mathcal{H}}_{\text{coll}}(\boldsymbol{a}) = -\frac{\hbar^2}{2\sqrt{\gamma(\boldsymbol{a})}} \sum_{kl} \frac{\partial}{\partial a_k} \sqrt{\gamma(\boldsymbol{a})}\, B_{kl}(\boldsymbol{a}) \frac{\partial}{\partial a_l} + V(\boldsymbol{a}), \tag{6.59}$$

with the collective inertia tensor $\boldsymbol{B}$ given by

$$B_{ij}(\boldsymbol{q}) = \frac{1}{2\hbar^2} \sum_{kl} G_{ik}^{-1}(\boldsymbol{q}) \left[h_{\boldsymbol{a}'\boldsymbol{a}'} - h_{\boldsymbol{a}\boldsymbol{a}'} + \Gamma_{kl}^{n}(\boldsymbol{a}) h_{\boldsymbol{a}} \right] G_{lj}^{-1}(\boldsymbol{q}). \tag{6.60}$$

Although there is no consensus on the vocabulary or the notation, we will use as much as possible the following conventions. The collective inertia is denoted by $B(\boldsymbol{q})$. The collective *mass* is the inverse tensor such that $B^{\alpha\beta} M_{\beta\gamma} = \delta_{\gamma}^{\beta}$ or, for one-dimensional collective spaces $K = 1$, $M(q^1) = 1/B(q^1)$. In the previous sections, the same symbol $\boldsymbol{B}$ was used for both inertia and mass, the distinction between the two coming from the use of contravariant (inertia) or covariant (mass) components; see the note in section 6.2.4.

The notation Γ_{kl}^{n} stands for the Christoffel symbol. We already encountered it in the context of the ASCC theory of section 6.3. It is related to the metric tensor $\boldsymbol{G}(\boldsymbol{a})$ through the relation

$$\Gamma_{kl}^{n}(\boldsymbol{a}) = \frac{1}{2} \sum_{i} G_{ni}^{-1}(\boldsymbol{a}) \left(\frac{\partial G_{ki}}{\partial a_l} + \frac{\partial G_{il}}{\partial a_k} - \frac{\partial G_{lk}}{\partial a_i} \right). \tag{6.61}$$

For a constant metric, $\Gamma_{kl}^{n} = 0$, by definition. All derivatives in the previous equation are evaluated at $\boldsymbol{a} = \boldsymbol{a}' = \boldsymbol{q}$. The last term in the expression of the collective inertia comes from the dependence of $\boldsymbol{G}$ on the collective space. This term is not present in the 'constant width' version of the GCM+GOA.

The potential part of the collective Hamiltonian reads

$$V_{\text{coll}}(\boldsymbol{q}) = V(\boldsymbol{q}) - \frac{1}{2} \boldsymbol{G}^{-1} h_{\boldsymbol{a}\boldsymbol{a}'} + \frac{1}{8} \boldsymbol{G}^{-1} \frac{\partial^2 h_{\boldsymbol{a}\boldsymbol{a}}}{\partial \boldsymbol{q}^2}, \tag{6.62}$$

where $V(\boldsymbol{q})$ is the HFB energy at the point $\boldsymbol{q}$ of the collective variables. The notation $\boldsymbol{G}^{-1} h_{\boldsymbol{a}\boldsymbol{a}'}$ is a short-hand notation for $\text{Tr} G_{ij}^{-1} h_{a_j a_i'}$. The zero-point energy is defined by

$$\epsilon_{\text{ZPE}}(\boldsymbol{q}) = -\frac{1}{2} \boldsymbol{G}^{-1} h_{\boldsymbol{a}\boldsymbol{a}'}. \tag{6.63}$$

It represents the energy associated with quantum fluctuations of the collective variable $\boldsymbol{q}$. The last term in equation (6.62) is rarely considered [66].

6.4.3 Collective Hamiltonian with HFB generator states

Expressions (6.60) for the inertia tensor, (6.62) for the collective potential and (6.63) for the zero-point collective energy all contain first and second derivatives of the reduced Hamiltonian. Although one could, of course, compute these derivatives numerically, one can exploit the properties of the HFB states $|\phi(\boldsymbol{q})\rangle$ to obtain an analytical expression. In order to achieve this we recall here a special case of the Thouless theorem proved by Ring and Schuck in [67].

Theorem 3 (Ring and Schuck). *If we consider a quasiparticle vacuum* $|\Phi\rangle$*, then, for any many-body state* $|\Psi\rangle$*, there exists a unitary transformation* $\hat{T}$ *such that*

$$1.\ |\Psi\rangle = e^{i\hat{T}}|\Phi\rangle$$
$$2.\ \hat{T} = \sum_{\mu<\nu} T_{\mu\nu}\beta^{\dagger}_{\mu}\beta^{\dagger}_{\nu} - \sum_{\mu<\nu} T^{*}_{\mu\nu}\beta_{\mu}\beta_{\nu}.$$

In other words, the matrix of the transformation $\hat{T}$ *in the q.p. basis associated with the state* $|\Phi\rangle$ *takes the generic form*

$$\tilde{T} = \begin{pmatrix} 0 & T \\ -T^{*} & 0 \end{pmatrix}.$$

In combination with a few other tools, this theorem will be invoked to write derivatives of the reduced Hamiltonian as a function of *local* values at point $\boldsymbol{q}$. In the rest of this section, we will show how the components of the collective Hamiltonian (6.59) can be computed in practice. We give a bottom-up style presentation: we first introduce a collective momentum operator related to the operation $\partial/\partial\boldsymbol{a}$; then express the matrix of the operator in the q.p. basis as a function of the generalized density; using the small amplitude limit of the time-dependent Hartree–Fock–Bogoliubov (TDHFB) equation, we can express $\partial\mathcal{R}/\partial\boldsymbol{q}$ as a function of the matrix elements of the constraint operator associated with $\boldsymbol{q}$.

Collective momentum

Let us consider the transformation $\hat{T}$ between the two states $|\Phi\rangle \equiv |\Phi(\boldsymbol{a})\rangle$ and $|\Psi\rangle \equiv |\Phi(\boldsymbol{a} + \delta\boldsymbol{a})\rangle$. Since we must have $\lim_{\delta\boldsymbol{a}\to 0}|\Psi\rangle = |\Phi\rangle$, we can always choose the transformation $\hat{T}$ in the form $\hat{T} = \delta\boldsymbol{a} \cdot \hat{\boldsymbol{P}}_{\boldsymbol{a}}/\hbar$. Here $\hbar$ is introduced for convenience: it will be used to show that $\hat{\boldsymbol{P}}_{\boldsymbol{a}}$ in this definition represent the set of canonical momenta associated with the collective variables $\boldsymbol{a}$. Remember that $\boldsymbol{a} = (a_1,\dots, a_N)$, so that we note $\hat{\boldsymbol{P}}_{\boldsymbol{a}} = (\hat{P}_1,\dots, \hat{P}_N)$. We apply the Ring and Schuck theorem and expand up to first order,

$$|\Phi(\boldsymbol{a} + \delta\boldsymbol{a})\rangle = e^{i\frac{\delta\boldsymbol{a}}{\hbar}\cdot\hat{\boldsymbol{P}}_{\boldsymbol{a}}}|\Phi(\boldsymbol{a})\rangle = |\Phi(\boldsymbol{a})\rangle + i\frac{\delta\boldsymbol{a}}{\hbar}\cdot\hat{\boldsymbol{P}}_{\boldsymbol{a}}|\Phi(\boldsymbol{a})\rangle + O(\delta\boldsymbol{a}^2).$$

Therefore, we have formally

$$\lim_{\delta a\to 0}\left(\frac{|\Phi(\boldsymbol{a}+\delta\boldsymbol{a})\rangle - |\Phi(\boldsymbol{a})\rangle}{\delta\boldsymbol{a}}\right) \equiv \frac{\partial}{\partial\boldsymbol{a}}|\Phi(\boldsymbol{a})\rangle = \frac{i}{\hbar}\hat{\boldsymbol{P}}_{\boldsymbol{a}}|\Phi(\boldsymbol{a})\rangle.$$

More rigorously, we should introduce the vector $\delta\boldsymbol{a} = (0\,,...,\delta a_k\,,...,0)$, and define each component of the collective momentum as

$$\lim_{\delta a\to 0}\left(\frac{|\Phi(\boldsymbol{a}+\delta\boldsymbol{a})\rangle - |\Phi(\boldsymbol{a})\rangle}{\delta a_k}\right) \equiv \frac{\partial}{\partial a_k}|\Phi(\boldsymbol{a})\rangle = \frac{i}{\hbar}\hat{P}_k|\Phi(\boldsymbol{a})\rangle.$$

From this, we see that $\hat{\boldsymbol{P}}_{\boldsymbol{a}} = -i\hbar\frac{\partial}{\partial\boldsymbol{a}}$, and we can easily show that $[\hat{\boldsymbol{a}}, \hat{\boldsymbol{P}}_{\boldsymbol{a}}] = i\hbar$, which shows that $\hat{\boldsymbol{P}}_{\boldsymbol{a}}$ indeed represents the set of canonical momenta associated with $\boldsymbol{a}$. Thus far, we have shown that the derivative operators $\partial/\partial\boldsymbol{a}$ are related to the canonical momentum operator $\hat{\boldsymbol{P}}_{\boldsymbol{a}}$. We now determine the relation between the canonical momenta and the generalized density.

The transformation $\hat{U} = e^{i\frac{\delta\boldsymbol{a}}{\hbar}\cdot\hat{\boldsymbol{P}}_{\boldsymbol{a}}}$ is unitary and acts in Fock space. It induces a transformation of the single-particle states, and we shall denote by U the matrix of that transformation. The key properties are summarized by

$$c'^{\dagger}_i = \hat{U}c_i^{\dagger}\hat{U}^{\dagger} = \sum_k U_{ki}c_k^{\dagger} \qquad c'_i = \hat{U}c_i\hat{U}^{\dagger} = \sum_k U_{ki}^{*}c_k.$$

The first part of each equality can easily be demonstrated by considering the action of $\hat{U}$ on Slater determinants, using the property that the particle vacuum must be invariant under $\hat{U}$ and that any many-body state can always be written as a linear superposition of Slater determinants. The second part of each equality can be deduced from looking at the action of a single $c_i^{\dagger}$ operator on the particle vacuum. The consequence of these two properties is that the matrix of the generalized density in the s.p. basis, formally

$$\mathcal{R}_{ij}(\boldsymbol{a}) = \langle\Phi(\boldsymbol{a})|\begin{pmatrix} c_j^{\dagger}c_i & c_jc_i \\ c_j^{\dagger}c_i^{\dagger} & c_jc_i^{\dagger}\end{pmatrix}|\Phi(\boldsymbol{a})\rangle,$$

can be written, after introducing the transformation $\hat{U}$,

$$\begin{aligned}\mathcal{R}_{ij}(\boldsymbol{a}) &= \langle\Phi(\boldsymbol{a}+\delta\boldsymbol{a})|\hat{U}\begin{pmatrix} c_j^{\dagger}c_i & c_jc_i \\ c_j^{\dagger}c_i^{\dagger} & c_jc_i^{\dagger}\end{pmatrix}\hat{U}^{\dagger}|\Phi(\boldsymbol{a}+\delta\boldsymbol{a})\rangle \\ &= \sum_{kl}\langle\Phi(\boldsymbol{a}+\delta\boldsymbol{a})|\begin{pmatrix} U_{ki}^{*}U_{lj}c_l^{\dagger}c_k & U_{ki}^{*}U_{lj}^{*}c_lc_k \\ U_{ki}U_{lj}c_l^{\dagger}c_k^{\dagger} & U_{ki}^{*}U_{lj}c_lc_k^{\dagger}\end{pmatrix}|\Phi(\boldsymbol{a}+\delta\boldsymbol{a})\rangle.\end{aligned}$$

If we introduce the matrix of the transformation in the doubled s.p. basis

$$\mathcal{U} = \begin{pmatrix} U & 0 \\ 0 & U^{*}\end{pmatrix},$$

we find $\mathcal{R}(\boldsymbol{a}) = \mathcal{U}^{\dagger}\mathcal{R}(\boldsymbol{a} + \delta\boldsymbol{a})\mathcal{U}$. Now recalling the form of the transformation $\hat{U}$ and expanding it to first order in $\delta\boldsymbol{a}$, we obtain

$$\mathcal{U} = 1 + \frac{i}{\hbar}\delta\boldsymbol{a} \cdot \boldsymbol{P}_a + O(\delta\boldsymbol{a}^2),$$

which leads to

$$\mathcal{R}(\boldsymbol{a} + \delta\boldsymbol{a}) = \mathcal{R}(\boldsymbol{a}) + \frac{i}{\hbar}\delta\boldsymbol{a} \cdot [\boldsymbol{P}_a, \mathcal{R}(\boldsymbol{a})]$$

and therefore, at the limit $\boldsymbol{a} \to 0$,

$$\frac{\partial\mathcal{R}}{\partial\boldsymbol{a}} = \frac{i}{\hbar}[\boldsymbol{P}_a, \mathcal{R}(\boldsymbol{a})].$$

As before, we use a loose shorthand notation for partial derivatives with respect to a full vector of collective variables. In the q.p. basis where $\mathcal{R}(\boldsymbol{a})$ is diagonal, we have

$$\mathcal{R}(\boldsymbol{a}) = \begin{pmatrix} 0 & 0 \\ 0 & 1 \end{pmatrix} \qquad \mathcal{P}(\boldsymbol{a}) = \begin{pmatrix} P_a^{11} & P_a^{12} \\ P_a^{21} & P_a^{22} \end{pmatrix},$$

which immediately leads to

$$\frac{\partial\mathcal{R}}{\partial\boldsymbol{a}} = \frac{i}{\hbar}\begin{pmatrix} 0 & P_a^{12} \\ -P_a^{21} & 0 \end{pmatrix}. \tag{6.64}$$

In this expression, $P_{k;\nu\mu}^{12*} = P_{k;\mu\nu}^{21}$ since $\hat{\boldsymbol{P}}$ is Hermitian. We thus have an expression that relates the matrix of the canonical momenta $\hat{\boldsymbol{P}}_a$ with the matrix of the derivative of the generalized density with respect to the collective variables $\boldsymbol{a}$. The next (and final) step is to replace the matrix elements of $\partial\mathcal{R}/\partial\boldsymbol{a}$ by local quantities at point $\boldsymbol{a}$.

Matrix elements of $\partial\mathcal{R}/\partial\boldsymbol{q}$

We recall that the HFB equations at point $\boldsymbol{a}$ take the form of the commutator

$$[\mathcal{H}(\boldsymbol{a}) - \sum_a \lambda_a \mathcal{Q}_a, \mathcal{R}(\boldsymbol{a})] = 0,$$

where $\mathcal{Q}_a$ is the matrix of the constraint operator $\hat{Q}_a$ in the double s.p. basis and λ_a is the Lagrange multiplier for the collective variable a ($a \equiv a_k$ for $k = 1, \dots, N$). This commutator can be written either in the double s.p. basis or in the q.p. basis associated with point $\boldsymbol{a}$. Next, we assume a small variation of the generalized density, leading to variations of the HFB matrix and the Lagrange multipliers,

$$\begin{aligned}
\mathcal{H}(\boldsymbol{a} + \delta\boldsymbol{a}) &= \mathcal{H}(\boldsymbol{a}) + \mathcal{H}_1 \\
\mathcal{R}(\boldsymbol{a} + \delta\boldsymbol{a}) &= \mathcal{R}(\boldsymbol{a}) + \mathcal{R}_1 \\
\lambda_a(\boldsymbol{a} + \delta\boldsymbol{a}) &= \lambda_a(\boldsymbol{a}) + \delta\lambda_a.
\end{aligned}$$

We impose that the HFB equations are still obeyed at point $\boldsymbol{a} + \delta\boldsymbol{a}$. Therefore,

$$\left[\mathcal{H}(\boldsymbol{a} + \delta\boldsymbol{a}) - \sum_a (\lambda_a + \delta\lambda_a)\mathcal{Q}_a,\ \mathcal{R}(\boldsymbol{a} + \delta\boldsymbol{a})\right] = 0.$$

Then of course, we expand the commutator and only keep terms up to first order in $\delta\boldsymbol{a}$. Taking into account the HFB equations at point $\boldsymbol{a}$, we find immediately

$$\left[\mathcal{R}_1,\ \mathcal{H}(\boldsymbol{a}) - \sum_a \lambda_a \mathcal{Q}_a\right] + [\mathcal{R}(\boldsymbol{a}), \mathcal{H}_1] = \sum_a \delta\lambda_a [\mathcal{R}(\boldsymbol{a}), \mathcal{Q}_a]. \tag{6.65}$$

We have already encountered equations very similar to equation (6.65) in the context of the QRPA theory, which is also obtained from the small amplitude limit of the TDHFB equation. We will also see later in section 8.3.1 that the same equation can be used to obtain a convenient formula to adjust the Lagrange parameters of linear constraints on the HFB solutions.

This equation is nothing other than the small amplitude limit of the TDHFB equation. It was introduced in a slightly different form (5.32) in chapter 5 to derive the QRPA equation of section 5.4. If we linearize all non-zero matrix elements (i.e. stack them in vector form), we can write it as

$$\mathcal{M}\begin{pmatrix} R^{12} \\ R^{12*} \end{pmatrix} = \sum_a \delta\lambda_a \begin{pmatrix} Q_a^{12} \\ Q_a^{12*} \end{pmatrix}, \tag{6.66}$$

with $\mathcal{M}$ the QRPA matrix (5.29). To go further, we recall the expectation value of the constraint associated with the collective variable

$$a = \mathrm{Tr}(Q_a\rho) = \mathrm{Tr}\, Q_a + \frac{1}{2}\mathrm{Tr}\, \mathcal{Q}_a \mathcal{R}(\boldsymbol{a}),$$

which leads to

$$\delta a = \frac{1}{2}\sum_{\mu\nu}\left(Q_{a;\mu\nu}^{12} R_{\mu\nu}^{12*} + Q_{a;\mu\nu}^{12*} R_{\mu\nu}^{12}\right).$$

Introducing the QRPA matrix, we find

$$\delta a = \frac{1}{2}(Q_a^{12*},\ Q_a^{12})\sum_b \delta\lambda_b \mathcal{M}^{-1}\begin{pmatrix} Q_b^{12} \\ Q_b^{12*} \end{pmatrix}.$$

Let us define the matrix of the moments $M^{(K)}$ by

$$M_{ab}^{(K)} = \frac{1}{2}(Q_a^{12*},\ Q_a^{12})\mathcal{M}^{-K}\begin{pmatrix} Q_b^{12} \\ Q_b^{12*} \end{pmatrix}, \tag{6.67}$$

where $\mathcal{M}^{-K} = \mathcal{M}^{-1} \times ... \times \mathcal{M}^{-1}$ K-times. Remember that Q_a^{12} is a block matrix in the q.p. basis, and $\mathcal{M}$ is a matrix in the q.p. basis. It may thus be worth recalling here that when we write

$$\mathcal{M}\begin{pmatrix} X^{12} \\ X^{12*} \end{pmatrix} = \begin{pmatrix} Y^{12} \\ Y^{12*} \end{pmatrix},$$

both X and Y are $n \times n$ matrices, where n is the size of the s.p. basis. Therefore, we have in fact something like

$$\begin{pmatrix} Y_{ij}^{12} \\ Y_{ij}^{12*} \end{pmatrix} = \sum_{\mu\nu} \mathcal{M}_{ij\mu\nu} \begin{pmatrix} X_{\mu\nu}^{12} \\ X_{\mu\nu}^{12*} \end{pmatrix},$$

where the summation runs over all indices μ and ν. Keeping this convention in mind is very important to obtain consistent formulas, and it will be recalled several times later. The variations of the constraint read

$$\delta a = \sum_b M_{ab}^{(1)} \delta\lambda_b.$$

This can be condensed into a matrix-vector notation,

$$\delta \boldsymbol{a} = \boldsymbol{M}^{(1)} \delta\boldsymbol{\lambda} \Rightarrow \delta\boldsymbol{\lambda} = [\boldsymbol{M}^{(1)}]^{-1} \delta\boldsymbol{a} \Rightarrow \delta\lambda_b = \sum_c [M^{(1)}]_{cb}^{-1} \delta c.$$

In terms of variations, $\delta\lambda_b/\delta a = (M^{(1)})_{ab}^{-1}$ since $\delta c/\delta a = \delta_{ac}$ (recall that a, b and c refer to expectation values of independent collective variables). Going back to the QRPA-like equations (6.66), we note that

$$\frac{\partial \mathcal{R}}{\partial a} = \sum_b \frac{\delta \mathcal{R}}{\delta\lambda_b} \frac{\delta\lambda_b}{\delta a} = \sum_b \frac{\mathcal{R}_1}{\delta\lambda_b} \frac{\delta\lambda_b}{\delta a}.$$

Since

$$\mathcal{R}_1 \equiv \begin{pmatrix} R^{12} \\ R^{12*} \end{pmatrix} = \sum_a \delta\lambda_a \mathcal{M}^{-1} \begin{pmatrix} Q_a^{12} \\ Q_a^{12*} \end{pmatrix}$$

we find

$$\frac{\mathcal{R}_1}{\delta\lambda_b} \equiv \begin{pmatrix} \dfrac{\partial \mathcal{R}^{12}}{\partial\lambda_b} \\ \dfrac{\partial \mathcal{R}^{12*}}{\partial\lambda_b} \end{pmatrix} = \sum_a \frac{\delta\lambda_a}{\delta\lambda_b} \mathcal{M}^{-1} \begin{pmatrix} Q_a^{12} \\ Q_a^{12*} \end{pmatrix} = \mathcal{M}^{-1} \begin{pmatrix} Q_b^{12} \\ Q_b^{12*} \end{pmatrix}$$

and therefore we arrive at our final result

$$\frac{\partial \mathcal{R}}{\partial a} \equiv \begin{pmatrix} \dfrac{\partial \mathcal{R}^{12}}{\partial a} \\ \dfrac{\partial \mathcal{R}^{12*}}{\partial a} \end{pmatrix} = \sum_b [M^{(1)}]_{ab}^{-1} \mathcal{M}^{-1} \begin{pmatrix} Q_b^{12} \\ Q_b^{12*} \end{pmatrix}.$$

If we now recall the relation (6.64) between the matrix of the momentum operator and the generalized density, we obtain

$$\begin{pmatrix} P_a^{12} \\ -P_a^{12*} \end{pmatrix} = \sum_b [M^{(1)}]_{ab}^{-1} \mathcal{M}^{-1} \begin{pmatrix} Q_b^{12} \\ Q_b^{12*} \end{pmatrix}. \tag{6.68}$$

Derivatives of overlaps

In order to evaluate the derivatives at point $\boldsymbol{q}$, we use again the Ring and Schuck theorem to express all HFB states as functions of the local HFB state $|\Phi_q\rangle$ at point $\boldsymbol{q}$,

$$|\Phi(\boldsymbol{a})\rangle = e^{i(\boldsymbol{a}-\boldsymbol{q})\hat{\boldsymbol{P}}_q/\hbar}|\Phi(\boldsymbol{q})\rangle.$$

Since $\hat{\boldsymbol{P}}_q$ is Hermitian, the norm kernel is thus given by

$$\mathcal{N}(\boldsymbol{a}, \boldsymbol{a}') = \langle\Phi_q|e^{-i(\boldsymbol{a}-\boldsymbol{q})\hat{\boldsymbol{P}}_q/\hbar}e^{i(\boldsymbol{a}'-\boldsymbol{q})\hat{\boldsymbol{P}}_q/\hbar}|\Phi_q\rangle.$$

Norm kernel

Derivatives with respect to $\boldsymbol{a}$ (or $\boldsymbol{a}'$) are now straightforward and lead to the following results,

$$\left.\frac{\partial^2 \mathcal{N}(\mathbf{a}, \mathbf{a}')}{\partial a_k \partial a'_l}\right|_{\mathbf{a}=\mathbf{a}'=\mathbf{q}} = +\frac{1}{\hbar^2}\langle\Phi_q|\hat{P}_k\hat{P}_l|\Phi_q\rangle \tag{6.69a}$$

$$\left.\frac{\partial^2 \mathcal{N}(\mathbf{a}, \mathbf{a}')}{\partial a_k \partial a'_l}\right|_{\mathbf{a}=\mathbf{a}'=\mathbf{q}} = -\frac{1}{\hbar^2}\langle\Phi_q|\hat{P}_k\hat{P}_l|\Phi_q\rangle. \tag{6.69b}$$

Note that first-order derivatives vanish since $|\Phi_q\rangle$ is a time-even state (by definition), while $\hat{\boldsymbol{P}}_q$ is a time-odd operator. Because the norm is given by equation (6.56), we also find

$$G_{kl} = \frac{1}{\hbar^2}\langle\Phi_q|\hat{P}_k\hat{P}_l|\Phi_q\rangle. \tag{6.70}$$

We now introduce the definition of the momentum in terms of quasiparticle operators. Using the fact that $|\Phi_q\rangle$ is a quasiparticle vacuum and that $G_{ab} = G_{ba}$, it is relatively easy to obtain

$$G_{ab} = \frac{1}{4}\left(P_{a;\mu\nu}^{12*}, -P_{a;\mu\nu}^{12}\right)\begin{pmatrix} P_{b;\mu\nu}^{12} \\ -P_{b;\mu\nu}^{12*} \end{pmatrix}. \tag{6.71}$$

Energy kernel

Owing to the fact that $\mathcal{N}(\boldsymbol{q}, \boldsymbol{q}) = 1$ and that $\mathcal{H}(\boldsymbol{q}, \boldsymbol{q}) = \mathcal{E}(\boldsymbol{q})$, the derivatives of the reduced Hamiltonian read

$$
\begin{aligned}
h_{aa'} &= \frac{\partial^2\mathcal{H}(\boldsymbol{a}, \boldsymbol{a}')}{\partial a_k \partial a_l}\bigg|_{\boldsymbol{a}=\boldsymbol{a}'=\boldsymbol{q}} - \mathcal{E}(\boldsymbol{q})\frac{\partial^2\mathcal{N}(\boldsymbol{a}, \boldsymbol{a}')}{\partial a_k \partial a_l}\bigg|_{\boldsymbol{a}=\boldsymbol{a}'=\boldsymbol{q}} \\
h_{aa'} &= \frac{\partial^2\mathcal{H}(\boldsymbol{a}, \boldsymbol{a}')}{\partial a_k \partial a'_l}\bigg|_{\boldsymbol{a}=\boldsymbol{a}'=\boldsymbol{q}} - \mathcal{E}(\boldsymbol{q})\frac{\partial^2\mathcal{N}(\boldsymbol{a}, \boldsymbol{a}')}{\partial a_k \partial a'_l}\bigg|_{\boldsymbol{a}=\boldsymbol{a}'=\boldsymbol{q}}.
\end{aligned}
$$

The partial derivatives of the energy kernel are a little more involved. First, it is important to note that differentiation should be performed *first* on $\boldsymbol{a}$ and $\boldsymbol{a}'$ and only then should we take the value of these derivatives at point $\boldsymbol{q}$. Since we want to express everything as a function of the characteristics at point $\boldsymbol{q}$, it is useful to use the Ring and Schuck theorem again. We find immediately

$$
\begin{aligned}
\frac{\partial^2\mathcal{H}(\boldsymbol{a}, \boldsymbol{a}')}{\partial a_k \partial a'_l}\bigg|_{\boldsymbol{a}=\boldsymbol{a}'=\boldsymbol{q}} &= +\frac{1}{\hbar^2}\langle\phi_{\boldsymbol{q}}|\hat{P}_k\hat{H}\hat{P}_l|\phi_{\boldsymbol{q}}\rangle \\
\frac{\partial^2\mathcal{H}(\boldsymbol{a}, \boldsymbol{a})}{\partial a_k \partial a_l}\bigg|_{\boldsymbol{a}=\boldsymbol{a}'=\boldsymbol{q}} &= -\frac{1}{\hbar^2}\langle\phi_{\boldsymbol{q}}|\hat{P}_k\hat{P}_l\hat{H}|\phi_{\boldsymbol{q}}\rangle.
\end{aligned}
$$

Using the definition of the operator $\boldsymbol{P}$ in terms of quasiparticle operators, we find for the first term

$$
\begin{aligned}
h_{aa'} = \frac{\partial^2 h(\boldsymbol{a}, \boldsymbol{a}')}{\partial a_k \partial a'_l}\bigg|_{\boldsymbol{a}=\boldsymbol{a}'=\boldsymbol{q}} &= \sum_{i<j,\mu<\nu} P^{12*}_{k,\,ij}P^{12}_{l,\,\mu\nu}\langle\phi_{\boldsymbol{q}}|\beta_j\beta_i\hat{H}\beta^\dagger_\mu\beta^\dagger_\nu|\phi_{\boldsymbol{q}}\rangle \\
&- \sum_{i<j,\mu<\nu} P^{12*}_{k,\,ij}P^{12}_{l,\,\mu\nu}\langle\phi_{\boldsymbol{q}}|\beta_j\beta_i\beta^\dagger_\mu\beta^\dagger_\nu|\phi_{\boldsymbol{q}}\rangle\mathcal{E}(\boldsymbol{q}).
\end{aligned}
$$

Notice the minus sign that appears in the second line. At this point, it is important to recall that the components of the QRPA matrix can be written

$$
\begin{aligned}
A_{ij\mu\nu} &= \langle\phi_{\boldsymbol{q}}|\beta_j\beta_i[\hat{H} - \mathcal{E}(\boldsymbol{q})]\beta^\dagger_\mu\beta^\dagger_\nu|\phi_{\boldsymbol{q}}\rangle \\
B_{ij\mu\nu} &= \langle\phi_{\boldsymbol{q}}|\beta_j\beta_i\beta_\nu\beta_\mu\hat{H}|\phi_{\boldsymbol{q}}\rangle,
\end{aligned}
\tag{6.72}
$$

so that we have

$$
h_{aa'} = \sum_{i<j,\mu<\nu} P^{12*}_{k,\,ij}P^{12}_{l,\,\mu\nu}A_{ij\mu\nu}. \tag{6.73}
$$

The calculation of the second term is done similarly. We first find

$$
\begin{aligned}
h_{aa} = \frac{\partial^2 h(\boldsymbol{a}, \boldsymbol{a})}{\partial a_k \partial a_l}\bigg|_{\boldsymbol{a}=\boldsymbol{a}'=\boldsymbol{q}} &= -\sum_{i<j,\mu<\nu} P^{12*}_{k,\,ij}P^{12}_{l,\,\mu\nu}\langle\phi_{\boldsymbol{q}}|\beta_j\beta_i\beta^\dagger_\mu\beta^\dagger_\nu\hat{H}|\phi_{\boldsymbol{q}}\rangle \\
&- \sum_{i<j,\mu<\nu} P^{12*}_{k,\,ij}P^{21}_{l,\,\mu\nu}\langle\phi_{\boldsymbol{q}}|\beta_j\beta_i\beta_\mu\beta_\nu\hat{H}|\phi_{\boldsymbol{q}}\rangle \\
&+ \sum_{i<j,\mu<\nu} P^{12*}_{k,\,ij}P^{12}_{l,\,\mu\nu}\langle\phi_{\boldsymbol{q}}|\beta_j\beta_i\beta^\dagger_\mu\beta^\dagger_\nu|\phi_{\boldsymbol{q}}\rangle\mathcal{E}(\boldsymbol{q}).
\end{aligned}
$$

In this expression, the last term corresponds to $\mathcal{E}(\boldsymbol{q})\partial^2\mathcal{N}/\partial a_k \partial a_l$, which is equal to $-G_{kl}$ as shown by combining equations (6.69b) and (6.70). The first and second terms come simply from applying the definition of the momentum operator. Note the presence of the first term, which is often forgotten in derivations, but essential: by bringing the creation operator to the left and using the fact that $\langle \Phi_{\boldsymbol{q}} | \beta_n^\dagger = 0, \forall n$, we can show easily that it exactly cancels the third term. Therefore, and given the definition (6.72), we find

$$h_{aa} = - \sum_{i<j,\mu<\nu} P^{12*}_{k,\,ij} P^{21}_{l,\,\mu\nu} B_{ij\nu\mu}. \tag{6.74}$$

Now, we use the fact that $A_{ij\mu\nu} = A^*_{\mu\nu ij}$ (the same for B), and also that $B_{ij\mu\nu} = -B_{ij\nu\mu}$. These relations are immediate from the definitions of these matrices. In addition, we use the fact that $P^{12}_{k,\,ij} = -P^{21*}_{k,\,ij}$, which is also straightforward to obtain from the hermiticity of the momentum operator. Finally (this is important), we also remove the restrictions on the summation indices, which adds an extra factor 1/2. All these manipulations lead to

$$h_{aa'} - h_{aa} = \frac{1}{4}\begin{pmatrix} P^{12*}_k, & P^{12}_k \end{pmatrix} \begin{pmatrix} A & B \\ B^* & A^* \end{pmatrix} \begin{pmatrix} P^{12}_l \\ P^{12*}_l \end{pmatrix}. \tag{6.75}$$

Closed form expression for the collective Hamiltonian

We can arrive at a closed-form expression for the metric, the collective inertia tensor and the zero-point by introducing equation (6.68) into (6.71) (metric), (6.75) (inertia tensor). This leads to

$$\boldsymbol{G} = \frac{1}{2}[\boldsymbol{M}^{(1)}]^{-1}\boldsymbol{M}^{(2)}[\boldsymbol{M}^{(1)}]^{-1} \tag{6.76}$$

for the metric. For the inertia tensor, we first need to define the matrix $\tilde{\mathcal{M}}$ by

$$\tilde{\mathcal{M}} = \begin{pmatrix} A & -B \\ -B^* & A^* \end{pmatrix} = \begin{pmatrix} 1 & 0 \\ 0 & -1 \end{pmatrix} \mathcal{M} \begin{pmatrix} 1 & 0 \\ 0 & -1 \end{pmatrix}$$

and we then find

$$h_{aa'} - h_{aa} = \frac{1}{2}[\boldsymbol{M}^{(1)}]^{-1}\tilde{\boldsymbol{M}}^{(1)}[\boldsymbol{M}^{(1)}]^{-1}.$$

Combining this with equation (6.76) and the definition (6.60), we find

$$\boldsymbol{B} = \boldsymbol{M}^{(1)}[\boldsymbol{M}^{(2)}]^{-1}\tilde{\boldsymbol{M}}^{(1)}[\boldsymbol{M}^{(2)}]^{-1}\boldsymbol{M}^{(1)}.$$

Looking more closely at the terms involved in the collective inertia, we find that they only depend on local quantities at point $\boldsymbol{q}$. Once the HFB solution is known at this point, computing the moments is rather trivial, and both the GCM metric G and the collective inertia tensor B only depend on them.

6.4.4 Cranking approximation

At the cranking approximation of the QRPA matrix, we have by definition $\mathcal{M}_{ij\mu\nu} = \delta_{i\mu}\delta_{j\nu}(E_\mu + E_\nu)$. Therefore, the moments $\boldsymbol{M}^{(K)}$ take the simpler form

$$M_{ab}^{(K)} = \frac{1}{2}\sum_{\mu\nu} Q_{a,\mu\nu}^{12*}\frac{1}{E_\mu + E_\nu}Q_{b,\mu\nu}^{12} + \frac{1}{2}\sum_{\mu\nu} Q_{a,\mu\nu}^{12}\frac{1}{E_\mu + E_\nu}Q_{b,\mu\nu}^{12*}.$$

Also recall that by definition $Q_{a,\mu\nu}^{12}$ is the upper right corner of the matrix Q of the constraint operator $\hat{Q}_a$ in the q.p. basis. Hence, it corresponds to

$$Q_{a,\mu\nu}^{12} = \langle\mu\nu|\hat{Q}_a|0\rangle,$$

where $|0\rangle = |\Phi(\boldsymbol{q})\rangle$ is the q.p. vacuum and $|\mu\nu\rangle$ is a two-q.p. excitation. This property can easily be obtained from the definition of the $\hat{Q}_a$ operator,

$$\hat{Q}_a = \sum_{\mu\nu} Q_{a,\mu\nu}^{12}\beta_\mu^\dagger\beta_\nu^\dagger - \sum_{\mu\nu} Q_{a,\mu\nu}^{21}\beta_\mu\beta_\nu,$$

by examining the action of such an operator on the vacuum. Similarly, $Q_{a,\mu\nu}^{21} = \langle 0|\hat{Q}_a|\mu\nu\rangle$ so that we have

$$\begin{aligned} M_{ab}^{(K)} = &\frac{1}{2}\sum_{\mu\nu}\langle\mu\nu|\hat{Q}_a|0\rangle\frac{1}{(E_\mu + E_\nu)^K}\langle 0|\hat{Q}_b|\mu\nu\rangle \\ &+ \frac{1}{2}\sum_{\mu\nu}\langle 0|\hat{Q}_a|\mu\nu\rangle\frac{1}{(E_\mu + E_\nu)^K}\langle\mu\nu|\hat{Q}_b|0\rangle. \end{aligned}$$

An alternative form for the moments often given in the literature is thus

$$M_{ab}^{(K)} = \mathfrak{Re}\sum_{\mu\nu}\frac{\langle\mu\nu|\hat{Q}_a|0\rangle\langle 0|\hat{Q}_b|\mu\nu\rangle}{(E_\mu + E_\nu)^K}.$$

In addition, the matrix $\tilde{\mathcal{M}} = \mathcal{M}$ at this approximation, hence $\tilde{\boldsymbol{M}}^{(K)} = \boldsymbol{M}^{(K)}$ as well. The inertia tensor simplifies into

$$\boldsymbol{B} = \boldsymbol{M}^{(1)}[\boldsymbol{M}^{(2)}]^{-1}\boldsymbol{M}^{(1)}[\boldsymbol{M}^{(2)}]^{-1}\boldsymbol{M}^{(1)}.$$

An alternative expression of the inertia tensor involving the metric is

$$\boldsymbol{B} = \frac{1}{4}\boldsymbol{G}^{-1}[\boldsymbol{M}^{(1)}]^{-1}\boldsymbol{G}^{-1}. \quad (6.77)$$

These two formulas give what is known as the Yoccoz formula for the collective mass [68]. It differs from the Thouless–Valatin collective mass, which is the one given by the ATDHFB theory [16, 69].

6.4.5 Problems in the GCM+GOA approach

The most appealing feature of the GOA is its numerical simplicity. The collective Hamiltonian does not require anything more than a set of HFB calculations for a range of collective variables $\boldsymbol{q}$, since everything can be expressed as a function of the moments $\boldsymbol{M}^{(K)}$. In addition, the GOA bypasses the formal problems of the GCM caused by density dependencies in the nuclear Hamiltonian.

However, the GOA is an approximation of the GCM which is based on the superposition principle of quantum mechanics with configuration mixing. In this sense, even if the starting point is fully quantum mechanical, and the GOA provides the zero-point-correction term for the potential, $V_{\mathrm{ZPE}}(q)$, one should remain cautious. The superposition is determined by off-diagonal matrix elements, or transition amplitudes, of the nuclear Hamiltonian. These off-diagonal elements are still ill-defined for conventional EDF, even if the numerical divergences observed in applying the exact GCM do not appear [56–59, 70].

It is known that the Yoccoz inertia does not reproduce the total mass for the particular case of translational motion, where the collective coordinate is the displacement of the center-of-mass [69]. In fact, the GOA inertial mass is even smaller than the Inglis–Belyaev cranking mass which is too small for realistic applications. This is another disadvantage of the GOA+GCM. In order to reproduce the exact mass for translational motion, it is necessary to adopt both $\boldsymbol{q}$ (collective coordinate) and $\boldsymbol{p}$ (conjugate collective momentum) as generator coordinates. This implies that the GCM wave function is not invariant with respect to time-reversal symmetry [71]. In contrast, both the ATDHFB and ASCC theories (see equation (6.51)) reproduce the exact translational mass, even though the collective subspace Σ_K is spanned by only time-even states defined at $\boldsymbol{p} = 0$. This is because both theories are derived from the full TDDFT dynamics and takes into account time-odd components in wave functions. Since the GOA inertial mass (6.60) is known to have this problem, the application of the GOA is often tweaked by replacing the GOA collective mass by the ATDHFB in the collective Hamiltonian.

Another difficulty with the GCM+GOA is the choice of the collective coordinates. We have shown in section 6.3 that the ASCC provide a method to determine the optimal collective coordinates. In the GCM, the variational principle dictates that we should also include the variation with respect to the path $|\boldsymbol{q}\rangle$ itself. This is known as the double variational method [72]. However, in nuclear physics, we often start from an EDF based on an effective pseudo-Hamiltonian with zero-range, two-body pseudo-potentials, $\propto v_0\delta(\boldsymbol{r}_1 - \boldsymbol{r}_2)$. In this case, the exact many-body ground state does not exist and the double variation may lead to a singular solution. This is a conceptual problem of the GCM+GOA.

References

[1] Negele J W 1982 The mean-field theory of nuclear structure and dynamics *Rev. Mod. Phys.* **54** 913–1015

[2] Bohr A and Mottelson B 1998 *Nuclear Deformations* Nuclear Structure vol 2 (Singapore: World Scientific)

[3] Runge E and Gross E K U 1984 Density-functional theory for time-dependent systems *Phys. Rev. Lett.* **52** 997–1000
[4] van Leeuwen R 1999 Mapping from densities to potentials in time-dependent density-functional theory *Phys. Rev. Lett.* **82** 3863–6
[5] Do Dang G u, Klein A and Walet N R 2000 Self-consistent theory of large-amplitude collective motion: applications to approximate quantization of nonseparable systems and to nuclear physics *Phys. Rep.* **335** 93–274
[6] Blaizot J-P and Ripka G 1985 *Quantum Theory of Finite Systems* (Cambridge, MA: MIT Press)
[7] Baranger M and Kumar K 1965 Nuclear deformations in the pairing-plus-quadrupole model: (I). The single-*j* shell *Nucl. Phys.* **62** 113–32
[8] Baranger M and Kumar K 1968 Nuclear deformations in the pairing-plus-quadrupole model: (II). Discussion of validity of the model *Nucl. Phys.* A **110** 490–528
[9] Kumar K and Baranger M 1968 Nuclear deformations in the pairing-plus-quadrupole model: (III). Static nuclear shapes in the rare-earth region *Nucl. Phys.* A **110** 529–54
[10] Baranger M and Kumar K 1968 Nuclear deformations in the pairing-plus-quadrupole model: (VI). Theory of collective motion *Nucl. Phys.* A **122** 241–72
[11] Kumar K and Baranger M 1968 Nuclear deformations in the pairing-plus-quadrupole model: (V). Energy levels and electromagnetic moments of the W, Os and Pt nuclei *Nucl. Phys.* A **122** 273–324
[12] Krieger S J and Goeke K 1974 Application of the adiabatic, time-dependent Hartree–Bogolyubov approximation to a solvable model *Nucl. Phys.* A **234** 269
[13] Brink D M, Giannoni M J and Veneroni M 1976 Derivation of an adiabatic time-dependent Hartree–Fock formalism from a variational principle *Nucl. Phys.* A **258** 237–56
[14] Giannoni M J *et al* 1976 A method for calculating adiabatic mass parameters: application to isoscalar quadrupole modes in light nuclei *Phys. Lett.* B **65** 305
[15] Villars F 1977 Adiabatic time-dependent Hartree–Fock theory in nuclear physics *Nucl. Phys.* A **285** 269
[16] Baranger M and Vénéroni M 1978 An adiabatic time-dependent Hartree–Fock theory of collective motion in finite systems *Ann. Phys.* **114** 123–200
[17] Goeke K and Reinhard P-G 1978 A consistent microscopic theory of collective motion in the framework of an ATDHF approach *Ann. Phys.* **112** 328
[18] Bonche P and Quentin P 1978 Adiabaticity of time-dependent Hartree–Fock solutions *Phys. Rev.* C **18** 1891
[19] Reinhard P-G and Goeke K 1978 The concept of a collective path and its range of validity *Nucl. Phys.* A **312** 121
[20] Giannoni M J and Quentin P 1980 Mass parameters in the adiabatic time-dependent Hartree–Fock approximation. I. Theoretical aspects; the case of a single collective variable *Phys. Rev.* C **21** 2060
[21] Giannoni M J and Quentin P 1980 Mass parameters in the adiabatic time-dependent Hartree–Fock approximation. II. Results for the isoscalar quadrupole mode *Phys. Rev.* C **21** 2076
[22] Mukherjee A K and Pal M K 1981 Analytical proof of the non-uniqueness of the ATDHF path *Phys. Lett.* B **100** 457
[23] Mukherjee A K and Pal M K 1982 Evaluation of the optimal path in ATDHF theory *Nucl. Phys.* A **373** 289

[24] Yuldashbaeva E Kh *et al* 1999 Mass parameters for large amplitude collective motion: a perturbative microscopic approach *Phys. Lett.* B **461** 1
[25] Schunck N and Robledo L M 2016 Microscopic theory of nuclear fission: a review *Rep. Prog. Phys.* **79** 116301
[26] Reinhard P G, Maruhn J and Goeke K 1980 Adiabatic time-dependent Hartree–Fock calculations of the optimal path, the potential, and the mass parameter for large-amplitude collective motion *Phys. Rev. Lett.* **44** 1740–3
[27] Nakatsukasa T *et al* 2016 Time-dependent density-functional description of nuclear dynamics *Rev. Mod. Phys.* **88** 045004
[28] Mukherjee A K and Pal M K 1982 Evaluation of the optimal path in ATDHF theory *Nucl. Phys.* A **373** 289–304.
[29] Klein A, Walet N R and Do Dang G 1991 Classical theory of collective motion in the large amplitude, small velocity regime *Ann. Phys.* **208** 90
[30] Giannoni M J 1984 An introduction to the adiabatic time-dependent Hartree–Fock method *Nucl. Phys.* A **428** 63
[31] Goeke K, Reinhard P-G and Rowe D J 1981 A study of collective paths in the time-dependent Hartree–Fock approach to large amplitude collective nuclear motion *Nucl. Phys.* A **359** 408
[32] Marumori T *et al* 1980 Self-consistent collective-coordinate method for the large-amplitude nuclear collective motion *Prog. Theor. Phys.* **64** 1294–314
[33] Yamamura M and Kuriyama A 1987 Time-dependent Hartree–Fock method and its extension *Prog. Theor. Phys. Suppl.* **93** 1–175
[34] Goeke K, Reinhard P-G and Reinhardt H 1982 The generator coordinate method, path integrals, and quantized time-dependent mean field motion *Phys. Lett.* B **118** 1–4
[35] Goeke K, Grümmer F and Reinhard P-G 1983 Three-dimensional nuclear dynamics in the quantized ATDHF approach *Ann. Phys.* **150** 504–51
[36] Matsuo M, Nakatsukasa T and Matsuyanagi K 2000 Adiabatic selfconsistent collective coordinate method for large amplitude collective motion in superconducting nuclei *Prog. Theor. Phys.* **103** 959–79
[37] Borisenko A I and Tarapov I E 1968 *Vector and Tensor Analysis with Applications* (New York: Dover)
[38] Landau L D and Lifshitz E M 1980 *The Classical Theory of Fields* 4th edn (Oxford: Butterworth-Heinemann)
[39] Hinohara N *et al* 2007 Gauge-invariant formulation of adiabatic self-consistent collective coordinate method *Prog. Theor. Phys.* **117** 451–78
[40] Hinohara N *et al* 2008 Microscopic derivation of collective Hamiltonian by means of the adiabatic self-consistent collective coordinate method *Prog. Theor. Phys.* **119** 59–101
[41] Hinohara N *et al* 2009 Microscopic description of oblate-prolate shape mixing in proton-rich Se isotopes *Phys. Rev.* C **80** 014305
[42] de Voigt M J A, Dudek J and Szymański Z 1983 High-spin phenomena in atomic nuclei *Rev. Mod. Phys.* **55** 949
[43] Nilsson S G and Ragnarsson I 1995 *Shapes and Shells in Nuclear Structure* (Cambridge: Cambridge University Press)
[44] Ullrich C A 2012 *Time-Dependent Density Functional Theory: Concepts and Applications* (New York: Oxford University Press)
[45] Griffin J J and Wheeler J A 1957 Collective motions in nuclei by the method of generator coordinates *Phys. Rev.* **108** 311

[46] Verrière M 2017 *Description de la dynamique de la fission dans le formalisme de la méthode de la coordonnée génératrice dépendante du temps* PhD thesisUniversité Paris-Saclay https://tel.archives-ouvertes.fr/tel-01559158
[47] Levit S, Negele J W and Paltiel Z 1980 Barrier penetration and spontaneous fission in the time-dependent mean-field approximation *Phys. Rev.* C **22** 1979
[48] Levit S, Negele J W and Paltiel Z 1980 Time-dependent mean-field theory and quantized bound states *Phys. Rev.* C **21** 1603–25
[49] Levit S 1980 Time-dependent mean-field approximation for nuclear dynamical problems *Phys. Rev.* C **21** 1594–602
[50] Negele J W 1989 Microscopic theory of fission dynamics *Nucl. Phys.* A **502** 371
[51] Bonche P, Koonin S and Negele J W 1976 One-dimensional nuclear dynamics in the time-dependent Hartree–Fock approximation *Phys. Rev.* C **13** 1226
[52] Wen K and Nakatsukasa T 2016 Self-consistent collective coordinate for reaction path and inertial mass *Phys. Rev.* C **94** 054618
[53] Hinohara N *et al* 2010 Microscopic description of large-amplitude shape-mixing dynamics with inertial functions derived in local quasiparticle random-phase approximation *Phys. Rev.* C **82** 064313
[54] Ring P and Schuck P 2004 *The Nuclear Many-Body ProblemTexts and Monographs in Physics* (Berlin: Springer)
[55] Goeke K and Reinhard P-G 1980 The generator-coordinate-method with conjugate parameters and the unification of microscopic theories for large amplitude collective motion *Ann. Phys.* **124** 249
[56] Stoitsov M V *et al* 2007 Variation after particle-number projection for the Hartree–Fock–Bogoliubov method with the Skyrme energy density functional *Phys. Rev.* C **76** 014308
[57] Duguet T *et al* 2009 Particle-number restoration within the energy density functional formalism: nonviability of terms depending on noninteger powers of the density matrices *Phys. Rev.* C **79** 044320
[58] Bender M, Duguet T and Lacroix D 2009 Particle-number restoration within the energy density functional formalism *Phys. Rev.* C **79** 044319
[59] Lacroix D, Duguet T and Bender M 2009 Configuration mixing within the energy density functional formalism: removing spurious contributions from nondiagonal energy kernels *Phys. Rev.* C **79** 044318
[60] Brink D M and Weiguny. A 1968 The generator coordinate theory of collective motion *Nucl. Phys.* A **120** 59
[61] Onishi N and Une T 1975 Local Gaussian approximation in the generator coordinate method *Prog. Theor. Phys.* **53** 504
[62] Banerjee B and Brink D M 1973 A Schrödinger equation for collective motion from the generator coordinate method *Z. Phys.* A 46
[63] Góźdź A 1985 An extended Gaussian overlap approximation in the generator coordinate method *Phys. Lett.* B **152** 281
[64] Hofmann H and Dietrich K 1971 Effects of variable inertia on collective dynamics *Nucl. Phys.* A **165** 1
[65] Góźdź A *et al* 1985 The mass parameters for the average mean-field potential *Nucl. Phys.* A **442** 26
[66] Berger J-F 1985 *Approche microscopique auto-consistante des processus nucléaires collectifs de grande amplitude à basse énergie. Application à la diffusion d'ions lourds et à la fission* PhD thesisCentre d'Orsay: Université Paris-Sud

[67] Ring P and Schuck P 1977 On the decomposition of the single-particle density in a time-even and a time-odd part *Nucl. Phys.* A **292** 20
[68] Peierls R E and Yoccoz J 1957 The collective model of nuclear motion *Proc. Phys. Soc.* A **70** 381
[69] Peierls R E and Thouless D J 1962 Variational approach to collective motion *Nucl. Phys.* **38** 154
[70] Anguiano M, Egido J L and Robledo L M 2001 Particle number projection with effective forces *Nucl. Phys.* A **696** 467
[71] Peierls R E and Thouless D J 1962 Variational approach to collective motion *Nucl. Phys.* **38** 154–76
[72] Holzwarth G and Yukawa T 1974 Choice of the constraining operator in the constrained Hartree-Fock method *Nucl. Phys.* A **219** 125–40

IOP Publishing

Energy Density Functional Methods for Atomic Nuclei

Nicolas Schunck

Chapter 7

Finite temperature

Nicolas Schunck

The linear response theory within either the random-phase approximation (RPA) or quasiparticle random-phase approximation (QRPA) frameworks, as presented in chapter 5, is the tool of choice to compute properties of low-lying excited states and/or transitions between such states. As the derivation of the RPA shows, it builds excited states by a variational exploration of particle–hole (p.h.) configurations around the ground state of a nucleus; the QRPA is simply the extension of this approach to superfluid systems, where p.h. configurations are replaced with quasiparticle (q.p.) configurations. The RPA and QRPA are examples of the vertical philosophy for nuclear structure discussed in section 1.1.3.

In contrast, the various methods to tackle the large-amplitude collective motion of a nucleus, discussed in chapter 6, are all based on the adiabatic exploration of some collective space—typically defined by a few deformations. While the energy of, say, the generator states $|\phi(\boldsymbol{q})\rangle$ in the generator coordinate method (GCM) + Gaussian overlap approximation (GOA), may be substantially higher than that of the ground state, only states with the lowest energy for a given value of $\boldsymbol{q}$ are often considered in practice. In general, these approaches are designed to describe mostly a very restricted class of excited states, the pure collective states of lowest energy. There are few exceptions such as [1], where the GCM+GOA formalism is extended to include two-q.p. excitations; [2], where several blocking configurations in odd-mass nuclei are used as generator states for the direct GCM; and [3] where collective rotational bands are described within the projected GCM by including states at non-zero rotational frequency as generators

As the excitation energy U of the nucleus increases, the level density grows exponentially. For example, in the Fermi gas model of the nucleus, nucleons are confined in a box and their wave functions are plane waves whose wave number $\boldsymbol{k}$ is determined by box boundary conditions. In these conditions, the density of states scales with the excitation energy U according to $\rho(U) \propto \exp(\sqrt{U})/U^{5/4}$; see [4] for detailed derivations. It follows that the very concept of discrete, intrinsic states loses its meaning. As a result, we should expect linear response theory to break down. Similarly, the few attempts to extend the scope of collective theories to higher excitation energies by considering generator states with various types of intrinsic excited states quickly become impractical as the number of individual configurations to track grows out of control. In such cases, it may be more relevant to adopt a statistical treatment of the nucleus. The goal of this chapter is, therefore, to present the extension of the Hartree–Fock–Bogoliubov (HFB) theory of chapter 3 at finite temperature. This is achieved by constructing suitable approximations of the statistical density operator used in statistical quantum mechanics. The material presented in this section is inspired to a large extent by the textbooks [5, 6] and a few particularly relevant articles [7, 8].

7.1 A reminder of statistical quantum mechanics

Before describing in more detail the finite-temperature extension of the Hartree–Fock (HF) and HFB theories, it may be relevant to recall a few important aspects of standard statistics. We will keep this presentation very simple by only discussing some very basic concepts of thermodynamics and recalling the definition of a partition function. An excellent textbook of (mostly) classical statistical mechanics is by Reichl [9].

7.1.1 Basic concepts of thermodynamics

The thermodynamic state of the nucleus is described by a certain number of 'state variables'. Some of them are 'extensive', i.e. they are directly proportional to the size of the system (volume and quantity of matter). Traditional examples of extensive variables are the volume V, the number of particles A and the entropy S. All other extensive variables will be denoted generically by the label X. An (important) example of such extensive variables in the energy density functional (EDF) approach of atomic nuclei is the expectation value of multipole moment operators $\langle \hat{Q}_{\ell m} \rangle$. To every extensive variable, we can associate the conjugate 'intensive' variable, which does not explicitly depend on the size of the system. Table 7.1 lists the relevant extensive and intensive variables for the thermodynamic description of a nucleus in the HFB framework.

There are traditionally three types of thermodynamic systems: (i) 'isolated' systems cannot exchange heat nor matter with the exterior, (ii) 'closed' systems

Table 7.1. List of conjugate variables in thermodynamics.

Extensive variable		Intensive variable	
V	Volume	P	Pressure
A	Number of particles	λ_F	Fermi energy
S	Entropy	T	Temperature
$\langle \hat{Q}_{\ell m} \rangle$	Multipole moment	$\lambda_{\ell m}$	Lagrange multipliers

Table 7.2. List of thermodynamic potentials, the corresponding independent state variables and the type of thermodynamic processes they apply to.

Potential	Variables	Processes
Internal energy	$E(V, S, X)$	Constant V, S, X
Helmholtz free energy	$F(V, T, X)$	Constant V, T, X
Enthalpy	$H(P, S, X)$	Constant P, S, X
Gibbs free energy	$G(P, T, X)$	Constant P, T, X
Grand potential	$\Omega(V, T, \lambda_F, X)$	Constant V, T, λ_F, X

may exchange heat, but not matter, and (iii) 'open' systems may exchange both heat and matter with the environment. In principle, a nucleus should be treated as an open quantum system; gamma-ray emission, for example, may cool down an excited nucleus, while coupling to the continuum is equivalent to exchanging particles. In the framework of the HFB theory, however, we will assume the nucleus is an isolated system. Furthermore, we suppose it is at 'thermal equilibrium', i.e. the value of state variables do not change with time.

Not all state variables that may describe the system are independent of one another. Equations of state relate such dependent variables. For each set of independent variables, we can associate a thermodynamic potential. Table 7.2 gives the five most important thermodynamic potentials. In nuclear physics, the volume V of the nucleus is constant. The most useful thermodynamic potentials are therefore the total internal energy E, the Helmholtz free energy F and the grand potential Ω. The latter should only be used when the nucleus is treated as an open quantum system. In practice, this hypothesis is made only to derive the HFB equation at finite temperature.

7.1.2 Partition functions

In statistical quantum mechanics, the expectation values of operators are calculated through the use of the statistical density operator $\hat{D}$; see, for example, the textbook [10] by N Bogoliubov, which gives an excellent introduction to some basic concepts of quantum statistical mechanics. In particular, it shows how the concept of statistical density operator naturally emerges when modeling a quantum system with many degrees of freedom. For arbitrary quantum many-body system characterized by a Hamiltonian $\hat{H}$, the density operator is found by minimization of the grand potential,

$$\Omega = E - TS - \lambda_F N. \tag{7.1}$$

The grand potential involves the internal energy E, the entropy S and the particle number N. They are given by the following relations

$$E = \mathrm{Tr}\,[\hat{D}\hat{H}] \qquad S = -k\,\mathrm{Tr}\,[\hat{D}\ln\hat{D}] \qquad N = \mathrm{Tr}\,[\hat{D}\hat{N}]. \tag{7.2}$$

Here, $\hat{H}$ is the true Hamiltonian of the system and $\hat{N} = \sum_i c_i^\dagger c_i$ the particle-number operator. The trace is taken on a convenient basis of the Fock space, to be determined. The entropy involves the logarithm of the density operator. At this stage, nothing has been assumed about the form of the Hamiltonian, or of the statistical density operator, which remains undetermined.

Because of equation (7.2), the grand potential Ω is obviously a functional of the density operator, $\Omega = \Omega[\hat{D}]$, and it reaches a minimum at equilibrium. We can use this condition to obtain an expression for the statistical density operator $\hat{D}$. More precisely, we will seek the operator $\hat{D}$ such that (i) $\mathrm{Tr}\,\hat{D} = 1$ and (ii) Ω is minimal. These two conditions can be satisfied simultaneously by minimizing $\Omega' = \Omega - \lambda(\mathrm{Tr}\,\hat{D} - 1)$, with λ a Lagrange multiplier to be determined. Such variations read

$$\delta\Omega' = \Omega(\hat{D} + \delta\hat{D}) - \Omega(\hat{D}) - \lambda\,\mathrm{Tr}\,\delta\hat{D},$$

which yields

$$\delta\Omega' = \mathrm{Tr}\left[\left(\hat{H} - \lambda_{\mathrm{F}}\hat{N}\right)\delta\hat{D} + kT\left(\hat{D} + \delta\hat{D}\right)\left(\ln\hat{D} + \ln\left(1 + \hat{D}\delta\hat{D}\right)\right) - \lambda\,\mathrm{Tr}\,\delta\hat{D}\right].$$

The exponential of a linear operator $\hat{O}$ such as the density operator is defined simply as $\exp\hat{O} = \sum_n \hat{O}^n/n!$. The logarithm $\hat{L}_O$ of the corresponding operator can then be defined implicitly by the requirement that $\hat{L}_O(\exp\hat{O}) = \hat{O}$. Note that determining if an arbitrary operator has a logarithm is not a trivial mathematical problem; see [11].

In the limit $\delta\hat{D} \to 0$, and retaining only the terms up to first order in $\delta\hat{D}$, the previous expression reduces to

$$\mathrm{Tr}\,[\hat{H} - \lambda_{\mathrm{F}}\hat{N} + kT\ln\hat{D} - \lambda] = 0.$$

Formally, this is solved by choosing

$$\hat{D} \equiv \hat{D}_{\mathrm{eq}} = \frac{1}{Z}e^{-\beta\left(\hat{H} - \lambda_{\mathrm{F}}\hat{N}\right)}, \tag{7.3}$$

where $\beta = 1/kT$ and Z is the coefficient that makes the density operator satisfy $\mathrm{Tr}\,\hat{D}_{\mathrm{eq}} = 1$. It is the partition function given by

$$Z = \mathrm{Tr}\left[^{-\beta\left(\hat{H} - \lambda_{\mathrm{F}}\hat{N}\right)}\right]. \tag{7.4}$$

At equilibrium, the value of the grand potential is

$$\Omega[\hat{D}] = \frac{1}{\beta}\ln Z \tag{7.5}$$

as can be computed easily by substituting equation (7.3) into the definition of the grand potential.

7.2 Finite-temperature Hartree–Fock theory

In equations (7.3) and (7.4), the Hamiltonian $\hat{H}$ is the 'true' Hamiltonian of the system. For the time being, we will only assume it is some effective, two-body pseudo-Hamiltonian. Although all demonstrations will be given assuming an effective pseudo-Hamiltonian, the general EDF approach does not require one as discussed in section 1.2. We will see later how to adapt our results to the case of an arbitrary energy density. In the 'mean-field' approximation of the density operator, we choose $\hat{D}$ as a quadratic form of c_i and $c_i^\dagger$. More specifically, in the HF mean-field approximation, we set

$$\hat{D}_{\rm HF} = \frac{1}{Z_{\rm HF}} e^{\beta \hat{K}} \qquad Z_{\rm HF} = {\rm Tr}[e^{-\beta \hat{K}}], \tag{7.6}$$

with

$$\hat{K} = \sum_{ij} K_{ij} c_i^\dagger c_j. \tag{7.7}$$

In other words, $\hat{K}$ is a one-body operator.

Statistical Wick theorem

The Wick theorem was introduced briefly in section 3.1.2 in the context of product states $|\Phi\rangle$, such as Slater determinants or HFB vacua. It is used to facilitate the calculation of expectation values of arbitrary operators onto such product states. The Wick theorem can be generalized to ensemble averages [12], where we now aim to compute the statistical average value of an operator $\hat{A}$ given a particular form of the statistical density operator. When the statistical density operator is of the form (7.7), the statistical Wick theorem states that there is a one-to-one correspondence between the density matrix ρ_{ij}, defined by $\rho_{ij} = {\rm Tr}\left[\hat{D}_{\rm HF} c_j^\dagger c_i\right]$, and the operator $\hat{K}$. Formally, we can write

$$\hat{\rho} = \frac{1}{1 + e^{\beta \hat{K}}}. \tag{7.8}$$

This implies, among other things, that in any given single-particle (s.p.) basis of the Hilbert space, variations of $\hat{\rho}$ and variations of $\hat{K}$ are equivalent. In addition, statistical traces can be computed as traces in the s.p. space

$${\rm Tr}[\hat{D}_{\rm HF}\hat{F}] = {\rm tr}[\hat{\rho}\hat{F}] = \sum_{ij} \rho_{ji} F_{ij}. \tag{7.9}$$

In the last equality, F_{ij} represents the matrix of the operator $\hat{F}$ in the s.p. basis $c_i, c_i^\dagger$. In the following, the notation 'tr' will always refer to the trace in the s.p. space.

Based on these results, it is not very difficult to rewrite the grand potential of equation (7.1) as

$$\begin{aligned}\Omega_{\mathrm{HF}} &= \mathrm{Tr}[\hat{D}_{\mathrm{HF}}(\hat{H} - \lambda\hat{N})] - \mathrm{Tr}[\hat{D}_{\mathrm{HF}}\hat{K}] - \frac{1}{\beta}\ln Z_{\mathrm{HF}} \\ &= \mathrm{tr}[\hat{h}\hat{\rho}] - \mathrm{tr}[\hat{K}\hat{\rho}] - \frac{1}{\beta}\ln Z_{\mathrm{HF}}\end{aligned} \tag{7.10}$$

where

$$h_{ij} = t_{ij} + \frac{1}{2}\sum_{kl}\bar{v}_{ijkl}\rho_{lj} \tag{7.11}$$

is the traditional HF Hamiltonian. In the most general EDF approach, the HF Hamiltonian should instead be called the Kohn–Sham (KS) Hamiltonian. It does not need to involve any two-body (or higher) potential but can be written in terms of the functional derivative of the energy with respect to the density matrix; see, e.g. the discussion in section 1.2 and the definition of the fields of the HFB matrix in equation (3.39).

Up to now, the operator $\hat{K}$ is arbitrary and is not known. We will determine it by requiring that the total energy, that is, the statistical average of the grand potential, be minimum with respect to variations of $\hat{K}$. In other words, the matrix elements K_{ij} of $\hat{K}$ play the role of variational parameters. We first note that

$$\delta Z_{\mathrm{HF}} = \delta\left[\mathrm{Tr}\left(e^{-\beta\hat{K}}\right)\right] = -\beta\mathrm{Tr}\left(e^{-\beta\hat{K}}\delta\hat{K}\right) = -\beta Z_{\mathrm{HF}}\mathrm{Tr}\left(\hat{D}_{\mathrm{HF}}\delta\hat{K}\right).$$

By virtue of the statistical Wick theorem, the last equality can be rewritten

$$\delta Z_{\mathrm{HF}} = -\beta Z_{\mathrm{HF}}\mathrm{tr}\left(\hat{\rho}\delta\hat{K}\right).$$

Starting from equation (7.10), we find

$$\begin{aligned}\delta\Omega_{HF} &= \mathrm{tr}(\hat{h}\delta\hat{\rho}) - \mathrm{tr}(\delta\hat{K}\hat{\rho}) - \mathrm{tr}(\hat{K}\delta\hat{\rho}) - \frac{1}{\beta}\frac{\delta Z_{\mathrm{HF}}}{Z_{\mathrm{HF}}} \\ &= \mathrm{tr}(\hat{h}\delta\hat{\rho}) - \mathrm{tr}(\delta\hat{K}\hat{\rho}) - \mathrm{tr}(\hat{K}\delta\hat{\rho}) + \mathrm{tr}(\hat{\rho}\delta\hat{K}).\end{aligned}$$

The cyclic invariance of the trace thus imposes that

$$\hat{K} = \hat{h}, \tag{7.12}$$

in other words, the operator $\hat{K}$ coincides with the HF Hamiltonian. It is essential to note that the variational principle does not give anything more than that. In particular, it does not require that we work in the basis that diagonalizes $\hat{h}$. We should also emphasize that the finite-temperature Hartree–Fock (FT-H) ground state, contrary to the $T = 0$ HF ground state, is *not* characterized by some product state $|\Psi\rangle = c_1^\dagger \cdots c_A^\dagger|-\rangle$, where $|-\rangle$ is the vacuum of particles. Such states, also called pure quantum states, can only represent a system at $T = 0$. In quantum statistical mechanics, systems are characterized instead by their density operator, here $\hat{D}_{\mathrm{HF}}$.

In practice, however, determining $\hat{h}$ requires knowing the density matrix $\hat{\rho}$, hence the density operator $\hat{K}$, see equation (7.8), hence the mean field $\hat{h}$, see equation (7.12). We are thus faced with the traditional self-consistency conundrum. A simple way to proceed is, therefore, to determine $\hat{\rho}$ by successive diagonalizations of $\hat{h}$. Indeed, starting with an initial guess for ρ_{ij} in the original s.p. basis of the c operators, we can construct the matrix h_{ij}; in the basis where this matrix is diagonal, $h_{ij} \equiv \delta_{ij} e_i$, the density becomes

$$\rho_{ij} = \frac{1}{1 + e^{\beta(e_i - \lambda)}} \delta_{ij}. \tag{7.13}$$

Transforming $\hat{\rho}$ back to the original basis yields a new iteration of the HF Hamiltonian h_{ij}, which we diagonalize again, etc. The procedure is repeated until the density matrix does not change any more. Note that this is not the only method to solve the FT-HF equations. Another popular alternative is to use the Thouless theorem and the conjugate gradient method. The FT-HF formalism was applied in studying atomic nuclei at high excitation energy in [13–18].

7.3 Finite-temperature Hartree–Fock–Bogoliubov theory

In this section, we first derive the HFB equations at $T > 0$. This is achieved by introducing an alternative approximation for the density operator which involves a quadratic form of the pair creation and annihilation operators. We then focus on the form of several quantities of interest in the HFB basis which diagonalizes the HFB Hamiltonian. Examples of full HFB calculations at finite temperature can be found in [19–24] and various extensions of the basic formalism such as the stability condition [25] and the inclusion of finite angular momentum [26, 27]. Other aspects going beyond the HFB level will be discussed briefly in sections 7.4 and 7.5.

7.3.1 Derivation of the FT-HFB equation

The considerations developed in the previous section can be extended to account for pairing correlations. In this case, we define the HFB mean-field approximation of the density operator as

$$\hat{D}_{\mathrm{HFB}} = \frac{1}{Z} e^{\beta \hat{K}} \qquad Z_{\mathrm{HFB}} = \mathrm{Tr}[e^{-\beta \hat{K}}], \tag{7.14}$$

with, this time,

$$\hat{K} = \frac{1}{2} \sum_{ij} K_{ij}^{11} c_i^\dagger c_j + \frac{1}{2} \sum_{ij} K_{ij}^{22} c_i c_j^\dagger + \frac{1}{2} \sum_{ij} K_{ij}^{20} c_i^\dagger c_j^\dagger + \frac{1}{2} \sum_{ij} K_{ij}^{02} c_i c_j. \tag{7.15}$$

We then introduce the 'doubled' matrix $\hat{\mathcal{K}}$ defined by

$$\hat{\mathcal{K}} = \begin{pmatrix} K^{11} & K^{20} \\ K^{02} & K^{22} \end{pmatrix} \tag{7.16}$$

and the generalized density matrix $\hat{\mathcal{R}}$, which, in the s.p. basis of the c operators, is defined as usual by

$$\mathcal{R} = \begin{pmatrix} \rho & \kappa \\ -\kappa^* & 1-\rho^* \end{pmatrix}. \tag{7.17}$$

Only this time, the components of the generalized density matrix are,

$$\begin{aligned} \rho_{ij} &= \mathrm{Tr}\left[\hat{D}_{\mathrm{HFB}} c_j^\dagger c_i\right] & \kappa_{ij} &= \mathrm{Tr}\left[\hat{D}_{\mathrm{HFB}} c_j c_i\right] \\ \kappa_{ij}^* &= \mathrm{Tr}\left[\hat{D}_{\mathrm{HFB}} c_i^\dagger c_j^\dagger\right] & (1-\rho)_{ij} &= \mathrm{Tr}\left[\hat{D}_{\mathrm{HFB}} c_i^\dagger c_j\right]. \end{aligned}$$

The two consequences of the generalized Wick theorem mentioned above for the HF approximation can be extended at the HFB approximation. In particular, we find that there is a one-to-one correspondence between the generalized density matrix $\hat{\mathcal{R}}$ and the operator $\hat{K}$, which formally is

$$\hat{\mathcal{R}} = \frac{1}{1+e^{\beta\hat{K}}}. \tag{7.18}$$

This means that variations of $\hat{K}$ and variations of $\hat{\mathcal{R}}$ are in fact equivalent. As before, we determine the operator $\hat{K}$ by requiring that the energy be minimal with respect to variations $\delta\hat{K}$, and as before, we find that a necessary condition to achieve this is to take

$$\hat{K} = \hat{\mathcal{H}}, \tag{7.19}$$

where the representation of $\hat{\mathcal{H}}$ in the basis of c operators is the familiar-looking HFB Hamiltonian

$$\mathcal{H} = \begin{pmatrix} h & \Delta \\ -\Delta^* & -h^* \end{pmatrix}. \tag{7.20}$$

Once again, we must insist that neither the mean-field approximation nor the variational principle requires that we be in a basis that diagonalizes $\mathcal{H}$. It is merely by convenience that one places oneself in such a basis—it is the most natural way to set up a system of self-consistent equations that can be solved by iterations.

> Recall that the HF and HFB equations at $T = 0$, as given by the variational principle, read
>
> $$\mathrm{HF}\ [\hat{h}, \hat{\rho}] = 0$$
> $$\mathrm{HFB}\ [\hat{\mathcal{H}}, \hat{\mathcal{R}}] = 0.$$
>
> Once again, these equations may be satisfied without necessarily going to the basis that simultaneously diagonalizes $\hat{h}$ and $\hat{\rho}$ (HF) or $\hat{\mathcal{H}}$ and $\hat{\mathcal{R}}$ (HFB). It just happens that going into this basis is a very simple way to ensure that these commutators are indeed zero.

Let us denote by β_μ, $\beta_\mu^\dagger$ the annihilation and creation operators that are associated with the basis that diagonalizes $\mathcal{H}$. These operators are related to the original c_i and $c_i^\dagger$ by the Bogoliubov transformation (3.6)

$$\beta_\mu = \sum_i \left[U_{\mu i}^\dagger \, c_i + V_{\mu i}^\dagger \, c_i^\dagger \right]$$

$$\beta_\mu^\dagger = \sum_i \left[V_{\mu i}^T \, c_i + U_{\mu i}^T \, c_i^\dagger \right].$$

The Bogoliubov transformation is nothing other than the transformation matrix that allows passing from the arbitrary basis of the c operators to the basis where $\mathcal{H}$ is diagonal. Since the new operators involve a linear combination of both creation and annihilation of particles, they are interpreted as q.p. operators. At $T = 0$, the ground state of the system is such that $\beta_\mu|\Psi\rangle = 0$ for all μ, i.e. the ground state is a q.p. vacuum; see equation (3.3). At $T > 0$, this is not true any more since the ground state is in fact a superposition of q.p. states characterized by the density operator $\hat{D}_{\text{HFB}}$, not by a pure quantum state $|\Psi\rangle$.

7.3.2 Statistical operators in the HFB basis

In the HFB basis (the one which diagonalizes the HFB Hamiltonian), the density operator can be written

$$\hat{D}_{\text{HFB}} = \frac{1}{Z_{\text{HFB}}} e^{-\beta\hat{\mathcal{H}}} = \frac{1}{Z_{\text{HFB}}} e^{-\beta\left(\sum_{\mu>0} E_\mu \beta_\mu^\dagger \beta_\mu - \sum_{\mu>0} E_\mu \beta_\mu \beta_\mu^\dagger\right)} = \frac{1}{Z_{\text{HFB}}} e^{\beta E_{\text{HFB}}} e^{-\beta \sum_\mu E_\mu \beta_\mu^\dagger \beta_\mu},$$

where the term $e^{\beta E_{\text{HFB}}}$ is nothing other than a normalization constant.

Partition function

The explicit expression of the partition function is

$$Z_{\text{HFB}} = \text{Tr}[e^{-\beta\hat{\mathcal{H}}}] = \sum_{\{n_\alpha\}} \langle \{n_\alpha\} | e^{-\beta\hat{\mathcal{H}}} | \{n_\alpha\} \rangle,$$

where $\{n_\alpha\} \equiv \{n_1, n_2, \ldots\}$ is the (infinite) set of all possible occupations of individual q.p.s in multi-q.p. states built on top of the HFB vacuum; n_k is the occupation of q.p. μ_k in level E_k in a multi-q.p. state $|\mu_1, \mu_2, \ldots\rangle$. We choose the kets $|\{n_\alpha\}\rangle$ to span the many-body Hilbert space simply because it is much easier to evaluate traces at the HFB approximation. At the HF approximation, we would choose many-particle many-hole states. The partition function therefore reads

$$Z_{\text{HFB}} = \text{Tr}[e^{-\beta\hat{\mathcal{H}}}] = \sum_{n_1, n_2, \ldots} \langle \{n_\alpha\} | e^{-\beta\hat{\mathcal{H}}} | \{n_\alpha\} \rangle = \sum_{n_1} \sum_{n_2} \ldots \langle \{n_\alpha\} | e^{-\beta\hat{\mathcal{H}}} | \{n_\alpha\} \rangle.$$

In practice, the occupation of Fermionic q.p. states is only 1 or 0,

$$\beta_\mu^\dagger \beta_\mu |\{n_\alpha\}\rangle = n_\mu |\{n_\alpha\}\rangle \quad \text{with: } n_\mu = \begin{cases} 1 \text{ if } \mu \in \{\alpha\} \\ 0 \text{ if } \mu \notin \{\alpha\} \end{cases},$$

and, therefore,

$$\langle \{n_\alpha\}| e^{-\beta \sum_\mu E_\mu \beta_\mu^\dagger \beta_\mu} |\{n_\alpha\}\rangle = e^{-\beta \sum_\mu E_\mu n_\mu},$$

which gives

$$\begin{aligned} Z_{\text{HFB}} &= e^{\beta E_{\text{HFB}}} \sum_{n_1} \sum_{n_2} \dots e^{-\beta \sum_\mu E_\mu n_\mu} \\ &= e^{\beta E_{\text{HFB}}} \sum_{n_1} \sum_{n_2} \dots e^{-\beta E_1 n_1} e^{-\beta E_2 n_2} \times \dots. \end{aligned}$$

It is clear from this expression that we can write

$$Z_{\text{HFB}} = e^{\beta E_{\text{HFB}}} \sum_{n_1} e^{-\beta E_1 n_1} \times \sum_{n_2} e^{-\beta E_2 n_2} \times \dots = e^{\beta E_{\text{HFB}}} \prod_\mu \sum_{n_\mu} e^{-\beta E_\mu n_\mu}.$$

Since $n_\mu = 0, 1$, we actually obtain

$$Z_{\text{HFB}} = e^{\beta E_{\text{HFB}}} \prod_\mu (1 + e^{-\beta E_\mu}). \tag{7.21}$$

Statistical density operator
In the HFB approximation, the density operator becomes

$$\hat{D}_{\text{HFB}} = \frac{1}{Z} e^{\beta E_{\text{HFB}}} e^{-\beta \sum_\mu E_\mu \beta_\mu^\dagger \beta_\mu}.$$

Define $\hat{n}_\mu = \beta_\mu^\dagger \beta_\mu$. In the q.p. basis, the operator $\hat{n}_\mu$ is a projector, $\hat{n}_\mu^2 = \hat{n}_\mu$. Therefore,

$$e^{-\beta \sum_\mu E_\mu \beta_\mu^\dagger \beta_\mu} = \prod_\mu e^{-\beta E_\mu \hat{n}_\mu} = \prod_\mu \sum_{n=0}^{+\infty} \frac{(-\beta E_\mu \hat{n}_\mu)^n}{n!} = \prod_\mu \left[1 + \hat{n}_\mu \sum_{n=1}^{+\infty} \frac{(-\beta E_\mu)^n}{n!} \right]$$

or

$$e^{-\beta \sum_\mu E_\mu \beta_\mu^\dagger \beta_\mu} = \prod_\mu \left[1 - \hat{n}_\mu + e^{-\beta E_\mu} \hat{n}_\mu \right].$$

Introducing the Fermi factors,

$$f_\mu = \frac{1}{1 + e^{\beta E_\mu}}, \tag{7.22}$$

we obtain immediately

$$\hat{D}_{\text{HFB}} = \prod_{\mu} \left[f_\mu \beta_\mu^\dagger \beta_\mu + (1 - f_\mu)(1 - \beta_\mu^\dagger \beta_\mu) \right]. \tag{7.23}$$

We see that the density operator does not depend on the HFB energy.

7.3.3 Density matrix and pairing tensor

Quasiparticle density

We first proceed to the calculation of the 'q.p. density matrix' in the q.p. basis, that is, we define $\tilde{\rho}$ such that

$$\tilde{\rho}_{\mu\nu} = \text{Tr}\left(\hat{D}_{\text{HFB}} \beta_\nu^\dagger \beta_\mu\right). \tag{7.24}$$

In the zero-temperature HFB theory, the expectation value of $\beta_\nu^\dagger \beta_\mu$ in the ground state would be exactly zero by construction. At finite temperature, however, this expectation value is given by equation (7.24) and can take arbitrary values between 0 and 1. Computing the trace gives

$$\tilde{\rho}_{\mu\nu} = \sum_{\{n_\alpha\}} \langle \{n_\alpha\} | \hat{D}_{\text{HFB}} \beta_\nu^\dagger \beta_\mu | \{n_\alpha\} \rangle = \sum_{n_1, n_2, \ldots} \langle \{n_\alpha\} | \prod_{\epsilon} \left[f_\epsilon n_\epsilon + (1 - f_\epsilon)(1 - n_\epsilon) \right] \beta_\nu^\dagger \beta_\mu | \{n_\alpha\} \rangle.$$

We have

$$\beta_\nu^\dagger \beta_\mu | \{n_\alpha\} \rangle = \begin{cases} \delta_{\mu\nu} n_\mu | \{n_\alpha\} \rangle \text{ if } \mu \in \{\alpha\} \\ 0 \text{ if } \mu \notin \{\alpha\} \end{cases},$$

with $n_\mu = 1$ then. Therefore

$$\tilde{\rho}_{\mu\nu} = \sum_{n_1, n_2, \ldots} \prod_{\epsilon} \left[f_\epsilon n_\epsilon + (1 - f_\epsilon)(1 - n_\epsilon) \right] \delta_{\mu\nu} n_\mu,$$

Just as for the calculation of the partition function, we note that we can exchange the summations and product, since

$$\sum_{n_1, n_2, \ldots} \prod_{\epsilon} f_\epsilon n_\epsilon + (1 - f_\epsilon)(1 - n_\epsilon) = \sum_{n_1} \sum_{n_2} \cdots (f_1 n_1 + (1 - f_1)(1 - n_1)) \\ \times (f_2 n_2 + (1 - f_2)(1 - n_2)) \times \cdots.$$

Therefore,

$$\tilde{\rho}_{\mu\nu} = \sum_{n_1, n_2, \ldots} \prod_{\epsilon} \left[f_\epsilon n_\epsilon + (1 - f_\epsilon)(1 - n_\epsilon) \right] \delta_{\mu\nu} n_\mu = \prod_{\epsilon} \sum_{n_\epsilon} \left[f_\epsilon n_\epsilon + (1 - f_\epsilon)(1 - n_\epsilon) \right] \delta_{\mu\nu} n_\mu.$$

We can extract from the product over ϵ the term $\epsilon = \mu$, for which $n_\mu = 1$, and this gives us

$$\tilde{\rho}_{\mu\nu} = f_\mu \delta_{\mu\nu} \prod_{\epsilon \neq \mu} \sum_{n_\epsilon} \left[f_\epsilon n_\epsilon + (1 - f_\epsilon)(1 - n_\epsilon) \right].$$

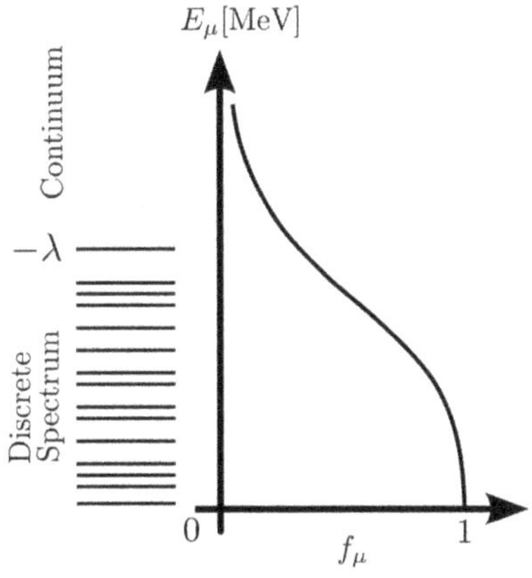

Figure 7.1. Schematic illustration of the occupation probability (7.25). Recall that the q.p. spectrum involves both discrete states and a continuum. At large temperatures, this continuum can be occupied. (Figure created by N Schunck.)

Since $n_\epsilon = 0, 1$, each sum is equal to 1 and we thus find (figure 7.1)

$$\tilde{\rho}_{\mu\nu} = f_\mu \delta_{\mu\nu}. \tag{7.25}$$

Density matrix

We can easily obtain the expression for the density matrix by going back from the s. p. to the q.p. basis. Noting that

$$\rho_{kl} = \mathrm{Tr}\left(\hat{D}_{\mathrm{HFB}} c_l^\dagger c_k\right), \tag{7.26}$$

we use the Bogoliubov transformation to write

$$\rho_{kl} = \mathrm{Tr}\left[\hat{D}_{\mathrm{HFB}} \sum_{\mu\nu} \left(V_{l\mu} U_{k\nu}\, \beta_\mu \beta_\nu + V_{l\mu} V_{k\nu}^*\, \beta_\mu \beta_\nu^\dagger + U_{l\mu}^* U_{k\nu}\, \beta_\mu^\dagger \beta_\nu + U_{l\mu}^* V_{k\nu}^*\, \beta_\mu^\dagger \beta_\nu^\dagger \right)\right].$$

Again, the trace here involves summing over all (identical) many-q.p. states; only the contributions with the same number of creation and annihilation operators will give non-zero contributions. Therefore,

$$\rho_{kl} = \mathrm{Tr}\left[\hat{D}_{\mathrm{HFB}} \sum_{\mu\nu} \left(V_{l\mu} V_{k\nu}^*\, \beta_\mu \beta_\nu^\dagger + U_{l\mu}^* U_{k\nu}\, \beta_\mu^\dagger \beta_\nu \right)\right].$$

Putting all operators in normal-order form and using equation (7.3.3), we find

$$\rho_{kl} = \sum_{\mu\nu} V_{l\mu} V_{k\nu}^* \mathrm{Tr}\left[\hat{D}_{\mathrm{HFB}}(\delta_{\mu\nu} - \beta_\nu^\dagger \beta_\mu)\right] + \sum_{\mu\nu} U_{l\mu}^* U_{k\nu} \mathrm{Tr}\left[\hat{D}_{\mathrm{HFB}} \beta_\mu^\dagger \beta_\nu\right].$$

Since $\mathrm{Tr}\, \hat{D}_{\mathrm{HFB}} = 1$ by definition, this gives us

$$\rho_{kl} = \sum_{\mu\nu} V_{l\mu} V^*_{k\nu} (1 - f_\mu) \delta_{\mu\nu} + \sum_{\mu\nu} U^*_{l\mu} U_{k\nu} f_\mu \delta_{\mu\nu},$$

hence

$$\rho_{kl} = \left(V^* (1 - f) V^T \right)_{kl} + \left(U f U^\dagger \right)_{kl}. \tag{7.27}$$

Pairing tensor

The pairing tensor is obtained in a similar way. We begin by introducing the quantity $\tilde{\kappa}_{\mu\nu} = \mathrm{Tr}(\hat{D}_{\mathrm{HFB}} \beta_\nu \beta_\mu)$, representing the statistical average of the pairing tensor expressed in the q.p. basis. Since the trace involves the bracket over multi-q.p. states with the same q.p. number on both sides (bra and ket), we see that $\tilde{\kappa}_{\mu\nu} = 0$ necessarily. We then introduce

$$\kappa_{kl} = \mathrm{Tr}\left(\hat{D}_{\mathrm{HFB}} c_l c_k \right) \tag{7.28}$$

and move into the q.p. basis,

$$\kappa_{kl} = \mathrm{Tr}\left[\hat{D}_{\mathrm{HFB}} \sum_{\mu\nu} \left(U_{l\mu} U_{k\nu}\, \beta_\mu \beta_\nu + U_{l\mu} V^*_{k\nu}\, \beta_\mu \beta^\dagger_\nu + V^*_{l\mu} U_{k\nu}\, \beta^\dagger_\mu \beta_\nu + V^*_{l\mu} V^*_{k\nu}\, \beta^\dagger_\mu \beta^\dagger_\nu \right) \right].$$

At this stage, the calculation is identical as for the density matrix. We only retain the terms that have identical numbers of q.p.s,

$$\kappa_{kl} = \mathrm{Tr}\left[\hat{D}_{\mathrm{HFB}} \sum_{\mu\nu} \left(U_{l\mu} V^*_{k\nu}\, \beta_\mu \beta^\dagger_\nu + V^*_{l\mu} U_{k\nu}\, \beta^\dagger_\mu \beta_\nu \right) \right],$$

and finally find

$$\kappa_{kl} = \left(V^* (1 - f) U^T \right)_{kl} + \left(U f V^\dagger \right)_{kl}. \tag{7.29}$$

Equations (7.22), (7.27) and (7.29) are nearly all we need to solve the finite-temperature Hartree–Fock–Bogoliubov (FT-HFB) equation, since both the HFB matrix and generalized density take a similar form as the $T = 0$ case. The algorithms used to solve the $T = 0$ HFB equation, which we present in detail in chapter 8, can be adapted in a straightforward manner.

While we can show that the column vectors V_μ are always localized, this is not the case for the vectors U_μ for μ such that $E_\mu > -\lambda$ [28]. Therefore, we find that the density matrix may contain a contribution that is unbound depending on the value of f_μ; see the discussion in [29]. In contrast to the density matrix, the pairing tensor is fully localized since both terms in equation (7.29) involve the vectors V_μ, which are localized for all μ.

7.4 Finite-temperature RPA

We will conclude this short chapter by giving a brief overview of the extension of the RPA theory at finite temperature. As before, we will simplify things by assuming we have a pseudo-Hamiltonian $\hat{H}$ rather than a general EDF. Let us assume a statistical ensemble now described by a time-dependent statistical density operator $\hat{D}(t)$. This operator obeys the Liouville equation

$$i\frac{\partial \hat{D}}{\partial t} = [\hat{H}, \hat{D}(t)]. \tag{7.30}$$

The Liouville equation can easily be obtained by introducing the formal definition of the density operator as $\hat{D} = \sum_n w_n |\Phi_n\rangle\langle\Phi_n|$, with w_n the probability that the system is in the (many-body) state $|\Phi_n\rangle$, and using the time-dependent Schrödinger equation for $|\Phi_n\rangle$. In this section, we will use a small amplitude limit around the static HF solution given by

$$\hat{D}_0 = \frac{1}{Z}e^{-\beta\hat{K}},$$

where $\hat{K}$ is given by equation (7.7). We will perform all calculations in the HF basis at finite temperature denoted by the operators $a_i^\dagger$ and a_j. As in section 5.1.1, we use the notation $(a, a^\dagger)$ for the HF s.p. operators to distinguish them from arbitrary s.p. operators, denoted by $(c, c^\dagger)$ throughout this book.

7.4.1 The time-dependent density operator

As in section 5.1.3, we want to explore what happens to the system when we perturb it from its FT-HF solution. Only this time, we will assume that the system is at non-zero temperature T. To this end, we assume that the full time-dependent density operator solution to the Liouville equation (7.30) can be expanded around the FT-HF solution $\hat{D}_0$ according to

$$\hat{D}(t) = e^{i\hat{\chi}(t)}\hat{D}_0 e^{-i\hat{\chi}(t)},$$

with $\hat{\chi}(t)$ a general time-dependent one-body operator to be determined. We will only assume that $\hat{\chi}$ is Hermitian and that it corresponds to a small perturbation, that is,

$$\hat{\chi}(t) = \sum_{ij}\chi_{ij}(t)a_i^\dagger a_j \qquad \chi_{ij}(t) = \chi_{ji}^*(t) \qquad |\chi_{ij}(t)| \ll 1 \quad \forall(i, j).$$

Let us expand $\hat{D}(t)$ in powers of $\hat{\chi}$. We find

$$\hat{D}(t) = \hat{D}_0 + \hat{D}_1(t) + \hat{D}_2(t) + \mathcal{O}(\hat{\chi}^3), \tag{7.31}$$

with

$$\hat{D}_1(t) = i\left[\hat{\chi}(t), \hat{D}_0\right] \tag{7.32a}$$

$$\hat{D}_2(t) = \frac{1}{2}\left[\left[\hat{\chi}(t), \hat{D}_0\right], \hat{\chi}(t)\right]. \tag{7.32b}$$

This is the exact analog of the perturbation expansion of the time-dependent Hartree–Fock (TDHF) equation, only it is applied to the statistical density operator rather than the one-body density matrix.

7.4.2 The time-dependent density matrix

To first order in $\hat{\chi}$, the full time-dependent one-body density matrix is given by the general expression

$$\rho_{ji}(t) = \mathrm{Tr}\left(\hat{D}(t)a_i^\dagger a_j\right) = \mathrm{Tr}\left(\hat{D}_0 a_i^\dagger a_j\right) + i\mathrm{Tr}\left([\hat{\chi}(t), \hat{D}_0]a_i^\dagger a_j\right).$$

Recall that we work in the FT-HF basis characterized by the s.p. creation and annihilation operators $a^\dagger$ and a, which are all time-independent. We can arrange this expression so that it reads as the trace of the density operator applied to some other operator, $\mathrm{Tr}(\hat{D}_0[..])$. This will allow us to use the statistical Wick theorem. As mentioned earlier, the notation 'Tr' refers to traces in the Fock space, and the statistical Wick theorem allows replacing these complicated traces by traces 'tr' in the s.p. space. To this purpose, we use the cyclic invariance of the trace and find

$$\rho_{ji}(t) = \mathrm{Tr}\left(\hat{D}_0 a_i^\dagger a_j\right) + i\mathrm{Tr}\left(\hat{D}_0\left[a_i^\dagger a_j, \hat{\chi}(t)\right]\right).$$

Since we operate in the FT-HF basis ($\rho_{ij} = f_i\,\delta_{ij}$), we obtain

$$\rho_{ji}(t) = f_i\,\delta_{ij} + i\chi_{ji}(t)(f_i - f_j). \tag{7.33}$$

It is clear from this expression that once the perturbation operator $\hat{\chi}(t)$ is known, then so is the density matrix. We can also see that the density matrix oscillates around its temperature-dependent, static HF solution. To be more precise, at each instant t, the solution of the time-dependent problem (in this small amplitude limit) can always be viewed as built on top of the initial HF solution, because of the first diagonal term in the previous equation. If this term was not present, one could not necessarily recognize the initial HF solution in $\rho(t)$. Also note that the diagonal term of the one-body density matrix, ρ_{ii}, is time-independent.

7.4.3 RPA equation at finite temperature

In section 5.1.3, we started from the TDHF equation to obtain a time-dependent differential equation for $\hat{\chi}$ alone. Here, we will adopt the same strategy, only we have to start from the Liouville equation. An alternative expression for the full time-dependent one-body density matrix is

$$i\frac{\partial \rho_{ji}}{\partial t} = i\frac{\partial}{\partial t}\mathrm{Tr}\left(\hat{D}(t)a_i^\dagger a_j\right) = \mathrm{Tr}\left(i\frac{\partial \hat{D}(t)}{\partial t}a_i^\dagger a_j\right) = \mathrm{Tr}\left([\hat{H},\, \hat{D}(t)]a_i^\dagger a_j\right).$$

By developing the commutator and using again the cyclic invariance of the trace, we find

$$i\frac{\partial \rho_{ji}}{\partial t} = \mathrm{Tr}\left(\hat{D}(t)\left[a_i^\dagger a_j,\, \hat{H}\right]\right).$$

Replacing $\hat{D}(t)$ by its expression (7.31), we find

$$i\frac{\partial \rho_{ji}}{\partial t} = \mathrm{Tr}\left(\hat{D}_0\left[a_i^\dagger a_j,\, \hat{H}\right]\right) + i\mathrm{Tr}\left(\left[\hat{\chi}(t),\, \hat{D}_0\right]\left[a_i^\dagger a_j,\, \hat{H}\right]\right).$$

Expanding the first commutator of the second term and using (again) the cyclic invariance of the trace, we obtain the following equation of motion for the time-dependent one-body density matrix

$$i\frac{\partial \rho_{ji}}{\partial t} = \mathrm{Tr}\left(\hat{D}_0\left[a_i^\dagger a_j,\, \hat{H}\right]\right) + i\sum_{kl}\chi_{kl}(t)\mathrm{Tr}\left(\hat{D}_0\left[\left[a_i^\dagger a_j,\, \hat{H}\right],\, a_k^\dagger a_l\right]\right). \tag{7.34}$$

Note that in this equation, the operators a are known (they correspond to the FT-HF basis), the operator $\hat{H}$ is also known (it is the original Hamiltonian), and so is the density operator $\hat{D}_0$ (it is the density operator at the HF approximation). The unknowns are the coefficients $\chi_{\mu\nu}(t)$ of the perturbation operator, and the time-dependent density matrix itself, which is, however, related to this perturbation operator via equation (7.33). The next phase thus consists in computing all the relevant commutators and putting equation (7.34) explicitly in the form of a time-differential equation for $\hat{\chi}$.

It is relatively simple to show that the first trace in equation (7.34) vanishes. Since we work in the FT-HF basis throughout, we can take advantage of the fact that $\rho_{ij} = f_i\,\delta_{ij}$ and that the mean-field Hamiltonian h reads

$$h_{ij} = t_{ij} + \sum_b \bar{v}_{ibjb}f_b = e_i\delta_{ij}.$$

The statistical Wick theorem then gives the announced result.

We can return to the equation of motion (7.34) for the time-dependent density matrix. Since the first trace gives zero, we have

$$\frac{\partial \rho_{ji}}{\partial t} = \sum_{kl}\chi_{kl}(t)\mathrm{Tr}\left(\hat{D}_0\left[\left[a_i^\dagger a_j,\, \hat{H}\right],\, a_k^\dagger a_l\right]\right).$$

The calculation of the double commutator is a little time consuming, but very straightforward. Throughout, we can take into account the properties of the FT-HF basis, so that after a little bit of algebra, we obtain

$$\frac{\partial \rho_{ji}}{\partial t} = \sum_{kl} \chi_{kl}(t)\bar{v}_{jlik}(f_i - f_j)(f_l - f_k) + \chi_{ji}(t)(e_j - e_i)(f_i - f_j).$$

The final step is to recall equation (7.33) where the time-dependent density matrix is simply expressed as function of the perturbation operator. Computing the time-derivative, we find the final equation of motion

$$i(f_i - f_j)\frac{\partial \chi_{ji}}{\partial t} = (e_j - e_i)(f_i - f_j)\chi_{ji}(t) + \sum_{kl} \chi_{kl}(t)\bar{v}_{jlik}(f_i - f_j)(f_l - f_k). \quad (7.35)$$

The next step is to distinguish between indices $i > j$ and $i < j$ (diagonal terms are time-independent, hence do not contribute to this equation), and use the hermiticity of $\hat{\chi}$. We can then rewrite the equation of motion for the perturbation operator $\hat{\chi}$ as a system of equations involving the two independent degrees of freedom χ_{ji} and χ_{ji}^* $(i < j)$,

$$i(f_i - f_j)\frac{\partial}{\partial t}\begin{bmatrix} \chi_{ji} \\ \chi_{ji}^* \end{bmatrix} = \begin{bmatrix} A_{ji\mu\nu} & B_{ji\mu\nu} \\ -B_{ji\mu\nu}^* & -A_{ji\mu\nu}^* \end{bmatrix}\begin{bmatrix} +\chi_{\mu\nu} \\ +\chi_{\mu\nu}^* \end{bmatrix}, \quad (7.36)$$

with

$$\begin{aligned} A_{ji\mu\nu} &= (e_j - e_i)(f_i - f_j)\delta_{j\mu}\delta_{i\nu} + \bar{v}_{j\nu i\mu}(f_i - f_j)(f_\nu - f_\mu) \\ B_{ji\mu\nu} &= \bar{v}_{j\mu i\nu}(f_i - f_j)(f_\nu - f_\mu). \end{aligned} \quad (7.37)$$

RPA equation from TDHF

From equations (7.36) and (7.37), we can obtain the traditional RPA equation at $T = 0$ by taking the following limits

$$\begin{aligned} f_h &\to 1 \; h \leqslant \lambda_F \\ f_p &\to 0 \; p > \lambda_F, \end{aligned}$$

which simply translates the fact that in the HF approximation at $T = 0$, all levels below the Fermi level are fully occupied, while all levels above are empty. It is easy to check that this leads to exactly the same RPA matrices as equation (5.4), and that equation (7.36) turns into

$$i\frac{\partial}{\partial t}\begin{bmatrix} \chi_{ph} \\ \chi_{ph}^* \end{bmatrix} = \begin{bmatrix} A_{php'h'} & B_{php'h'} \\ -B_{php'h'}^* & -A_{php'h'}^* \end{bmatrix}\begin{bmatrix} +\chi_{p'h'} \\ +\chi_{p'h'}^* \end{bmatrix}.$$

We can then seek solutions to these equations in the form

$$\chi_\nu(t) = X^\nu e^{-i\Omega^\nu t} + Y^{\nu *} e^{+i\Omega^\nu t},$$

where X^ν and Y^ν are matrices of size $N_{sp} \times N_{sp}$. This leads directly to the equation (5.3).

As the form of equation (7.36) shows, the matrices A and B of equation (7.37) are in fact obtained by linearizing the pairs of indices i, j and μ, ν: $ij \to m$ and $\mu\nu \to n$, with $1 \leqslant i, j, \mu, \nu \leqslant N_{\text{sp}}$ and $1 \leqslant m, n \leqslant N_{\text{sp}}^2$, where N_{sp} is the size of the s.p. FT-HF basis. Formally, the RPA matrix at finite temperature thus resembles the one at $T = 0$. The main difference comes from the presence of the statistical occupation factors in the definition of the A and B blocks. In fact, this is a source of ambiguity, since it leads to different conventions for the definition of the matrices A and B (hence of the RPA matrix itself). The two main conventions are are listed in table 7.3.

The Vautherin prescription of [30] comes from simplifying by $(f_i - f_j)$ on both sides of the RPA equation. Mathematically, it is of course only valid when $f_i \neq f_j$. This convention leads to defining an RPA matrix that is not Hermitian with respect to the change of indexes $(j, i) \leftrightarrow (\mu, \nu)$, and which has an asymmetry in the occupation probabilities. Because the RPA matrix is not Hermitian, there may be complex values in its spectrum ($\Omega^\nu \in \mathbb{C}$). In addition, such a matrix is not consistent with the stability matrix that could be derived from second-order expansion of the energy, in full analogy of what was presented in section 3.1.4. In spite of these limitations, this prescription has been used, e.g. in [8, 32–35].

In the Sommermann solution of [31], the RPA matrix is both Hermitian and compatible with the stability matrix, and leads to a solution that can be expanded on the RPA eigenmodes. However, the square roots require special care when combining protons and neutrons, and the proper sign must be assigned to each element of the RPA matrix. This is the prescription used in [36, 37]. A proper comparison of the two prescriptions is still missing.

7.5 Beyond mean field

One of the well-known effects of finite-temperature systems is the existence of statistical fluctuations around the mean value of observables [9]. In quantum mechanical systems described by symmetry-breaking theories, these fluctuations add up to the ones, of quantum nature, resulting from the non-conservation of quantum numbers [38]. Statistical fluctuations increase with temperature. However, as discussed briefly in the introduction of this chapter, the finite-temperature formalism is only a tool used to describe highly excited states: even at very high excitation energy E^*, a nucleus is still characterized by a well-defined number of

Table 7.3. Prescription for the RPA matrix at finite temperature $T > 0$.

Prescription	RPA matrices
Vautherin [30]	$A_{ji,\,\mu v} = (e_j - e_i)\delta_{j\mu}\delta_{iv} + \bar{v}_{jvi\mu}(f_v - f_\mu)$
	$B_{ji,\,\mu v} = \bar{v}_{j\mu iv}(f_\mu - f_v)$
Sommermann [31]	$A_{ji,\,\mu v} = (e_j - e_i)\sqrt{f_i - f_j}\,\delta_{j\mu}\delta_{iv} + \bar{v}_{jvi\mu}\sqrt{f_i - f_j}\,\sqrt{f_v - f_\mu}$
	$B_{ji,\,\mu v} = \bar{v}_{j\mu iv}\sqrt{f_i - f_j}\,\sqrt{f_\mu - f_v}$

particles, angular momentum, parity, etc. Just like the zero temperature, restoring broken symmetries may sometimes be essential to connect with experimental data.

> For a system described by a Hamiltonian $\hat{H}$, let us consider an operator $\hat{A}$ such that $[\hat{H}, \hat{A}] = 0$. This could be, for example, the projection of angular momentum on some quantization axis or the parity operator; see chapter 8 or section 3.1.1 for additional discussions. Symmetry-breaking states $|\Phi\rangle$ are such that $\hat{A}|\Phi\rangle \neq A|\Phi\rangle$, and, therefore $\Delta A = \sqrt{\langle \hat{A}^2 \rangle - \langle A \rangle^2} \neq 0$, where $\langle\rangle$ refers to the expectation value on the state $|\Phi\rangle$, $\langle A \rangle = \langle \Phi | \hat{A} \Phi \rangle$. At finite temperature, the expectation value becomes $\langle A \rangle = \mathrm{Tr}(\hat{D}\hat{A})$ and one can show that it is a sum of two terms, one of quantum origin that remains even at $T = 0$, the other one of statistical nature, which is not zero only at $T > 0$ [38].

Projected statistical quantum mechanics begins with the definition of the projected density operator. Let us consider a symmetry S and a projector operator on S. For a system at $T > 0$ characterized by a density operator $\hat{D}$, the projected density operator becomes [39]

$$\hat{D}_{\mathrm{proj}} = \frac{1}{Z_{\mathrm{proj}}} \hat{P}\hat{D}\hat{P}, \tag{7.38}$$

where Z_{proj} is the projected partition function

$$Z_{\mathrm{proj}} = \mathrm{Tr}(\hat{P}\hat{D}\hat{P}).$$

It is straightforward to show that the effect of the form (7.38) is to filter out states with the wrong quantum numbers when taking the statistical trace. As usual, the calculation of the projected partition function and of the projected density operator is extremely challenging. In practice, one can apply the same type of approximations as for unprojected statistics, that is, replace $\hat{D}$ by $\hat{D}_{\mathrm{HF}}$ or $\hat{D}_{\mathrm{HFB}}$. The important case of particle-number-projected statistics was treated directly in [39–41] and within the thermofield formalism in [42] (see section 7.5 for a very brief introduction to thermofield dynamics). A general framework for symmetry restoration in the case of independent particle or quasiparticle density operators was presented in [43], with application to angular momentum restoration and a comparison with a direct method to compute partition functions based on the static path approximation. Note that most practical applications of these techniques so far have been made with schematic nuclear Hamiltonians.

In contrast to small amplitude collective motion (RPA and QRPA of chapter 5) and symmetry restoration (see section 3.2), the extension of the techniques presented in chapter 6 to describe large-amplitude collective motion poses more challenges. There have been a few applications involving the extension of the adiabatic time-dependent Hartree–Fock–Bogoliubov (ATDHFB) theory of collective inertia at finite temperature [22, 44, 45]. The inertia tensor in this case is given by a similar formula as at $T = 0$, namely,

$$\mathrm{M}_{\mu\nu} = 2\hbar^2\left[M^{(1)}\right]^{-1}M^{(3)}\left[M^{(1)}\right]^{-1}, \tag{7.39}$$

but this time the moments are given,

$$\begin{aligned}\mathrm{M}_{\mu\nu}^{(K)} = \sum_{ij}\Bigg[& Q_{\mu;ij}^{11*}\frac{f_j - f_i}{(E_i - E_j)^K}Q_{\nu;ij}^{11} + Q_{\mu;ij}^{12*}\frac{1 - f_j - f_i}{(E_i + E_j)^K}Q_{\nu;ij}^{12} \\ & + Q_{\mu;ij}^{21*}\frac{1 - f_j - f_i}{(E_i + E_j)^K}Q_{\nu;ij}^{21} + Q_{\mu;ij}^{22*}\frac{f_j - f_i}{(E_i - E_j)^K}Q_{\nu;ij}^{22}\Bigg],\end{aligned} \tag{7.40}$$

where $\hat{Q}_\mu$ is the operator for the collective variable μ and $f_i = 1/(1 + e^{\beta E_i})$ is the Fermi–Dirac statistical occupation of the quasiparticle E_i. This approximation can be used to estimate, e.g. spontaneous fission half-lives at finite temperature. However, note that there is no study of the validity of this approach in the current literature.

We do not show a demonstration for the formula (7.39) and (7.40), since they correspond to an approximation that has not been validated. The interested reader could obtain these equations by starting from the finite-temperature time-dependent Hartree–Fock–Bogoliubov (TDHFB) equation (see [6]), which is simply $i\hbar\dot{\mathcal{R}} = [\mathcal{H}, \mathcal{R}]$ with $\mathcal{H}$ given by equation (7.20) and $\mathcal{R}$ by equation (7.17). The next step is to expand $\mathcal{R}$ up to second order around the FT-HFB solution. Note that this implies that we assume the collective path is confined to a sequence of FT-HFB solutions. This expansion leads to $i\hbar\dot{\mathcal{R}}_0 = [\mathcal{H}_0, \mathcal{R}_1] + [\mathcal{H}_1, \mathcal{R}_0]$, which is nothing other than the QRPA at finite temperature. Neglecting the first trace (residual interaction) leads to a relationship between $\mathcal{R}_1$ and the perturbation operator $\hat{\chi}$. Writing the energy up to second order yields the requested formula.

The extension of configuration mixing techniques following, e.g., the GCM approach discussed in sections 3.2 and 6.4, is even more challenging. Here, the difficulty is really conceptual; the GCM most fundamentally relies on the ansatz (3.79), that is,

$$|\Psi\rangle = \int d^N\boldsymbol{q}\, f(\boldsymbol{q})|\Phi(\boldsymbol{q})\rangle,$$

where $|\Psi\rangle$ is the many-body state, $|\Phi(\boldsymbol{q})\rangle$ is a convenient basis of known many-body states and $f(\boldsymbol{q})$ is the weight functions—to be determined from the variational principle. At finite temperature, this ansatz simply becomes meaningless, since the system is not described by a state vector but by a density operator. At the moment, there seems to have been only one attempt to combine the GCM with statistical quantum mechanics through an extended transport theory [46]. An alternative, as of yet unexplored, option could be to use the thermofield dynamics, in which finite-temperature systems are described by a state vector—at the price of doubling the size of the Hilbert space.

Thermofield dynamics

One of the major technical difficulties of statistical quantum mechanics is that a system is not described by a state vector $|\Psi\rangle$, but by a statistical density operator $\hat{D}$. The thermofield dynamics proposed by Umezawa and others in quantum field theory provides a mathematical recipe to recover a description of $T > 0$ systems in terms of state vectors; see, e.g. [47, 48]. The main idea is to embed the system of interest with its environment. This process is known in the theory of quantum entanglement as *purification*. If the system is maintained at temperature T ($\beta = 1/T$), the density operator in $\mathcal{H}$ reads

$$\hat{D} = \sum_n d_{nn}|n\rangle\langle n| = \frac{1}{Z}e^{-\beta\hat{H}},$$

so that $d_{nn} = e^{-\beta e_n}/Z$ where e_n are the eigenvalues of $\hat{H}$ and Z the partition function. The system is in a mixed state characterized by the density operator $\hat{D}$. We purify this state by doubling the Hilbert space. We thus introduce a new space $\tilde{\mathcal{H}}$. The Schmidt theorem tells us that, given a basis $|n\rangle \in \mathcal{H}$, we can always find a basis $|\tilde{n}\rangle \in \tilde{\mathcal{H}}$ such that any pure state in $|\Psi\rangle \in \mathcal{H} \otimes \tilde{\mathcal{H}}$ reads

$$|\Psi\rangle = \sum_n \sqrt{d_{nn}}\,|n\rangle \otimes |\tilde{n}\rangle.$$

If we choose for $|n\rangle$ the basis made of the eigenstates of the Hamiltonian, we find that the states

$$|0(\beta)\rangle = \frac{1}{\sqrt{Z}}\sum_n e^{-\beta e_n/2}|n\rangle \otimes |\tilde{n}\rangle \tag{7.41}$$

obey the requirement

$$\mathrm{Tr}(\hat{D}\hat{O}) = \langle 0(\beta)|\hat{O}|0(\beta)\rangle.$$

In other words, the statistical trace that defines the value of an observable at $T > 0$ is expressed in terms of a regular expectation value. This general formalism can be applied with any of the popular approximations to the density operator. In particular, it was applied to study small amplitude collective motion with an extension of the RPA and second RPA in hot nuclei [49–52]; to derive an extension of the Bloch–Messiah theorem for $T > 0$ nuclei [53]; and to formulate an alternative derivation of particle-number projection [42].

As an example, we can choose the HFB approximation of the density operator and neglect any interaction between q.p.s. The Hamiltonian is $\hat{K} = E_{\text{HFB}} + \sum_\mu \beta_\mu^\dagger \beta_\mu$. We choose a basis of multi-quasiparticles states

$$|n\rangle = \prod_{\mu \in I_n} \beta_\mu^\dagger |0\rangle,$$

where $|0\rangle$ is the FT-HFB ground state. The energy of these basis states is $e_n = \sum_{\mu \in I_n} E_\mu$, where the E_μ are the q.p. energies at $T > 0$. The basis of the doubled space is formally

$$|n, \tilde{n}\rangle = \prod_{\mu \in I_n} \beta_\mu^\dagger \tilde{\beta}_\mu^\dagger |0, \tilde{0}\rangle$$

and verifies

$$\hat{H}|n, \tilde{n}\rangle = e_n|n, \tilde{n}\rangle.$$

From there, it is quite easy to recover the results of the FT-HFB theory.

References

[1] Bernard R *et al* 2011 Microscopic and nonadiabatic Schrödinger equation derived from the generator coordinate method based on zero- and two-quasiparticle states *Phys. Rev.* C **84** 044308

[2] Bally B 2014 *Description des noyaux impairs à l'aide d'une méthode de fonctionnelle énergie de la densité à plusieurs états de référence* PhD thesis Université de Bordeaux

[3] Borrajo M, Rodríguez T and Egido J 2015 Symmetry conserving configuration mixing method with cranked states *Phys. Lett.* B **746** 341

[4] Bohr A and Mottelson B R 1998 *Single-Particle MotionNuclear Structure*vol 1 (Singapore: World Scientific)

[5] des Cloizeaux J 1968 Approximation de Hartree–Fock et approximation de phase aléatoire à température finie *Many-Body Physics* ed C DeWitt and R Balian (Philadelphia, PA: Gordon and Breach)

[6] Blaizot J-P and Ripka G 1985 *Quantum Theory of Finite Systems* (Cambridge, MA: MIT Press)

[7] Goodman A L 1981 Finite-temperature HFB theory *Nucl. Phys.* A **352** 30

[8] Egido J L and Ring P 1993 The decay of hot nuclei *J. Phys. G: Nucl. Part. Phys.* **19** 1

[9] Reichl L E 1988 *A Modern Course in Statistical Physics.* (New York: Wiley)

[10] Bogolubov N N and Bogolubov N N Jr. 1982 *Introduction to Quantum Statistical Mechanics* (Singapore: World Scientific)

[11] Conway J B 1987 Roots and logarithms of bounded operators on Hilbert space *J. Funct. Anal.* **70** 171

[12] Gaudin M 1960 Une démonstration simplifiée du théorème de Wick en mécanique statistique *Nucl. Phys.* **15** 89

[13] Brack M and Quentin P 1974 Self-consistent calculations of highly excited nuclei *Phys. Lett.* B **52** 159

[14] Mosel U, Zint P-G and Passler K H 1974 Self-consistent calculations for highly excited compound nuclei *Nucl. Phys.* A **236** 252

[15] Sauer G, Chandra H and Mosel U 1976 Thermal properties of nuclei *Nucl. Phys.* A **264** 221
[16] Bonche P, Levit S and Vautherin D 1984 Properties of highly excited nuclei *Nucl. Phys.* A **427** 278
[17] Bonche P, Levit S and Vautherin D 1985 Statistical properties and stability of hot nuclei *Nucl. Phys.* A **436** 265
[18] Bonche P, Vautherin D and Vénéroni M 1986 Limiting temperatures of excited nuclei: static and dynamical aspects *J. Phys. Coll.* **47** 339
[19] Egido J L, Robledo L M and Martin V 2000 Behavior of shell effects with the excitation energy in atomic nuclei *Phys. Rev. Lett.* **85** 26
[20] Martin V, Egido J L and Robledo L M 2003 Thermal shape fluctuation effects in the description of hot nuclei *Phys. Rev.* C **68** 034327
[21] Pei J C *et al* 2009 Fission barriers of compound superheavy nuclei *Phys. Rev. Lett.* **102** 192501
[22] Martin V and Robledo L M 2009 Fission barriers at finite temperature: a theoretical description with the Gogny force *Int. J. Mod. Phys.* E **18** 861
[23] Sheikh J A, Nazarewicz W and Pei J C 2009 Systematic study of fission barriers of excited superheavy nuclei *Phys. Rev.* C **80** 011302
[24] Schunck N *et al* 2014 Quantification of uncertainties in nuclear density functional theory *Nucl. Data Sheets* **123** 115
[25] Lee H C and Das Gupta S 1979 Nuclear shape transitions at finite temperature *Phys. Rev.* C **19** 2369
[26] Tanabe K, Sugawara-Tanabe K and Mang H J 1981 Theory of the cranked temperature-dependent Hartree–Fock–Bogoliubov approximation and parity projected statistics *Nucl. Phys.* A **357** 20
[27] Goodman A L 1983 Landau–Ginzburg theory of phase transitions in heated rotating nuclei *Nucl. Phys.* A **406** 94
[28] Dobaczewski J *et al* 1996 Mean-field description of ground-state properties of drip-line nuclei: pairing and continuum effects *Phys. Rev.* C **53** 2809
[29] Schunck N, Duke D and Carr H 2015 Description of induced nuclear fission with Skyrme energy functionals. II. Finite temperature effects *Phys. Rev.* C **91** 034327
[30] Vautherin D and Vinh Mau N 1984 Temperature dependence of collective states in the random-phase approximation *Nucl. Phys.* A **422** 140
[31] Sommermann H M 1983 Microscopic description of giant resonances in highly excited nuclei *Ann. Phys.* **151** 163
[32] Ring P *et al* 1984 Microscopic theory of the isovector dipole resonance at high angular momenta *Nucl. Phys.* A **419** 261
[33] Lacroix D, Chomaz P and Ayik S 1998 Finite temperature nuclear response in the extended random phase approximation *Phys. Rev.* C **58** 2154
[34] Canosa N, Matera J and Rossignoli R 2007 Description of thermal entanglement with the static path plus random-phase approximation *Phys. Rev.* A **76** 022310
[35] Paar N, Khan E and Vretenar D 2009 Calculation of stellar electron-capture cross sections on nuclei based on microscopic Skyrme functionals *Phys. Rev.* C **80** 055801
[36] Dang N and Tanabe K 2006 Self-consistent random-phase approximation at finite temperature within the Richardson model *Phys. Rev.* C **74** 034326
[37] Minato F and Hagino K 2009 β-decay half-lives at finite temperatures for $N = 82$ isotones *Phys. Rev.* C **80** 065808

[38] Egido J L 1988 Quantum versus statistical fluctuations in mean-field theories *Phys. Rev. Lett.* **61** 767
[39] Esebbag C and Egido J L 1993 Number projected statistics and the pairing correlations at high excitation energies *Nucl. Phys.* A **552** 205
[40] Fanto P 2017 Projection after variation in the finite-temperature Hartree–Fock–Bogoliubov approximation *Phys. Rev.* C **96** 051301
[41] Fanto P, Alhassid Y and Bertsch G F 2017 Particle-number projection in the finite-temperature mean-field approximation *Phys. Rev.* C **96** 014305
[42] Tanabe K and Nakada H 2005 Quantum number projection at finite temperature via thermofield dynamics *Phys. Rev.* C **71** 024314
[43] Rossignoli R and Ring P 1994 Projection at finite temperature *Ann. Phys.* **235** 350
[44] Egido J L *et al* 1986 The nuclear deformation parameters at high excitation energies *Phys. Lett.* B **178** 139
[45] Alhassid Y *et al* 2005 Nuclear moment of inertia and spin distribution of nuclear levels *Phys. Rev.* C **72** 064326
[46] Dietrich K, Niez J-J and Berger J-F 2010 Microscopic transport theory of nuclear processes *Nucl. Phys.* A **832** 249
[47] Umezawa H, Matsumoto H and Tachiki M 1982 *Thermofield Dynamics and Condensed States.* (Amsterdam: North-Holland)
[48] Khanna F C *et al* 2009 *Thermal Quantum Field Theory.* (Singapore: World Scientific)
[49] Tanabe K and Sugawara-Tanabe K 1986 On the variational derivation of the thermal RPA equation *Phys. Lett.* B **172** 129
[50] Tanabe K 1988 Extended thermal random phase approximation equation for nuclear collective excitation at finite temperature *Phys. Rev.* C **37** 2802
[51] Dzhioev A A and Vdovin A I 2009 On the TFD treatment of collective vibrations in hot nuclei *Int. J. Mod. Phys.* E **18** 1–535
[52] Alan A and Dzhioev *et al* 2016 Thermal quasiparticle random-phase approximation with Skyrme interactions and supernova neutral-current neutrino–nucleus reactions *Phys. Rev.* C **94** 015805
[53] Tanabe K and Sugawara-Tanabe K 1990 Proof of the extended Bloch–Messiah theorem in the thermal Hartree–Fock–Bogoliubov theory *Phys. Lett.* B **247.2** 202

Chapter 8

Numerical implementations

Nicolas Schunck

The previous chapters have given an overview of the ensemble of methods used to compute nuclear properties within the general framework of the energy density functional (EDF) theory. For practical applications, all these methods must be implemented in a computer program. As may have become clear, the Hartree–Fock–Bogoliubov (HFB) equation is key to most of them: many ground-state properties of nuclei are computed at the single-reference energy density functional (SR-EDF) approximation of section 3.1; low-lying excited states and transitions are often obtained with the linear response theory of section 5.1.3, which requires knowledge of the HFB solution; collective excited states can be computed with the direct generator coordinate method (GCM) or projected GCM calculation of section 3.2, which again, is expanded on a basis of HFB states; and the description of large-amplitude collective motion such as fission is often based on the cranking approximation of the adiabatic time-dependent Hartree–Fock–Bogoliubov (ATDHFB) theory or the Gaussian overlap approximation (GOA) of the GCM of section 6.4—both need only quasiparticle (q.p.) energies and HFB solutions.

In this chapter, we will thus describe the various techniques that have been developed to solve HFB equation. One of the most popular methods to represent the HFB equation, used among others in the published codes HOSPHE [1], HFBTHO [2], HFODD [3] and DIRHB [4], is based on expanding the HFB solution on a convenient basis of the single-particle (s.p.) Hilbert space, which turns the HFB equation into a non-linear eigenvalue problem. We will present this approach in detail in section 8.1. The alternative to basis expansion methods is the use of lattice techniques, e.g. in SKY [5], EV [6, 7] and HFBRAD [8], which we will describe in section 8.2. Note that SKY [5] solves the time-dependent Hartree–Fock (TDHF) equation, and pairing correlations are treated at the Bardeen–Cooper–Schrieffer (BCS) level for the initial, static solution only. The suite of codes EV of [6, 7] solve the static Hartree–Fock (HF)+BCS equation. In our presentation, we will focus on

doi:10.1088/2053-2563/aae0edch8

those techniques implemented in publicly available codes. Whenever relevant, we will indicate alternative methods that have been employed in specific studies. Since the HFB equation is non-linear, two main techniques are used to handle this non-linearity: successive diagonalizations, and the gradient method. Both will be presented in section 8.3. Finally, we will say a few words about specific algorithms used in time-dependent extensions.

8.1 Configuration space and basis expansions

In an arbitrary basis of the s.p. Hilbert space, we recall that the HFB equation derived in section 3.1.3 can be recast into a non-linear eigenvalue problem

$$\begin{pmatrix} h & \Delta \\ -\Delta^* & -h^* \end{pmatrix}\begin{pmatrix} U_\mu \\ V_\mu \end{pmatrix} = E_\mu \begin{pmatrix} U_\mu \\ V_\mu \end{pmatrix}, \tag{8.1}$$

with the HFB matrix $\mathcal{H}$ given by

$$\mathcal{H} = \begin{pmatrix} h & \Delta \\ -\Delta^* & -h^* \end{pmatrix}.$$

We will establish this equivalence between the original HFB equation (3.43) and a non-linear eigenvalue problem in section 8.3.1. The HFB matrix contains the mean field $h = t + \Gamma$ (t the kinetic energy and Γ the HF potential) and pairing field Δ. When the HFB equation is derived from some Hamiltonian $\hat{H}$, the mean field and pairing field read

$$h_{ij} = t_{ij} + \sum_{kl} \bar{v}_{ikjl}\rho_{lk} \qquad \Delta_{ij} = \frac{1}{2}\sum_{kl} \bar{v}_{ijkl}\kappa_{kl}. \tag{8.2}$$

If the HFB equation is derived from a general energy functional $\mathcal{E}[\rho, \kappa, \kappa^*]$, then they are given by equation (3.39),

$$h_{ij} = \frac{\partial \mathcal{E}}{\partial \rho_{ji}} \qquad \Delta_{ij} = \frac{\partial \mathcal{E}}{\partial \kappa_{ij}^*}.$$

In practice, the implementation of the HFB theory in configuration space thus begins with the calculation of the matrix elements of the mean field and pairing field in a given basis. We refer to section 1.1.1, for a reminder of some basic properties of the s.p. Hilbert space. Here, we will first discuss in more detail the harmonic oscillator (HO) basis in section 8.1.1, since it plays a special role in nuclear physics. Finally, we will show how matrix elements can be computed in section 8.1.2.

8.1.1 The harmonic oscillator basis

The harmonic oscillator basis plays a very special role in nuclear physics. In this section, we will recall some of the basic definitions of this basis in order to illustrate how to implement the HFB equation in configuration space. Following the notation in section 1.1.1, we first write the HO basis functions as

$$\psi_i(\boldsymbol{r}\sigma) \equiv \psi_i(\boldsymbol{r})\chi_{\frac{1}{2}\sigma}, \tag{8.3}$$

with i a set of relevant, spatial quantum numbers. As we will see below, these quantum numbers depend on the underlying symmetries and on the system of coordinates of the HO. Note that in equation (8.3), the quantum numbers relative to spatial coordinates and to spin coordinates are not coupled. There are several important cases where it is advantageous to couple them explicitly together. This section summarizes information on the HO basis that can be found in several nuclear physics textbooks [9–11], or in the texts accompanying specific HFB solvers [1, 12, 13].

In this section, we do not explicitly include the isospin degree of freedom in our discussion. In implementations of the HFB theory that consider isospin mixing such as presented in [14–19], it may be advantageous to consider basis states of the form

$$\psi_i(\boldsymbol{r}\sigma\tau) \equiv \psi_i(\boldsymbol{r})\chi_{\frac{1}{2}\sigma}\zeta_{\frac{1}{2}\tau}.$$

Coordinate systems

We choose spherical coordinates, $\boldsymbol{r} \equiv (r, \theta, \varphi)$, to describe an isotropic 3D quantum HO for a particle of mass m,

$$\hat{H}_{\mathrm{HO}} = \frac{\boldsymbol{p}^2}{2m} + \frac{1}{2}m\omega^2\boldsymbol{r}^2. \tag{8.4}$$

In this case, the Laplacian $\Delta = -\boldsymbol{p}^2/\hbar^2$ is separable into a purely radial and purely angular part,

$$\hat{H}_{\mathrm{HO}} = \frac{\hat{\ell}^2}{2m} - \frac{\hbar^2}{2m}\frac{1}{r}\frac{d^2}{dr^2} + \frac{1}{2}m\omega^2 r^2,$$

and the spatial HO eigenfunctions thus take the form

$$\psi_{nlm_\ell}(r, \theta, \varphi) = g_{nl}(r)\,Y_{lm_\ell}(\theta, \varphi), \tag{8.5}$$

where $Y_{lm_\ell}(\theta, \varphi)$ is the spherical harmonic of order l; see [20] for a comprehensive presentation of their properties. The radial wave function is

$$g_{nl}(r) = \sqrt{\frac{2n!}{b^3\Gamma\left(n + l + \frac{3}{2}\right)}}\left(\frac{r}{b}\right)^l \mathrm{e}^{-r^2/2b^2} L_n^{(l+1/2)}\left(\frac{r^2}{b^2}\right),$$

where $L_n^{(l+1/2)}$ is the associated Laguerre polynomial and Γ is the Γ-function; see [21] for definitions. The oscillator length b (typically given in Fermis) is related to the oscillator frequency according to

$$b = \sqrt{\frac{\hbar}{m_{\mathrm{N}}\omega}},$$

with m_N the nucleon mass. Following the generic definition of equation (8.3), intrinsic spin degrees of freedom can be added by simply considering the tensor product

$$\psi_{nlm_\ell\sigma}(r, \theta, \varphi) = \psi_{nlm_\ell}(r, \theta, \varphi)\chi_{\frac{1}{2}\sigma}.$$

Alternatively, one may introduce the coupled representation by coupling the orbital and intrinsic spin. Formally, we write

$$|\psi_{nljm}\rangle = \left[|\psi_{nlm_\ell}\rangle \otimes |\chi_{\frac{1}{2}\sigma}\rangle\right]_j^m,$$

that is,

$$|\psi_{nljm}\rangle = \sum_{m_\ell\sigma} C^{jm}_{lm_\ell\frac{1}{2}\sigma}|\psi_{nlm_\ell}\rangle \otimes |\chi_{\frac{1}{2}\sigma}\rangle, \tag{8.6}$$

with $C^{jm}_{lm_\ell\frac{1}{2}\sigma}$ the Clebsch–Gordan coefficient [20]. Choosing the coupled representation can be advantageous when spherical symmetry is explicitly built into the code. Basis functions are characterized by the quantum numbers n, ℓ, jm in the coupled representation. As is well known, see, e.g. [9, 10], the energy levels of the three-dimensional HO are given by $E_N = \hbar\omega(N + 3/2)$, with $N = 2n + \ell$, $0 \leqslant \ell \leqslant N$.

As discussed in chapter 3, the EDF approach naturally leads to the concept of symmetry-breaking. In this context, it is, therefore, natural to introduce basis functions that somehow are adapted to non-isotropic density distributions. Most atomic nuclei are axially deformed in their ground state. In this case, we may choose the cylindrical system of coordinates $\boldsymbol{r} = (\rho, \theta, z)$ and use the eigenstates of the axially symmetric HO Hamiltonian,

$$\hat{H}_{\mathrm{HO}} = \frac{\boldsymbol{p}^2}{2m} + \frac{1}{2}m(\omega_\perp^2\rho^2 + \omega_z z^2),$$

which are given by

$$\psi_{n_\rho\Lambda n_z}(\rho, \varphi, z) = \frac{1}{\sqrt{2\pi}}\phi^{|\Lambda|}_{n_\rho}(\rho)\psi_{n_z}(z)e^{i\Lambda\varphi},$$

where $\psi^{|\Lambda|}_{n_\rho}(\rho)$ is the perpendicular part and $\psi_{n_z}(z)$ is the longitudinal part of the wave function. Furthermore, Λ is the orbital angular momentum projection on the z-direction and can take negative values. Here $\psi_{n_z}(z)$ is the one-dimensional harmonic oscillator function (see below) and

$$\psi^{|\Lambda|}_{n_\rho}(\rho) = N^\Lambda_{n_\rho}\frac{\sqrt{2}}{b_\perp}\left(\frac{\rho}{b_\perp}\right)^{|\Lambda|}e^{-\rho^2/2b_\perp}L^{(|\Lambda|)}_{n_\rho}\left(\frac{\rho}{b_\perp}\right),$$

with

$$N_{n_\rho}^{\Lambda} = \left(\frac{n_\rho!}{(n_\rho + |\Lambda|)!} \right)^{1/2},$$

where $L_n^{(\Lambda)}(x)$ is the associated Laguerre polynomial and $b_\perp$ is the oscillator length parameter corresponding to $\omega_\perp$. Again, the intrinsic spin degree of freedom can be added by simple tensor coupling. In cylindrical coordinates, it is customary to note the z-projection of the intrinsic spin Σ. Hence,

$$\psi_{n_\rho \Lambda n_z \Sigma}(\rho, \varphi, z) = \psi_{n_\rho \Lambda n_z}(\rho, \varphi, z)\chi_{\frac{1}{2}\Sigma}.$$

The quantum numbers of the axially symmetric HO are now Λ, n_ρ, which characterize the 'perpendicular' wave function, and n_z which characterizes the 'longitudinal' wave function. When including spin, we note $\Omega = \Lambda + \Sigma$ with $\Sigma = \pm 1/2$. The main quantum number of the HO is $N = (2n_\rho + |\Lambda| + n_z + 3/2)$; see [22] for details.

In the most general case, where no particular symmetry is present, we may use the Cartesian coordinates, $\boldsymbol{r} \equiv (x, y, z)$ and the eigenfunctions of the anisotropic harmonic oscillator,

$$\hat{H}_{\mathrm{HO}} = \frac{\boldsymbol{p}^2}{2m} + \frac{1}{2}m\left(\omega_x^2 x^2 + \omega_y^2 y^2 + \omega_z z^2\right).$$

These eigenfunctions can be factorized along each Cartesian direction,

$$\psi_{n_x n_y n_z}(x, y, z) = \psi_{n_x}(x)\psi_{n_y}(y)\psi_{n_z}(z),$$

where each $\psi_{n_i}(x)$ is a one-dimensional harmonic oscillator wave function towards the i-direction,

$$\psi_{n_i}(x) = (\sqrt{\pi}2^{n_i}n_i!)^{-1/2} b_i^{-1/2} H_{n_i}\left(\frac{x}{b_i}\right) e^{-x^2/2b_i^2}, \tag{8.7}$$

with $b_i = \sqrt{\hbar/m_{\mathrm{N}}\omega_i}$ and $H_n(z)$ is the Hermite polynomial of order n. The intrinsic spin degree of freedom is incorporated by tensor product,

$$\psi_{n_x n_y n_z \sigma}(x, y, z) = \psi_{n_x n_y n_z}(x, y, z)\chi_{\frac{1}{2}\sigma}. \tag{8.8}$$

Moshinsky transformation and Talmi coefficients

One of the many advantages of the HO basis is its analytical properties. In particular, the HO wave functions in any coordinate system can be transformed exactly into another coordinate system [23]. In addition, the one-dimensional HO wave functions (8.7) satisfy a number of relations that prove very useful in practical applications. Some of the properties of the HO that we will discuss below are most conveniently established using a second quantization representation of the HO wave functions with creation and annihilation of 'quanta' in each relevant direction of the coordinate system. For instance, one may write $|n_x\rangle \approx \eta_x^{n_x}|-\rangle$; see [10, 23, 24] for a more detailed discussion.

The first of these properties is related to the expansion of the product of two Hermite polynomials [25]

$$H_{n'}^{(0)}(\xi)H_{n'}^{(0)}(\xi) = \sum_{k=0}^{n'+n} C_{n'n}^{k} H_{k}^{(0)}(\xi). \tag{8.9}$$

Here, the notation $H_n^{(0)}(x)$ refers to the normalized Hermite polynomial of order n,

$$\int_{-\infty}^{+\infty} dx H_n^{(0)}(x) H_m^{(0)}(x) e^{-x^2} = \delta_{nm}.$$

This implies that the product of two full, one-dimensional, HO wave functions (8.7) can also be expanded according to

$$\varphi_{n'}^{(b)}(x)\varphi_n^{(b)}(x) = be^{-\xi^2}\sum_k C_{n'n}^k H_k^{(0)}(\xi) = \sqrt{b}\,e^{-\xi^2/2}\sum_k C_{n'n}^k \varphi_k^{(b)}(x).$$

The coefficients $C_{n'n}^k$ can be computed easily by quadrature. The second interesting property of the HO wave functions comes from the scaling transformation of the normalized Hermite polynomial,

$$H_n^{(0)}\left(\frac{x}{b}\right) = \sqrt{\frac{b}{b'}} \sum_{n'=0}^{n} D_{nn'}\left(\frac{b'}{b}\right) H_n^{(0)}\left(\frac{x}{b'}\right). \tag{8.10}$$

The coefficients $D_{nn'}(\eta)$ are defined by

$$D_{nn'}(\eta) = \begin{cases} (-1)^{\frac{n-n'}{2}}\left(\frac{n!}{n'!}\right)^{1/2} \dfrac{\eta^{n'+1/2}(1-\eta^2)^{\frac{n-n'}{2}}}{2^{\frac{n-n'}{2}}\left(\frac{n-n'}{2}\right)!} & n-n' \text{ even} \\ 0 \quad n-n' \text{ odd} \end{cases}.$$

Perhaps the most important property of HO wave functions is the Moshinsky transformation. It consists in introducing the variables

$$\begin{cases} U = \dfrac{1}{\sqrt{2}}(x + x') \\ u = \dfrac{1}{\sqrt{2}}(x - x') \end{cases}.$$

The Jacobian of this transformation is 1. Note that the new variables (U, u) always have the same units as the old ones (x, x'). In fact, the Moshinsky transformation below can be applied either on the normalized Hermite polynomials, $H_n^{(0)}$, or on the

Hermite functions, φ_n, irrespective of whether the latter are properly normalized by the $\sqrt{b}$ scale, in exactly the same way. Here, we present the Moshinski transformation on fully normalized Hermite functions,

$$\varphi_k^{(b)}(x)\varphi_l^{(b)}(x') = \sum_{N,n} M_{kl}^{Nn}\varphi_N^{(b)}(U)\varphi_n^{(b)}(u). \tag{8.11}$$

In this expression, the relation $N + n = k + l$ must be satisfied, which provides limits to the summation. We note that the inverse transformation is given by

$$\begin{cases} x = \dfrac{1}{\sqrt{2}}(U + u) \\ x' = \dfrac{1}{\sqrt{2}}(U - u) \end{cases} \Rightarrow x^2 + x'^2 = U^2 + u^2.$$

The Talmi–Moshinsky coefficients are defined by

$$M_{kl}^{Nn} = \delta_{k+l,N+n}\left(\frac{k!l!}{2^{k+l}N!n!}\right)^{1/2}\sum_m(-1)^m\binom{N}{k-n+m}\binom{n}{m},$$

with $\binom{n}{m}$ denoting the usual binomial coefficients. For $N = 0$, this reduces to

$$M_{kl}^{0n} = \delta_{k+l,n}\left(\frac{k!l!}{2^{k+l}n!}\right)^{1/2}\sum_m(-1)^m\binom{0}{k-n+m}\binom{n}{m}.$$

By definition,

$$\binom{0}{k-n+m} \Rightarrow k - n + m = 0 \Rightarrow m = n - k = l.$$

Hence

$$M_{kl}^{0n} = \delta_{k+l,n}\left(\frac{k!l!}{2^{k+l}n!}\right)^{1/2}(-1)^l\binom{n}{l} = \delta_{k+l,n}\left(\frac{k!l!}{2^{k+l}n!}\right)^{1/2}(-1)^l\frac{n!}{l!k!}.$$

Therefore

$$M_{kl}^{0n} = \delta_{k+l,n}(-1)^l\sqrt{\frac{n!}{2^{k+l}k!l!}}.$$

As we will see later, the Moshinski transformation is very useful to compute the matrix elements of a Gaussian potential. In a totally different context, it is also a key element of *ab initio* theories of nuclear structure, since it is nothing other than a transformation in a center-of-mass frame, with U playing the role of the coordinate of the center-of-mass, and u that of the relative motion.

Symmetries

Symmetries in the nuclear EDF approach were discussed extensively in chapter 3. However, this discussion was focused more on general aspects related to spontaneous symmetry-breaking as a mechanism to include many-body correlations in the nuclear wave function. Here, we will look instead at specific examples of common symmetries and show how to build such symmetries into the *basis* functions used to expand the solution of the HFB equation. Taking advantage of built-in symmetries of the basis to enforce the same symmetries into the HFB solution often considerably accelerates a HFB solver. The price for this speed-up is a loss of generality, since the same conserved symmetries may not give the lowest energy for a given configuration. We will consider here both continuous symmetries, i.e. operators $\hat{R}(\alpha)$ associated with one or several continuous parameters α, and well as discrete symmetries, i.e. operators associated with finite groups. Rotations are examples of the former, parity or time-reversal are examples of the latter.

Let us assume that the basis that defines the configuration space is the set of eigenvectors of an s.p. Hamiltonian $\hat{h}$, for instance the three-dimensional harmonic oscillator discussed earlier. We first introduce a group $\mathcal{G}$ of transformations that leave the system described by $\hat{h}$ invariant. In other words, while any transformation $\hat{R} \in \mathcal{G}$ may change the coordinates of the system, it does not change the Hamiltonian $\hat{h}$. For example, the isotropic 3D HO given by equation (8.4) remains obviously unchanged if $\boldsymbol{r} \rightarrow -\boldsymbol{r}$ or under any rotation of the coordinates. If $\mathcal{G}$ is a compact Lie group of dimension d, any transformation $\hat{R} \in \mathcal{G}$ can be expressed as a function of the generators $\hat{\boldsymbol{C}} = (\hat{C}_1, \dots, \hat{C}_d)$ of the group according to

$$\hat{R} = e^{-i\boldsymbol{\alpha}\cdot\hat{\boldsymbol{C}}}.$$

Therefore, the group $\mathcal{G}$ is a symmetry group if its generators commute with the Hamiltonian,

$$[\hat{h}, \hat{\boldsymbol{C}}] = 0,$$

where we have adopted a short-hand notation, $[\hat{h}, \hat{\boldsymbol{C}}] \equiv [\hat{h}, \hat{C}_i], \forall i \in [1, d]$. See section 2.1 for a much more extensive presentation of symmetries, Lie groups, their algebra and their representations in the context of special relativity. Most of these considerations still apply here.

Point groups are another important example of symmetries [26]. They correspond to geometrical transformations that leave at least one point of the system invariant. If $\mathcal{G}$ is a point group of dimension d, it is, by definition, made of d transformations $\hat{R}_i$. In order for this group to be a symmetry group for $\hat{h}$, we must thus have

$$[\hat{h}, \hat{R}_i] = 0 \quad \forall i \in [1, d].$$

Note that point symmetries are exclusively geometric in nature.

The theory of point groups is presented in great detail in [26, 27]. Let us give just one simple example of such a group. Suppose we have a system (a nucleus, a molecule, etc) that has a symmetry axis of order n, i.e. which remains invariant under a rotation of angles $2\pi/n$ about said axis. The ensemble of such transformations forms a group of dimension n, and is denoted by $\mathcal{C}_n$. There is only a finite number of such groups, which have been comprehensively studied, in particular in the context of molecular symmetries.

For a symmetry group, we can thus find a basis that simultaneously diagonalizes both $\hat{h}$ and the various transformations $\hat{\boldsymbol{C}}$ (compact Lie group) or $\hat{R}_i$ (point group). We have already encountered an example of such a basis: in the coupled representation of the spherical, isotropic HO, the basis functions are both eigenfunctions of the HO Hamiltonian and of the angular momentum operator $\hat{j}^2$ and $\hat{j}_z$. The advantage of using such symmetries becomes obvious if we examine the structure of the matrix of an arbitrary operator $\hat{O}$. To illustrate this point, let us take the simple case of a symmetry operator $\hat{R}$ with only two eigenvalues, $r = \pm 1$. Let us now consider an operator $\hat{O}$ that also commutes with $\hat{R}$. We look at the matrix elements of the type

$$O_{+-} = \langle + | \hat{O} | - \rangle$$

with $|\pm\rangle$ some generic eigenstate of $\hat{h}$ corresponding to the eigenvalue ± 1 of $\hat{R}$. We know that, by definition,

$$\hat{R}|\pm\rangle = \pm| \pm \rangle \Rightarrow \langle \pm | \hat{R}^{\dagger} = \pm \langle \pm |.$$

Hence,

$$O_{+-} = \langle + | \hat{O} | - \rangle = \left(+ \langle + | \hat{R}^{\dagger} \right) \hat{O} \left(- \hat{R} | - \rangle \right) = -\langle + | \hat{R}^{\dagger} \hat{O} \hat{R} | - \rangle = -\langle + | \hat{O} | - \rangle = 0,$$

where we have use the property that for unitary operators, $[\hat{R}, \hat{O}] = 0 \Leftrightarrow \hat{R}\hat{O}\hat{R}^{\dagger} = \hat{O}$. Therefore, if we order the basis functions by putting first the states $|+\rangle$, then the states $|-\rangle$, we find that the matrix of the operator $\hat{O}$ has the following block structure

$$O = \begin{pmatrix} O_{++} & 0 \\ 0 & O_{--} \end{pmatrix}.$$

If $\hat{R}$ is a symmetry of the s.p. Hamiltonian $\hat{h}$, the block structure also applies to the s. p. levels: we can thus label the eigenstates of $\hat{h}$ with the quantum numbers r.

The block structure of matrices induced by symmetry transformations explains why it is computationally attractive to take advantage of such symmetries. Recall that, e.g. the diagonalization of a square matrix of size $N \times N$ scales like N^3—if the matrix is block diagonal, one only has to diagonalize two matrices of size $\frac{N}{2} \times \frac{N}{2}$, each with 1/8 of the cost of the full diagonalization.

Self-consistent symmetries

Let us take a group of transformation $\mathcal{G}$ characterized by n generators $\hat{C}$ (we drop the index for convenience). We make the important assumption that $\mathcal{G}$ is a group of symmetry of the nuclear many-body Hamiltonian $\hat{H}$. Under the action of $\hat{C}$, the states of a given s.p. basis of the Hilbert space transform according to

$$|n\rangle \to |n'\rangle = \sum_m C_{nm}|m\rangle,$$

where C is the matrix of the operator $\hat{C}$. We have seen in section 3.1.5 that the one-body density matrix and anomalous density that characterize the HFB solution transform according to

$$\rho' = C^\dagger \rho C \qquad \kappa' = C^\dagger \kappa C^*$$

and have deduced from these relations that the mean field Γ and pairing field Δ will then transform similarly,

$$\Gamma' = C^\dagger \Gamma C \qquad \Delta' = C^\dagger \Delta C^*.$$

It is thus possible to define the double matrix

$$\mathcal{C} = \begin{pmatrix} C & 0 \\ 0 & C^* \end{pmatrix}$$

such that the generalized density $\mathcal{R}$ and the HFB matrix $\mathcal{H}$ transform according to

$$\mathcal{R} \to \mathcal{R}' = \mathcal{C}^\dagger \mathcal{R} \mathcal{C} \qquad \mathcal{H} \to \mathcal{H}' = \mathcal{C}^\dagger \mathcal{H} \mathcal{C}.$$

Let us now assume that the generalized density $\mathcal{R}^{(0)}$ that characterizes the system at the first iteration of the HFB equation is invariant under the transformations of $\mathcal{G}$. For example, we may choose the *initial* density matrix and pairing tensor constructed out of the solutions $\varphi_n(\boldsymbol{r})$ to the one-body Schrödinger equation for some spherical potential. The invariance of the generalized density is characterized by the property $\mathcal{R}^{(0)'} = \mathcal{R}^{(0)}$. Since $\mathcal{G}$ is a symmetry group for the nuclear Hamiltonian, the many-body nuclear potential $\hat{V}$ is invariant under any transformation of $\mathcal{G}$, $\hat{V} = \hat{C}^\dagger \hat{V} \hat{C}$, which guarantees that $\mathcal{H}^{(0)'} = \mathcal{H}^{(0)}$. Therefore, the eigenvectors $(U^{(0)}, V^{(0)})$ of $\mathcal{H}^{(0)}$ are also eigenstates of $\hat{C}$. Since the density matrix at the next iteration, $\mathcal{R}^{(1)}$, is precisely constructed out of these eigenvectors, it will also be invariant under $\mathcal{G}$ and this property will be thus also be valid at the end of the self-consistent process.

Even though the concept of symmetry-breaking is at the core of the EDF approach, not all symmetries of the nuclear Hamiltonian are broken in actual calculations. For instance, large-scale surveys of nuclear properties have shown that nearly all nuclei (with a very few possible exceptions) are most likely either spherical or axially deformed in their ground states. In such cases, the HFB solution is invariant under a number of geometrical rotations that form a (point) group of symmetry and there exist a number of relevant self-consistent symmetries, see [16, 28, 29] for a complete description. When solving the HFB equation, it would thus be advantageous to work in a basis that possesses the same symmetries as the physical solution. This will guarantee that all operators of interest, in particular the HFB Hamiltonian, are block diagonal.

> Although we emphasize that we can choose the characteristics of the basis based on what know of the solutions, the reverse is also true: practitioners often use codes with built-in self-consistent symmetries and apply them even in nuclei where we know these symmetries are probably broken.

Below is a (not complete) list of symmetry operators commonly used in HFB solvers. Additional discussion of important symmetries relevant for nuclear structure can be found in, e.g. the textbooks by Bohr and Mottelson [9] and in review articles such as [30].

Angular momentum. We have already seen in section 8.1.1 how quantum numbers associated with both orbital and total angular momentum could naturally appear as characteristics of HO basis functions. In particular, spherically symmetric systems are most conveniently described with the eigenfunctions of the isotropic, 3D HO. Following equation (8.5), we recall that the spatial part of the HO eigenfunctions,

$$\psi_{nlm_\ell}(r, \theta, \varphi) = g_{nl}(r) Y_{lm_\ell}(\theta, \varphi),$$

are eigenfunctions of the s.p. orbital angular momentum

$$\hat{\ell}^2 Y_{lm_\ell}(\theta, \varphi) = \ell(\ell + 1) Y_{lm_\ell}(\theta, \varphi)$$
$$\hat{\ell}_z Y_{lm_\ell}(\theta, \varphi) = m_\ell Y_{lm_\ell}(\theta, \varphi).$$

The basis functions in the coupled representation of the spherical HO basis are, similarly, eigenfunctions of the total angular momentum operator $\hat{j}^2$ and $\hat{J}_z$. For a spherical system, the matrix elements of any operator $\hat{O}$ that commutes $\hat{j}^2$ and $\hat{J}_z$ will thus acquire a particular block structure—it will be block diagonal, each block being labeled by j and ℓ.

Spherical tensor and the Wigner–Eckart theorem

As mentioned several times, spherical systems are most naturally described in a spherical coordinate system characterized by the coordinates (r, θ, φ) and the basis vectors $(e_r, e_\theta, e_\varphi)$. In fact, it is even more convenient to introduce a spherical basis characterized by the vectors (e_{+1}, e_{-1}, e_0) defined by

$$e_{+1} = -\frac{1}{2}(e_x + ie_y) \qquad e_0 = +e_z \qquad e_{-1} = +\frac{1}{2}(e_x - ie_y).$$

The interest of such a representation is that the components of a vector V in this basis can be treated as the components of what is called an irreducible tensor V_1 of rank 1. Irreducible (spherical) tensors or rank n are objects with $2n + 1$ components that transform like the eigenfunctions of angular momentum under a rotation of the coordinate system.

Theorem 4 (Wigner–Eckart). *For a given irreducible tensor T of rank k, the matrix elements of its $2k + 1$ components $\hat{T}_{k\mu}$ between the states $|njm\rangle$ and $|n'j'm'\rangle$ read*

$$\langle n'j'm'|\hat{T}_{k\mu}|njm\rangle = (-1)^{2\mu} C_{jmk\mu}^{j'm'} \frac{\langle n'j'||\hat{T}_k||nj\rangle}{\sqrt{2j' + 1}}. \tag{8.12}$$

The interest of this theorem is the following: instead of calculating all matrix elements for all states $|njm\rangle$ and all components μ, one can choose the most convenient to calculate—hence fixing m, m' and μ—wherefrom we obtain the reduced matrix elements $\langle n'j'm'||\hat{T}_k||njm\rangle$. For the other m, m' and μ terms, we have only to compute a Clebsch–Gordan coefficient. In addition, the Wigner–Eckart theorem immediately shows that the matrix elements of scalar operators, which are zero-rank tensors ($k = \mu = 0$), are

$$\langle n'j'm'|\hat{T}_{00}|njm\rangle = \delta_{jj'}\delta_{mm'} \frac{\langle n'j||\hat{T}_k||nj\rangle}{\sqrt{2j + 1}}$$

by virtue of the properties of Clebsch–Gordan coefficients. A comprehensive presentation of spherical tensors and their algebra can be found in [20].

Parity. The parity operator is a reflection with respect to the origin of the reference frame. It is a purely spatial transformation, that is, it leaves intrinsic spin vectors invariant. In Cartesian coordinates, it can be summarized by the following rules

$$\hat{P}: \begin{cases} x \to x' = -x \\ y \to y' = -y. \\ z \to z' = -z \end{cases}$$

Note that the aforementioned reflection is defined in the intrinsic reference frame of the nucleus. The parity eigenvalues are ± 1.

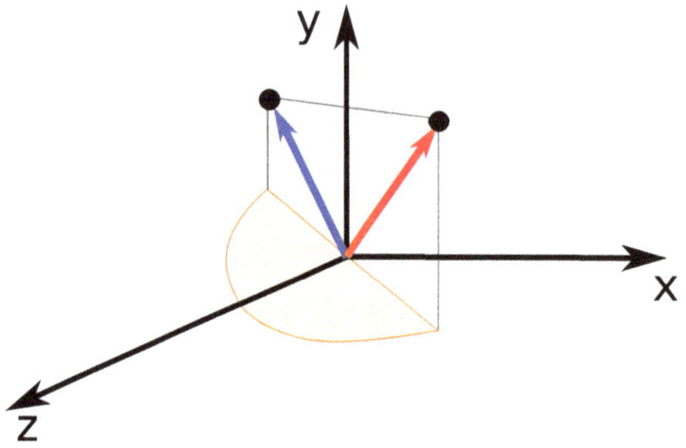

Figure 8.1. Schematic illustration of the y-signature symmetry, a rotation by an angle π around the y-axis. The x- and z-signatures are obtained similarly. (Figure created by Nicolas Schunck.)

Signature. The signature operator $\hat{R}_\mu$, $\mu = x, y, z$ is a rotation of angle π about the μ-axis. For example, the $\hat{R}_{xy}$ operation corresponds to the transformation

$$\hat{R}_y: \begin{cases} x \to x' = -x \\ y \to y' = +y \\ z \to z' = -z \end{cases}.$$

The operator performing this operation reads

$$\hat{R}_\mu = e^{-i\pi\hat{j}_\mu}, \tag{8.13}$$

where $\hat{j}_\mu$ is the μ-component of the total angular momentum (figure 8.1). A more convenient notation consists in using the properties of the Pauli matrices, and rewriting the signature operator as

$$\hat{S}_\mu = e^{-i\pi\hat{\ell}_\mu}(-i\sigma_\mu).$$

The eigenvalues of the signature operator are $\pm i$.

Simplex. The simplex operator operator $\hat{S}_\mu$, $\mu = x, y, z$ is a rotation of angle π about the μ-axis followed by an inversion. It corresponds to a reflection about the plane perpendicular to the μ-axis. For example, the $\hat{S}_y$ operation corresponds to a reflection about the x–z plane, i.e. it corresponds to the transformation

$$\hat{S}_y: \begin{cases} x \to x' = +x \\ y \to y' = -y \\ z \to z' = +z \end{cases}.$$

The operator performing this operation reads

$$\hat{S}_\mu = e^{-i\pi\hat{j}_\mu}\hat{P} = \hat{R}_\mu\hat{P}, \tag{8.14}$$

where $\hat{j}_\mu$ is, as before, the μ-component of the total angular momentum, and $\hat{P}$ is the parity operator (figure 8.2). Just as for the signature, we can rewrite it as

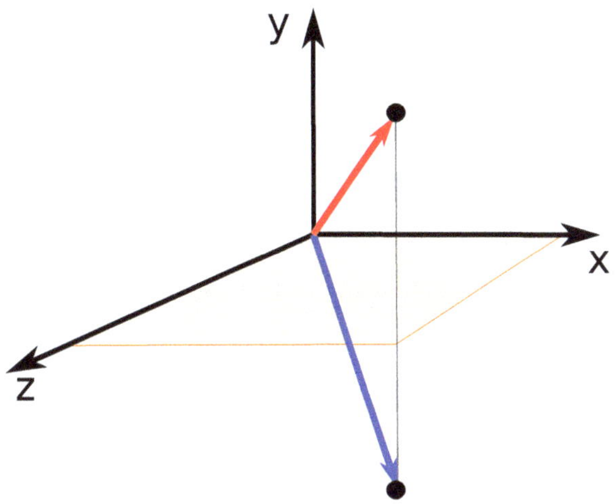

Figure 8.2. Schematic illustration of the y-simplex symmetry, a rotation by an angle π around the y-axis followed by an inversion. This is equivalent to a reflection with respect to the x–z plane. The x- and z-simplexes are obtained similarly. (Figure created by Nicolas Schunck.)

$$\hat{S}_\mu = e^{-i\pi\hat{\ell}_\mu}(-i\sigma_\mu)\hat{P}.$$

Because of this relation, the eigenvalues of the simplex operator are also $\pm i$ like those of the signature. The simplex operator is anti-symmetric and its inverse is given by $\hat{S}_\mu^{-1} = -\hat{S}_\mu$.

Time-reversal. There are usually two ways to define the time-reversal operator,

$$\hat{T} = \pm i\sigma_y\hat{K}, \tag{8.15}$$

where σ_y is the second Pauli matrix, and $\hat{K}$ is the complex conjugation operator. Among all the symmetry operators, time-reversal is the only one that is not unitary but anti-unitary. An anti-unitary operator verifies the following properties [31]: it is anti-linear, hence transforms kets as

$$\hat{T}(\alpha|u\rangle + \beta|\nu\rangle) = \alpha^*\hat{T}|u\rangle + \beta^*|\nu\rangle$$

and it also verifies

$$\langle\hat{T}u|\hat{T}\nu\rangle = \langle u|\nu\rangle^*.$$

Its action on spin 1/2 states is summarized by the two relations

$$\begin{aligned}\hat{T}|+\rangle &= -|-\rangle \\ \hat{T}|-\rangle &= +|+\rangle,\end{aligned} \tag{8.16}$$

which is sometimes condensed into $\hat{T}|\sigma\rangle = (-1)^{1/2+\sigma}|-\sigma\rangle$ with $\sigma = \pm 1/2$.

The spherical HO basis is the most natural example of a basis that can be used when rotational invariance is a self-consistent symmetry. In the case of nuclei that are axially deformed in their ground state, we note that the eigenfunctions of the axial HO are particularly adapted, since they are also eigenstates of $\hat{j}_z$. This means that we can order the basis by grouping together all the basis functions with the same Ω (the usual notation for the eigenvalue of $\hat{j}_z$). The matrix of all operators will thus be block diagonal with as many Ω-blocks as the maximum value of Ω included in the basis. The case of the Cartesian HO, however, is not so straightforward and requires a little bit more creativity.

Goodman basis

Although the Goodman transformation was originally introduced for axially symmetric rotating nuclei (where time-reversal symmetry is broken) [32], it has come to refer to a generic type of transformation that builds specific discrete symmetries in basis functions. Here, we are going to take the example of the Goodman basis for the y-simplex operator. We define the following set of s.p. states,

$$|t_+\rangle = \frac{1}{\sqrt{2}}(i^{n_y}|\boldsymbol{n}\rangle|+\rangle - i^{-n_y+1}|\boldsymbol{n}\rangle|-\rangle) \tag{8.17}$$

$$|t_-\rangle = \frac{1}{\sqrt{2}}(-i^{n_y+1}|\boldsymbol{n}\rangle|+\rangle + i^{-n_y}|\boldsymbol{n}\rangle|-\rangle), \tag{8.18}$$

where $|\boldsymbol{n}\rangle \equiv |n_x n_y n_z\rangle$ is a short-hand notation for an eigenstate of the Cartesian HO where the quantum numbers n_x, n_y, n_z and $|\pm\rangle$ refer, as before, to an eigenstate of the z-projection of the intrinsic spin $\hat{s}_z$. The transformation from the original HO states to the t-basis is unitary and has a matrix

$$U = \frac{1}{\sqrt{2}}\begin{pmatrix} +1 & +i \\ -i & -1 \end{pmatrix}.$$

It is not very difficult to show that the vectors $|t_+\rangle$ and $|t_-\rangle$ are eigenstates of both the time-reversal and y-simplex operators. Specifically,

$$\begin{aligned} \hat{T}|t_+\rangle &= -|t_-\rangle & \hat{T}|t_-\rangle &= +|t_+\rangle \\ \hat{S}_y|t_+\rangle &= +i|t_+\rangle & \hat{S}_y|t_-\rangle &= -i|t_-\rangle. \end{aligned}$$

These properties can be used to establish the particular structure of a one-body operator in the t-basis depending on their commutation or anti-commutation with the time-reversal and y-simplex operator. For example, if an operator $\hat{O}$ commutes with both $\hat{T}$ and $\hat{S}_y$, then we find that the block structure of the matrix of $\hat{O}$ reads

$$O = \begin{pmatrix} \langle t'_+|\hat{O}|t_+\rangle & 0 \\ 0 & \langle t'_+|\hat{O}|t_+\rangle^* \end{pmatrix}.$$

8.1.2 Calculation of matrix elements

Once a basis of the s.p. Hilbert space has been chosen, implementing the EDF theory requires computing matrix elements of various operators. Even though the EDF approach has been formulated to recast the original many-body problem in the form of an effective one-body problem, it still involves matrix elements of two- or three-body operators, see, e.g. the term $\bar{v}_{abcd}$ of equation (8.2). In spherical symmetry, it is possible to compute and store all these matrix elements explicitly and use them when needed. In more general calculations of deformed nuclei, however, the number of matrix elements becomes prohibitive and such a direct approach is often not possible. As a result, one often tries to compute the matrix elements of the mean field and pairing field on the fly at each iteration of the self-consistent loop. When the EDF is derived from a finite-range, two-body, effective pseudo-potential (or, equivalently, if it is directly built out of the non-local density), such calculations involve double integrals; if the EDF derives from a three-body potential, then triple integrals are needed. Since such calculations must be carried out at every iteration, special techniques—most of them specific to the HO basis—have been designed in to mitigate the computational cost.

It is easy to understand why computing matrix elements of N-body potentials in symmetry-breaking schemes can quickly become challenging. Let us take the simplest example of a two-body potential in the Cartesian HO basis characterized by the quantum numbers n_x, n_y, n_z. Before anti-symmetrization, matrix elements are therefore of the type

$$v_{abcd} \equiv \langle \boldsymbol{n}_a \boldsymbol{n}_b | \hat{v} | \boldsymbol{n}_c \boldsymbol{n}_d \rangle,$$

with $\boldsymbol{n}_a \equiv (n_{x,\,a}, n_{y,\,a}, n_{z,\,a})$. Let us first assume naively that each quantum number goes from 0 to $N_0 = 20$. The total number of states in the basis is thus $N_0^{12} = 20^{12} \approx 4 \times 10^{15}$. Arguably, in this back-of-the-envelop estimate, we did not take into account properties such as the hermiticity of $\hat{v}$, but we have also not considered spin degrees of freedom. Let us assume instead that we require the major quantum number to be bound, $N = n_x + n_y + n_z \leqslant N_0$. The total number of combinations of (n_x, n_y, n_z) that satisfy this property is $A_{N_0+2}^{N_0} = (N_0 + 2)!/N_0! = (N_0 + 2)(N_0 + 1) = 462$ for $N_0 = 20$. The total number of matrix elements is now about $462^4 \approx 4 \times 10^{10}$. Assuming each matrix element is a double-precision complex number, hence requires 16 bits of storage, the memory requirements are of the order of 100 GB. The problem here is not so much the amount of data (which can easily be stored on today's supercomputers and tomorrow's computers) but the fact that these data need to be accessed at every iteration of the HFB self-consistent loop, see section 8.3, and contracted with objects such as the density matrix or pairing tensor, see, e.g. equation (3.36). Evidently, the situation deteriorates quickly if we need to increase N_0 (for instance in fission calculations [33]) or want to consider three-body potentials, see, e.g. [34] for additional discussion.

From wave functions to densities

As we have seen in chapter 3, the one-body density matrix and pairing tensors in configuration space are obtained from the eigenvectors of the HFB equation according to

$$\rho_{mn} = (V^* V^T)_{mn} = \sum_\mu V^*_{m\mu} V_{n\mu} \qquad \kappa_{mn} = (V^* U^T)_{mn} = \sum_\mu V^*_{m\mu} U_{n\mu}. \tag{8.19}$$

Therefore, once the HFB matrix has been diagonalized, simple matrix multiplications are all that is needed to compute the density matrix and pairing tensor. In practice, even implementations of the HFB equation in configuration space often require the value of these two objects in coordinate space, e.g. on the quadrature mesh used in the code. This is obtained by writing

$$\rho(x, x') = \sum_{mn} \rho_{mn} \psi_m(x) \psi^*_n(x') \qquad \kappa(x, x') = \sum_{mn} \kappa_{mn} \psi_m(x) \psi_n(x'), \tag{8.20}$$

with $x \equiv (\boldsymbol{r}, \sigma)$ as before, and $\psi_m(x)$ the spinors for the s.p. Hilbert space. These relations are obtained simply by going back to the definition (3.1) of the density matrix and pairing tensor and using the relations (1.1) to introduce creation/annihilation operators in coordinate space. By substituting equation (8.19), we find naturally the relations

$$\rho(x, x') = \sum_\mu \left(\sum_m V^*_{m\mu} \psi_m(x) \right) \left(\sum_m V_{n\mu} \psi^*_n(x') \right) = \sum_\mu V^*_\mu(x) V_\mu(x'). \tag{8.21}$$

Together with similar relations for κ, this leads us to pose

$$V_\mu(x) = \sum_m V_{m\mu} \psi^*_m(x) \qquad U_\mu(x) = \sum_n U_{m\mu} \psi_n(x). \tag{8.22}$$

A slightly different way of introducing $V_\mu(x)$ and $U_\mu(x)$ is also described in section 3.1.2. In addition to being used when computing the density in coordinate space (8.21), these relations are also particularly useful to determine the transformation rules of the coefficients $V_{m\mu}$ and $U_{m\mu}$ under various symmetry operations.

Knowledge of the density in coordinate space is especially useful when working with EDFs of the local density, such as those derived from zero-range pseudo-potentials such as the Skyrme potential, see section 1.2.1, since the energy can be expressed as an integral over the space of a local energy density.

Symmetries and matrix block structure

Let us continue with the example of the Goodman basis of equations (8.17) and (8.18). Recall that, in our example, this basis is made of eigenstates of the y-simplex operator. We first assume that the HFB solution also conserves the y-simplex. This implies that the U and V matrices take the following block structure

$$U = \begin{pmatrix} 0 & U_{+-} \\ U_{-+} & 0 \end{pmatrix} V = \begin{pmatrix} V_{++} & 0 \\ 0 & V_{--} \end{pmatrix}. \tag{8.23}$$

It is simple to show that the density matrix ρ, and consequently the matrix of the mean field h, will have the same kind of block structure as V, and that the pairing tensor, hence the matrix of the pairing field Δ, will be similar to U. If we further assume that time-reversal symmetry is also conserved by the HFB solution, then the following relations between blocks hold: $V_{--}^{*} = V_{++}$ and $U_{-+} = -U_{+-}^{*}$.

Spin–isospin expansion
One of the often overlooked aspects of computing matrix elements is the need to handle spin and isospin degrees of freedom. A particularly useful technique, called the spin–isospin expansion, is to express any one- or two-body operator as the product of a purely spatial part and some Pauli matrices representing the spin and isospin quantum numbers. In this section, we present the general recipes to work out such expansions.

One-body operators. Let us take the example of the one-body density matrix. Its matrix elements ρ_{ij} can be interpreted as the matrix elements of a one-body operator $\hat{\rho}$,

$$\rho_{ij} = \langle i|\hat{\rho}|j\rangle.$$

Similarly, we can write in coordinate space (see equation (1.21) for additional details)

$$\rho(x, x') = \langle \sigma\tau|\hat{\rho}(\boldsymbol{r}, \boldsymbol{r}')|\sigma'\tau'\rangle,$$

where, in the latter expression, the operator $\hat{\rho}(\boldsymbol{r}, \boldsymbol{r}')$ acts in spin $\otimes$ isospin space (and is a function of spatial coordinates). Operators acting in spin space can be represented by 2×2 unitary matrices and the Pauli matrices (together with the identity) form a basis for the space $SU(2)$ of such matrices. Also recall that the isospin space is formally identical to the spin space. In both cases, an operator is thus represented by a matrix of $SU(2)$, and we may write a generic operator acting in spin $\otimes$ isospin in the form

$$\hat{O} = \hat{O}_\sigma \otimes \hat{O}_\tau \equiv \hat{O}_\sigma \hat{O}_\tau.$$

This leads to writing the density matrix s.p. operator

$$\hat{\rho}(\boldsymbol{r}, \boldsymbol{r}') = \sum_{uk} \rho_{\mu k}(\boldsymbol{r}, \boldsymbol{r}')\hat{\sigma}_\mu \hat{\tau}_k.$$

Hence we have

$$\rho(x, x') = \sum_{uk} \rho_{\mu k}(\boldsymbol{r}, \boldsymbol{r}')\langle\sigma|\hat{\sigma}_\mu|\sigma'\rangle\langle\tau|\hat{\tau}_k|\tau'\rangle. \tag{8.24}$$

This is the spin–isospin expansion of the one-body density matrix $\rho(x, x')$. The coefficients of the expansion, the functions $\rho_{\mu k}(\boldsymbol{r}, \boldsymbol{r}')$ can be calculated by taking advantage of the resolution of the identity relation in spin space,

$$\begin{aligned}
\sum_{\sigma\sigma'}\langle\sigma|\hat{\sigma}_\mu|\sigma'\rangle\langle\sigma'|\hat{\sigma}_\nu|\sigma\rangle &= \sum_{\sigma}\langle\sigma|\hat{\sigma}_\mu\hat{\sigma}_\nu|\sigma\rangle \\
&= \sum_{\sigma}\langle\sigma|\left(\delta_{\mu\nu} + i\sum_{c}\varepsilon_{\mu\nu c}\hat{\sigma}_c\right)|\sigma\rangle \\
&= \delta_{\mu\nu} + i\varepsilon_{\mu\nu z}\sum_{\sigma}\langle\sigma|\hat{\sigma}_c|\sigma\rangle = \delta_{\mu\nu}.
\end{aligned}$$

In the spin–isospin expansion, we thus multiply both terms by $\langle\sigma|\hat{\sigma}_\nu|\sigma'\rangle\langle\tau|\hat{\tau}_l|\tau'\rangle$ and sum over all quantum numbers,

$$\sum_{\sigma\sigma'\tau\tau'} \rho(x, x')\langle\sigma|\hat{\sigma}_\nu|\sigma'\rangle\langle\tau|\hat{\tau}_l|\tau'\rangle = \sum_{\mu k}\rho_{\mu k}(\boldsymbol{r}, \boldsymbol{r}')\sum_{\sigma\sigma'\tau\tau'}\langle\sigma|\hat{\sigma}_\mu|\sigma'\rangle\langle\tau|\hat{\tau}_k|\tau'\rangle\langle\sigma|\hat{\sigma}_\nu|\sigma'\rangle\langle\tau|\hat{\tau}_l|\tau'\rangle,$$

which simply gives us

$$\rho_{\mu k}(\boldsymbol{r}, \boldsymbol{r}') = \sum_{\sigma\sigma'\tau\tau'} \rho(x, x')\langle\sigma'|\hat{\sigma}_\mu|\sigma\rangle\langle\tau'|\hat{\tau}_k|\tau\rangle.$$

The numbers $\langle\sigma|\hat{\sigma}_\mu|\sigma'\rangle$ and $\langle\tau|\hat{\tau}_k|\tau'\rangle$ are simply the relevant matrix elements of the Pauli matrices. The density matrix in full coordinate space, $\rho(x, x')$, can be computed by transforming from its expression in configuration space ρ_{bd} with equation (8.20).

Let us take the case of σ_z,

$$\sigma_z = \begin{pmatrix} 1 & 0 \\ 0 & -1 \end{pmatrix}.$$

Let us denote the eigenvalues of the z-component of the spin $s_z = +1/2 = \sigma/2$ and $s_z = -1/2 = \sigma'/2$. The matrix elements are

$$\begin{aligned}
\langle\sigma|\hat{\sigma}_z|\sigma\rangle &\equiv \langle+|\sigma_z|+\rangle = 1 \\
\langle\sigma'|\hat{\sigma}_z|\sigma'\rangle &\equiv \langle+|\sigma_z|+\rangle = 1 \\
\langle\sigma|\hat{\sigma}_z|\sigma'\rangle &\equiv \langle+|\sigma_z|-\rangle = 0 \\
\langle\sigma'|\hat{\sigma}_z|\sigma\rangle &\equiv \langle-|\sigma_z|+\rangle = 0.
\end{aligned}$$

Two-body operators. We can also introduce a spin–isospin expansion for a generic, spin- and isospin-dependent two-body operator by generalizing the framework developed for one-body operators. Therefore, we write

$$\hat{v}(\boldsymbol{r}_1, \boldsymbol{r}_2, \boldsymbol{r}'_1, \boldsymbol{r}'_2) = \sum_{\mu k} v_{\mu k}(\boldsymbol{r}_1, \boldsymbol{r}_2, \boldsymbol{r}'_1, \boldsymbol{r}'_2)\hat{\sigma}_\mu^{(1)}\hat{\sigma}_\mu^{(2)}\hat{\tau}_k^{(1)}\hat{\tau}_k^{(2)},$$

where $\hat{\sigma}_\mu^{(1)}\hat{\sigma}_\mu^{(2)}$ is the operator that is the tensor product of $\hat{\sigma}_\mu^{(1)}$ and $\hat{\sigma}_\mu^{(2)}$ (similarly for $\hat{\tau}_k^{(1)}\hat{\tau}_k^{(2)}$). The functions $v_{\mu k}(\boldsymbol{r}_1, \boldsymbol{r}_2, \boldsymbol{r}'_1, \boldsymbol{r}'_2)$ are purely spatial. They can be computed from the matrix elements $v(x_1 x_2, x'_1 x'_2)$ of the two-body potential,

$$v(x_1, x_2, x'_1, x'_2) = \sum_{\mu k} v_{\mu k}(\boldsymbol{r}_1, \boldsymbol{r}_2, \boldsymbol{r}'_1, \boldsymbol{r}'_2)\langle\sigma_1|\hat{\sigma}_\mu^{(1)}|\sigma_1'\rangle\langle\tau_1|\hat{\tau}_k^{(1)}|\tau_1'\rangle\langle\sigma_2|\hat{\sigma}_\mu^{(2)}|\sigma_2'\rangle\langle\tau_2|\hat{\tau}_k^{(2)}|\tau_2'\rangle,$$

which, for the special case of a spatially local interaction, reduces to

$$v(x_1, x_2, x'_1, x'_2) = \sum_{\mu k} v_{\mu k}(\boldsymbol{r}_1, \boldsymbol{r}_2)\langle\sigma_1|\hat{\sigma}_\mu^{(1)}|\sigma_1'\rangle\langle\tau_1|\hat{\tau}_k^{(1)}|\tau_1'\rangle\langle\sigma_2|\hat{\sigma}_\mu^{(2)}|\sigma_2'\rangle\langle\tau_2|\hat{\tau}_k^{(2)}|\tau_2'\rangle. \tag{8.25}$$

The components of the spin–isospin expansion can then be expressed as

$$\begin{aligned} v_{\mu k}(\boldsymbol{r}_1, \boldsymbol{r}_2) = \sum_{\sigma_1\sigma_1'\sigma_2\sigma_2'} \sum_{\tau_1\tau_1'\tau_2\tau_2'} & v(x'_1, x'_2, x_1, x_2)\langle\sigma_1'|\hat{\sigma}_\mu^{(1)}|\sigma_1\rangle\langle\tau_1'|\hat{\tau}_k^{(1)}|\tau_1\rangle \\ & \times\langle\sigma_2'|\hat{\sigma}_\mu^{(2)}|\sigma_2\rangle\langle\tau_2'|\hat{\tau}_k^{(2)}|\tau_2\rangle. \end{aligned} \tag{8.26}$$

Paradoxically, the spin–isospin expansion is relevant even for purely spatial operators. This is simply because the matrix elements of such operators must always be anti-symmetrized when used in nuclear structure calculations. We recall (see section 1.1.1 of chapter 1) that

$$\bar{v}_{abcd} = \langle ab|\hat{v}|cd\rangle - \langle ab|\hat{v}|dc\rangle = \langle ab|\hat{V}\hat{\mathcal{P}}|cd\rangle,$$

where $\hat{\mathcal{P}} = 1 - \hat{P}_x\hat{P}_\sigma\hat{P}_\tau$ is the full anti-symmetrizer operator. Since $\hat{P}_\sigma = \frac{1}{2}(1 + \boldsymbol{\sigma}^{(1)} \cdot \boldsymbol{\sigma}^{(2)})$ (and similarly for $\hat{P}_\tau$), even if $\hat{v}$ does not involve spin and/or isospin operators, its anti-symmetrized version will, and the spin–isospin expansion will be relevant. A good example is the Coulomb potential, see section 1.3.2

Application to the calculation of the energy. To illustrate how such expansions can be applied in practice, let us look at the mean-field contribution for the interaction energy,

$$\mathcal{E}_{\text{int}} = \frac{1}{2}\int dx_1 dx_2 dx'_1 dx'_2\ \bar{v}(x_1, x_2, x'_1, x'_2)\rho(x'_1, x_1)\rho(x'_2, x_2).$$

In a Hamiltonian picture where the energy is simply given by $\mathcal{E} = \langle\Phi|\hat{H}|\Phi\rangle/\langle\Phi|\Phi\rangle$, the interaction energy is the part of the total energy that only involves the potential. The total energy is obtained by also adding to it the kinetic energy. We can use the spin–isospin expansion of the two-body potential, (8.25) and (8.26), to express the direct and exchange terms in the potential energy. This gives us

$$\mathcal{E}_{\text{int}} = \frac{1}{2}\int d^3\boldsymbol{r}_1 \int d^3\boldsymbol{r}_2 \sum_{\mu k} \nu^D_{\mu k}(\boldsymbol{r}_1, \boldsymbol{r}_2) \sum_{\sigma\sigma', \tau\tau'} \langle\sigma_1|\hat{\sigma}^{(1)}_\mu|\sigma_1'\rangle\langle\tau_1|\hat{\tau}^{(1)}_k|\tau_1'\rangle\langle\sigma_2|\hat{\sigma}^{(2)}_\mu|\sigma_2'\rangle\langle\tau_2|\hat{\tau}^{(2)}_k|\tau_2'\rangle$$
$$\times \rho(\boldsymbol{r}_1\sigma'_1\tau'_1, \boldsymbol{r}_1\sigma_1\tau_1)\rho(\boldsymbol{r}_2\sigma'_2\tau'_2, \boldsymbol{r}_2\sigma_2\tau_2)$$
$$- \frac{1}{2}\int d^3\boldsymbol{r}_1 \int d^3\boldsymbol{r}_2 \sum_{\mu k} \nu^E_{\mu k}(\boldsymbol{r}_1, \boldsymbol{r}_2) \sum_{\sigma\sigma', \tau\tau'} \langle\sigma_1|\hat{\sigma}^{(1)}_\mu|\sigma_1'\rangle\langle\tau_1|\hat{\tau}^{(1)}_k|\tau_1'\rangle\langle\sigma_2|\hat{\sigma}^{(2)}_\mu|\sigma_2'\rangle\langle\tau_2|\hat{\tau}^{(2)}_k|\tau_2'\rangle$$
$$\times \rho(\boldsymbol{r}_2\sigma'_1\tau'_1, \boldsymbol{r}_1\sigma_1\tau_1)\rho(\boldsymbol{r}_1\sigma'_2\tau'_2, \boldsymbol{r}_2\sigma_2\tau_2).$$

We now recognize the summations that define the spin–isospin expansion of the density matrix, hence

$$\begin{aligned}\mathcal{E}_{\text{int}} &= \frac{1}{2}\int d^3\boldsymbol{r}_1 \int d^3\boldsymbol{r}_2 \sum_{\mu k} \nu^D_{\mu k}(\boldsymbol{r}_1, \boldsymbol{r}_2)\rho_{uk}(\boldsymbol{r}_1)\rho_{uk}(\boldsymbol{r}_2) \\ &- \frac{1}{2}\int d^3\boldsymbol{r}_1 \int d^3\boldsymbol{r}_2 \sum_{\mu k} \nu^E_{\mu k}(\boldsymbol{r}_1, \boldsymbol{r}_2)\rho_{uk}(\boldsymbol{r}_2, \boldsymbol{r}_1)\rho_{uk}(\boldsymbol{r}_1, \boldsymbol{r}_2).\end{aligned} \tag{8.27}$$

The main advantage of equation (8.27) is that the spin and isospin degrees of freedom are now integrated out into the spin- and isospin-dependent functions $\nu^D_{\mu k}$, $\nu^E_{\mu k}$ and $\rho_{\mu k}$. Computing the interaction energy reduces to simple integrations.

Gaussian potential

Many popular EDFs are constructed from the expectation value of some local, effective pseudo-potential $\hat{V}$ on a Slater determinant. This is the case, for instance, of the Skyrme and Gogny functionals. Even when building the EDF directly by a systematic expansion of density couplings up to a given order, the resulting EDF still requires a Coulomb term. At the simplest HF approximation, this Coulomb term also originates from a local two-body potential. In many cases, the potentials of interest are local, central and have a finite range. Computing matrix elements of such potentials in an arbitrary basis can quickly become cumbersome and computationally expensive. In the special case of the HO basis, however, it is advantageous to expand the potential on a series of Gaussians, for which the matrix elements possess a number of interesting properties. In the following, we illustrate how the properties of the HO basis functions and the Moshinsky transformation can be used to considerably simplify the calculation of matrix elements of a Gaussian potential. Although we focus on the special (while also most general) case of Cartesian coordinates, similar considerations also apply to spherical or cylindrical HO functions.

Gaussian expansions have been used to approximate the Yukawa spatial form factor $\exp(-\mu r)/(\mu r)$ [35, 36] or the Coulomb potential itself through the Gaussian substitution method, i.e. the property

$$\frac{1}{|\boldsymbol{r} - \boldsymbol{r}'|} = \frac{2}{\sqrt{\pi}}\int_0^\infty e^{-(\boldsymbol{r}-\boldsymbol{r}')^2/\mu^2}\frac{d\mu}{\mu^2}.$$

The discretization of the integral leads to a Legendre quadrature that corresponds to a finite sum of Gaussian terms [37–39].

Let us consider a sum of Gaussian potentials with different ranges and scales,

$$\hat{V}(\boldsymbol{r}_1, \boldsymbol{r}_2) = \sum_{w=1}^{N_w} \alpha_w e^{-\beta_w(\boldsymbol{r}_1-\boldsymbol{r}_2)^2},$$

where the dimensions are $[\alpha_w]$ = MeV and $[\beta_w] = \text{fm}^{-2}$. We look at the non anti-symmetrized matrix elements of each individual term w of the sum on Cartesian HO functions,

$$\nu_{\boldsymbol{n}'\boldsymbol{m}'\boldsymbol{n}\boldsymbol{m}} = \langle \boldsymbol{n}'\boldsymbol{m}' | \alpha_w e^{-\beta_w(\boldsymbol{r}_1-\boldsymbol{r}_2)^2} | \boldsymbol{n}\boldsymbol{m} \rangle. \tag{8.28}$$

Recall that the notation $|\boldsymbol{n}\rangle$ refers to a spinor of the type (8.8), that is, for the spatial part only,

$$\langle \boldsymbol{r} | \boldsymbol{n} \rangle \equiv \varphi_{n_x}^{(b_x)}(x)\varphi_{n_y}^{(b_y)}(y)\varphi_{n_z}^{(b_z)}(z),$$

with each of the $\varphi_{n_\mu}^{(b_\mu)}(x_\mu)$ given by equation (8.7). Because of the separability of both the exponential and the basis functions, this matrix element can be factorized as

$$\nu_{\boldsymbol{n}'\boldsymbol{m}'\boldsymbol{n}\boldsymbol{m}} = \nu^{(x)}_{n_x'm_x'n_xm_x}\nu^{(y)}_{n_y'm_y'n_ym_y}\nu^{(z)}_{n_z'm'_zn_zm_z},$$

with

$$\nu^{(\mu)}_{n'm'nm} = \alpha_w^{1/3} \int dx_\mu \int dx_\mu' \varphi_{n_\mu'}^{(b_\mu)*}(x_\mu)\varphi_{m_\mu'}^{(b_\mu)*}(x_\mu')e^{-\beta_w(x_\mu-x_\mu')^2}\varphi_{n_\mu}^{(b_\mu)}(x_\mu)\varphi_{m_\mu}^{(b_\mu)}(x_\mu'). \tag{8.29}$$

From now on, we drop the index μ for convenience.

Inserting the expansion (8.9) of the product of two Hermite polynomials into the matrix element (8.29), we find

$$\nu^{(\mu)}_{n'm'nm} = \alpha_w^{1/3} b \sum_{kl} C^k_{n'n} C^l_{m'm'} \int dx \int dx' e^{-\frac{1}{2}b^2(x^2+x'^2)} e^{-\beta_w(x-x')^2} \varphi_k^{(b)}(x)\varphi_l^{(b)}(x').$$

We now introduce the Moshinsky transformation (8.11) and obtain

$$\nu^{(\mu)}_{n'm'nm} = \alpha_w^{1/3} b \sum_{kl} C^k_{n'n} C^l_{m'm'} \sum_{N,n} M^{Nn}_{kl} \int dU \int du e^{-\frac{1}{2}b^2(U^2+u^2)} e^{-2\beta_w u^2} \varphi_N^{(b)}(U)\varphi_n^{(b)}(u),$$

which can be factorized in an integral over the center-of-mass coordinate U and another one over the relative motion coordinate u as follows

$$\nu^{(\mu)}_{n'm'nm} = \alpha_w^{1/3} b \sum_{kl} C^k_{n'n} C^l_{m'm'} \sum_{N,n} M^{Nn}_{kl} \int dU e^{-\frac{1}{2}b^2U^2} \varphi_N^{(b)}(U) \int du e^{-\frac{1}{2}b^2u^2-2\beta_w u^2} \varphi_n^{(b)}(u).$$

The first integral over U is easily computed, since the exponential factor is nothing other than the Hermite function of order 0 (within some normalization coefficient). Hence

$$\int dU e^{-\frac{1}{2}b^2U^2} \varphi_N^{(b)}(U) = \frac{\pi^{1/4}}{\sqrt{b}} \int dU \varphi_0^{(b)}(U)\varphi_N^{(b)}(U) = \frac{\pi^{1/4}}{\sqrt{b}} \delta_{N0}.$$

Therefore,

$$\nu^{(\mu)}_{n'm'nm} = \alpha_w^{1/3}\pi^{1/4} b \sum_{kl} C^k_{n'n} C^l_{m'm'} \sum_n M^{0n}_{kl} \int du e^{-b^2u^2-2\beta_w u^2} H^{(0)}_n(bu).$$

Next, we change the integration variable to $\theta = u\sqrt{b^2 + 2\beta_w}$ in order to put the integral into tabularized form compiled in, e.g. [40]. Straightforward mathematics leads to the final result

$$\nu^{(\mu)}_{n'm'nm} = \alpha_w^{1/3}\pi^{1/2} \frac{b}{\sqrt{b^2 + 2\beta_w}} \sum_{kl} C^k_{n'n} C^l_{m'm'} \sum_n M^{0n}_{kl} \left(\frac{-\beta_w}{b^2 + 2\beta_w} \right)^{n/2} \frac{\sqrt{n!}}{(n/2)!}. \qquad (8.30)$$

Here, recall that each of the indices n', m', n, m stand for one Cartesian direction. In a large 'stretched' basis with the elongation along the z-axis, this implies that $0 \leqslant n \leqslant N_{\max}$ with $N_{\max}$ possibly as high as 30. The summation over k, l typically extends to twice that value. From a computational point of view, the matrix element (8.30) involves only very cheap operations and minimal storage.

Computational efficiency

The main difficulty when implementing these formulas is to handle the large number of nested loops; since equation (8.30) applies only to one Cartesian direction, we see that the full calculation of *each* matrix element (8.28) would involve a 12-nested loop—and we still need to loop over $\boldsymbol{n}'$, $\boldsymbol{m}'$, $\boldsymbol{n}$ and $\boldsymbol{m}$ to fill the full matrix. For these reasons, the matrix elements are often calculated on-the-fly whenever the mean field potential Γ or pairing field Δ (see the definitions in equation (3.39)) are needed; see also the discussion at the beginning of this section. Below we sketch briefly how such strategies can be implemented. Let us take the example of a single Gaussian potential, which is separable,

$$V(\boldsymbol{r}, \boldsymbol{r}') = e^{-\frac{(x-x')^2}{\mu^2}} e^{-\frac{(y-y')^2}{\mu^2}} e^{-\frac{(z-z')^2}{\mu^2}}.$$

Therefore the mean field is given by the contraction

$$\Gamma_{\boldsymbol{n}'\boldsymbol{m}'} = \sum_{n_x'm_x'} V_{n_x n_x' m_x m_x'} \sum_{n_y'm_y'} V_{n_y n_y' m_y m_y'} \sum_{n_z'm_z'} V_{n_z n_z' m_z m_z'} \rho_{\boldsymbol{mn}}.$$

Instead of computing the naive, 12-dimensional loop, we can introduce separate contributions from each direction as follows (red indices imply summations, but not contractions)

$$
\begin{aligned}
Y^{n_z'm_z'}_{m_xn_xm_yn_y} &= \sum_{n_z'm_z'}\sum_{n_zm_z} V_{n_zn_z'm_zm_z'}\,\rho_{m_xm_ym_zn_xn_yn_z} \\
Z^{n_z'm_z'n_y'm_y'}_{m_xn_x} &= \sum_{n_y'm_y'}\sum_{n_ym_y} V_{n_yn_y'm_ym_y'}\,Y^{n_z'm_z'}_{m_xn_xm_yn_y} \\
\Gamma^{n_z'm_z'n_y'm_y'n_x'm_x'} &= \sum_{n_x'm_x'}\sum_{n_xm_x} V_{n_xn_x'm_xm_x'}\,Z^{n_z'm_z'n_y'm_y'}_{m_xn_x}.
\end{aligned}
$$

The resulting loops are of order N_0^9 instead of N_0^{12}, see the discussion in [2, 34, 38].

8.2 Lattice techniques

Numerical methods based on expanding the solutions of the HFB equation on a basis, especially the HO basis, are particularly convenient for computing, e.g., ground-state properties of nuclei or low-lying states. However, there are a few important physics problems where basis expansions are not very efficient. In particular, properties of very neutron-rich nuclei near the neutron drip lines, the structure of superheavy elements, or nuclear fission, are examples where expansions onto a one-center basis of spatially localized functions can lead to severe truncation errors; see figure 8.3 for an illustration.

As we will see in section 8.2.1, the HFB equation in coordinate space takes the form of a non-linear, integro-differential system of coupled equations. The numerical methods used to solve this system of equation depend on the symmetries one is willing to impose on the solution. For example, if spherical symmetry is assumed, one can show that the full HFB equation reduces to a system of coupled differential equations with respect to the radial coordinate r only, and all the dependence on angular variables is factorized out. The resulting system is simple enough that direct integration and differentiation techniques can be used. Since this special case is of interest to discuss some properties of the HFB solutions, we will present it in detail in section 8.2.2. In the most general case where no symmetry is assumed, solving the HFB equation requires explicit consideration of all three Cartesian coordinates. The cost of direct methods becomes prohibitive and special lattice techniques are often introduced, especially to improve the numerical efficiency of computing partial derivatives. We will give a brief review of these techniques in section 8.2.3. Finally, we will also discuss finite elements (FE) methods in section 8.2.4, which can be thought of as hybrid between basis expansions and lattice techniques.

8.2.1 HFB equation in coordinate space

Starting from the HFB equation in configuration space (8.1) and using the transformation (1.1) and (1.2), we can obtain a coordinate-space representation. Writing down explicitly the coordinate and spin degrees of freedom, we find an integro-differential equation of the kind

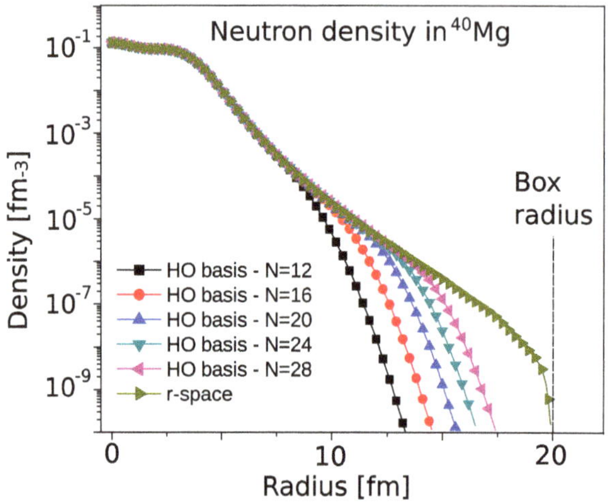

Figure 8.3. Profile of the radial density (8.32) in ^{40}Mg computed with the D1S Gogny interaction. The figure shows the density obtained when solving the spherical HFB equation in the HO basis (for different basis sizes) and in the basis made of eigenstates of the Woods–Saxon Hamiltonian; see [41, 42] for additional details. The HO basis is not adapted to describing the exponentially decreasing tail of the one-body density because of the incorrect asymptotic properties of HO basis functions. This is one of the reasons why coordinate-space, or lattice, methods are sometimes advantageous. (Figure created by Nicolas Schunck.)

$$\int d^3\boldsymbol{r}' \sum_{\sigma'} \begin{pmatrix} h(\boldsymbol{r}\sigma, \boldsymbol{r}'\sigma') - \lambda_{\mathrm{F}}\delta_{\sigma\sigma'} & \Delta(\boldsymbol{r}\sigma, \boldsymbol{r}'\sigma')\delta_{\sigma\sigma'} \\ \Delta(\boldsymbol{r}\sigma, \boldsymbol{r}'\sigma')\delta_{\sigma\sigma'} & -h(\boldsymbol{r}\sigma, \boldsymbol{r}'\sigma') + \lambda_{\mathrm{F}}\delta_{\sigma\sigma'} \end{pmatrix} \begin{pmatrix} U(\boldsymbol{r}'\sigma') \\ V(\boldsymbol{r}'\sigma') \end{pmatrix} = E \begin{pmatrix} U(\boldsymbol{r}\sigma) \\ V(\boldsymbol{r}\sigma) \end{pmatrix}, \tag{8.31}$$

where E are the q.p. energies, λ_{F} the Fermi energy, or chemical potential, used to ensure that the number of particles is conserved on average. In the most general case, both the mean field $h(\boldsymbol{r}\sigma, \boldsymbol{r}'\sigma')$ and pairing field $\Delta(\boldsymbol{r}\sigma, \boldsymbol{r}'\sigma')$ can be non-local. This is the case, for instance, of mean fields constructed from local, finite-range pseudo-potentials. If one makes the additional assumptions (most frequent in nuclear physics) that there is no proton–neutron mixing in the system, then this equation applies separately to protons and neutrons.

We show in equation (8.31) the HFB equation for one type of particle only, i.e. we do not consider proton–neutron mixing. If we did so, we would have to involve an additional summation over the isospin quantum number $\sum_{\sigma'} \rightarrow \sum_{\sigma'\tau'}$, and all fields would acquire an additional isospin dependence; for a presentation of the most general EDF formalism including to proton–neutron mixing and isospin projection, see [14–19, 43].

Locality: potentials versus densities

As was recalled earlier on, a one-body potential operator $\hat{V}$ is local if its action is of the form $\hat{V}: \psi(\boldsymbol{r}) \rightarrow V(\boldsymbol{r})\psi(\boldsymbol{r})$ and non-local if it is expressed as

$$\hat{V}: \psi(\boldsymbol{r}) \to \int d^3r' V(\boldsymbol{r}, \boldsymbol{r}')\psi(\boldsymbol{r}').$$

The same distinction applies to N-body potentials. An N-body, non-local operator has matrix elements of the type

$$\int d^3r_1 ... \int d^3r_N \int d^3r'_1 ... \int d^3r'_N \Psi^*(\boldsymbol{r}_1, \dots, \boldsymbol{r}_N)\hat{V}\Psi(\boldsymbol{r}'_1, \dots, \boldsymbol{r}'_N).$$

Confusion may arise when extending this discussion to energy densities. A 'local' two-body potential, such as, e.g. the Gogny or Coulomb potential, gives rise to energy densities that are functionals of the 'non-local density'. By extension, we often call such EDF non-local functionals. For example, the Coulomb potential discussed in section 1.3.2 reads

$$\hat{V}_{\text{Cou}} = \frac{e^2}{4\pi\epsilon_0}\frac{1}{|\boldsymbol{r}_1 - \boldsymbol{r}_2|}.$$

The direct part of the Coulomb EDF is a functional of the local (charge) density and reads

$$\mathcal{H}_{\text{dir}}(\boldsymbol{r}_1) = \frac{1}{2}\frac{e^2}{4\pi\epsilon_0}\rho_{\text{ch}}(\boldsymbol{r}_1)\int d^3r_2 \frac{\rho_{\text{ch}}(\boldsymbol{r}_2)}{|\boldsymbol{r}_1 - \boldsymbol{r}_2|}.$$

In contrast, the exchange part of the EDF is a functional of the non-local charge density,

$$\mathcal{H}_{\text{exc}}(\boldsymbol{r}_1) = -\frac{1}{2}\frac{e^2}{4\pi\epsilon_0}\int d^3r_2\left(\frac{\rho_{\text{ch}}(\boldsymbol{r}_1, \boldsymbol{r}_2)\rho_{\text{ch}}(\boldsymbol{r}_2, \boldsymbol{r}_1)}{|\boldsymbol{r}_1 - \boldsymbol{r}_2|} - \frac{\boldsymbol{s}_{\text{ch}}(\boldsymbol{r}_1, \boldsymbol{r}_2)\cdot\boldsymbol{s}_{\text{ch}}(\boldsymbol{r}_2, \boldsymbol{r}_1)}{|\boldsymbol{r}_1 - \boldsymbol{r}_2|}\right),$$

where the last term is only non-zero in systems where time-reversal invariance is internally broken such as, e.g. odd nuclei.

We define the energy functional $\mathcal{H}(\boldsymbol{r})$ as a local function of a single coordinate vector $\boldsymbol{r}$ such that the energy of the system is given by

$$\mathcal{E} = \int d^3r \mathcal{H}(\boldsymbol{r}).$$

As discussed in chapter 3, the spectrum of the HFB equation is unbounded from above and below. More precisely, the spectrum is continuous for $-\infty \leqslant E \leqslant \lambda_{\text{F}}$ and $-\lambda_{\text{F}} \leqslant E \leqslant +\infty$ and discrete for $\lambda_{\text{F}} \leqslant E_\mu \leqslant -\lambda_{\text{F}}$ (recall that $\lambda_{\text{F}} < 0$). As a result, the orthonormalization condition for the wave functions takes slightly different forms for both cases. For the discrete eigenvalues, we have

$$\sum_\sigma \int d^3r\left[U_\mu^*(\boldsymbol{r}\sigma)U_\nu(\boldsymbol{r}\sigma) + V_\mu^*(\boldsymbol{r}\sigma)V_\nu(\boldsymbol{r}\sigma)\right] = \delta_{\mu\nu},$$

while for the continuous part of the spectrum we impose

$$\sum_{\sigma} \int d^3\mathbf{r}[U^*(E, \mathbf{r}\sigma)U(E', \mathbf{r}\sigma) + V^*(E, \mathbf{r}\sigma)V(E', \mathbf{r}\sigma)] = \delta(E - E').$$

The general expression of the one-body density and pairing tensor in coordinate space are

$$\rho(\mathbf{r}\sigma, \mathbf{r}'\sigma') = \sum_{0<|E_\mu|<-\lambda_F} V_\mu^*(\mathbf{r}\sigma)V_\mu(\mathbf{r}'\sigma') + \int_{|\lambda_F|<|E|} dE\ V^*(E, \mathbf{r}\sigma)V(E, \mathbf{r}'\sigma')$$
$$\kappa(\mathbf{r}\sigma, \mathbf{r}'\sigma') = \sum_{0<|E_\mu|<-\lambda_F} V_\mu^*(\mathbf{r}\sigma)U_\mu(\mathbf{r}'\sigma') + \int_{|\lambda_F|<|E|} dE\ V^*(E, \mathbf{r}\sigma)U(E, \mathbf{r}'\sigma').$$

As we will show in the next section, the direct resolution of the HFB equation in coordinate space requires imposing boundary conditions at the edges of the spatial domain. Most of the time, these boundary conditions correspond to fixing the value of the wave functions or their derivatives to 0.

More realistic boundary conditions would involve imposing that the asymptotic form of the wave functions corresponds to scattering waves. The analytical forms of such scattering wave solutions are well known: they correspond to the Schrödinger equation where the nuclear potential goes to 0. We give a few additional details in section 8.2.2 in the context of the spherical HFB equation.

As a result, the continuous part of the spectrum is effectively discretized, and solving the HFB equation only requires setting the maximum value of the q.p. energy E_{cut}. In this case, the densities become simply

$$\rho(\mathbf{r}\sigma, \mathbf{r}'\sigma') = \sum_{0<E_\mu<E_{\mathrm{cut}}} V_\mu^*(\mathbf{r}\sigma)V_\mu(\mathbf{r}'\sigma')$$

and

$$\kappa(\mathbf{r}\sigma, \mathbf{r}'\sigma') = \sum_{0<E_\mu<E_{\mathrm{cut}}} V_\mu^*(\mathbf{r}\sigma)U_\mu(\mathbf{r}'\sigma').$$

The determination of the Fermi energy λ_F can be done in different ways. One possibility is to treat the Fermi energy as a constraint on the expectation value of the (one-body) number operator. Standard methods to deal with constraints involve the linear method based on the cranking approximation of the quasiparticle random-phase approximation (QRPA) matrix [3, 44, 45], the augmented Lagrangian method [46] or the quadratic constraint method [25, 47]. An alternative method that is specific to the number of particle consists in using a BCS-like formulation. At each iteration, we first compute the norm

$$\mathcal{N}_\mu = \int d^3\mathbf{r} \sum_{\sigma} |V_\mu(E, \mathbf{r}\sigma)|^2.$$

This allows us to define effective s.p. energies according to

$$\tilde{e}_\mu = E_\mu\left[1 - 2\mathcal{N}_\mu^2\right] + \lambda_{\rm F} \qquad \Delta_\mu = E_\mu^2 - (\tilde{e}_\mu - \lambda_{\rm F})^2,$$

where we use a previously determined value for the Fermi energy $\lambda_{\rm F}$. In the next step, we use the BCS formula

$$\nu_\mu^2 = \frac{1}{2}\left(1 - \frac{\tilde{e}_\mu - \lambda_{\rm F}}{E_\mu^{\rm BCS}}\right) N = \sum_\mu \nu_\mu^2,$$

with $E_\mu^{\rm BCS} = \sqrt{(\tilde{e}_\mu - \lambda_{\rm F}^2 + \Delta_\mu^2}$ to determine the optimum value of $\lambda_{\rm F}$ at this iteration. This technique was invented in [48] and is used in the HFBTHO [2] and HFODD [3] HFB solvers.

8.2.2 The simplest case of spherical symmetry and local EDFs

Let us now assume that the HFB solutions are invariant under rotations. Then the eigenvectors of the HFB matrix can be chosen so that they are also eigenvectors of $\hat{j}^2$ and $\hat{\ell}^2$, and are thus characterized by the spherical quantum numbers (n, ℓ, j, m). We may then write

$$U_{n\ell jm}(\boldsymbol{r}) = \frac{u_{n\ell j}(r)}{r}\Omega^\ell_{jm}(\theta, \varphi)$$
$$V_{n\ell jm}(\boldsymbol{r}) = \frac{\nu_{n\ell j}(r)}{r}\Omega^\ell_{jm}(\theta, \varphi),$$

where $\Omega^\ell_{jm}(\theta, \varphi)$ are spinor spherical harmonics, the special case $s = 1/2$ of tensor spherical harmonics representing the coupling of the usual spherical harmonics with spin 1/2 spinors,

$$\Omega^\ell_{jm}(\theta, \varphi) = \sum_{m_\ell, m_s} C^{jm}_{\ell m_\ell s m_s} Y_{\ell m_\ell}(\theta, \varphi)\chi_{s m_s};$$

see [20] for details. With these notations, the local one-body density matrix becomes

$$\rho(\boldsymbol{r}) = \sum_{n\ell jm} V^*_{n\ell jm}(\boldsymbol{r}) V_{n\ell jm}(\boldsymbol{r}) = \sum_{n\ell jm} \frac{\nu^2_{n\ell}(r)}{r^2}\Omega^{\ell\,\dagger}_{jm}(\theta, \varphi)\Omega^\ell_{jm}(\theta, \varphi).$$

We then use the property that the quadratic form $\Omega^{\ell\,\dagger}_{jm}(\theta, \varphi)\Omega^\ell_{jm}(\theta, \varphi)$ is independent of ℓ and φ, which leads to

$$\rho(\boldsymbol{r}) = \sum_{n\ell} \frac{\nu^2_{n\ell}(r)}{r^2} \sum_{jm} \Omega^{\ell\,\dagger}_{jm}(\theta, \varphi)\Omega^\ell_{jm}(\theta, \varphi) = \sum_{n\ell} \frac{2j + 1}{4\pi r^2}\nu^2_{n\ell}(r). \tag{8.32}$$

This shows that the local density is rotationally invariant (as expected) and only depends on the radial coordinate r, all the dependence on angles being factorized out. We would obtain a similar result for the pairing density. In fact, one could show that the kinetic energy density $\tau(\boldsymbol{r})$ and the spin-current tensor $\mathrm{J}(\boldsymbol{r})$ also reduce to simple functions of r. These densities are the only ones needed to define the Skyrme EDF in spherical symmetry since among the densities listed in section 1.2.1, time-odd densities such as the spin density $\boldsymbol{s}(\boldsymbol{r})$, the current density $\boldsymbol{j}(\boldsymbol{r})$ or the vector part of the spin kinetic density $\boldsymbol{T}(\boldsymbol{r})$ are identically zero in spherically symmetric systems —which are particular cases of time-reversal invariant systems.

To further discuss numerical implementation methods, we must specify the form of the EDF, which determines the structure of the HFB matrix. As already mentioned, coordinate-space-based methods have been applied mostly to solve the HFB equation for Skyrme EDF [8, 48, 49]. In this case, the HFB equation becomes

$$\left[-\frac{d}{dr}\mathcal{M}(r)\frac{d}{dr} + \mathcal{M}\frac{\ell(\ell+1)}{r^2} + \frac{1}{r}\frac{d\mathcal{M}}{dr} + \mathcal{U}(r) + \mathcal{U}_{\mathrm{s.o.}}(r)\right]\begin{pmatrix} u_{n_\mu \ell_\mu} \\ v_{n_\mu \ell_\mu}\end{pmatrix} = E_\mu \begin{pmatrix} u_{n_\mu \ell_\mu} \\ v_{n_\mu \ell_\mu}\end{pmatrix}, \quad (8.33)$$

where $\mathcal{M}$, $\mathcal{U}$ and $\mathcal{U}_{\mathrm{s.o.}}$ are 2×2 matrices with the following structure

$$\mathcal{M} = \begin{pmatrix} M & 0 \\ 0 & -M \end{pmatrix} \quad \mathcal{U} = \begin{pmatrix} U - \lambda & \Delta \\ -\Delta & -U + \lambda \end{pmatrix} \quad \mathcal{U}_{\mathrm{s.o.}} = \begin{pmatrix} U_{\mathrm{s.o.}} & 0 \\ 0 & -U_{\mathrm{s.o.}} \end{pmatrix}.$$

By a proper scaling transformation

$$\begin{pmatrix} u_{n_\mu \ell_\mu} \\ v_{n_\mu \ell_\mu}\end{pmatrix} = \frac{1}{\sqrt{M(r)}}\begin{pmatrix} f_{n_\mu \ell_\mu} \\ g_{n_\mu \ell_\mu}\end{pmatrix}$$

we can eliminate first-order derivatives and put the HFB equation in the generic matrix form

$$\begin{pmatrix} -M\dfrac{d^2}{dr^2} + X(r) & Y(r) \\ Y(r) & +M\dfrac{d^2}{dr^2} + X(r) \end{pmatrix}\begin{pmatrix} f_{n_\mu \ell_\mu} \\ g_{n_\mu \ell_\mu}\end{pmatrix} = E_{n_\mu \ell_\mu}\begin{pmatrix} f_{n_\mu \ell_\mu} \\ g_{n_\mu \ell_\mu}\end{pmatrix}. \quad (8.34)$$

This form is particularly adapted to using the Numerov method for second-order differential equations. For details about the expression of the blocks $X(r)$ and $Y(r)$, see [8].

Numerov method for ordinary differential equation

Recall that a generic second-order differential equation for the function $y(r)$ reads

$$\frac{d^2y}{dr^2} + p(r)y(r) = 0.$$

The Numerov method to solve numerically this equation is easily obtained by combining the Taylor expansion of the function $y(r)$ at points $r \pm h$, with h

small. In practice, it is implemented by considering a regular grid of points. Given the value of the function $y(r)$ at two previous points, y_n and y_{n-1}, the value y_{n+1} of the function at point $n+1$ is obtained by recurrence from the relation

$$\left(1 + \frac{1}{12}h^2 p_{n+1}\right)y_{n+1} - 2\left(1 - \frac{5}{12}h^2 p_n\right)y_n + \left(1 + \frac{1}{12}h^2 p_{n-1}\right)y_{n-1} = 0.$$

The method is of order h^6 and thus converges faster with the grid size h than the standard Runge–Kutta algorithm. The previous formula can easily be generalized to a system of coupled second-order differential equations. Let us denote by $\boldsymbol{y}$ the vector of unknown functions $y(r)$ and by P a matrix of known functions $p(r)$. We construct the auxiliary matrices A and vectors $\boldsymbol{g}$ as follows

$$A = 1 + \frac{1}{12}h^2 P \qquad \boldsymbol{g} = A\boldsymbol{y}.$$

With these notations, we can rewrite the differential equation as

$$\boldsymbol{g}_{n+1} = 12\boldsymbol{y}_n - 10\boldsymbol{g}_n - \boldsymbol{g}_{n-1} \tag{8.35a}$$

$$\boldsymbol{y}_{n+1} = (A_{n+1})^{-1}\boldsymbol{g}_{n+1}. \tag{8.35b}$$

The numerical implementation of this set of equations often begins by specifying a radius R_b for the (spherical) box and a spacing h for the grid in the box. For each q. p. energy E_μ, we then consider the two solutions obtained by forward (starting at $r = 0$) and backward (starting at $r = R_b$) integration. These two solutions are matched at some radius R_m. In practice, the system of equations (8.35) is solved by expanding the solutions of equation (8.34) on a basis of linearly independent vectors,

$$\boldsymbol{F}(r) \equiv \begin{pmatrix} f(r) \\ 0 \end{pmatrix} \quad \boldsymbol{G}(r) \equiv \begin{pmatrix} 0 \\ g(r) \end{pmatrix},$$

and using these 'basis' vectors to search for the most general solution of equation (8.34) in the form

$$\begin{aligned} \boldsymbol{y}^{(+)}(r) &= A\boldsymbol{F}^{(+)}(r) + B\boldsymbol{G}^{(+)}(r) \qquad \text{(forward solution)} \\ \boldsymbol{y}^{(-)}(r) &= C\boldsymbol{F}^{(-)}(r) + D\boldsymbol{G}^{(-)}(r) \qquad \text{(backward solution).} \end{aligned} \tag{8.36}$$

The notation $\boldsymbol{y}$ encapsulates the fact that we have expanded our solution on a basis of independent vectors. In practice, we will use the algorithm (8.35) on each of the two components of that vector. As usual when solving differential equations, the constants A, B, C and D are found by using boundary conditions on the edges of the box. Like all similar methods, the Numerov algorithm requires boundary conditions, that is, initial values of the functions (Dirichlet boundary conditions) or their first derivatives (Neumann boundary conditions).

Boundary conditions are deduced from the following general observations. For $r \to 0$, the centrifugal term dominates over other terms in equation (8.33). At large distances from the center of the nucleus, $r \to +\infty$, nuclear forces vanish and the HFB equations (8.33) decouple [50]. One can easily show that the radial part of the asymptotic solutions takes the familiar form of scattering wave functions

$$u_\ell(k, r) \sim C_u^+ H_\ell^+(k_u r) + C_u^- H_\ell^-(k_u r)$$
$$v_\ell(k, r) \sim C_v^+ H_\ell^+(k_v r),$$

where $k_u = \sqrt{2m(\lambda + E)/\hbar^2}$, $k_v = \sqrt{2m(\lambda - E)/\hbar^2}$ are the generalized momenta, H_ℓ^+ the Hankel functions of the first kind and H_ℓ^- the Hankel functions of the second kind [26].

Specifically, we need either the value of the function at $r = 0$ or $r = R_b$. These conditions are obtained by considering the limits $r \to 0$ and $r \to +\infty$ of the q.p. spinors. Let us take the example of Dirichlet boundary conditions. We find for $r \to 0$,

$$\boldsymbol{F}^{(+)}(r) \underset{r\to 0}{\sim} \begin{pmatrix} r^{\ell+1} \\ 0 \end{pmatrix} \qquad \boldsymbol{G}^{(+)}(r) \underset{r\to 0}{\sim} \begin{pmatrix} 0 \\ r^{\ell+1} \end{pmatrix}$$

for the forward solution and

$$\boldsymbol{F}^{(-)}(r) \underset{r\to 0}{\sim} \begin{pmatrix} r - r_N \\ 0 \end{pmatrix} \qquad \boldsymbol{G}^{(-)}(r) \underset{r\to 0}{\sim} \begin{pmatrix} 0 \\ r - r_N \end{pmatrix}$$

for the backward solution. For the forward solution, this gives the boundary conditions on the grid

$$\boldsymbol{F}_0^{(+)} = \begin{pmatrix} 0 \\ 0 \end{pmatrix} \qquad \boldsymbol{F}_1^{(+)} = \begin{pmatrix} h^{\ell+1} \\ 0 \end{pmatrix}$$
$$\boldsymbol{G}_0^{(+)} = \begin{pmatrix} 0 \\ 0 \end{pmatrix} \qquad \boldsymbol{G}_1^{(+)} = \begin{pmatrix} 0 \\ h^{\ell+1} \end{pmatrix}$$

and for the backward solution

$$\boldsymbol{F}_N^{(-)} = \begin{pmatrix} 0 \\ 0 \end{pmatrix} \qquad \boldsymbol{F}_{N-1}^{(-)} = \begin{pmatrix} h \\ 0 \end{pmatrix}$$
$$\boldsymbol{G}_N^{(-)} = \begin{pmatrix} 0 \\ 0 \end{pmatrix} \qquad \boldsymbol{G}_{N-1}^{(-)} = \begin{pmatrix} 0 \\ h \end{pmatrix}.$$

With the Dirichlet boundary conditions, the value of the function is specified, but that of its first derivative is not—since the original partial differential equation only specifies the link between the function and its second derivative. We then require that the full solutions (8.36) are equal at $r = R_m$, and that the value of their first derivatives are also equal. This set of conditions can be expressed by the linear system

$$\begin{pmatrix} \boldsymbol{F}_m^{(+)} & \boldsymbol{G}_m^{(+)} & -\boldsymbol{F}_m^{(-)} & \boldsymbol{G}_m^{(-)} \\ \boldsymbol{F}'^{(+)}_m & \boldsymbol{G}'^{(+)}_m & -\boldsymbol{F}'^{(-)}_m & \boldsymbol{G}'^{(-)}_m \end{pmatrix} \begin{pmatrix} A \\ B \\ C \\ D \end{pmatrix} = 0. \tag{8.37}$$

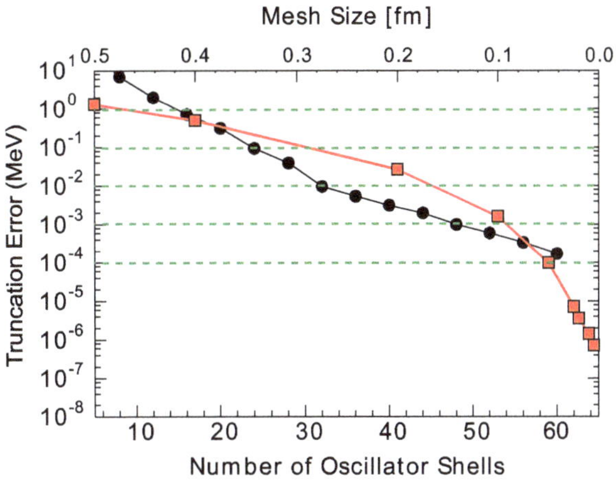

Figure 8.4. Numerical error when solving the spherical HFB equation for ^{208}Pb for the coordinate-space solver HFBRAD of [8], red squares, and the HO-based solver HOSPHE of [1], black circles. For HFBRAD, calculations were performed in a box of 20 fm. Refer to [51] for additional details about the comparison. (Adapted from [51]. Copyright 2015 IOP Publishing.)

Note that since $\boldsymbol{F}$ and $\boldsymbol{G}$ are 2×1 vectors, the matrix in this system is a 4× matrix. Solving it gives the coefficients A, B, C and D, which fully determines the initial conditions for the HFB spinors.

Figure 8.4 illustrates the precision of coordinate-space methods against basis expansion methods. The figure shows the truncation error estimated by extrapolating results to infinite basis size (HO basis) and zero lattice spacing (coordinate space). In order to achieve an absolute truncation error of 1 keV on the total energy, we need a basis made of 48 full shells (this corresponds to 20 825 states). In coordinate-space implementations, the same precision requires a lattice spacing of about 0.1 fm.

8.2.3 Broken symmetries and special lattices

Extensions of direct methods for solving the HFB equation to the case of a three-dimensional geometry are faced with a number of challenges. First, the number of grid points needed to achieve reasonable accuracy quickly becomes very large. For example, let us consider a cubic box $L \times L \times L$. With $L = 20$ fm and a grid of spacing $h = 0.1$ fm (these are typical values in spherical calculations), the total number of grid points in the volume becomes 8 120 601. Second, the number of coupled partial differential equations to solve in the HFB case is much larger than in the HF case, or even the HF+BCS case. This difficulty is made even worse in coordinate space—the density of continuum states in a box is much larger than the density of discretized continuum states in localized bases such as the HO. As a consequence, it is not unusual to have to consider a few thousands of q.p.s to achieve convergence for the HFB solution. In time-dependent density functional theory (TDDFT) applications based on a coordinate-space solver, which are discussed in chapter 4, the number of q.p.s to track is even larger. For example, the TDDFT calculations of induced nuclear fission reported in [52] involved tracking over 50 000 q.p.s at each time step. As mentioned above, each of these few thousands of q.p. would involve solving a partial differential equation (PDE) on a grid of more than eight million points. Finally, one of the main reasons to use a 3D geometry is to break symmetries. In particular, breaking time-reversal invariance allows solving the HFB equation for odd nuclei and is also needed to solve the time-dependent HFB equation. When time-reversal symmetry is broken, additional densities need to be constructed, and contribute to the mean field.

Several strategies can be adopted to reduce this computational cost. The simplest one is to take advantage of conserved symmetries. For many applications of the static HFB equation, such as computation of ground-state properties, axial and reflection symmetries are most often conserved. This means that only 1/8 of the full $\mathbb{R}^3$ needs be considered. A more general technique to reduce the size of the grid consists in using special meshes that allow for very accurate computation of derivatives. In the rest of this section, we will briefly prevent several such methods that are mostly based on interpolation techniques.

Let us consider a generic spinor $\Phi(\boldsymbol{r}, \sigma)$ in coordinate space. We set a box of size $L_x \times L_y \times L_z$ with n_μ $(\mu = x, y, z)$ points along each Cartesian directions. Discretization of the field at the collocation points (x_i, y_j, z_k) yields the set $\{\Phi_{ijk}\}$: these values will be taken as variational parameters. The goal of lattice techniques is thus to choose the grid of collocation points in such a way as to maximize the numerical precision of the calculations.

The Lagrange mesh method is a general approach to obtain a variational expression for the eigenfunction $\Phi(\boldsymbol{r}, \sigma)$ of a quantum Hamiltonian $\hat{h} = \hat{t} + \hat{v}$; see [53] for a comprehensive review. Its main advantage is that it only requires matrix elements of the kinetic energy operator (which, depending on the choice of the mesh, can be computed analytically) and of the values of the potential at the collocation points. In simple terms, it is based on expanding the unknown spinor $\Phi(\boldsymbol{r}, \sigma)$ on a basis of Lagrange functions. In 1D, we thus write

$$\Phi(x, \sigma) = \sum_{n=1}^{n_x} c_n f_n(x),$$

where c_n are the variational parameters (which are determined by solving a linear system of equation) and the family $\{f_i(x)\}_i$ of Lagrange functions verifies the two important properties

$$f_n(x_i) = \frac{\delta_{ni}}{\sqrt{\lambda_i}}$$

$$\int_a^b f_n(x) f_m(x) dx = \sum_{i=1}^{n_x} \lambda_i f_n(x_i) f_m(x_i).$$

The last equality expresses the fact that the integral of a product of two Lagrange functions over the domain is *exactly* represented by a Gauss-like quadrature formula. In fact, one can derive Lagrange meshes from the quadrature formula of orthogonal polynomials by setting $\lambda_i = \omega_i/w(x_i)$ with ω_i the weight of the Gauss quadrature and $w(x)$ the weight function associated with a set of orthogonal polynomials. Based on these properties of the Lagrange functions, we find that the unknown spinor can instead be written

$$\Phi(x, \sigma) = \sum_{n=1}^{n_x} \Phi_n f_n(x),$$

with Φ_n the values of the spinor on the grid—in a coordinate-based approach, these are the true variational parameters.

One particularly interesting variant of the Lagrange mesh is the one obtained when considering the family of functions

$$\varphi_n(x) = \frac{1}{\sqrt{N}} e^{i2\pi nx/N} \qquad n = -\frac{1}{2}(N-1), \ldots, +\frac{1}{2}(N-1)$$

for a regular mesh $x_i = -\frac{1}{2}(N-1), \ldots, +\frac{1}{2}(N-1)$. For these functions, the weights are $\lambda_i = 1$. The actual Lagrange functions are defined by

$$f_n(x) = \frac{\sin \pi(x - x_n)}{N \sin \frac{\pi}{N}(x - x_n)}.$$

The advantage of this particular mesh is that the matrix elements of the kinetic energy operator (in fact of the operator d^p/dx^p for any p) can be computed exactly. For the first-order derivative, we find

$$D_{ij}^{(1)} = (-1)^{i-j} \frac{\pi}{2Nh} \frac{1}{\sin\left[\pi(x_i - x_j)/(2N)\right]} \qquad D_{ii}^{(1)} = 0$$

and, therefore,

$$D_{ij}^{(2)} = (-1)^{i-j+1} 2 \left(\frac{\pi}{2Nh}\right)^2 \frac{\cos\left[\pi(x_i - x_j)/(2N)\right]}{\sin^2\left[\pi(x_i - x_j)/(2N)\right]}$$

$$D_{ii}^{(2)} = -\frac{\pi^2}{3h^2}\left(1 - \frac{1}{(2N)^2}\right).$$

An alternative method to compute spatial derivatives, which is partially available in the code SKY and is the method of choice in the TDDFT solvers discussed in chapter 4, is based on the fast Fourier transform (FFT) and uses the fact that $\nabla \to \boldsymbol{k}\cdot$ in Fourier space. This method is both accurate and extremely efficient; see [52, 54, 55].

In Cartesian coordinates, three-dimensional Lagrange functions are most conveniently obtained by setting the tensor product of 1D functions,

$$F_{lmn}(x, y, z) = f_l(x) f_m(y) f_n(z).$$

This separability of the basis functions implies that the representation of the spinor and of its second derivative would be simply

$$\Phi(\boldsymbol{r}, \sigma) = \sum_{ijk} \Phi_{ijk} f_i(x) f_j(y) f_k(z)$$

$$\nabla^2 \Phi(\boldsymbol{r}, \sigma) = \sum_{ijk} \left(\sum_p D_{pi}^{(2)} \Phi_{ijk} + \sum_q D_{qj}^{(2)} \Phi_{ijk} + \sum_r D_{rk}^{(2)} \Phi_{ijk} \right) f_i(x) f_j(y) f_k(z).$$

To finish this section, we note that the method of Lagrange meshes encompasses what is known as the discrete variable representation of [56]. This class of methods has been used in the HF+BCS codes EV of [6, 7, 57] and in the TDHF code SKY of [5].

> To quote from [58]: 'calculations in medium size volume $32 \times 32 \times 48$ fm^3 with lattice resolution $dx = 1$ fm corresponding to $p_{\mathrm{cut}} = \hbar\pi/dx \approx 600$ MeV c^{-1}. Then, the $N \times N$ HFB matrix has $N^2 = (4 \times 32 \times 32 \times 48)^2 \approx 2\,000\,002$ matrix elements, and requires more than 0.5 TB memory just to store it.' For this reason, static lattice-based solvers such as EV or SKY do not solve the full HFB equation but the HF+BCS equation instead.

Scaling behaviors of HFB implementation

In computer science, 'strong scaling' is the property that the cost of solving a given problem size is decreased by a factor k if the number of processing units is increased by the same factor k. In other words, the more processing units one has, the faster the calculation. It is different from 'weak scaling', where the time of running a problem of size s on p processing units is the same as one of size $k \times s$ on $k \times p$ processing units—more processing units means one can solve larger problems at the same cost.

In the context of HFB calculations, both properties present advantages. Having fast and thrifty HFB solvers imply one can use the large number of processing units of a supercomputer to run many HFB calculations in parallel (i.e. weak scaling). This is typically the case for configuration space implementations of the HFB equation, which are particularly suited to computing, e.g. potential energy surfaces for deformed nuclei. On the other hand, the potential speed-up obtained from parallelizing these codes is relatively limited; see [3].

Naively, one might think that lattice-based methods could potentially be more amenable to exhibiting strong scaling properties, since every variational parameter (i.e. the value of the wave function at each point on the lattice) is in principle independent of the others. However, a direct approach where one would simply construct the HFB matrix (with row and column indices referring to positions on the grid) and diagonalize it would be extremely challenging computationally speaking. However, in the spirit of density functional theory (DFT), wave functions are only needed to reconstruct the density matrix and pairing tensors which in principle encapsulate all the relevant information. The method proposed in [58] uses Green function techniques to bypass entirely the calculation of wave functions and compute the density directly at point $\boldsymbol{r}$ by contour integration. This guarantees both weak (if one uses more processors to solve problems in larger boxes at constant grid size) and strong (if one uses these processors to increase the density of the grid at constant size) scaling.

8.2.4 Hybrid methods

While configuration space methods based on the expansion of HFB solutions are fast and economical in terms of computer resources (memory, storage, CPU speed), their precision when describing systems with, e.g. very large spatial extension such as fissioning nuclei or neutron star interiors, is very limited. Conversely, the precision of lattice methods is nearly independent of the characteristics of the system, but these approaches require a larger amount of computational resources and do not lend themselves to implementations with non-local EDFs.

FE methods offer an appealing alternative. While they still cast the HFB problem in the form of a non-linear eigenvalue problem, FE methods rely on introducing some kind of local basis of the coordinate system for a specific cell of the domain—a segment in 1D, a N-dimensional volume in $\mathbb{R}^N$. One of their interesting features is that they can be generalized relatively easily to an arbitrary number of dimensions. As a consequence, they could be used to solve the spherical, axial or triaxial HFB equation, but could also be used, e.g. to diagonalize collective Hamiltonians. In nuclear physics, finite elements have been used to solve the equation of the relativistic mean field and relativistic Hartree–Bogoliubov theory in spherical symmetry [59–63]; the non-relativistic Skyrme HFB equation in [64, 65] (the latter for a special case of finite elements called the basis spline collocation method); and for the collective Schrödinger equation from the GCM+GOA in [66, 67].

Another emerging technology that has been recently applied to nuclear physics problems is multi-resolution analysis (MRA) [68–73]. The main idea of MRA is to introduce the space of functions, here the space $\mathcal{L}_2(\mathbb{C})$ of square-integrable wave functions, into a sequence of subspaces of increasing refinement, $\mathcal{V}_0 \subset \mathcal{V}_1 \subset \ldots \mathcal{V}_k \subset \mathcal{L}_2$, with k denoting the level of refinement. We have

$$\bigcup_{k\geqslant 0}^{\infty} \mathcal{V}_k = \mathcal{L}_2(\mathbb{C}).$$

The unique advantage of such methods is that they allow an arbitrary level of numerical precision, which can be preset by the user. This is particularly relevant when computing properties of very deformed nuclei, as in fission, or very weakly bound nuclei. The downside of MRA at the moment is that the computational cost can be relatively high.

In this section, we will briefly sketch some of the basic concepts of this approach. We start with the space $\mathbb{R}^N$ and consider the Hilbert space $\mathcal{L}_2(\mathbb{R}^N, \mathbb{C})$ of square-integrable functions in that space, that is, the set of complex-valued functions ϕ such that

$$\int d^N\boldsymbol{q}|\phi(\boldsymbol{q})|^2 \in \mathbb{R}_+.$$

We now consider a differential operator $\hat{D}$: $\mathcal{L}_2(\mathbb{R}^N, \mathbb{C}) \mapsto \mathcal{L}_2(\mathbb{R}^N, \mathbb{C})$. For $N = 3$, the s.p. Hamiltonian h is an example of such an operator. In $\mathcal{L}_2(\mathbb{R}^N, \mathbb{C})$, we have the property that

$$\forall \varphi \in \mathcal{L}_2(\mathbb{R}^N, \mathbb{C}) \qquad (\varphi, \psi) = 0 \Rightarrow \psi = 0,$$

where the notation $(\cdot, \cdot)$ refers to the usual scalar product in $\mathcal{L}_2(\mathbb{R}^N, \mathbb{C})$. As a reminder, the scalar product for two arbitrary functions φ and ϕ of $\mathcal{L}_2(\mathbb{R}^N, \mathbb{C})$ is defined by the integral

$$(\varphi, \phi) = \int d^N\boldsymbol{q}\varphi^*(\boldsymbol{q})\phi(\boldsymbol{q}).$$

In other words, if the scalar product between a given function ψ and *any* other function φ vanishes, then ψ must be identically zero. The Galerkin finite element method takes advantage of this property to look for the solution of $\hat{D}\psi = 0$ by expanding the unknown function ψ onto a piece-wise polynomial basis of the domain $\mathbb{R}^N$ or, more often, a subdomain $\Omega \in \mathbb{R}^N$. For example, if $\hat{D} = \hat{h} - e$, then the method would provide the HF solutions; if $\hat{D} = \hat{\mathcal{H}} - E$ ($\mathcal{H}$ the HFB matrix), then it provides the HFB solution; etc. Interestingly, the method can also be applied to time-dependent operators. For example,

$$\hat{D} = i\hbar\frac{\partial}{\partial t} - \sum_{ij=1}^{N}\frac{\partial}{\partial q_i}B_{ij}\frac{\partial}{\partial q_j} + V(\boldsymbol{q})$$

would provide the solution of the collective Schrödinger equation extracted from the time-dependent generator coordinate under the GOA.

The static GOA has been discussed extensively in section 6.4. The time-dependent GOA is obtained by using the ansatz

$$|\Psi(t)\rangle = \int d^N\boldsymbol{q}\, f(\boldsymbol{q}, t)|\Phi(\boldsymbol{q})\rangle$$

instead of the one given by equation (3.79), see section 3.2.1. The time-dependent Hill–Wheeler equation is obtained from minimizing the action constructed with this ansatz [74].

The set up of the Garlerkin finite element method starts with the definition of a grid in Ω. A popular choice consists in partitioning Ω into simplices. In 1D, a simplex is simply an interval $[a, b]$, with a and b referred to as the vertices of the simplex; in 2D, a simplex is a triangle defined by three vertices; in 3D, it is a tetrahedron defined by four vertices; etc. In N dimensions, a simplex is defined by $N + 1$ vertices. If we denote by $\mathcal{S}$ an arbitrary simplex, then we pose that

$$\forall \boldsymbol{q} \in \mathcal{S} \qquad \psi(\boldsymbol{q}) = P_{d_s}(\boldsymbol{q}),$$

where P_{d_s} is a polynomial of degree d_S. We can thus attach to each simplex the vector space $\mathcal{P}_s$ of all such polynomials. Next, we define a number of nodes $\boldsymbol{q}_i$ inside the simplex and require our basis functions $\phi_{s,i}(\boldsymbol{q})$ to obey the conditions

$$\phi_{s,\,i}(\boldsymbol{q}_j) = \delta_{ij}.$$

This means that the functions $\phi_{s,i}$ are simply the Lagrange polynomials for the set of nodes in the simplex $\mathcal{S}$. We can obtain in this way a basis of the whole domain Ω by stacking together each simplex,

$$\phi_i(\boldsymbol{q}) = \begin{cases} \phi_{s,i}(\boldsymbol{q}) & \text{if } \boldsymbol{q} \in \mathcal{S} \text{ and } i \text{ is a node of } \mathcal{S} \\ 0 & \text{otherwise} \end{cases}.$$

The solutions ψ to our problem are then expanded in that basis,

$$\psi(\boldsymbol{q}) = \sum_{i=1}^{m} \psi(\boldsymbol{q}_i)\phi_i(\boldsymbol{q}), \tag{8.38}$$

where the sum runs over all the nodes of all the simplices of the domain Ω. Equation (8.38) is simply the expansion of an arbitrary function of $\mathcal{L}_2(\mathbb{R}^2, \mathbb{C})$ on a basis that is built out of a juxtaposition of bases of all the simplices that pave the domain Ω. We then use equation (8.38) to solve the set of linear equations

$$\left(\phi_i(\boldsymbol{q})|\hat{D}\psi\right) = 0 \quad \forall i \in [1, m].$$

This can be recast into a set of coupled equations,

$$D_{ij} = 0 \quad \forall (i, j) \in [1, m]^2, \tag{8.39}$$

where the matrix D_{ij} is given by

$$D_{ij} = (\phi_i(\boldsymbol{q})|\hat{D}\phi_j(\boldsymbol{q})).$$

The characteristics of this matrix depend on the operator $\hat{D}$. If it is a static (i.e. time-independent) operator such as, e.g. the HF or HFB Hamiltonian, the system (8.39) is nothing other than a linear system. If the operator is time-dependent as in the time-dependent HF, HFB or GCM theories, then we obtain a system of coupled time-differential equations.

Lattice formulation of Hartree–Fock equations

Let us illustrate this method with the concrete example of the HF s.p. Hamiltonian for a simple EDF of the local density (possibly corresponding to a two-body potential of zero-range). In this case, the operator $\hat{D}$ is simply $-\frac{\hbar^2}{2m}\nabla^2 + \frac{1}{2}\Gamma(\boldsymbol{r})$, with

$$\Gamma(\boldsymbol{r}) = \frac{\partial \mathcal{E}}{\partial \rho(\boldsymbol{r})}.$$

The system (8.39) thus becomes

$$(\mathrm{H} - e\mathrm{N})\Phi = 0,$$

with $\Phi = [\phi_1, \ldots, \phi_m]$ the column of all the unknown coefficients $\phi_i \equiv \phi(\boldsymbol{q}_i)$ in equation (8.38) and

$$\mathrm{N} \equiv N_{ij} = (\phi_i | \phi_j)$$

$$\mathrm{H} \equiv H_{ij} = \left(\phi_i \middle| \left[-\frac{\hbar^2}{2m}\nabla^2 + \frac{1}{2}\Gamma(\boldsymbol{r}) \right] \middle| \phi_j \right).$$

Contrary to traditional basis expansion methods, the resulting system of equations is a non-linear generalized eigenvalue problem, akin to the equations of the GCM. The system is non-linear because of the self-consistent mean field $\Gamma(\boldsymbol{r})$ which, as usual, depends on the density, hence on the solutions.

8.3 The self-consistent loop

In the HFB theory, we use the variational principle to determine the optimal transformation (U, V) that minimizes the energy. There are essentially two common strategies to implement this theory. The first one takes advantage of the fact that the variational principle leads to the condition $[\mathcal{H}, \mathcal{R}] = 0$, with $\mathcal{H}$ the HFB matrix and $\mathcal{R}$ the generalized density. As we will show in section 8.3.1, one may find the solution to such an equation by a series of successive diagonalization of a (highly) non-linear problem. The second approach is based on a convenient parametrization of the total energy which allows its direct minimization with gradient-like methods. This second approach will be summarized in section 8.3.2.

8.3.1 Non-linear eigenvalue problems

A majority of HFB solvers on the market use successive diagonalizations of the HFB matrix to solve the HFB equation [1–4, 8, 65]. A practical advantage of this approach is that most of the linear algebra involved, matrix–vector and matrix–matrix multiplication, as well as diagonalization, can be outsourced to specialized software libraries. The method is also conceptually simple to implement and very flexible. Special care should be taken to ensure the convergence of the calculation. The efficient handling of constraints requires specialized algorithms.

Matrix diagonalization
Let us recall the original form of the HFB equation (8.1),

$$[\mathcal{H}, \mathcal{R}] = 0,$$

where the HFB and generalized densities are given by

$$\mathcal{H} = \begin{pmatrix} h & \Delta \\ -\Delta^* & -h^* \end{pmatrix} \quad \mathcal{R} = \begin{pmatrix} \rho & \kappa \\ -\kappa^* & 1-\rho^* \end{pmatrix}. \tag{8.40}$$

The density matrix ρ and pairing tensor κ entering the generalized density are given by

$$\rho_{mn} = \frac{\langle\Phi|c_n^\dagger c_m|\Phi\rangle}{\langle\Phi|\Phi\rangle} \qquad \kappa_{mn} = \frac{\langle\Phi|c_n c_m|\Phi\rangle}{\langle\Phi|\Phi\rangle},$$

where $|\Phi\rangle$ is the q.p. vacuum (3.3). Recall that these formal expressions are valid both in configuration space, that is, in a basis, and in coordinate space. In the former case, h, Δ, ρ and κ are $N \times N$ matrices, where N is the actual number of states in the basis. In the latter case, h, Δ, ρ and κ are of the type $h(\boldsymbol{r}\sigma\tau, \boldsymbol{r}'\sigma'\tau')$ where σ and τ refer to the projection of the spin and isospin on the quantization axis. In this section, we show how we can solve the HFB equation by performing successive diagonalizations of the HFB matrix. To this end, we first recall some properties of both the HFB matrix $\mathcal{H}$ and the generalized density $\mathcal{R}$.

We saw in section 3.1.3 that the matrix $\mathcal{H}$ was Hermitian. It is straightforward to show that it also verifies $\mathcal{H}^T = \mathcal{H}^* = -\gamma\mathcal{H}\gamma$ with

$$\gamma = \begin{pmatrix} 0 & 1 \\ 1 & 0 \end{pmatrix}.$$

Since $\mathcal{H}$ is Hermitian, its eigenvalues are real. If E_μ is the eigenvalue associated with the eigenvector $\boldsymbol{W}_\mu$, then

$$\mathcal{H}\boldsymbol{W}_\mu = E_\mu \boldsymbol{W}_\mu \Rightarrow -\gamma\mathcal{H}\gamma\boldsymbol{W}_\mu^* = E_\mu\boldsymbol{W}_\mu^* \Rightarrow \mathcal{H}(\gamma\boldsymbol{W}_\mu^*) = -E_\mu(\gamma\boldsymbol{W}_\mu^*).$$

Hence, the eigenvalues of $\mathcal{H}$ come in pairs $\pm E_\mu$ and if

$$\mathcal{H}\boldsymbol{W}_\mu = E_\mu\boldsymbol{W}_\mu \quad \text{with} \quad \boldsymbol{W}_\mu = \begin{pmatrix} U_\mu \\ V_\mu \end{pmatrix}$$

then $\mathcal{H}\tilde{\boldsymbol{W}}_\mu = -E_\mu\tilde{\boldsymbol{W}}_\mu$ with

$$\tilde{\boldsymbol{W}}_\mu = \gamma\boldsymbol{W}_\mu^* = \begin{pmatrix} V_\mu^* \\ U_\mu^* \end{pmatrix}.$$

Since the eigenvectors $\boldsymbol{W}$ and $\tilde{\boldsymbol{W}}$ are orthonormal, $\boldsymbol{W}_\mu^\dagger \cdot \boldsymbol{W}_\nu = \tilde{\boldsymbol{W}}_\mu^\dagger \cdot \tilde{\boldsymbol{W}}_\nu = \delta_{\mu\nu}$ and $\boldsymbol{W}_\mu^\dagger \cdot \tilde{\boldsymbol{W}}_\nu = 0$, it is easy to show then that these eigenvectors verify the property

$$\sum_\mu \left(\boldsymbol{W}_\mu \boldsymbol{W}_\mu^\dagger + \tilde{\boldsymbol{W}}_\mu \tilde{\boldsymbol{W}}_\mu^\dagger \right) = 1.$$

These various properties of $\mathcal{H}$ are key to prove how to turn the HFB equation into a (non-linear) eigenvalue problem. First, we note that since the matrices $\mathcal{H}$ and $\mathcal{R}$ commute and are diagonalizable (being both Hermitian), it is possible to find a basis that simultaneously diagonalizes the two matrices. We will now show that we can determine such a basis by constructing the generalized density out of the eigenvectors of the HFB matrix. First, let us use the set of eigenvectors $\tilde{\boldsymbol{W}}_\mu$ of $\mathcal{H}$ to form the square matrix

$$\mathcal{R} = \tilde{\boldsymbol{W}} \tilde{\boldsymbol{W}}^\dagger \Rightarrow \mathcal{R}_{kl} = \sum_\mu \tilde{\boldsymbol{W}}_{k\mu} \tilde{\boldsymbol{W}}_{\mu l}^* = \begin{pmatrix} V^* V^T & V^* U^T \\ U^* V^T & U^* U^T \end{pmatrix}. \tag{8.41}$$

Although we denote by $\mathcal{R}$ the matrix that we form with the eigenvectors of the HFB matrix $\mathcal{H}$, we have not yet proven that it corresponds to the actual generalized density defined by equation (8.40) or, more to the point, by the formula (3.17) with the densities defined by equation (3.1), the many-body wave function by equation (3.3) and the q.p. operators by equation (3.6). Note that the matrix $\mathcal{R}$ thus formed is written in the original (doubled) basis of the s.p. Hilbert space. A direct and straightforward calculation shows that $\mathcal{R}^\dagger = \mathcal{R}$ and $\gamma \mathcal{R} \gamma = 1 - \mathcal{R}^*$. In addition, $\mathcal{R}^2 = \mathcal{R}$ since the vectors $\boldsymbol{W}_\mu$ are orthonormal, that is, $\sum_i \tilde{W}_{i\mu} \tilde{W}_{i\nu}^* = \delta_{\mu\nu}$. These conditions are sufficient to guarantee that the matrix $\mathcal{R}$ can be associated with a product state $|\Phi\rangle$ which is a q.p. vacuum for the q.p.s defined by the transformation

$$\beta_\mu^\dagger = \sum_n \left(U_{n\mu} c_n^\dagger + V_{n\mu} c_n \right)$$

$$\beta_\mu = \sum_n \left(V_{n\mu}^* c_n^\dagger + U_{n\mu}^* c_n \right).$$

> It is shown in [74] that 'a necessary and sufficient condition for a completely antisymmetrized state $|\Phi\rangle$ to be a q.p. vacuum is that the associated generalized density matrix $\mathcal{R}$ satisfies'
>
> $$R^2 = R \qquad \gamma R \gamma = 1 - R^* \qquad R^\dagger = R'.$$

In turn, this guarantees that $\mathcal{R}$ has the canonical form

$$\tilde{\mathcal{R}} = \begin{pmatrix} 0 & 0 \\ 0 & 1 \end{pmatrix} \tag{8.42}$$

in the basis of the $(\beta, \beta^\dagger)$ operators, that is, in the eigenbasis of the HFB matrix. It also follows from equation (8.41) that the density matrix and pairing tensor are obtained from the matrix of the Bogoliubov transformation with

$$\rho = V^* V^T \qquad \kappa = V^* U^T.$$

These formula are identical to the ones already shown in section 3.1.2.

We thus showed that the basis that diagonalizes the HFB matrix also diagonalizes the generalized density. However, the determination of this basis is not as straightforward as it may look, since the HFB matrix depends on ρ and κ, hence on the generalized density $\mathcal{R}$. The simplest strategy consists in setting an iterative process, or self-consistent loop. One possibility of such a loop could be:

1. Take an initial guess for the generalized density, say $\mathcal{R}^{(0)}$. This initial guess can be constructed by solving the Schrödinger equation for some realistic nuclear potential and construct the density matrix from the resulting s.p. wave functions.
2. Construct the HFB matrix $\mathcal{H}^{(0)}$ at the first iteration by computing the mean field $h^{(0)}$ and pairing field $\Delta^{(0)}$ out of the components $\rho^{(0)}$ and $\kappa^{(0)}$ of the generalized density.
3. Diagonalize $\mathcal{H}^{(0)}$. The eigenvectors can be ordered by increasing eigenvalue and form the matrix

$$\mathcal{W}^{(0)} = \begin{pmatrix} U^{(0)} & V^{(0)*} \\ V^{(0)} & U^{(0)*} \end{pmatrix}.$$

4. Reconstruct the density matrix and pairing tensor at the next iteration according to

$$\rho^{(1)} = V^{(0)*} V^{(0)T} \qquad \kappa^{(1)} = V^{(0)*} U^{(0)T}.$$

5. Go back to point 1 and continue until the generalized density does not change too much from one iteration to the next,

$$||\mathcal{R}^{(n)} - \mathcal{R}^{(n+1)}|| \leqslant \varepsilon.$$

Note that the self-consistent loop can be initialized in many different ways. Instead of taking a guess for the generalized density, one can initialize directly the mean field and pairing field, or the Bogoliubov matrix. Similarly, the convergence criterion listed in point 5 could involve the HFB matrix, the Bogoliubov matrix, etc. The structure of the self-consistent loop will affect the definition of acceleration schemes discussed in section 8.3.1.

Acceleration schemes

The self-consistent loop can be represented formally as the transformation $\mathcal{S}$ of a generic N-dimensional vector $\boldsymbol{V}_{\text{in}}^{(n)}$ into a new vector $\boldsymbol{V}_{\text{out}}^{(n)}$,

$$\boldsymbol{V}_{\text{out}}^{(n)} = \mathcal{S}(\boldsymbol{V}_{\text{in}}^{(n)}).$$

There is no unique choice to define the vector $\boldsymbol{V}$. The only requirement is that it contains the complete information about the current iteration of the self-consistent loop. For instance, $\boldsymbol{V}$ can contain the generalized density, the HFB matrix, or the Bogoliubov matrix. In the case of the HFB with Skyrme-like EDFs, it could contain the density matrix and pairing tensor either in coordinate space or in configuration space—in which case, one obviously only needs to include the independent matrix elements. Depending on the method used to fix the particle number, one may include also in $\boldsymbol{V}$ the Fermi energies λ_{F} for neutrons and protons. Similarly, when the Lipkin–Nogami prescription is used, one typically adds λ_{2N} and λ_{2P}. If constraints are imposed on the HFB solution, for instance through the use of Lagrange parameters as described in section 8.3.1, then these parameters should also be included in the vector $\boldsymbol{V}$. The definition of $\boldsymbol{V}$ is thus entirely dependent on the particular implementation of the HFB equation.

By definition, the convergence of the self-consistent loop requires that the final solution $\boldsymbol{V}^*$ be a fixed-point of the iteration process. Formally, this implies that $\mathcal{S}(\boldsymbol{V}^*) = \boldsymbol{V}^*$. Numerically, this equation is only satisfied approximately. Program developers must choose a metric to define the convergence of the iterative process, typically, some particular norm $||.||$, and must set some criterion ε such that the iterations are stopped when

$$||\boldsymbol{V}_{\mathrm{out}}^{(n)} - \boldsymbol{V}_{\mathrm{in}}^{(n)}|| \leqslant \varepsilon. \tag{8.43}$$

If the self-consistent loop is implemented by simply using the output of iteration n as input for iteration $n + 1$, that is, $\boldsymbol{V}_{\mathrm{in}}^{(n+1)} = \boldsymbol{V}_{\mathrm{out}}^{(n)}$, the iterations often cannot converge to a solution. To ensure convergence, it is necessary to 'slow down' the iterative process. This is a similar problem to what is observed in the gradient method discussed in section 8.3.2, where full steps in the direction of the gradient can lead to chaotic divergences. The simplest and most robust method to slow down iterations consists in simply mixing the input and output at iteration n to define the actual input at iteration $n + 1$. We thus introduce a linear mixing coefficient α such that

$$\boldsymbol{V}_{\mathrm{in}}^{(n+1)} = \alpha \boldsymbol{V}_{\mathrm{out}}^{(n)} + (1 - \alpha)\boldsymbol{V}_{\mathrm{in}}^{(n)}. \tag{8.44}$$

This simple 'linear mixing' effectively slows down the iterations. The mixing parameter $\alpha \in [0, 1]$ is a user-defined parameter for the calculations. Values $\alpha \approx 0.5$ lead to converged results in most simple cases such as solutions near the ground state. However, it is very frequent to find cases where the convergence, while occurring, is very slow. In the context of nuclear physics, this can typically occur when computing potential energy surfaces in heavy nuclei. When these potential energy surfaces are characterized by coexisting minima and/or when they are extended to large deformations, the pace of convergence tends to slow down noticeably. For these reasons, it is sometimes advantageous to use acceleration schemes that exploit the fixed-point nature of the self-consistent loop.

Original Broyden method

If the vector field $\mathcal{S}(\boldsymbol{V})$ is differentiable, one may improve the pace of convergence by exploiting information about the derivatives of $\mathcal{S}(\boldsymbol{V})$. Let us note $\boldsymbol{F}^{(n)} = \boldsymbol{V}_{\mathrm{out}}^{(n)} - \boldsymbol{V}_{\mathrm{in}}^{(n)}$.

We can regard the self-consistent condition $||\boldsymbol{F}(\boldsymbol{V}^*)|| = 0$ as equivalent to finding the roots of the corresponding system of non-linear equations. This can be achieved with the multidimensional Newton–Raphson method. In this general iterative method for root finding, the value of the root at iteration $n+1$ is approximated by

$$\boldsymbol{V}^{(n+1)} \equiv \boldsymbol{V}_{\text{in}}^{(n+1)} = \boldsymbol{V}_{\text{in}}^{(n)} - \mathrm{B}^{(n)}\boldsymbol{F}^{(n)}, \tag{8.45}$$

where $\mathrm{B}^{(n)} = (\mathrm{J}^{(n)})^{-1}$ and $\mathrm{J} \equiv J_{jk}^{(n)} = \partial F_j^{(n)}(\boldsymbol{V})/\partial V_k$ is the Jacobian matrix of the non-linear equations at $\boldsymbol{V} = \boldsymbol{V}_{\text{in}}^{(n)}$. This is the Broyden method, which was originally proposed in [75]. While it converges quadratically for sufficiently smooth functions, it can become very expensive since it requires explicit evaluation of the Jacobian matrix. In particular, as mentioned above, the size N of the vectors $\boldsymbol{V}$ is, by construction, proportional to the number of independent variables that characterize the HFB equation. When $\boldsymbol{V}$ contains matrix elements in a basis, N scales like the square of the size of the basis; when $\boldsymbol{V}$ contains value of fields on the lattice, N scales like the number of mesh points, which itself depends on the built-in symmetries of the code. In any case, when N is large, storing $N \times N$ matrix elements of the inverse Jacobian and performing the $N \times N$ matrix multiplications of equation (8.45) is computationally expensive.

Modified Broyden method

The modified version of the Broyden method originally proposed in [76] for quantum chemistry applications builds an approximation of the inverse Jacobian by using information collected over M previous iterations. Details can be found in [76] and in [77], which presents nuclear physics applications. Here, we only recall the final result,

$$\boldsymbol{V}_{\text{in}}^{(n+1)} = \boldsymbol{V}_{\text{in}}^{(n)} + \alpha \boldsymbol{F}^{(n)} - \sum_{k=k_{\min}}^{n-1} w_k \gamma_{nk} \boldsymbol{u}^{(k)}, \tag{8.46}$$

with $k_{\min} = \max(1, n - M)$, $\boldsymbol{u}^{(n)} = \alpha \Delta \boldsymbol{F}^{(n)} + \Delta \boldsymbol{V}^{(n)}$ and

$$\Delta \boldsymbol{V}^{(n)} = \frac{\boldsymbol{V}_{\text{in}}^{(n+1)} - \boldsymbol{V}_{\text{in}}^{(n)}}{|\boldsymbol{F}^{(n+1)} - \boldsymbol{F}^{(n)}|} \qquad \Delta \boldsymbol{F}^{(n)} = \frac{\boldsymbol{F}^{(n+1)} - \boldsymbol{F}^{(n)}}{|\boldsymbol{F}^{(n+1)} - \boldsymbol{F}^{(n)}|}. \tag{8.47}$$

The matrix γ_{kn} depends on $\Delta \boldsymbol{V}^{(n)}$ and $\Delta \boldsymbol{V}^{(n)}$; see [77] for details. The weights w_k ($k = 1, \dots, M$) are associated with each previous iteration and one often takes $w_k = 1$. The weight w_0 is assigned to the error in the evaluation of the inverse Jacobian. Note that the first two terms in equation (8.46) are simply the linear mixing of equation (8.44) with the mixing parameter α. The last term is an additional correction. We see that the modified Broyden method only involves a single $M \times M$ matrix and M vectors of length N. Since M is typically less than 10, the storage cost remains small. Figure 8.5 shows how techniques such as the Broyden method can accelerate the convergence of non-linear self-consistent calculations. The example depicted in the figure shows constrained and unconstrained HF calculations with Skyrme functionals. The paper [77] gives additional examples such as coupled-

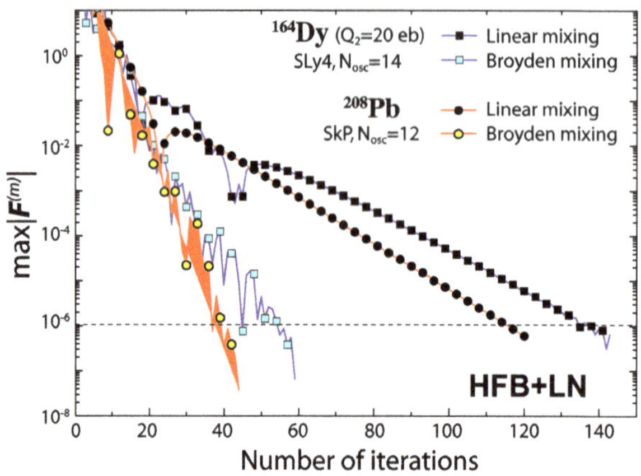

Figure 8.5. Comparison of HFODD convergence speed between linear mixing with $\alpha = 0.5$ (black symbols) and the modified Broyden's method ($\alpha = 0.7$, $M = 7$). We plot a HFB calculation in the superdeformed minimum of ^{152}Dy at the quadrupole moment $\langle \hat{Q}_{20} \rangle = 20$ b with the SLy4 Skyrme functional and a full HO basis with $N = 14$, as well as a HF calculation for the spherical ground state of ^{208}Pb with the SkP Skyrme functional in a smaller basis of $N = 12$. (Figure reproduced from [77], courtesy of Nazarewicz. Copyright 2008 The American Physical Society.)

cluster calculations. The Broyden method has also been implemented to solve the finite-amplitude method (FAM) equation of the QRPA method discussed in section 5.4.3, since the latter are also non-linear equations.

Constrained calculations

In contrast to gradient methods, the handling of constraints on the HFB solution requires dedicated algorithms when the HFB equation is solved by successive iterations. Here, we will describe a powerful method originally introduced in [44] and presented in more detail in [3, 45]. The method is a variant of the standard Lagrange multiplier for linear constraints, only the update of said multipliers uses additional information specifically related to the structure of the HFB solution.

> Several methods to handle constraints have been considered in the past. For a long time, quadratic constraints, where a term of the kind $\frac{1}{2}\mu\big(\langle \hat{Q} \rangle - Q_0\big)^2$ is added to the energy, were hailed as one of the most robust methods [78]. Its drawbacks were that the parameter μ has to be set by hand and that the method rarely converges to the requested value Q_0. Combining the linear and quadratic constraints within what is called the augmented Lagrangian method (ALM) solved the latter deficiency of the quadratic method, but not the former [46]. We show in figure 8.6 a comparison between the ALM and the random-phase approximation (RPA)-based approach described here.

Let us consider a set of one-body constraint operators $\hat{F}_a$. Their expression in second quantization is

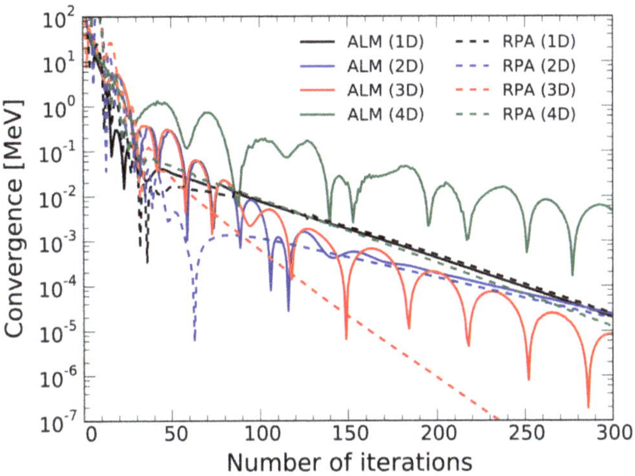

Figure 8.6. Convergence of the HFB calculation as function of the number of iterations in ^{240}Pu for four different sets of constraints: (i) a 1D case with only a constraint on $\langle \hat{Q}_{20} \rangle = 60$ b, (ii) a 2D case with a constraint on both $\langle \hat{Q}_{20} \rangle = 60$ b and $\langle \hat{Q}_{22} \rangle = 0$ b, (iii) a 3D case with the constraints $\langle \hat{Q}_{20} \rangle = 60$ b, $\langle \hat{Q}_{22} \rangle = 0$ b b, and $\langle \hat{Q}_{40} \rangle = 5$ b^2, and finally (iv) a 4D case with the constraints $\langle \hat{Q}_{10} \rangle = 0$ fm, $\langle \hat{Q}_{20} \rangle = 60$ b, $\langle \hat{Q}_{22} \rangle = 0$ b, and $\langle \hat{Q}_{30} \rangle = 10$ b$^{3/2}$. (Figure reproduced from [3], courtesy of Schunck. Copyright 2017 with permission from Elsevier.)

$$\hat{F}_a = \sum_{ij} F_{a;ij} c_i^{\dagger} c_j = \sum_{ij} \langle i | \hat{F}_a | j \rangle c_i^{\dagger} c_j.$$

The problem is to obtain solutions $|\Phi\rangle$ to the HFB equation such that the following constraints are satisfied

$$\forall a, \ \langle \hat{F}_a \rangle = \langle \Phi | \hat{F}_a | \Phi \rangle = \bar{F}_a,$$

where $|\Phi\rangle$ is the q.p. vacuum. This is a classic problem of optimization under constraints. For Hamiltonian-based EDF, one can introduce the Hamiltonian

$$\hat{H} = \hat{H}_0 - \sum_a \lambda_a (\hat{F}_a - \bar{F}_a) \tag{8.48}$$

and minimize $\langle \Phi | \hat{H}_0 - \sum_a \lambda_a (\hat{F}_a - \bar{F}_a) | \Phi \rangle / \langle \Phi | \Phi \rangle$. By following the same line of argument as in chapter 3, we easily obtain the HFB equation with constraints

$$\left[\mathcal{R}, \mathcal{H}_0 - \sum_a \lambda_a \mathcal{F}_a \right] = 0, \tag{8.49}$$

where $\mathcal{H}_0$ is the 'free' HFB matrix (without the constraints) and $\mathcal{F}_a$ is the matrix of the operator $\hat{F}_a$ in the double s.p. basis.

If F_a is the matrix of an arbitrary, Hermitian one-body operator in the s.p. basis, then the matrix $\mathcal{F}_a$ of the same operator in the double s.p. basis reads

$$\mathcal{F}_a = \begin{pmatrix} F_a & 0 \\ 0 & -F_a^* \end{pmatrix}.$$

This can easily be obtained by using the definition (1.4) of a one-body operator in second quantization and recalling the form (3.24) of the generalized density.

We now assume that we have a generalized density matrix $\mathcal{R}^{(0)}$ that diagonalizes the HFB matrix $\mathcal{H}^{(0)}$ under the set of constraints $\hat{F}_a$. We are interested in the effect of a small perturbation $\delta\mathcal{R}$ of the density matrix, on the HFB solutions. Under such a perturbation, we can write

$$\mathcal{R} = \mathcal{R}^{(0)} + \delta\mathcal{R} \qquad \mathcal{H} = \mathcal{H}^{(0)} + \delta\mathcal{H} \qquad \lambda_a = \lambda_a^{(0)} + \delta\lambda_a \quad \forall\, a. \tag{8.50}$$

The variation of the HFB matrix can be obtained by Taylor expansion

$$\delta\mathcal{H} = \delta\mathcal{R}\frac{\partial\mathcal{H}}{\partial\mathcal{R}} = \sum_{a\le b} \delta\mathcal{R}_{ab}\frac{\partial\mathcal{H}}{\partial\mathcal{R}_{ab}}.$$

Note that since $\mathcal{H}$ is a $(2M \times 2M)$ matrix, $\delta\mathcal{H}$ is in fact a rank-4 tensor: it has four running indices. We impose that the perturbed HFB matrix still commutes with the perturbed generalized density matrix,

$$\left[\mathcal{R}, \mathcal{H} - \sum_a \lambda_a \mathcal{F}_a\right] = 0.$$

Substituting $\mathcal{R}$, $\mathcal{H}$ and λ_a by their expressions given in equation (8.50), we obtain to first order in $\delta\mathcal{R}$,

$$\left[\delta\mathcal{R}, \mathcal{H}^{(0)} - \sum_a \lambda_a^{(0)}\mathcal{F}_a\right] + \left[\mathcal{R}^{(0)}, \delta\mathcal{R}\frac{\partial\mathcal{H}}{\partial\mathcal{R}}\right] - \sum_a \delta\lambda_a[\mathcal{R}^{(0)}, \mathcal{F}_a] = 0. \tag{8.51}$$

These considerations are analogous to the RPA response function discussed in section 5.1.3. In particular, by putting $G = 0$ in equation (5.10) and assuming purely static solutions and perturbations $d(\delta\rho)/dt = 0$, we obtain very similar equations.

We now make the additional assumption to neglect the term $\partial\mathcal{H}/\partial\mathcal{R}$. This is equivalent to working at the so-called cranking approximation. The equation for the linear constraint becomes

$$\left[\delta\mathcal{R}, \mathcal{H}^{(0)} - \sum_a \lambda_a^{(0)}\mathcal{F}_a\right] - \sum_a \delta\lambda_a[\mathcal{R}^{(0)}, \mathcal{F}_a] = 0. \tag{8.52}$$

Let us obtain additional information on the matrix of $\delta\mathcal{R}$. To first order, the condition that both $\mathcal{R}$ and $\mathcal{R}^{(0)}$ are projectors reads

$$\mathcal{R}^2 = (\mathcal{R}^{(0)} + \delta\mathcal{R})^2 = \mathcal{R}^{(0)} + \mathcal{R}^{(0)}\delta\mathcal{R} + \delta\mathcal{R}\mathcal{R}^{(0)} = \mathcal{R}^{(0)} + \delta\mathcal{R}.$$

Hence,

$$\delta\mathcal{R} = \mathcal{R}^{(0)}\delta\mathcal{R} + \delta\mathcal{R}\mathcal{R}^{(0)}.$$

We write down this equation in the q.p. basis. From now on, and by convention, all matrices of operators in the q.p. basis will have a 'tilde' sign, all matrices in the s.p. basis, e.g. the HO basis, will have no sign. We have

$$\tilde{\mathcal{R}}^{(0)} = \begin{pmatrix} 0 & 0 \\ 0 & 1 \end{pmatrix} \qquad \delta\tilde{\mathcal{R}} = \begin{pmatrix} \tilde{R}^{11} & \tilde{R}^{12} \\ \tilde{R}^{21} & \tilde{R}^{22} \end{pmatrix} \qquad \tilde{\mathcal{F}}_a = \begin{pmatrix} \tilde{F}_a^{11} & \tilde{F}_a^{12} \\ \tilde{F}_a^{21} & \tilde{F}_a^{22} \end{pmatrix}$$

and, since the q.p. is the basis that diagonalizes the HFB matrix at zero order,

$$\tilde{\mathcal{H}}^{(0)} - \sum_a \lambda_a^{(0)}\tilde{\mathcal{F}}_a = \begin{pmatrix} E & 0 \\ 0 & -E \end{pmatrix}.$$

It is then straightforward to show that, in the q.p. basis, the matrix $\delta\tilde{\mathcal{R}}$ is anti-diagonal

$$\delta\tilde{\mathcal{R}} = \begin{pmatrix} 0 & \tilde{R}^{12} \\ \tilde{R}^{21} & 0 \end{pmatrix}. \tag{8.53}$$

Based on the form (8.53) and of the matrix of the constraint in the q.p. basis, we can compute each commutator in equation (8.51) and express the unknown matrices $\tilde{R}^{12}$ as a function of the known matrix F_a^{12} and the q.p. energies. After some elementary algebra, we find

$$\tilde{R}_{ij}^{12} = \sum_a \delta\lambda_a \frac{1}{E_i + E_j}\tilde{F}_{a;ij}^{12}$$

$$\tilde{R}_{ij}^{21} = \sum_a \delta\lambda_a \frac{1}{E_i + E_j}\tilde{F}_{a;ij}^{21}.$$

At this point, we can obtain an additional simplification by computing explicitly the expression of $\tilde{\mathcal{F}}_a$, i.e. the matrix of the constraint in the q.p. basis, in terms of the matrix in the s.p. basis and the Bogoliubov transformation. This explicit calculation leads to

$$\tilde{\mathcal{F}}_a = \begin{pmatrix} U^\dagger F_a U - V^\dagger F_a^* V & U^\dagger F_a V^* - V^\dagger F_a^* U^* \\ V^T F_a U - U^T F_a^* V & V^T F_a V^* - U^T F_a^* U^* \end{pmatrix} = \begin{pmatrix} \tilde{F}_a^{11} & \tilde{F}_a^{12} \\ -\tilde{F}_a^{12\,*} & -\tilde{F}_a^{11\,*} \end{pmatrix}. \tag{8.54}$$

Therefore,

$$\tilde{R}_{ij}^{12} = +\sum_a \delta\lambda_a \frac{1}{E_i + E_j}\tilde{F}_{a;ij}^{12} \tag{8.55a}$$

$$\tilde{R}_{ij}^{21} = -\sum_a \delta\lambda_a \frac{1}{E_i + E_j}\tilde{F}_{a;ij}^{12\,*}. \tag{8.55b}$$

This equation relates the variation of the generalized density induced by a small change $\delta\lambda_a$ of the Lagrange parameters associated with the constraints. Conversely, we can use it to optimize the value of the Lagrange parameter at each iteration in order to satisfy the requested constraints. Let us assume that at iteration n of the self-consistent loop, the deviation between the actual and requested value of the constraint operators are

$$\delta F_a = \bar{F}_a - \langle \hat{F}_a \rangle^{(n)} \qquad \forall a.$$

It will be convenient to introduce the vector of constraint operators $\hat{\boldsymbol{F}}$ and the related vector of linear constraints values $\boldsymbol{\lambda}$,

$$\hat{\boldsymbol{F}} = \begin{pmatrix} \hat{F}_1 \\ \vdots \\ \hat{F}_N \end{pmatrix} \qquad \boldsymbol{\lambda} = \begin{pmatrix} \lambda_1 \\ \vdots \\ \lambda_N \end{pmatrix}.$$

Let us now define the vector of perturbations $\delta\boldsymbol{\lambda}$ such that

$$\langle \hat{\boldsymbol{F}}(\boldsymbol{\lambda} + \delta\boldsymbol{\lambda}) \rangle = \langle \Phi(\boldsymbol{\lambda} + \delta\boldsymbol{\lambda}) | \hat{\boldsymbol{F}} | \Phi(\boldsymbol{\lambda} + \delta\boldsymbol{\lambda}) \rangle.$$

We find

$$\delta \boldsymbol{F} = \frac{1}{2}\mathrm{Tr}(\boldsymbol{F}\delta\mathcal{R}) = \frac{1}{2}\sum_{ij} \mathcal{F}_{a;ij}\delta\mathcal{R}_{ji} \quad \forall a. \tag{8.56}$$

In the q.p. basis, the trace thus becomes

$$\delta F_a = \frac{1}{2}\sum_{\mu\nu}\left(\tilde{F}^{12}_{a;\mu\nu}\tilde{R}^{21}_{\nu\mu} - \tilde{F}^{12\,*}_{a;\mu\nu}\tilde{R}^{12}_{\nu\mu}\right).$$

Let us now introduce relations (8.55) into this latter result. We obtain

$$\delta F_a = -\frac{1}{2}\sum_{\mu\nu}\tilde{F}^{12}_{a;\mu\nu}\sum_b \delta\lambda_b \frac{1}{E_\mu + E_\nu}\tilde{F}^{12\,*}_{b;\nu\mu} - \frac{1}{2}\sum_{\mu\nu}\tilde{F}^{12\,*}_{a;\mu\nu}\sum_b \delta\lambda_b \frac{1}{E_\mu + E_\nu}\tilde{F}^{12}_{b;\nu\mu}.$$

Let us denote by M_{ab} the matrix made of the elements

$$M^{(K)}_{ab} = -\sum_{\mu\nu}\tilde{F}^{12}_{a;\mu\nu}\frac{1}{(E_\mu + E_\nu)^K}\tilde{F}^{12\,*}_{b;\nu\mu}. \tag{8.57}$$

We already encountered the moments $M^{(K)}_{ab}$ in section 6.4.3, equation (6.67), in the context of the GOA approximation of the GCM. The main difference is that equation (6.67) did not include the approximation $\delta\mathcal{H}/\delta\mathcal{R} = 0$ but the inverse of the full QRPA matrix. With this notation, we can then write

$$\delta\boldsymbol{\lambda} = [M^{(1)}]^{-1}\delta\boldsymbol{F}. \tag{8.58}$$

Therefore, we have obtained an expression to determine at each iteration the change needed in the value of the Lagrange parameters to satisfy the constraints. This expression is in closed form: it only depends on the difference between the actual expectation value of the constraint operator and its target value, and the matrix of the moments $M^{(K)}$, which itself depends on the q.p. energies and the matrix of the constraint operator.

The interpretation of the formula (8.58) is straightforward. It relies on exploiting the knowledge of an approximation of the Hessian matrix associated with the function to minimize. In the context of the HFB theory, the function is the energy and the Hessian matrix is the QRPA matrix or, equivalently, the stability matrix. The cranking approximation is a particularly convenient approximation of the QRPA matrix, since it puts it into diagonal form. This approximation is also used in the gradient method to optimize the choice of the Thouless matrix giving the next iteration; see equation (8.68) in section 8.3.2. It is also a very popular approximation to compute the collective mass for large-amplitude collective motion, either in the ATDHFB theory or in the GCM+GOA; see equation (6.77) in section 6.4.4.

8.3.2 Gradient methods

In spite of its popularity, the non-linear eigenvalue problem formulation of the HFB equation suffers from the need to properly perform a mixing of consecutive iterations to ensure convergence of the self-consistent loop. This sometimes leads to numerical instabilities and is typically observed in constrained HFB calculations in 'flat' collective spaces. Gradient methods provide an alternative that is less sensitive to such effects. They are based on the powerful Thouless theorem, which provides a formula to relate two independent q.p. vacua.

Thouless theorem

Theorem 5 (Thouless). *Let us consider two arbitrary q.p. vacua $|\Phi_0\rangle$ and $|\Phi_1\rangle$. Then there exists a Hermitian one-body operator $\hat{Z}$ such that*

$$|\Phi_1\rangle = e^{i\hat{Z}}|\Phi_0\rangle.$$

We choose to present the Thouless theorem in terms of a unitary transformation ($\propto e^{i\hat{Z}}$). In [74, 11], we also find the traditional form of the theorem,

$$|\Phi_1\rangle = \mathcal{N}e^{\frac{1}{2}\sum_{\mu\nu} Z_{\mu\nu}\beta^\dagger_\mu\beta^\dagger_\nu}|\Phi_0\rangle$$

with the Z matrix defined as skew-symmetric: $Z^T = -Z$. Evidently, the Thouless matrix iZ is thus Hermitian, and we fall back to the definition given here.

Recall that the expression of a Hermitian one-body operator $\hat{Z}$ in second quantization is

$$\hat{Z} = \sum_{ij} Z_{ij} c_i^\dagger c_j, \qquad Z_{ij} = Z_{ji}^*.$$

By applying the Bogoliubov transformation $\mathcal{W}_0$, it can be written in terms of the q.p. operators $\beta_{0,\mu}$, $\beta_{0,\mu}^\dagger$

$$\hat{Z} = \sum_{\mu\nu} Z_{\mu\nu}^{11} \beta_{0,\mu}^\dagger \beta_{0,\nu} + \frac{1}{2}\sum_{\mu\nu} Z_{\mu\nu}^{20} \beta_{0,\mu}^\dagger \beta_{0,\nu}^\dagger - \frac{1}{2}\sum_{\mu\nu} Z_{\mu\nu}^{20*} \beta_{0,\mu} \beta_{0,\nu} \tag{8.59}$$

The conventions implicitly adopted here, normal ordering and factors 1/2, will become clear below. Note the properties of the matrices Z^{11} and Z^{20}, which can be obtained simply from the hermiticity of Z and the properties of the Bogoliubov transformation: $Z^{11} = Z^{11\dagger}$ (Hermitian) and $Z^{20} = -Z^{20T}$ (skew-symmetric). Also note that the first and last terms of equation (8.59) are somewhat redundant since when acting on the vacuum $|\Phi_0\rangle$, they will give 0. In many presentations of the applications of the Thouless theorem, these terms are often omitted.

Let us use the properties of the q.p. vacuua to infer a transformation rule for the q.p. operators. We start with the definitions of said vacuua

$$\beta_{0,\mu}|\Phi_0\rangle = 0 \quad \beta_{1,\mu}|\Phi_1\rangle = 0 \quad \forall \mu.$$

Therefore,

$$e^{-i\hat{Z}} e^{+i\hat{Z}} \beta_{0,\mu} e^{-i\hat{Z}} e^{+i\hat{Z}} |\Phi_0\rangle = 0.$$

Since $\hat{Z}$ is Hermitian, the transformation

$$\mathcal{Z}\colon \beta_{0,\mu} \mapsto \gamma_{1,\mu} = e^{+i\hat{Z}} \beta_{0,\mu} e^{-i\hat{Z}}$$

is unitary and preserves the anti-commutation relations. In other words, the operators $\gamma_{1,\mu}$, $\gamma_{1,\mu}^\dagger$ obey the same Fermionic anti-commutation relation as the $\beta_{0,\mu}$, $\beta_{0,\mu}^\dagger$. Moreover, we find that

$$e^{-i\hat{Z}} \gamma_{1,\mu} |\Phi_1\rangle = 0 \qquad \forall \mu.$$

This is only possible if $\gamma_{1,\mu}|\Phi_1\rangle = 0$, $\forall \mu$, which implies that the operators γ coincide with the operators β_1 associated with the q.p. vacuum $|\Phi_1\rangle$. We thus can write

$$\begin{pmatrix} \beta_1 \\ \beta_1^\dagger \end{pmatrix} = e^{+i\hat{Z}} \begin{pmatrix} \beta_0 \\ \beta_0^\dagger \end{pmatrix} e^{-i\hat{Z}}. \tag{8.60}$$

We can obtain a more convenient form of the transformation (8.60) by using the Baker–Campbell–Hausdorff formula for the operators $\beta_{0,\alpha}$ and $\beta_{0,\alpha}^\dagger$,

$$e^{-i\hat{Z}} \hat{A} e^{i\hat{Z}} = \hat{A} - i[\hat{Z}, \hat{A}] - \frac{1}{2}[\hat{Z}, [\hat{Z}, \hat{A}]] + \ldots.$$

The commutator $[\hat{Z}, \beta_{0,\alpha}]$ gives

$$[\hat{Z}, \beta_{0,\alpha}] = -\sum_{\nu} Z^{11}_{\alpha\nu}\beta_{0,\nu} - \sum_{\nu} Z^{20}_{\alpha\nu}\beta^{\dagger}_{0,\nu}$$

and for $[\hat{Z}, \beta^{\dagger}_{0,\alpha}]$ we find

$$[\hat{Z}, \beta_{0,\alpha}] = +\sum_{\nu} Z^{11*}_{\alpha\nu}\beta^{\dagger}_{0,\nu} + \sum_{\nu} Z^{20*}_{\alpha\nu}\beta_{0,\nu},$$

which can be cast into the concise form

$$\left[\hat{Z}, \begin{pmatrix} \beta_0 \\ \beta_0^{\dagger} \end{pmatrix}\right] = i\begin{pmatrix} Z^{11} & Z^{20} \\ -Z^{20*} & -Z^{11*} \end{pmatrix}\begin{pmatrix} \beta_0 \\ \beta_0^{\dagger} \end{pmatrix}.$$

Let us denote

$$\mathcal{Z} = \begin{pmatrix} Z^{11} & Z^{20} \\ -Z^{20*} & -Z^{11*} \end{pmatrix}. \tag{8.61}$$

Similar calculations for the other terms of the Baker–Hausdorff–Campbell expansion lead to the following result

$$\begin{pmatrix} \beta_1 \\ \beta_1^{\dagger} \end{pmatrix} = e^{+i\mathcal{Z}}\begin{pmatrix} \beta_0 \\ \beta_0^{\dagger} \end{pmatrix}.$$

Note the differences between this latter expression and equation (8.60): here, $\mathcal{Z}$ is a matrix in the original q.p. basis, which means that the q.p. operators β_1, $\beta_1^{\dagger}$ are in fact linear combinations of the β_0 and $\beta_0^{\dagger}$ operators. By using the definition of the two Bogoliubov transformations involved here,

$$\begin{pmatrix} \beta_0 \\ \beta_0^{\dagger} \end{pmatrix} = W_0^{\dagger}\begin{pmatrix} c \\ c^{\dagger} \end{pmatrix} \quad \begin{pmatrix} \beta_1 \\ \beta_1^{\dagger} \end{pmatrix} = W_1^{\dagger}\begin{pmatrix} c \\ c^{\dagger} \end{pmatrix}, \tag{8.62}$$

we find immediately the following transformation rule between the two Bogoliubov matrices

$$W_1^{\dagger} = e^{+i\mathcal{Z}}W_0^{\dagger}. \tag{8.63}$$

The idea behind the Thouless derivation of the HFB equation is to look at the variations of the generalized density when going from an initial basis $(\beta_0, \beta_0^{\dagger})$ to a new basis $(\beta_1, \beta_1^{\dagger})$: $\mathcal{R}_1 = \mathcal{R}_0 + \delta\mathcal{R}$. Using the transformation (8.63) to first order in powers of $\mathcal{Z}$, we find

$$\mathcal{R}_1 = (W_0 - iW_0\mathcal{Z})\tilde{\mathcal{R}}_0(W_0^{\dagger} + i\mathcal{Z}W_0^{\dagger}),$$

which leads directly to an expression for $\delta\mathcal{R}$.

HFB equation from the Thouless theorem

The Thouless theorem can be used to obtain an alternative derivation of the HFB equation. Let us consider the expansion of equation (8.63) in terms of powers of $\mathcal{Z}$. To first order, it reads

$$W_1 = W_0 - iW_0\mathcal{Z}.$$

Remember that the generalized density in the s.p. basis $\mathcal{R}$ is obtained from the generalized density in the q.p. basis $\tilde{\mathcal{R}}$ through

$$\mathcal{R} = W\tilde{\mathcal{R}}W^{\dagger}, \qquad \tilde{\mathcal{R}} = \begin{pmatrix} 0 & 0 \\ 0 & 1 \end{pmatrix}.$$

From this last expression, variations of $\mathcal{R}$ can be expressed as

$$\delta\mathcal{R} = -iW_0(\mathcal{Z}\tilde{\mathcal{R}}-\tilde{\mathcal{R}}\mathcal{Z})W_0^{\dagger}.$$

The matrix $\tilde{\mathcal{R}}$ is expressed in q.p. space. If we transform it to the s.p. basis, we find

$$\delta\mathcal{R} = i(W_0\mathcal{Z}W_0^{\dagger}\mathcal{R} - \mathcal{R}W_0\mathcal{Z}W_0^{\dagger}),$$

that is

$$\delta\mathcal{R} = i[Z, \mathcal{R}].$$

Since the variations of the energy for a q.p. vacuum associated with $\mathcal{R}$ are given by

$$\delta E = \frac{1}{2}\mathrm{tr}(\mathcal{H}\delta\mathcal{R})$$

the use of the cyclic invariance of the trace naturally yields

$$[\mathcal{H}, \mathcal{R}] = 0,$$

which is nothing other than the HFB equation.

Let us obtain a more convenient relation between the original Bogoliubov transformation characterized by the matrices (U_0, V_0) and the new transformation characterized by (U_1, V_1). Going back to equation (8.62) we can express the relationship between the q.p. operators in the matrix form

$$\begin{pmatrix} \beta_1 \\ \beta_1^{\dagger} \end{pmatrix} = W_1^{\dagger}W_0\begin{pmatrix} \beta_0 \\ \beta_0^{\dagger} \end{pmatrix} = \mathcal{W}^{\dagger}\begin{pmatrix} \beta_0 \\ \beta_0^{\dagger} \end{pmatrix}, \tag{8.64}$$

where the matrix of the Bogoliubov transformation $\mathcal{W}$ between the two q.p. vacua is denoted by

$$W = \begin{pmatrix} U & V^* \\ V & U^* \end{pmatrix}. \tag{8.65}$$

Since both W_0 and W_1 are Hermitian, so is W. Therefore, it also defines a Bogoliubov transformation (only, between two sets of q.p. operators). Remember that we know W_0 but not W_1 (hence U_1 and V_1). Since the Onishi theorem states that $|\langle\Phi_0|\Phi_1\rangle|^2 \propto |\det(U)|^2$ (see equation (3.105) in section 3.2.2), the non-orthogonality of the two states implies that $\det(U)$ is non-zero and the matrix U can be inverted. Therefore, we can introduce a new set of operators γ_1, $\gamma_1^\dagger$ that verify

$$\gamma_{1,\mu}^\dagger = \sum_n (U^{-1})_{n\mu}\beta_{1,n}^\dagger = \beta_{0,\mu}^\dagger + \sum_n (VU^{-1})_{n\mu}\beta_{0,n}.$$

This comes directly from equation (8.65),

$$\begin{aligned}\beta_{1,n}^\dagger &= \sum_m V_{nm}^T\beta_{0,m} + \sum_m U_{nm}^T\beta_{0,m}^\dagger \\ &= \sum_{mn}\left[V_{mn}(U^{-1})_{n\mu}\beta_{0,m} + U_{mn}(U^{-1})_{n\mu}\beta_{0,m}^\dagger\right] = \sum_m (VU^{-1})_{m\mu}\beta_{0,m} + \beta_{0,\mu}^\dagger.\end{aligned}$$

Similarly,

$$\gamma_{1,\mu} = \beta_{0,\mu} + \sum_n (VU^{-1})_{n\mu}^\dagger\beta_{0,n}^\dagger.$$

By construction, we must have $\gamma_{1,\mu}|\Phi_1\rangle = 0$ since the γ_1 operators are linear combinations of β_1. Given the particular form of $|\Phi_1\rangle$ given by the Thouless theorem, we can show that a necessary condition for γ_1 to satisfy $\gamma_{1,\mu}|\Phi_1\rangle = 0$ is that $VU^{-1} = Z^{20}$. Although the operators γ_1 properly annihilate the vacuum $|\Phi_1\rangle$, their anti-commutation relation is

$$\{\gamma_{1,\mu}, \gamma_{1,\nu}^\dagger\} = \delta_{\mu\nu} + (Z^{20\,T}Z^{20*})_{\mu\nu}$$

and they can, therefore, not be used to define our new q.p. In order to ensure that the anti-commutation relation are satisfied, we thus introduce the Cholesky decomposition of the matrix $I + Z^{20\,T}Z^{20*}$, that is, the matrix L such that $LL^\dagger = I + Z^{20\,T}Z^{20*}$. We then consider the q.p. operators defined by

$$\eta_{1,\mu} = \sum_\nu (L^{-1})_{\mu\nu}\gamma_{1,\nu}$$

and similarly for $\eta_{1,\mu}^\dagger$. By construction, these operators satisfy the anti-commutation rules of Fermion operators, annihilate the vacuum $|\Phi_1\rangle$ and their Bogoliubov transformation is expressed as a function of the original one through

$$U_1 = (U_0 + V_0^* Z^{20*})[L^{-1}]^\dagger \tag{8.66a}$$

$$V_1 = (V_0 + U_0^* Z^{20*})[L^{-1}]^\dagger. \tag{8.66b}$$

Relations (8.66a) can be obtained by combining the definition of the operators $\eta_{1,\mu}$ and $\gamma_{1,\nu}$ as a function of the operators $(\beta_0, \beta_0^\dagger)$ and then recalling the definition (3.6) of the Bogoliubov transformation to express $\eta_{1,\mu}$ as a function of the s.p. creation and annihilation operators. Therefore, if we know the Bogoliubov matrices (U_0, V_0) and the Thouless matrix Z^{20}, then we can easily compute the new Bogoliubov matrices (U_1, V_1) through a series of standard linear algebra operations. We will see in the next section that the gradient method provides a very simple expression for the Thouless matrix.

Minimization of the energy

The first key ingredient for gradient methods is equation (8.66), which gives the Bogoliubov matrices associated with a new ket $|\Phi_1\rangle$ given an old ket $|\Phi_0\rangle$ and a Thouless matrix Z. Note: to make the notation a little less dense, we now choose to denote by $Z \equiv iZ^{20}$, with Z^{20} the upper-right block of the full Thouless matrix $\mathcal{Z}$ of equation (8.61). By definition, we have $Z^T = -Z$. This choice of notation should make it easier to connect with the existing literature, in particular [11, 74] and papers [79, 80]. The second key ingredient is, quite naturally, the expression for the gradient $\partial E/\partial Z$ (formally). Let us parametrize the energy associated with an arbitrary ket $|\Phi_1\rangle$ in the neighborhood of $|\Phi_0\rangle$,

$$E(Z, Z^*) = \frac{\langle\Phi_1|\hat{H}|\Phi_1\rangle}{\langle\Phi_1|\Phi_1\rangle} = E_0 + \left.\frac{\partial E}{\partial Z}\right|_{Z=0} \delta Z + \left.\frac{\partial E}{\partial Z^*}\right|_{Z^*=0} \delta Z^*.$$

Using the definition of $|\Phi_1\rangle$, we find

$$E(Z, Z^*) = \frac{\langle\Phi_0|e^{\frac{1}{2}\sum_{\gamma\delta} Z^*_{\gamma\delta}\beta_{0,\delta}\beta_{0,\gamma}} \hat{H} e^{\frac{1}{2}\sum_{\mu\nu} Z_{\mu\nu}\beta^\dagger_{0,\mu}\beta^\dagger_{0,\nu}}|\Phi_0\rangle}{\langle\Phi_0|e^{\frac{1}{2}\sum_{\gamma\delta} Z^*_{\gamma\delta}\beta_{0,\delta}\beta_{0,\gamma}} e^{\frac{1}{2}\sum_{\mu\nu} Z_{\mu\nu}\beta^\dagger_{0,\mu}\beta^\dagger_{0,\nu}}|\Phi_0\rangle}. \tag{8.67}$$

Hence the derivative with respect to Z^* reads

$$\left.\frac{\partial E}{\partial Z^*_{\gamma\delta}}\right|_{Z=0} = \left.\frac{1}{2}\frac{\langle\Phi_1|\beta_{0,\delta}\beta_{0,\gamma}[\hat{H} - E(Z, Z^*)]|\Phi_1\rangle}{\langle\Phi_1|\Phi_1\rangle}\right|_{Z=0} = -\frac{1}{2}H^{20}_{\delta\gamma} = \frac{1}{2}H^{20}_{\gamma\delta}.$$

We define the variation of the energy E under variations δZ^* of the Thouless matrix as

$$\delta E = \sum_{\mu\nu} \frac{\partial E}{\partial Z^*_{\mu\nu}} \delta Z^*_{\mu\nu}$$

and similarly for δZ. Both the numerator and the denominator of the function $E(Z, Z^*)$ depend on Z^*, hence we find

$$\frac{\partial E}{\partial Z^*_{\mu\nu}} = \frac{1}{2}\frac{\langle\Phi_1|\beta_{0,\delta}\beta_{0,\gamma}\hat{H}|\Phi_1\rangle}{\langle\Phi_1|\Phi_1\rangle} - \frac{1}{2}\frac{\langle\Phi_1|\hat{H}|\Phi_1\rangle\langle\Phi_1|\beta_{0,\delta}\beta_{0,\gamma}|\Phi_1\rangle}{\langle\Phi_1|\Phi_1\rangle^2},$$

which gives the desired result. Similar calculations give the derivative with respect to $Z_{\mu\nu}$.

We thus find that the gradient of the energy is proportional to the term H^{20} in the expansion of the Hamiltonian into normal-ordered products of q.p. operators $(\beta_0, \beta_0^\dagger)$, see equation (3.35). A naive application of the gradient method would thus consist in the following algorithm:

1. Start with a given q.p. vacuum $|\Phi_0\rangle$ and the corresponding Bogoliubov matrix W_0.
2. Compute the matrix of the gradient H^{20}, set the Thouless matrix to a fraction η along the line of steepest descent, $Z = -\eta H^{20}$, and determine the optimal value of η such that the energy is minimum (line minimization).
3. Update the Bogoliubov matrix to W_1 using the Thouless matrix and the formula (8.66).
4. Start over until convergence, defined by some criterion ε such that $||H^{20}|| \leqslant \varepsilon$.

This algorithm can be considerably improved by informing the calculation of the gradient with the local information about the curvature of the (hyper-)surface $E(Z)$ around the local point. Let us go back to the expression of the energy (8.67) and expand it up to second-order in Z. The Taylor expansion of the function $E(Z, Z^*)$ up to second order reads

$$E(Z, Z^*) = E_0 + \frac{\partial E}{\partial Z}\delta Z + \frac{\partial E}{\partial Z^*}\delta Z^* + \frac{1}{2}\left[\frac{\partial^2 E}{\partial Z^2}\delta Z^2 + \frac{\partial^2 E}{\partial Z^{*2}}\delta Z^{*2}\right] + \frac{1}{2}\frac{\partial^2 E}{\partial Z \partial Z^*}\delta Z \delta Z^*.$$

When doing the actual calculation, terms proportional to the energy E_0 that contain only q.p. creation or annihilation operators will vanish. Only terms of the kind $\propto\langle\beta E\beta^\dagger\rangle$, $\propto\langle\beta\beta E\beta^\dagger\beta^\dagger\rangle$, etc, will contribute. Only focusing on the numerator, we find

$$\begin{aligned} E(Z, Z^*) \propto E_0 &+ \frac{1}{2}\sum_{\gamma\delta} Z^*_{\gamma\delta}\langle\Phi_0|\beta_{0,\delta}\beta_{0,\gamma}\hat{H}|\Phi_0\rangle + \frac{1}{2}\sum_{\mu\nu}\langle\Phi_0|\hat{H}\beta^\dagger_{0,\mu}\beta^\dagger_{0,\nu}|\Phi_0\rangle Z_{\mu\nu} \\ &+ \frac{1}{4}\sum_{\gamma\delta\mu\nu} Z^*_{\gamma\delta}\langle\Phi_0|\beta_{0,\delta}\,\beta_{0,\gamma}(\hat{H} - E_0)\beta^\dagger_{0,\mu}\beta^\dagger_{0,\nu}|\Phi_0\rangle Z_{\mu\nu} \\ &+ \frac{1}{4}\sum_{\gamma\delta\mu\nu} Z^*_{\gamma\delta}Z^*_{\mu\nu}\langle\Phi_0|\beta_{0,\delta}\beta_{0,\gamma}\beta_{0,\nu}\beta_{0,\mu}\hat{H}|\Phi_0\rangle \\ &+ \frac{1}{4}\sum_{\gamma\delta\mu\nu} Z_{\gamma\delta}Z_{\mu\nu}\langle\Phi_0|\hat{H}\beta^\dagger_{0,\gamma}\beta^\dagger_{0,\delta}\beta^\dagger_{0,\mu}\beta^\dagger_{0,\nu}|\Phi_0\rangle. \end{aligned}$$

This can be recast into the following compact form

$$E(Z, Z^*) = E_0 + \frac{1}{2}(Z^*, Z)\begin{pmatrix} H^{20} \\ H^{20\,*} \end{pmatrix} + \frac{1}{4}(Z^*, Z)\begin{pmatrix} A & B \\ B^* & A^* \end{pmatrix}\begin{pmatrix} Z \\ Z^* \end{pmatrix},$$

with

$$A_{\mu\nu\gamma\delta} = \langle\Phi_0|\beta_{0,\delta}\beta_{0,\gamma}(\hat{H} - E_0)\beta^{\dagger}_{0,\mu}\beta^{\dagger}_{0,\nu}|\Phi_0\rangle$$
$$B_{\mu\nu\gamma\delta} = \langle\Phi_0|\beta_{0,\delta}\beta_{0,\gamma}\beta_{0,\nu}\beta_{0,\mu}\hat{H}|\Phi_0\rangle$$

and, after playing a bit with the anti-commutation relations of the β operators, we recognize the QRPA matrix $\mathcal{M}$. This is the same matrix as in equation (6.72) obtained in the derivation of the collective Hamiltonian within the GCM+GOA in section 6.4.3. The explicit expression of A and B is given by equation (5.36) in section 5.4.3. Requiring that the derivative of the energy with respect to Z and Z^* be zero, we find now the matrix equation

$$H^{20} + \mathcal{M}Z = 0 \Rightarrow Z = -\mathcal{M}^{-1}H^{20}.$$

While solving this equation rigorously is a formidable computational task, since it requires inverting the full QRPA matrix, it can be simplified by adopting the usual cranking approximation such that the residual interaction in $\mathcal{M}$ vanishes, i.e. $B = 0$ and $A_{\mu\nu,\,\gamma\delta} = (E_\mu + E_\nu)\delta_{\mu\gamma}\delta_{\nu\delta}$. This leads to the following formula

$$Z_{\mu\nu} = \frac{H^{20}_{\mu\nu}}{E_\mu + E_\nu}, \tag{8.68}$$

where E_μ and E_ν are the q.p. energies associated with the vacuum $|\Phi_0\rangle$. This version of the gradient method is used, e.g. in [80].

Conjugate gradient method

Let us consider a scalar function f of the vector $\boldsymbol{x}$,

$$f: \begin{array}{l}\mathbb{R}^N \to \mathbb{R}\\ \boldsymbol{x} \mapsto f(\boldsymbol{x}).\end{array}$$

We can perform a Taylor expansion around an arbitrary point $\boldsymbol{x}_0$ with

$$f(\boldsymbol{x}) = f(\boldsymbol{x}_0) + \boldsymbol{b}\cdot\boldsymbol{x} + \frac{1}{2}\boldsymbol{x}^T\mathrm{H}\boldsymbol{x},$$

with $\boldsymbol{b} = \nabla f|_{\boldsymbol{x}_0}$ the gradient of the function at point $\boldsymbol{x}_0$ and H the Hessian matrix of the function f at that point,

$$\mathrm{H} \equiv H_{ij} = \frac{\partial^2 f}{\partial x_i \partial x_j}|_{x_0}.$$

By taking partial derivatives of this Taylor expansion, we can express the gradient of the function f at another point $\boldsymbol{x}'$ as $\nabla f| = \mathrm{H}\boldsymbol{x} - \boldsymbol{b}$. Therefore, minimizing the function f is equivalent to solving the linear system of equation $\mathrm{H}\boldsymbol{x} - \boldsymbol{b} = 0$. The conjugate gradient method provides an effective way to do this. If we can find a set $\mathcal{P} = (\boldsymbol{p}_1, \dots, \boldsymbol{p}_N)$ of vectors that verify $\boldsymbol{p}_i^T\mathrm{H}\boldsymbol{p}_j = 0$ for

$i \neq j$, then the solution to the linear system of equation $\mathrm{H}\boldsymbol{x} - \boldsymbol{b} = 0$ can be expanded as

$$x_* = \sum_{i-1}^{N} \alpha_i \boldsymbol{p}_i \qquad \alpha_i = \frac{\boldsymbol{p}_i^T \boldsymbol{b}}{\boldsymbol{p}_i^T \mathrm{H} \boldsymbol{p}_i}.$$

The conjugate gradient method is implemented by building the set $\mathcal{P}$ iteratively. We start at the point $\boldsymbol{x}_0$ with the first vector $\boldsymbol{p}_0$ usually chosen as the gradient of f at $\boldsymbol{x}_0$, that is $\boldsymbol{p}_0 = \boldsymbol{b} - H\boldsymbol{x}_0$. We then move through a sequence of points $\boldsymbol{x}_k = \boldsymbol{x}_{k-1} + \alpha_{k-1}k\boldsymbol{p}_{k-1}$ with

$$\alpha_k = \frac{\boldsymbol{p}_k^T \boldsymbol{r}_k}{\boldsymbol{p}_k^T \mathrm{H} \boldsymbol{p}_k}$$

and the residual $\boldsymbol{r}_k = \boldsymbol{b} - H\boldsymbol{x}_k$. For a function of N variables, N steps are in principle needed to obtain the solution.

The conjugate gradient method can be applied to the minimization of the HFB energy viewed as the scalar function $E(\boldsymbol{Z})$ with $\boldsymbol{Z}$ the linearized version of the Thouless matrix Z (i.e. the vector formed by stacking together all independent elements of the Z matrix). The number of iterations needed to find the solution is in principle equal to the number of independent elements of $\boldsymbol{Z}$, which is itself $N(N+1)/2$, where N is the size of the basis used. Additional details can be found in [79].

8.4 Time-evolution algorithms

In the previous sections, we have reviewed the most common methods to solve the static HFB equation. This emphasis is justified by the fundamental role played by the HFB equation in the EDF approach. In the SR-EDF approach, the HFB solution is used to compute ground-state nuclear properties, see chapter 9 for additional discussion. In the multi-reference energy density functional (MR-EDF) approach, HFB solutions for different values of $\boldsymbol{q}$, which is a generic symbol to refer to either a set of collective coordinates (GCM) or a set of gauge angles (symmetry restoration), provide a basis of states used in configuration mixing techniques.

Chapter 4 contains a brief presentation of TDDFT methods. While the discussion in section 6.4 on the GCM+GOA was focussed on static aspects, the theory can easily be extended to describe time-dependent collective phenomena [66, 67, 81–85]. In both cases, one needs to find a time-differential equation. In this section, we will very briefly describe the numerical techniques used to solve time-differential equations.

All the time-dependent theories we consider here can be cast into the following generic form,

$$\phi_\mu(t + \Delta t) = \hat{U}(t,\, t,\, +\Delta t)\phi_\mu(t),$$

where $\phi_\mu(t)$ refers to either a single-particle wave function (TDHF), a HFB spinor time-dependent Hartree–Fock–Bogoliubov (TDHFB) or the collective amplitude (time-dependent generator coordinate method (TDGCM) with the GOA approximation). The evolution operator can be written as

$$\hat{U}(t,\, t + \Delta t) = \hat{\mathcal{T}} \exp\left(-\frac{i}{\hbar}\int_t^{t+\Delta t} \hat{H}(t')dt'\right),$$

with $\hat{\mathcal{T}}$ the time-ordering operator, and $\hat{H}$ the relevant Hamiltonian, which can now be time-dependent. For TDHF, $\hat{H}$ coincides with the time-dependent single-particle Hamiltonian (or mean-field potential, or HF potential) given by equation (5.6), and for TDHFB, it is the time-dependent HFB matrix of equation (8.1). In both cases, the time dependence comes from the fact that these Hamiltonians are density-dependent, and it is the density matrix and generalized density that carry the time dependence. In the case of the TDGCM+GOA approximation, the relevant operator is the time-independent collective Hamiltonian (6.59).

Most of the time-dependent codes on the market such as SKY or FELIX use an adapted version of the general class of predictor–corrector methods to determine the solution to the time-differential equation [86–88]. Given the Hamiltonian at time t, the method is based on the following steps.

Predictor. Determine an estimate of the Hamiltonian at mid-time $t + \Delta t/2$ by computing the guess wave functions

$$\tilde{\phi}_\mu = \exp\left(-\frac{i}{\hbar}\hat{H}(t')\Delta t\right)\phi_\mu(t).$$

In the case of time-dependent energy density functional (TDEDF), these intermediate wave functions allow computing densities at half-step $\Delta t/2$: for TDHF, only the one-body density $\tilde{\rho}_{\text{pre}}$ is needed; for TDHFB we need the one-body density and anomalous density to form the estimate $\tilde{\mathcal{R}}_{\text{pre}}$ of the generalized density. The predictor step then consists in computing an estimate $\tilde{H}_{\text{pre}}$ of the Hamiltonian at mid-step using the average densities $(\rho(t) + \tilde{\rho}_{\text{pre}})/2$ or $(\mathcal{R}(t) + \tilde{\mathcal{R}}_{\text{pre}})/2$.

Corrector. Use the Hamiltonian at mid-step $\tilde{H}_{\text{pre}}$ to compute the wave functions at $t + \Delta t$. The calculation is based on the same formulas as the predictor phase after substituting $\hat{H} \to H_{\text{pre}} = H(t + \Delta t/2)$. Therefore, we obtain the wave functions at step $t + \Delta t$ as

$$\phi_\mu(t + \Delta t) = \exp\left(-\frac{i}{\hbar}\hat{H}_{\text{pre}}\Delta t\right)\phi_\mu(t).$$

In both cases, the operator exponential is simply approximated by the first m terms of its Taylor expansion,

$$\exp\left(-\frac{i}{\hbar}\hat{H}_{\text{pre}}\Delta t\right) \approx \sum_{n=0}^{m} \frac{(-i\Delta t)^n}{\hbar^n n!}\hat{H}^n.$$

In the TDHF solver SKY, the first $m = 4$ terms are used. This code uses a coordinate-space implementation of the HF equation and is restricted to EDF of the local density. As a result, the s.p. Hamiltonian contains only local potentials and terms involving derivative operators. The calculation of terms such as $\hat{H}^2\phi_\nu(\boldsymbol{r}, t)$ is then rather straightforward.

References

[1] Carlsson B G *et al* 2010 Solution of self-consistent equations for the N^{3}LO nuclear energy density functional in spherical symmetry. The program HOSPHE (v1.02) *Comput. Phys. Commun.* **181** 1641

[2] Navarro Perez R *et al* 2017 Axially deformed solution of the Skyrme–Hartree–Fock–Bogolyubov equations using the transformed harmonic oscillator basis (III) HFBTHO (v3.00): a new version of the program *Comput. Phys. commun.* **220** 363–75

[3] Schunck N *et al* 2017 Solution of the Skyrme–Hartree–Fock–Bogolyubov equations in the Cartesian deformed harmonic-oscillator basis. (VIII) HFODD (v2.73y): a new version of the program *Comput. Phys. Commun.* **216** 145

[4] Nikšić T *et al* 2014 DIRHB—a relativistic self-consistent mean-field framework for atomic nuclei *Comput. Phys. Commun.* **185** 1808

[5] Maruhn J A *et al* 2014 The TDHF code Sky3D *Comput. Phys. Commun.* **185** 2195–216

[6] Bonche P, Flocard H and Heenen P H 2005 Solution of the Skyrme HF + BCS equation on a 3D mesh *Comput. Phys. Commun.* **171** 49

[7] Ryssens W *et al* 2015 Solution of the Skyrme–HF+BCS equation on a 3D mesh, II: a new version of the Ev8 code *Comput. Phys. Commun.* **187** 175

[8] Bennaceur K and Dobaczewski J 2005 Coordinate-space solution of the Skyrme–Hartree–Fock–Bogolyubov equations within spherical symmetry. The program HFBRAD (v1.00) *Comput. Phys. Commun.* **168** 96

[9] Bohr A and Mottelson B R 1998 *Nuclear StructureSingle-Particle Motion* vol 1 (Singapore: World Scientific)

[10] Nilsson S G and Ragnarsson I 1995 *Shapes and Shells in Nuclear Structure* (Cambridge: Cambridge University Press)

[11] Ring P and Schuck P 2004 *The Nuclear Many-Body ProblemTexts and Monographs in Physics* (Berlin: Springer)

[12] Stoitsov M V *et al* 2005 Axially deformed solution of the Skyrme-Hartree-Fock-Bogolyubov equations using the transformed harmonic oscillator basis. The program HFBTHO (v1.66p) *Comput. Phys. Commun.* **167** 43–63

[13] Dobaczewski J and Dudek J 1997 Solution of the Skyrme–Hartree–Fock equations in the Cartesian deformed harmonic oscillator basis I. The method *Comput. Phys. Commun.* **102** 166

[14] Perlińska E *et al* 2004 Local density approximation for proton–neutron pairing correlations: formalism *Phys. Rev.* C **69** 014316

[15] Satuła W *et al* 2009 Isospin mixing in nuclei within the nuclear density functional theory *Phys. Rev. Lett.* **103** 012502
[16] Rohoziński S G, Dobaczewski J and Nazarewicz W 2010 Self-consistent symmetries in the proton–neutron Hartree–Fock–Bogoliubov approach *Phys. Rev.* C **81** 014313
[17] Satuła W *et al* 2010 Isospin-symmetry restoration within the nuclear density functional theory: formalism and applications *Phys. Rev.* C **81** 054310
[18] Sato K *et al* 2013 Energy-density-functional calculations including proton–neutron mixing *Phys. Rev.* C **88** 061301
[19] Sheikh J A *et al* 2014 Isospin-invariant Skyrme energy-density-functional approach with axial symmetry *Phys. Rev.* C **89** 054317
[20] Varshalovich D A, Moskalev A N and Khersonskii V K 1988 *Quantum Theory of Angular Momentum* (Singapore: World Scientific)
[21] Abramowitz M and Stegun I 1964 *Handbook of Mathematical Functions with Formulas, Graphs, and Mathematical Tables* (New York: Dover)
[22] Vautherin D 1973 Hartree–Fock calculations with Skyrme's interaction. II. Axially deformed nuclei *Phys. Rev.* C **7** 296
[23] Davies K T R and Krieger S J 1991 Harmonic-oscillator transformation coefficients *Can. J. Phys.* **69** 62
[24] Moshinsky M and Smirnov Y F 1996 The harmonic oscillator in modern physics *Contemporary Concepts in Physics* (Boca Raton, FL: CRC Press)
[25] Dobaczewski J and Dudek J 1997 Solution of the Skyrme–Hartree–Fock equations in the Cartesian deformed harmonic oscillator basis II. The program HFODD *Comput. Phys. Commun.* **102** 183
[26] Landau L D and Lifshitz E M 1981 *Quantum Mechanics: Nonrelativistic Theory* 3rd edn (Oxford: Butterworth-Heinemann)
[27] Carter R 1998 *Molecular Symmetry and Group Theory* (New York: Wiley)
[28] Dobaczewski J *et al* 2000 Point symmetries in the Hartree–Fock approach. I. Densities, shapes, and currents *Phys. Rev.* C **62** 014310
[29] Dobaczewski J *et al* 2000 Point symmetries in the Hartree–Fock approach. II. Symmetry-breaking schemes *Phys. Rev.* C **62** 014311
[30] Frauendorf S 2000 Description of multi-quasiparticle bands by the tilted axis cranking method *Nucl. Phys.* A **677** 115
[31] Messiah A 1962 *Quantum Mechanics* vol 2 (Amsterdam: North-Holland)
[32] Goodman A L 1974 Self-consistent symmetries of the Hartree–Fock–Bogoliubov equations in a rotating frame *Nucl. Phys.* A **230** 466
[33] Schunck N and Robledo L M 2016 Microscopic theory of nuclear fission: a review *Rep. Prog. Phys.* **79** 116301
[34] Parrish R M *et al* 2013 Exact tensor hypercontraction: a universal technique for the resolution of matrix elements of local finite-range *N*-body potentials in many-body quantum problems *Phys. Rev. Lett.* **111** 32505
[35] Engel J *et al* 2003 Time-reversal violating Schiff moment of 225Ra *Phys. Rev.* C **68** 025501
[36] Dobaczewski J and Olbratowski P 2005 Solution of the Skyrme–Hartree–Fock–Bogolyubov equations in the Cartesian deformed harmonic-oscillator basis. (V) HFODD (v2.08k) *Comput. Phys. Commun.* **167** 214
[37] Girod M and Grammaticos B 1983 Triaxial Hartree–Fock–Bogolyubov calculations with D1 effective interaction *Phys. Rev.* C **27** 2317

[38] Dobaczewski J *et al* 2009 Solution of the Skyrme–Hartree–Fock–Bogolyubov equations in the Cartesian deformed harmonic-oscillator basis. (VI) HFODD (v.240h): a new version of the program *Comput. Phys. Commun.* **180** 2361

[39] Stoitsov M V *et al* 2013 Axially deformed solution of the Skyrme–Hartree–Fock–Bogoliubov equations using the transformed harmonic oscillator basis (II) HFBTHO v2.00d: a new version of the program *Comput. Phys. Commun.* **184** 1592–604

[40] Gradshteyn I S and Ryzhik I M 2007 *Table of Integrals, Series, and Products* (New York: Academic)

[41] Schunck N and Egido J L 2008 Continuum and symmetry-conserving effects in drip-line nuclei using finite-range forces *Phys. Rev.* C **77** 011301(R)

[42] Schunck N and Egido J L 2008 Nuclear halos and drip lines in symmetry-conserving continuum Hartree–Fock–Bogoliubov theory *Phys. Rev.* C **78** 064305

[43] Satuła W *et al* 2012 Isospin-breaking corrections to superallowed Fermi β decay in isospin- and angular-momentum-projected nuclear density functional theory *Phys. Rev.* C **86** 054316

[44] Dechargé J and Gogny D 1980 Hartree–Fock–Bogolyubov calculations with the D1 effective interaction on spherical nuclei *Phys. Rev.* C **21** 1568–93

[45] Younes W and Gogny D 2009 Microscopic calculation of ^{240}Pu scission with a finite-range effective force *Phys. Rev.* C **80** 054313

[46] Staszczak A *et al* 2010 Augmented Lagrangian method for constrained nuclear density functional theory *Eur. Phys. J.* A **46** 85

[47] Flocard H *et al* 1974 Self-consistent calculation of the fission barrier of 240Pu *Nucl. Phys.* A **231** 176

[48] Dobaczewski J, Flocard H and Treiner J 1984 Hartree–Fock–Bogolyubov description of nuclei near the neutron-drip line *Nucl. Phys.* A **422** 103–39

[49] Dobaczewski J *et al* 1996 Mean-field description of ground-state properties of drip-line nuclei: pairing and continuum effects *Phys. Rev.* C **53** 2809

[50] Michel N, Matsuyanagi K and Stoitsov M 2008 Gamow–Hartree–Fock–Bogoliubov method: representation of quasiparticles with Berggren sets of wave functions *Phys. Rev.* C **78** 044319

[51] Schunck N *et al* 2015 Error analysis in nuclear density functional theory *J. Phys. G: Nucl. Part. Phys.* **42** 034024

[52] Bulgac A *et al* 2016 Induced fission of ^{240}Pu within a real-time microscopic framework *Phys. Rev. Lett.* **116** 122504

[53] Daniel B 2015 The Lagrange-mesh method *Phys. Rep.* **565** 1

[54] Bulgac A and Roche K J 2008 Time-dependent density functional theory applied to superfluid nuclei *J. Phys.: Conf. Ser.* **125** 012064

[55] Bulgac A *et al* 2013 *Quantum Friction: Cooling Quantum Systems with Unitary Time Evolution* arXiv: 1305.6891

[56] Bulgac A and Forbes M M 2013 *Time-Dependent Superfluid Local Density Approximation* ed N Proukakis *et al* (Singapore: World Scientific) ch 32, pp 397–406

[57] Ryssens W, Heenen P-H and Bender M 2015 Numerical accuracy of mean-field calculations in coordinate space *Phys. Rev.* C **92** 064318

[58] Jin S *et al* 2017 Coordinate-space solver for superfluid many-fermion systems with the shifted conjugate-orthogonal conjugate-gradient method *Phys. Rev.* C **95** 044302

[59] Pöschl W 1998 *B*-spline finite elements and their efficiency in solving relativistic mean field equations *Comput. Phys. Commun.* **112** 42

[60] Pöschl W, Vretenar D and Ring P 1996 Application of the finite element method in self-consistent relativistic mean field calculations *Comput. Phys. Commun.* **99** 128
[61] Pöschl W, Vretenar D and Ring P 1997 Relativistic Hartree–Bogoliubov theory in coordinate space: finite element solution for nuclear system with spherical symmetry *Comput. Phys. Commun.* **103** 217
[62] Pöschl W *et al* 1997 Relativistic Hartree–Bogoliubov theory with finite range pairing forces in coordinate space: neutron halo in light nuclei *Phys. Rev. Lett.* **79** 3841
[63] Pöschl W *et al* 1997 Application of finite element methods in relativistic mean-field theory: spherical nucleus *Comput. Phys. Commun.* **101** 75
[64] Blazkiewicz A *et al* 2005 Coordinate space Hartree–Fock–Bogoliubov calculations for the zirconium isotope chain up to the two-neutron drip line *Phys. Rev.* C **71** 054321
[65] Pei J *et al* 2008 Deformed coordinate-space Hartree–Fock–Bogoliubov approach to weakly bound nuclei and large deformations *Phys. Rev.* C **78** 064306
[66] Regnier D *et al* 2016 FELIX-1.0: a finite element solver for the time dependent generator coordinate method with the Gaussian overlap approximation *Comput. Phys. Commun.* **200** 350
[67] Regnier D *et al* 2018 FELIX-2.0: new version of the finite element solver for the time dependent generator coordinate method with the Gaussian overlap approximation *Comput. Phys. Commun.* **225** 180
[68] Fann G I *et al* 2009 Fast multiresolution methods for density functional theory in nuclear physics *J. Phys.: Conf. Ser.* **180** 012080
[69] Robert J and Harrison *et al* 2004 Multiresolution quantum chemistry: basic theory and initial applications *J. Chem. Phys.* **121** 11587
[70] Pei J C *et al* 2014 Adaptive multi-resolution 3D Hartree–Fock–Bogoliubov solver for nuclear structure *Phys. Rev.* C **90** 024317
[71] Sagert I *et al* 2016 Quantum simulations of nuclei and nuclear pasta with the multiresolution adaptive numerical environment for scientific simulations *Phys. Rev.* C **93** 055801
[72] Yanai T *et al* 2004 Multiresolution quantum chemistry in multiwavelet bases: analytic derivatives for Hartree–Fock and density functional theory *J. Chem. Phys.* **121** 2866
[73] Yanai T *et al* 2004 Multiresolution quantum chemistry in multiwavelet bases: Hartree–Fock exchange *J. Chem. Phys.* **121** 6680
[74] Blaizot J-P and Ripka G 1985 *Quantum Theory of Finite Systems* (Cambridge, MA: MIT Press)
[75] Broyden C G 1965 A class of methods for solving nonlinear simultaneous equations *Math. Comput.* **19** 577
[76] Johnson D D 1988 Modified Broyden's method for accelerating convergence in self-consistent calculations *Phys. Rev.* B **38** 12807
[77] Baran A *et al* 2008 Broyden's method in nuclear structure calculations *Phys. Rev.* C **78** 014318
[78] Flocard H *et al* 1973 Nuclear deformation energy curves with the constrained Hartree–Fock method *Nucl. Phys.* A **203** 433
[79] Egido J L *et al* 1995 On the solution of the Hartree–Fock–Bogoliubov equations by the conjugate gradient method *Nucl. Phys.* A **594** 70
[80] Robledo L M and Bertsch G F 2011 Application of the gradient method to Hartree–Fock–Bogoliubov theory *Phys. Rev.* C **84** 014312

[81] Berger J F, Girod M and Gogny D 1991 Time-dependent quantum collective dynamics applied to nuclear fission *Comput. Phys. Commun.* **63** 365–74
[82] Goutte H, Casoli P and Berger J-F 2004 Mass and kinetic energy distributions of fission fragments using the time dependent generator coordinate method *Nucl. Phys.* A **734** 217
[83] Goutte H *et al* 2005 Microscopic approach of fission dynamics applied to fragment kinetic energy and mass distributions in ^{238}U *Phys. Rev.* C **71** 024316
[84] Regnier D *et al* 2016 Fission fragment charge and mass distributions in ^{239}Pu(n,f) in the adiabatic nuclear energy density functional theory *Phys. Rev.* C **93** 054611
[85] Younes W and Gogny D 2012 Collective dissipation from saddle to scission in a microscopic approach *Technical report* LLNL-TR-586694 Lawrence Livermore National Laboratory (LLNL)
[86] Atkinson K, Han W and Stewart D 2009 *Numerical Solution of Ordinary Differential Equations* (New York: Wiley)
[87] Pang T 2006 *An Introduction to Computational Physics* (Cambridge: Cambridge University Press)
[88] Riley K F, Hobson M P and Bence S J 2006 *Mathematical Methods for Physics and Engineering* (Cambridge: Cambridge University Press)

IOP Publishing

Energy Density Functional Methods for Atomic Nuclei

Nicolas Schunck

Chapter 9

Calibration of energy functionals

Nicolas Schunck and Markus Kortelainen

It is a truth (not always) universally acknowledged, that all models in nuclear physics—including *ab initio* approaches—must be calibrated on experimental data. In the energy density functional (EDF) approach to nuclear structure, this problem is particularly relevant since most energy functionals used so far have been phenomenological in nature: while the form of the functional is often constrained by elementary requirements such as symmetries, there is no mathematical connection between the most fundamental theory of nuclear forces represented by chiral effective field theory (χEFT) and the structure of the EDF. In fact, as discussed in section 1.1.3, the real challenge is that the energy functional is supposed to encode many-body correlations that do not come from the form of the original nucleon–nucleon (NN) or three-body (NNN) force, but rather from the many-body methods used to solve the Schrödinger equation. This compounds the difficulty in establishing a firm, rigorous mapping between a fundamental theory of nuclear forces and the EDF framework. As a result, one must treat energy functionals as phenomenological objects.

> There is a fundamental difference between quantum chromodynamics (QCD) and low-energy nuclear theory. QCD relies in principle on a small set of parameters, namely the masses of elementary particles and the coupling constants of the theory, that can be directly measured experimentally. Since nuclear forces are not elementary forces but, in principle, derived from QCD, it is not possible to measure their characteristics. Instead, one has to determine the free parameters on the theory by comparing predictions of the model with experimental measurements. This is this process that we call calibration in this chapter. For example, χEFT is typically calibrated on scattering phase shifts [1, 2]

This has mainly two consequences. Given an energy functional, one must establish a procedure to determine its parameters, typically a small set of coupling constants. This procedure involves first identifying a set of pertinent experimental data. If, as is most often the case, this set of data involves more than one type, one must then

doi:10.1088/2053-2563/aae0edch9

decide the relative importance of each type so that a regression problem can be properly set up. Indeed, experimental data such as nuclear binding energies, charge radii or electromagnetic transition strengths, have very different scales and units and probe different physics. The final stage is to solve this regression problem. This seemingly simple program already teems with hidden challenges that have to do with the compatibility of the experimental data with the level (single-reference energy density functional (SR-EDF), multi-reference energy density functional (MR-EDF), quasiparticle random-phase approximation (QRPA), etc) at which the calibration is performed; with the need to properly constrain each coupling constant of the functional including, for instance, the isovector one, etc. In sections 9.1 and 9.2, we will discuss some of these issues.

It is important to realize the extent to which the parametrization resulting from this process is affected by multiple sources of uncertainties: there is a systematic uncertainty coming from the choice of the functional itself (Skyrme, Gogny, covariant, other) and the particular approximation level (SR-EDF, MR-EDF, etc); a statistical uncertainty caused by the choice of the data, their type, their number, etc; and even, in most cases, a numerical uncertainty due to the possible truncations (in the basis, in the resolution of the mesh, etc.) motivated by the computational cost. In recent years, there has been a concerted effort in the nuclear physics community to come up with rigorous statistical tools to quantify these uncertainties and propagate them to model predictions. These will be reviewed in sections 9.3 and 9.4.

Classifying uncertainties in systematic, statistical and numerical is somewhat arbitrary and reductive—all three sources are in fact entangled in any given calculation. For example, fitting the parameters of the energy functional on nuclear binding energies at the Hartree–Fock–Bogoliubov (HFB) level may lead to over-fitting and statistical bias, since variations of binding energies with Z and N involve correlation effects that go beyond HFB; see, e.g. [3, 4]. Similarly, calculations within a constant model space, say a unique harmonic oscillator (HO) basis, will lead to small truncation errors in light nuclei, but much larger errors in heavy nuclei. If both light and heavy nuclei are included in the fit, this will also introduce a statistical bias in the model parameters.

9.1 Parameters of energy functionals

The (only) common characteristics of all EDF, non-relativistic Skyrme-like, Gogny, or phenomenological, or covariant EDF, is that they depend on a relatively small set of parameters that need to be adjusted to experimental data. In the case of Hamiltonian-based EDF—energy functionals that can be strictly derived from the expectation value of some effective pseudo-Hamiltonian on a Slater determinant—one can fit either the parameters of the pseudo-potential itself, or of the EDF. This is sometimes a source of confusion. For example, the tensor term of Skyrme-like EDF is the term proportional to the spin density current $J \equiv J_{\mu\nu}$ which has the mathematical structure of a rank-2 tensor. This tensor term in the EDF is different from the tensor term in the two-body nucleon–nucleon pseudo-potential, which is, schematically, the term proportional to the spin tensor operator $\hat{S}_{12}(\hat{\boldsymbol{r}}) = 3(\boldsymbol{\sigma}_1 \cdot \hat{\boldsymbol{r}})(\boldsymbol{\sigma}_2 \cdot \hat{\boldsymbol{r}}) - \boldsymbol{\sigma}_1 \cdot \boldsymbol{\sigma}_2$. Studies have shown that the tensor interaction impacts both the tensor and spin–orbit terms of the EDF [5–7]. In section 9.1.1, we recall the transformation between the parameters of the EDF and that of an underlying Skyrme pseudo-potential.

One of the first validation tests of any EDF fit is to compute infinite nuclear matter (INM) properties. In the special case of functionals of the local density such as Skyrme-like EDF, there exists a one-to-one mapping between the volume parameters of the EDF and INM properties. This mapping has been used either to directly set Skyrme coupling constants [8] or to set tight constraints on them [9–11]. In section 9.1.2, we recall the expressions for INM at the Hartree–Fock (HF) for both Skyrme and Gogny EDF.

9.1.1 From nucleon–nucleon potentials to energy functionals

As mentioned in the introduction to this chapter, since Hamiltonian-based EDF are constructed from the expectation value of some effective pseudo-potential, usually a two-body only, it is always possible to relate the coupling constant of the EDF to those of the potential. As we will show later in the particular case of tensor coupling constants, there can still be some non-trivial technical issues in doing so.

Let us begin by recalling one more time the form of the Skyrme EDF. What we called the canonical form (1.23) of the Skyrme EDF presented in section 1.2.1, can be viewed as a special case of the slightly more general functional

$$\begin{aligned}
\mathcal{H}_t^{\text{even}} &= C_t^{\rho\rho}\rho_t^2 + C_t^{\rho\Delta\rho}\rho_t\Delta\rho_t + C_t^{\rho\tau}\rho_t\tau_t + C_t^{\rho\nabla J}\rho_t\nabla\cdot\boldsymbol{J}_t \\
&\quad + C_t^{JJ}\sum_{\mu\nu}J^t_{\mu\nu}J_{t,\,\mu\nu} + C_t^{J\bar{J}}\left[\left(\sum_{\mu}J_{t,\,\mu\mu}\right)\left(\sum_{\mu}J_{t,\,\mu\mu}\right) + \sum_{\mu\nu}J_{t,\,\mu\nu}J_{t,\,\nu\mu}\right] \\
\mathcal{H}_t^{\text{odd}} &= C_t^{ss}\boldsymbol{s}_t^2 + C_t^{s\Delta s}\boldsymbol{s}_t\cdot\Delta\boldsymbol{s}_t + C_t^{sj}\boldsymbol{j}_t^2 + C_t^{sT}\boldsymbol{s}_t\cdot\boldsymbol{T}_t \\
&\quad + C_t^{s\nabla j}\boldsymbol{s}_t\cdot\left(\nabla\times\boldsymbol{j}_t\right) + C_t^{s\nabla s}(\nabla\cdot\boldsymbol{s}_t)^2 + C_t^{sF}\boldsymbol{s}_t\cdot\boldsymbol{F}_t,
\end{aligned}$$

where $t = 0$ refers to the isoscalar channel and $t = 1$ to the isovector one. We recall that $\mathcal{H}_t^{\text{even}}$ only depends on time-even densities, while $\mathcal{H}_t^{\text{odd}}$ involves time-odd densities and thus automatically vanishes if time-reversal symmetry is conserved. The main difference to equation (1.23) is that we do not impose the local gauge invariance of the functional, see the side note below.

As noted in section 1.2.1, local gauge invariance of the energy functional under transformations of the type

$$\rho(x, x') \rightarrow \rho'(x, x') = e^{i[\phi(\boldsymbol{r})-\phi(\boldsymbol{r}')]}\rho(x, x')$$

for the full density matrix (where, as always, $x \equiv (\boldsymbol{r}, \sigma, \tau)$) imposes the following constraints on the time-odd coupling constants [12, 13]

$$
\begin{aligned}
C_t^{sj} &= -\,C_t^{\rho\tau} \\
C_t^{s\nabla j} &= +\,C_t^{\rho\nabla J} \\
C_t^{sT} &= -\,C_t^{JJ} \\
C_t^{sF} &= -\,2C_t^{J\bar{J}}.
\end{aligned}
$$

The remaining coupling constants C_t^{ss}, $C_t^{s\Delta s}$ are not constrained. The coupling constants $C_t^{s\nabla s}$ and C_t^{sF} are related to the tensor piece of the functional and are discussed separately.

There is an explicit one-to-one relation between the $C_t^{uu'}$ coupling constants and the (t, x) parameters of the Skyrme pseudo-potential of equations (1.26)–(1.28) [5, 12]. For the coupling constants of the time-even energy density, we find

$$
\begin{aligned}
C_0^{\rho\rho} &= +\frac{3t_0}{8} & C_1^{\rho\rho} &= -\frac{t_0}{4}\left(x_0 + \frac{1}{2}\right) \\
C_{0\mathrm{D}}^{\rho\rho} &= +\frac{t_3}{16} & C_{1\mathrm{D}}^{\rho\rho} &= -\frac{t_3}{24}\left(x_3 + \frac{1}{2}\right) \\
C_0^{\rho\Delta\rho} &= -\frac{9}{64}t_1 + \frac{t_2}{16}\left(x_2 + \frac{5}{4}\right) & C_1^{\rho\Delta\rho} &= \frac{3t_1}{32}\left(x_1 + \frac{1}{2}\right) + \frac{t_2}{32}\left(x_2 + \frac{1}{2}\right) \\
C_0^{\rho\tau} &= +\frac{3t_1}{16} + \frac{t_2}{4}\left(x_2 + \frac{5}{4}\right) & C_1^{\rho\tau} &= -\frac{t_1}{8}\left(x_1 + \frac{1}{2}\right) + \frac{t_2}{8}\left(x_2 + \frac{1}{2}\right) \\
C_0^{\rho\nabla J} &= -b_4 - \frac{1}{2}b_4' & C_1^{\rho\nabla J} &= -\frac{1}{2}b_4',
\end{aligned}
$$

while for the ones in the time-odd channel,

$$
\begin{aligned}
C_0^{ss} &= -\frac{t_0}{4}\left(\frac{1}{2} - x_0\right) & C_1^{ss} &= -\frac{t_0}{8} \\
C_{0\mathrm{D}}^{ss} &= -\frac{t_3}{24}\left(\frac{1}{2} - x_3\right) & C_{1\mathrm{D}}^{ss} &= -\frac{t_3}{48} \\
C_0^{s\Delta s} &= -\frac{3}{32}t_1\left(\frac{1}{2} - x_1\right) + \frac{t_2}{32}\left(\frac{1}{2} + x_2\right) & C_1^{s\Delta s} &= \frac{3t_1}{64} + \frac{t_2}{64} \\
C_0^{s\nabla s} &= -\frac{3}{32}(t_e - t_o) & C_1^{s\nabla s} &= -\frac{1}{32}(3t_e + t_o).
\end{aligned}
$$

Recall that $C_t^{\rho\rho}$ (and C_t^{ss}) is in fact density-dependent. Here we will take it in the form

$$
C_t^{\rho\rho} = C_{t0}^{\rho\rho} + C_{t\mathrm{D}}^{\rho\rho}\rho^{\alpha}(\boldsymbol{r}). \tag{9.1}
$$

The case of the tensor term deserves some further comments. In section 1.2.1, we have adopted a presentation of the Skyrme EDF that allows establishing a one-to-one matching between the Skyrme pseudo-potential and the Skyrme EDF.

However, there are other conventions for the tensor terms in the literature. Recall that in equation (1.23), the tensor component (including time-even and time-odd terms) is

$$\mathcal{H}_t^{\text{tens.}} = C_t^{JJ} \sum_{\mu\nu} J_{\mu\nu}^t J_{t,\,\mu\nu} - C_t^{sT} \mathbf{s}_t \cdot \mathbf{T}_t + C_t^{J\bar{J}}\left[\left(\sum_{\mu} J_{t,\,\mu\mu}\right)\left(\sum_{\mu} J_{t,\,\mu\mu}\right) + \sum_{\mu\nu} J_{t,\,\mu\nu} J_{t,\,\nu\mu}\right] - 2C_t^{sF} \mathbf{s}_t \cdot \mathbf{F}_t. \tag{9.2}$$

When using the $(C_t^{JJ},\, C_t^{J\bar{J}})$ representation of the Skyrme energy functional, the isoscalar $t = 0$ and isovector $t = 1$ coupling constants can be related to the parameters of the Skyrme pseudo-potential according to

$$\begin{aligned}
C_0^{JJ} &= -\frac{1}{16}[t_1(1 - 2x_1) - t_2(1 + 2x_2)] - \frac{1}{8}(t_e + 3t_o) \\
C_1^{JJ} &= -\frac{1}{16}(t_1 - t_2) - \frac{1}{8}(t_e - t_o) \\
C_0^{J\bar{J}} &= -\frac{3}{8}(t_e + 3t_o) \\
C_1^{J\bar{J}} &= -\frac{3}{8}(t_e - t_o).
\end{aligned}$$

We will focus only on the terms involving time-even densities only. We can recouple terms following equations (24) and (25) given in [5], so that we obtain

$$\mathcal{H}_t^{\text{tens.}} = \frac{1}{3}\left(-C_t^{JJ} - 2C_t^{J\bar{J}}\right)\left(J_t^0\right)^2 + \frac{1}{2}\left(-C_t^{JJ} + \frac{1}{2}C_t^{J\bar{J}}\right)\mathbf{J}_t^2 + \left(-C_t^{JJ} - \frac{1}{2}C_t^{J\bar{J}}\right)\mathsf{J}_t^2, \tag{9.3}$$

where J_t^0 is a pseudoscalar, $\mathbf{J}_t$ is a vector and J_t is a pseudotensor. These terms are given by the following definitions

$$\begin{aligned}
J_t^{(0)} &= \sum_{\mu} J_{\mu\mu,\,t} \\
\mathbf{J}_{\kappa,\,t} &= \sum_{\mu\nu} \epsilon_{\kappa\mu\nu} J_{\mu\nu,\,t} \\
\mathsf{J}_{\mu\nu,\,t} &= \frac{1}{2}\left(J_{\mu\nu,\,t} + J_{\nu\mu,\,t}\right) - \delta_{\mu\nu}\frac{1}{3}J_t^{(0)}
\end{aligned}$$

with $\epsilon_{\kappa\mu\nu}$ the Levi-Civita tensor. Since $J_t^{(0)}$ is a pseudoscalar, it changes sign under parity. Similarly, the pseudotensor $\mathsf{J}_{\mu\nu,\,t}$ also changes sign under parity transformation.

Consider a transformation of the (Cartesian) coordinate system according to the law

$$x_\alpha \to x'_\alpha = R_{\alpha\beta} x_\beta,$$

where we use the Einstein summation convention: repeated indices are summed over. Then, a (Cartesian) rank-n tensor $T_{\alpha_1 \dots \alpha_n}$ is an object characterized by 3^n real numbers which transform under such changes of the coordinate system according to

$$T_{\alpha_1 \dots \alpha_n} = R_{\alpha_1\beta_1} \dots R_{\alpha_n\beta_n} T_{\beta_1 \dots \beta_n}.$$

If, instead, the law of transformation is given by

$$T_{\alpha_1 \dots \alpha_n} = (\det R) R_{\alpha_1\beta_1} \dots R_{\alpha_n\beta_n} T_{\beta_1 \dots \beta_n},$$

with det referring to the determinant of the transformation, then we speak of pseudotensors. For $n = 0$, these are pseudoscalars; for $n = 1$, pseudovectors; and for $n > 1$, rank-n pseudotensors; see [14, 15] for additional discussion on tensor algebra. The Levi-Civita tensor, also called unit tensor, is a rank-3 pseudotensor. It changes sign under any odd number of permutations of any of its indices.

Contrast equation (9.3) with the expression found in [16],

$$\mathcal{H}_t^{\text{tens.}} = \frac{1}{3} C_t^J \left(J_t^0\right)^2 + \frac{1}{2} C_t^J \mathbf{J}_t^2 + C_t^J J_t^2, \tag{9.4}$$

and another choice in [17],

$$\mathcal{H}_t^{\text{tens.}} = C_t^{J0} \left(J_t^0\right)^2 + C_t^{J1} \mathbf{J}_t^2 + C_t^{J2} J_t^2. \tag{9.5}$$

The difference between the last two cases comes from relaxing the requirement that the coupling constants multiplying the pseudoscalar, vector and pseudotensor components are the same. The coupling constants C^{J0}, C^{J1} and C^{J2} of equation (9.5) can be related to the various coupling constants C_t^{JJ} and $C_t^{J\bar{J}}$ used in equation (9.3) according to

$$C_t^{J0} = \frac{1}{3}\left(- C_t^{JJ} - 2C_t^{J\bar{J}}\right) \quad C_t^{J1} = \frac{1}{2}\left(-C_t^{JJ} + \frac{1}{2}C_t^{J\bar{J}}\right) \quad C_t^{J2} = \left(-C_t^{JJ} - \frac{1}{2}C_t^{J\bar{J}}\right). \tag{9.6}$$

The advantage of using the $(C_t^{JJ}, C_t^{J\bar{J}})$ representation is that these coupling constants can be related to the (t, x) parameters of the Skyrme pseudo-potential and the parameters t_e, t_o of the tensor force. However, in the most general case where both $C_t^{J\bar{J}}$ and C_t^{JJ} are non-zero, there is no equivalence between the parameterization of equation (9.4) (where the same coupling constant C^J multiplies all three terms of the tensor energy density) and that of either equation (9.3) or (9.5), since we should then have simultaneously

$$C_t^J = \left(-C_t^{JJ} - 2C_t^{J\bar{J}}\right) = \left(-C_t^{JJ} + \frac{1}{2}C_t^{J\bar{J}}\right) = \left(-C_t^{JJ} - \frac{1}{2}C_t^{J\bar{J}}\right).$$

In the particular case where $C_t^{JJ} = 0$, $(t = 0, 1)$, all three prescriptions to define the tensor energy density coincide. Since in this case,

$$C_0^{J\bar{J}} = -\frac{3}{8}(t_e + 3t_o) \qquad C_1^{J\bar{J}} = -\frac{3}{8}(t_e - t_o),$$

this implies that $t_e = t_o = 0$, i.e. there can be no tensor *force* underlying the EDF. However, in this case, it would actually not be possible to relate the EDF to an underlying Skyrme pseudo-potential anyway: we have six coupling constants $C_t^{\rho\tau}$, $C_t^{\rho\Delta\rho}$ and C_t^{JJ} to express as functions of only four Skyrme force parameters t_1, x_1, t_2 and x_2.

If $C_t^{JJ} \neq 0$, $(t = 0, 1)$, we can still reconcile the three prescriptions, but only if spherical symmetry is conserved. Indeed, the terms $J_t^{(0)}$ and J_t are then identically 0, hence we can set $C_t^J = -C_t^{JJ} + \frac{1}{2}C_t^{J\bar{J}} = 2C_t^{J1}$. It is then possible to relate the functional to a Skyrme pseudo-potential with an explicit tensor force, since this boils down to expressing t_e and t_o as functions of t_1, x_1, t_2, and x_2, which are themselves uniquely determined from $C_t^{\rho\tau}$ and $C_t^{\rho\Delta\rho}$. We find

$$C_0^J = \frac{1}{16}[t_1(1 - 2x_1) - t_2(1 + 2x_2)] - \frac{5}{16}(t_e + 3t_o)$$
$$C_1^J = \frac{1}{16}(t_1 - t_2) - \frac{5}{16}(t_e - t_o).$$

The resulting pseudo-potential, however, can only be used in applications where spherical symmetry is conserved. If it is not the case, then we can not have the same coupling constants C_t^J multiplying the pseudoscalar, vector and pseudotensor components of the tensor energy density. Instead, each of these components must be specifically parametrized by the relations (9.6). This implies that we can express each of the eight coupling constants $C_t^{\rho\tau}$, $C_t^{\rho\Delta\rho}$, C_t^{JJ} and $C_t^{J\bar{J}}$ as a function of t_1, x_1, t_2, x_2, t_e and t_o. However, eight coupling constants cannot be determined uniquely by only six parameters, and the resulting EDF would therefore be ill-defined.

9.1.2 Nuclear matter properties

We turn now to a discussion of how equilibrium properties of INM can be used to constrain some of the coupling constants of the EDF. In INM, the total energy per particle defines the saturation curve $W(\rho_n, \rho_p)$. Its derivation discards the Coulomb energy and all gradient terms that are zero for a homogeneous system, and substitutes the kinetic energy density with its Thomas–Fermi expression, which is exact in this case. As a reminder, the Thomas–Fermi model describes a homogeneous gas of non-interacting spin 1/2 fermions. The single-particle Hamiltonian is simply the kinetic energy operator. The expression for the density ρ^{TF} can be obtained with the tools of statistical mechanics and quantizing the system in a box [18–20]. Assuming spin saturation, one also disregards the terms involving the spin–orbit current density $\boldsymbol{J}_t$.

The expansion of $W(\rho_n, \rho_p)$ around the equilibrium density ρ_{sat} in a Taylor series in $\rho = \rho_n + \rho_p$ and $\beta = (\rho_n - \rho_p)/\rho$ yields

$$W(\rho, \beta) = W(\rho) + S_2(\rho)\beta^2 + S_4(\rho)\beta^4, \tag{9.7}$$

with

$$\begin{aligned} W(\rho) &= \frac{E^{\text{NM}}}{A} + \frac{P^{\text{NM}}}{\rho_{\text{sat}}}(\rho - \rho_{\text{sat}}) + \frac{K^{\text{NM}}}{18\rho_{\text{sat}}^2}(\rho - \rho_{\text{sat}})^2 \\ S_2(\rho) &= a_{\text{sym}}^{\text{NM}} + \frac{L^{\text{NM}}}{3\rho_{\text{sat}}}(\rho - \rho_{\text{sat}}) + \frac{\Delta K^{\text{NM}}}{18\rho_{\text{sat}}^2}(\rho - \rho_{\text{sat}})^2, \end{aligned}$$

where E^{NM}/A, P^{NM}, K^{NM}, $a_{\text{sym}}^{\text{NM}}$, L^{NM} and ΔK^{NM} denote the total energy per nucleon at equilibrium, the pressure, the nuclear matter incompressibility, the symmetry energy coefficient, the density derivative of $a_{\text{sym}}^{\text{NM}}$ and the isovector correction to the incompressibility at saturation density ρ_{sat} of nuclear matter, respectively. The quartic term in β can be safely neglected in equation (9.7) in practice (figure 9.1).

Skyrme functionals

For the Skyrme EDF, equation (9.7) takes the particularly simple form

$$W(\rho, \beta) = \frac{\hbar^2}{2m}\tau_0 + (C_{00}^{\rho} + C_{0D}^{\rho}\rho^{\gamma})\rho + (C_{10}^{\rho} + C_{1D}^{\rho}\rho^{\gamma})\beta^2\rho + C_0^{\tau}\rho\tau_0 + C_1^{\tau}\beta\rho\tau_1, \tag{9.8}$$

where τ_0 and τ_1 are, respectively, the isoscalar and isovector kinetic densities. These densities can be expressed analytically as

$$\tau_0 = \frac{1}{2}\frac{3}{5}\left(\frac{3\pi^2}{2}\right)^{2/3}\rho^{2/3}((1 + \beta)^{5/3} + (1 - \beta)^{5/3}) \tag{9.9a}$$

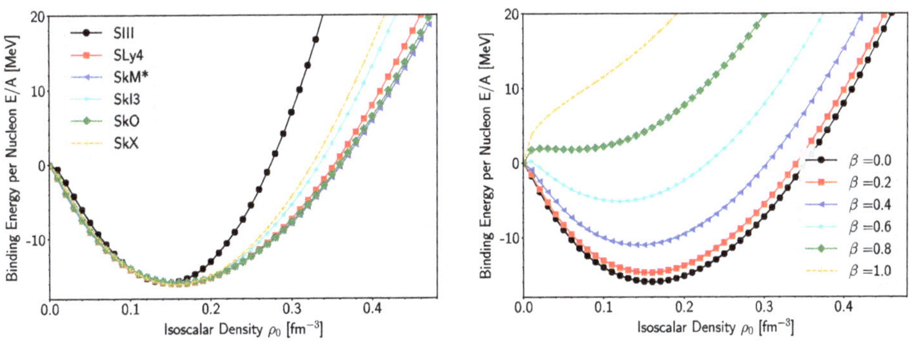

Figure 9.1. Left: Binding energy per nucleons (in MeV) in symmetric nuclear matter as a function of the isoscalar density ρ for several standard Skyrme EDFs. Right: Binding energy per nucleon (in MeV) in asymmetric nuclear matter as a function of the isoscalar density ρ for the SLy4 EDF for different values of the neutron–proton asymmetry parameter β. (Figure created by Nicolas Schunck.)

$$\tau_1 = \frac{1}{2}\frac{3}{5}\left(\frac{3\pi^2}{2}\right)^{2/3}\rho^{2/3}((1+\beta)^{5/3} - (1-\beta)^{5/3}). \tag{9.9b}$$

Owing to the particularly simple form of the energy per particle as a function of the EDF coupling constants, it is possible to obtain an analytical expression of the nuclear matter equilibrium quantities $\rho_{\rm sat}$, $E^{\rm NM}/A$, $K^{\rm NM}$, $a_{\rm sym}^{\rm NM}$, $L^{\rm NM}$, m_s^* and m_v^*, as a function of the Skyrme coupling constants. Here, m_s^* and m_v^* are the isoscalar and isovector effective masses, respectively, and are given by

$$m_s^* = \frac{2m}{\hbar^2}\frac{\partial W}{\partial \tau_0}\bigg|_{\rho_{\rm sat}} \qquad m_v^* = \frac{2m}{\hbar^2}\frac{\partial W}{\partial \tau_1}\bigg|_{\rho_{\rm sat}}.$$

The reason for performing such a transformation is that nuclear matter equilibrium properties are relatively tightly constrained by either experimental observations or by direct *ab initio* calculations.

Symmetric nuclear matter

In symmetric nuclear matter (SNM) the neutron and proton densities are equal, $\rho_n = \rho_p = \rho/2$, where ρ is the isoscalar density, hence $\beta = 0$. All isovector terms are thus zero and the isoscalar kinetic energy density per particle, denoted by τ_c, is obtained by setting $\rho = \rho_{\rm sat}$ in equation (9.9a). The parameters C_0^ρ, C_{0D}^ρ and C_0^τ are then expressed in terms of $E^{\rm NM}/A$, $\rho_{\rm sat}$, τ_c and m_s^* through the following set of equations

$$C_0^\rho = \frac{1}{3\gamma\rho_{\rm sat}}\left[3(\gamma+1)\frac{E^{\rm NM}}{A} - \frac{\hbar^2}{2m}(3-(2-3\gamma)m_s^{*-1})\tau_c\right] \tag{9.10a}$$

$$C_{0D}^\rho = \frac{1}{3\gamma\rho_c^{\gamma+1}}\left[-3\frac{E^{\rm NM}}{A} - \frac{\hbar^2}{2m}(2m_s^{*-1}-3)\tau_c\right] \tag{9.10b}$$

$$C_0^\tau = \frac{\hbar^2}{2m}(m_s^{*-1}-1)\frac{1}{\rho_c}. \tag{9.10c}$$

The parameter γ can also be eliminated by expressing it as a function of the incompressibility $K^{\rm NM}$,

$$\gamma = \frac{-K^{\rm NM} - 9E^{\rm NM}\big/A + \frac{\hbar^2}{2m}(4m_s^{*-1}-3)\tau_c}{9E^{\rm NM}\big/A + 3\frac{\hbar^2}{2m}(2m_s^{*-1}-3)\tau_c}.$$

Asymmetric nuclear matter

In the regime of asymmetric nuclear matter (ANM), neutron and proton densities are now different ($\rho_0 = \rho, \rho_1 = \beta\rho$) and the isovector terms contribute. One can therefore express parameters C_1^ρ, C_{1D}^ρ and C_1^τ in terms of $a_{\rm sym}^{\rm NM}$, $L^{\rm NM}$ and m_v^* through

$$C_1^\rho = \frac{27(\gamma+1)a_{\rm sym}^{\rm NM} - 9L^{\rm NM} + 20(2-3\gamma)C_0^{\rho\tau}\rho_c\tau_c}{27\gamma\rho_{\rm sat}} + \frac{\frac{\hbar^2}{2m}\left((9\gamma-6)m_v^{*-1} - 12\gamma + 5\right)\tau_c}{27\gamma\rho_{\rm sat}} \tag{9.11a}$$

$$C_{1D}^\rho = \frac{-27a_{\rm sym}^{\rm NM} + 9L^{\rm NM} - 40C_0^{\rho\tau}\rho_{\rm sat}\tau_c}{27\gamma\rho_{\rm sat}^{\gamma+1}} + \frac{\frac{\hbar^2}{2m}(30m_v^{*-1} - 25)\tau_c}{27\gamma\rho_{\rm sat}^{\gamma+1}} \tag{9.11b}$$

$$C_1^\tau = \frac{\hbar^2}{2m}\left(m_s^{*-1} - m_v^{*-1}\right)\frac{1}{\rho_{\rm sat}}, \tag{9.11c}$$

where C_0^τ has already been determined by equation (9.10).

Gogny functionals

In the case of the Gogny force, the expansion (9.7) for the energy per particle in ANM reads [21, 22]

$$W(\rho,\beta) = \frac{\hbar^2}{2m}\tau_0 + \left(C_{0D}^\rho + C_{1D}^\rho\beta^2\right)\rho^{\alpha_i+1} + \frac{1}{2}\sum_{i=1,2}(A_0^i + A_1^i\beta^2)\rho + \frac{1}{2}\sum_{i=1,2}\left\{B_{nn}^i\left[\frac{1+\beta}{2}g(\mu_i k_{\rm F}^n) + \frac{1-\beta}{2}g(\mu_i k_{\rm F}^p)\right] + B_{np}^i h(\mu_i k_{\rm F}^n, \mu_i k_{\rm F}^p)\right\}, \tag{9.12}$$

where ρ is the total density, $\beta = (\rho_n - \rho_p)/\rho$ is, as before, the isospin asymmetry parameter, and we have used the Skyrme convention for the coupling constants of the density-dependent term. This means that the coupling constants are expressed as a function of the (t_3, x_3) parameters of the Skyrme NN effective pseudo-potential according to the relations given earlier. For the Gogny pseudo-potential, we recall that the parameters of the density-dependent term are denoted by (t_0, x_0) and are related to the Skyrme one through $t_0 \equiv t_3/6$ and $x_0 = x_3$. The coefficients A and B are given by

$$A_0^i = \frac{\pi^{3/2}\mu_i^3}{4}(4W_i + 2B_i - 2H_i - M_i)$$

$$A_1^i = \frac{\pi^{3/2}\mu_i^3}{4}(-2H_i - M_i)$$

$$B_{nn}^i = -\frac{1}{\sqrt{\pi}}(W_i + 2B_i - H_i - 2M_i) = B_{pp}^i$$

$$B_{np}^i = \frac{1}{\sqrt{\pi}}(H_i + 2M_i),$$

and the functions $g(q)$ and $h(q_1, q_2)$ are the result of a double integration of the exchange matrix elements over the same Fermi surface for g and two different surfaces for h which result in

$$g(q) = \frac{2}{q^3} - \frac{3}{q} - \left(\frac{2}{q^3} - \frac{1}{q}\right)e^{-q^2} + \sqrt{\pi}\,\mathrm{erf}(q)$$

and

$$h(q_1, q_2) = 2\frac{q_1^2 - q_1q_2 + q_2^2 - 2}{q_1^3 + q_2^3}e^{-\frac{(q_1+q_2)^2}{4}} - 2\frac{q_1^2 + q_1q_2 + q_2^2 - 2}{q_1^3 + q_2^3}e^{-\frac{(q_1-q_2)^2}{4}}$$
$$- \sqrt{\pi}\frac{q_1^3 - q_2^3}{q_1^3 + q_2^3}\,\mathrm{erf}\left(\frac{q_1 - q_2}{2}\right) + \sqrt{\pi}\,\mathrm{erf}\left(\frac{q_1 + q_2}{2}\right).$$

Note that there are 11 coupling constants in total that characterize nuclear matter: the power of the density α, the two density-dependent coupling constants C_{tD}^{ρ} and the eight constants W_i, B_i, H_i, M_i (recall that $i = 1, 2$) entering the A and B coefficients. Therefore, it is not possible to express directly the parameters of the Gogny force as a function of INM characteristics unless we keep quartic terms in the expansion (9.7)—terms that are less constrained. In spite of this lack of one-to-one correspondence, it is useful to obtain expressions for INM properties, which can be used to validate fits of Gogny forces. Let us begin with the effective mass at the Fermi surface for the particle of type $q = p, n$. It is given by

$$\frac{m}{m_q^*} = 1 + \frac{m}{\hbar^2}\sum_i\left[B_{nn}^i\mathsf{m}(\mu_i k_F^{(n)}, \mu_i k_F^{(n)}) + B_{np}^i\mathsf{m}(\mu_i k_F^{(p)}, \mu_i k_F^{(p)})\right],$$

with [21]

$$\mathsf{m}(q_F, q_F) = +\frac{1}{q_F^3}\left[2(1 - e^{-q_F^2}) - q_F^2(1 + e^{-q_F^2})\right].$$

Let us recall in passing that the isoscalar and isovector effective masses denoted by m_s^* and m_v^*, respectively, are related to the proton and neutron effective masses through

$$\frac{m}{m_q^*} = \frac{m}{m_s^*} + 2\tau_z\beta\left(\frac{m}{m_s^*} - \frac{m}{m_v^*}\right),$$

where $\tau_z = \pm 1/2$ is the isospin, $\tau_z = +1/2$ for neutrons and $\tau_z = -1/2$ for protons. We also recall that m_v^* is not an isovector quantity in the sense of isovector couplings in the functional [5].

Symmetric nuclear matter

An important quantity to gauge the quality of the equation of state (EoS) for nuclear matter is the saturation density ρ_{sat}. In the case of the Gogny force, it can be obtained numerically by determining the position and value of minimum of the curve (9.12). The characteristics of SNM such as P^{NM}, K^{NM} and E^{NM}/A are then obtained easily by setting $\rho = \rho_{\text{sat}}$ in their definitions. By using the expansion (9.7) and the properties $h(q, q) = h(0, 2q) = h(2q, 0) = g(q)$ valid in SNM, we find the following expressions for the pressure,

$$P(\rho) = \frac{2}{5}\frac{\hbar^2 k_{\text{F}}^2}{2m}\rho + (\alpha + 1)C_{0D}^{\rho}\,\rho^{\alpha+2} + \frac{1}{2}\sum_{i=1,2} A_0^i\rho^2 + \sum_{i=1,2} B_0^i p(\mu_i k_{\text{F}})\rho,$$

and incompressibility,

$$K(\rho) = \frac{6}{5}\frac{\hbar^2 k_{\text{F}}^2}{2m} + 9\alpha(\alpha + 1)C_{0D}^{\rho}\,\rho^{\alpha+1} - 3\sum_{i=1,2} B_0^i k(\mu_i k_{\text{F}}),$$

with the functions

$$p(q) = -\frac{1}{q^3} + \frac{1}{2q} + \left(\frac{1}{q^3} + \frac{1}{2q}\right)e^{-q^2}$$

$$k(q) = -\frac{6}{q^3} + \frac{2}{q} + \left(\frac{6}{q^3} + \frac{4}{q} + q\right)e^{-q^2}$$

$$q(q) = -\frac{54}{q^3} + 14q + \left(\frac{54}{q^3} + \frac{40}{q} + 13q + 2q^2\right)e^{-q^2}.$$

Asymmetric nuclear matter

Once the saturation density ρ_{sat} has been determined, we can extract the properties of ANM very easily. For the symmetry energy at saturation $a_{\text{sym}}^{\text{NM}} \equiv S(\rho_{\text{sat}})$, we find

$$S(\rho) = \frac{1}{3}\frac{\hbar^2 k_{\text{F}}^2}{2m} + C_{1D}\rho^{\alpha+1} + \frac{1}{2}\sum_{i=1,2} A_1^i\rho + \frac{1}{6}\sum_{i=1,2}\left[B_{nn}^i s_1(\mu_i k_{\text{F}}) + B_{np}^i s_2(\mu_i k_{\text{F}})\right]$$

for the slope of the symmetry energy at saturation $L^{\text{NM}} \equiv L(\rho_{\text{sat}})$

$$L(\rho) = \frac{2}{3}\frac{\hbar^2 k_{\text{F}}^2}{2m} + 3(\alpha + 1)C_{1D}\rho^{\alpha+1} + \frac{3}{2}\sum_{i=1,2} A_1^i\rho + \frac{1}{6}\sum_{i=1,2}\left[B_{nn}^i l_1(\mu_i k_{\text{F}}) + B_{np}^i l_2(\mu_i k_{\text{F}})\right]$$

and for the isovector correction to the incompressibility,

$$\Delta K^{\text{NM}}(\rho) = -\frac{2}{3}\frac{\hbar^2 k_{\text{F}}^2}{2m} + 9(\alpha + 1)\alpha C_{1D}\rho^{\alpha+1} - \frac{2}{3}\sum_{i=1,2}\left[B_{nn}^i k_1(\mu_i k_{\text{F}}) + B_{np}^i k_2(\mu_i k_{\text{F}})\right].$$

The various functions of q entering the previous expressions read

$$s_1(q) = +\frac{1}{q} - \left(\frac{1}{q} + q\right)e^{-q^2}$$
$$s_2(q) = +\frac{1}{q} - q - \frac{1}{q}e^{-q^2}$$
$$l_1(q) = -\frac{1}{q} + \left(\frac{1}{q} + q + 2q^3\right)e^{-q^2}$$
$$l_2(q) = -\frac{1}{q} - q + \left(\frac{1}{q} + 2q\right)e^{-q^2}$$
$$k_1(q) = -\frac{1}{q} + \left(\frac{1}{q} + q + \frac{q^3}{2} + q^5\right)e^{-q^2}$$
$$k_2(q) = -\frac{1}{q} - \frac{q}{2} + \left(\frac{1}{q} + \frac{3}{2}q + q^3\right)e^{-q^2}.$$

9.1.3 Parameters of covariant functionals

The original formulation of covariant EDF was based on non-linear effective Lagrangians with explicit meson exchange, including self-couplings of the scalar meson σ, and was presented in section 2.4.1. The real parameters of these Lagrangians are the three nucleon–meson couplings constants and the two self-coupling constants, g_σ, g_ω, g_ρ, g_2, g_3 for the 'canonical version' of the theory. In addition, the mass of each of the mesons was often taken as an adjustable parameter, adding another three parameters. One of the most representative parametrizations of meson-exchange Lagrangians is NL3 [23].

The density-dependent effective Lagrangian is defined by equation (2.52). We recall that these couplings take the generic form (2.53)

$$g_i(\rho_v) = g_i(\rho_{\text{sat}})f_i(\xi) \qquad i = \sigma, \omega,$$

where ρ_{sat} is the saturation density and the functions $f(\xi)$ are given by equation (2.54)

$$f_i(\xi) = a_i\frac{1 + b_i(\xi + d_i)^2}{1 + c_i(\xi + d_i)^2}.$$

Together with the range of the exponential dependence of the ρ and π coupling constants, this means the free parameters involved in this phenomenological Lagrangian density are

$$m_\sigma,\ g_\sigma(\rho_{\text{sat}}),\ a_\sigma,\ b_\sigma,\ c_\sigma,\ d_\sigma,$$
$$m_\omega,\ g_\omega(\rho_{\text{sat}}),\ a_\omega,\ b_\omega,\ c_\omega,\ d_\omega,$$
$$m_\rho,\ g_\rho(0),\ a_\rho,$$
$$m_\pi,\ f_\pi(0),\ a_\pi.$$

The number of parameters can be reduced by fixing those that have a counterpart in free space to their measured value. In particular, we can use the measured masses of the mesons

$$m_\omega = 783\ \text{MeV}$$
$$m_\rho = 769\ \text{MeV} \quad g_\rho(0) = 2.629$$
$$m_\pi = 138\ \text{MeV} \quad f_\pi(0) = 1.0.$$

On the other hand, the eight parameters a_i, b_i, c_i, d_i ($i = \sigma$, ω) are not independent, since we must have for $\rho_v = \rho_{\text{sat}}$

$$f_i(\xi = 1) = 1.$$

One can further reduce the number of free parameters by imposing additional constraints for the functions f_i

$$\left.\frac{d^2 f_\sigma}{d\xi^2}\right|_{\xi=1} = \left.\frac{d^2 f_\omega}{d\xi^2}\right|_{\xi=1}$$
$$\left.\frac{d^2 f_i}{d\xi^2}\right|_{\xi=0} = 0.$$

In particular the last set of constraints prevents the rearrangement terms from diverging. These constraints reduce the eight free parameters a_i, b_i, c_i, d_i, to only three independent ones. We usually choose b_σ, d_σ and b_ω as free parameters to specify the density dependence associated with the σ and ω mesons. In total, the Lagrangian density is therefore characterized by eight free parameters

$$m_\sigma,\ g_\sigma(\rho_{\text{sat}}),\ b_\sigma,\ d_\sigma,\ g_\omega(\rho_{\text{sat}}),\ b_\omega,\ a_\rho,\ a_\pi,$$

adjusted to typical nuclear properties. DD-ME2 [24] and PKO3 [25] are representative parametrizations of this class of relativistic functionals.

Finally, the point-coupling Lagrangians of equation (2.56) involve the four functions α_i ($i = S$, V, TS, TV) and the parameter δ_S. Each of the α_i is a functional of the nucleon four-current j^μ—in practice, only the vector term ρ_v. One popular form of the coupling functions is [26, 27]

$$\alpha_i(\rho) = a_i + (b_i + c_i x)e^{-d_i x},$$

with $x = \rho/\rho_{\text{sat}}$. In practice, the scalar–isovector term of the Lagrangian is rarely considered. The reference parametrization of point-coupling Lagrangians is DD-PC1 [26], where $\alpha_{TS} = 0$ and the parameters $c_V = a_{TV} = c_{TV} = 0$. In the case of DD-PC1, the remaining eight parameters for the isoscalar channels S and V are the free parameters of the theory adjusted on experimental data, while the single isovector channel TV is fitted on Dirac–Brueckner–Hartree–Fock (DBHF) calculations of ANM.

Another popular realization of point-coupling Lagrangians is the PC-PK1 parametrization of [28]. In that work, the couplings are not density-dependent but actual constants, and the pairing channel is treated with a finite-range, separable interaction. The Lagrangian used also contains 'derivative' terms intended to mock-up the effect of finite-range couplings.

9.2 Physical observables

Given the form of the EDF, its parameters must be adjusted to a set of experimental data. The methods used to perform this adjustment will be detailed in the next section. Here, we want to briefly discuss some of the challenges that arise when choosing data for the fit.

One may distinguish two types of data. Observables are quantities directly extracted from an experiment and 'minimally' processed. Typical examples of such observables are atomic masses, γ-ray energies, cross-sections, half-lives, etc. The advantage of using such experimental data is that there is little ambiguity as to what quantity we are comparing the theoretical prediction with. As a result, observables truly test the predictive power of a theory or, in the case of EDF approaches, of the level of approximation where the EDF is used (SR-EDF, MR-EDF, QRPA, etc). Note that quantities such as nuclear binding energies or proton/charge radii are not as clean from processing as they may look: nuclear binding energies have to be extracted from atomic masses by subtracting the contribution of the electron cloud to the atomic mass [29]. Charge radii are extracted from electron scattering experiment using standard Fourier transforms, but deducing from them the proton radius requires additional input about the charge distribution of the proton [30].

When extracting nuclear binding energies from atomic mass measurements, the contribution of the electronic cloud to the mass can reach up to 1 MeV in heavy nuclei. Its expression if often parameterized as a function of Z; see [29, 31] and references therein for the procedures related to the 2016 Atomic Mass Evaluation. From a nuclear theory perspective, the nuclear binding energy can be computed at the SR-EDF approximation, i.e. using (3.37) for a given functional or at the MR-EDF approximation with/without projection on broken symmetries as in equation (3.112) or configuration mixing through equation (3.91). Each level of approximation gives a different actual number for the binding energy.

In contrast, pseudo-observables, also called meta-data, are obtained from the actual experimental results through a model-dependent analysis that involves specific assumptions and, often, a set of parameters. A number of pseudo-observables are used in the calibration of EDF, in spite of important limitations, which we briefly describe below.

Single-particle energies have been used, among others, in the fit of [8, 11, 32]. Single-particle energies are never measured directly in an experiment. Instead, experimentalists would typically measure the angular dependence of differential cross-sections in pick-up or stripping reactions of, e.g. the kind $(e, e'p)$ or (p, d), extract spectroscopic factors by fitting the observed patterns within the distorted-wave Born approximation (DBWA) model [33], and using said spectroscopic factors to assign single-particle levels within an independent-particle model such as the Woods–Saxon potential [34]. When the strength is fragmented, this assignment can be ambiguous at best. In any case, it is heavily model-dependent since it relies on the hypotheses and parameters of the DBWA model—which itself depends on input hypotheses and parameters such as the characteristics of the nuclear potential. From these considerations, it should be clear that using single-particle energies to calibrate EDF could be misleading because of the considerable uncertainties associated with the experimental procedures. But the situation is aggravated by the fact that on the theory side, the very concept of single-particle energies makes little sense except in the special case of HF calculations, where they emerge naturally as eigenvalues of the HF potential. In any many-body method that describes the nucleus as a fully correlated system, such as, e.g. MR-EDF calculations, single-particle levels are ill-defined; see the discussion in [35].

Were we to assume that the HF theory is the exact theory for atomic nuclei, then Koopman's theorem would allow us to identify one-particle separation energies with HF single-particle (s.p.) energies [36]. Starting from a nucleus $A = Z + N$ and considering the nucleus $A - 1$ where a hole was made in the s.p. level number i, then we can write [37]

$$E(Z + 1) - E_i(Z) = e_i$$

if we assume that the sets of s.p. levels are the same in the two nuclei. In practice, this condition is not verified (although differences are small) because of mass polarization [6]. In any case, the HF is a poor approximation to a nucleus; see the discussion in [35].

Pairing gaps are another popular example of an experimental quantity that is unambiguously defined neither experimentally nor theoretically. It has been used in [9–11, 38]. In a qualitative interpretation of pairing correlations, the amount of pairing in the system should be proportional, somewhat, to the difference in energy between a fully paired even–even nucleus A and its odd neighbors $A \pm 1$. What exactly this difference should be has been the subject of long debates; see for instance [39–44]. While a mild consensus has emerged on using the three-point odd–even staggering $\Delta^{(3)}$ defined in the nucleus (Z, N) by

$$\Delta^{(3)}(N) = \frac{(-1)^N}{2}[B(N + 1) - 2B(N) + B(N - 1)], \tag{9.13}$$

the main issue is that this indicator is by no means dependent *only* on pairing correlations. The difficulties posed in determining the parameters of the pairing

EDF are thus related to the fact that pairing correlations are both a correction to the bulk energy of a nucleus, which comes mostly from the particle–hole (p.h.) channel, and at the same time deeply correlated to mean-field types of correlations. This argument is one of the main reasons why EDF practitioners, especially for Skyrme or phenomenological EDF and covariant EDF, often choose a much simpler form of the pairing EDF; see the discussion and references in section 1.3.1.

In addition, the computational cost of computing an odd nucleus is much higher than for an even–even one. Nuclei with an odd number of particles are described within the blocking approximation, where the energy of the odd nucleus is given as a one-quasiparticle excitation of an even–even one [43, 45]. In a rigorous implementation of the blocking prescription, time-reversal symmetry is spontaneously broken, which can also induce triaxial shape polarization [46]. This implies that, at least in principle, blocking calculations should allow for the breaking of axial symmetry, which substantially increases the time of execution. In practice, however, detailed studies have shown that the effect of both time-reversal symmetry and axial symmetry-breaking are less than a few dozen of keV [45, 46]. As a result, most blocking calculations are performed within the equal filling approximation [47], which preserves these symmetries [10, 11]. However, the characteristics of the one-quasiparticle excitation that yields the correct grounds-state are not known. As a result, practitioners typically have to compute different candidates and retain the one with the lowest energy. As a consequence, the theoretical pairing indicator used for comparison with the experimental formula (9.13) is often taken as the average value of the pairing field of the HFB matrix, $\langle \Delta \rangle = \text{tr}\, \Delta\rho$. There is no good reason for doing this apart from computational convenience, since it transforms equation (9.13) into a pseudo-observable (from the theory side); see also the discussion in [44]. We refer the reader to section 3.1.3 in chapter 3 for a formal introduction of the blocking approximation and a discussion of its practical implementation.

Fission barriers are yet another example of a popular pseudo-observable. The Skyrme SkM* [48] and Gogny D1S [49] EDF have been fitted by adjusting the height of the first fission barrier in ^{240}Pu. The advantage of using such data is that the surface properties of the EDF are better constrained; see [50] and references therein. This benefits the description of both bulk nuclear properties and the shell structure. However, fission barriers are not obtained in direct measurements. Instead, they are often regarded as fitting parameters into a Hauser–Feshbach calculation of the cross-section for neutron-induced fission or particle transfer [51]. Hauser–Feshbach calculations are notoriously complex simulations that involve a lot of different models to describe level density, capture rates, excitation spectra, etc, each coming with its healthy dose of parameters. In all major Hauser–Feshbach codes on the market, for instance [52, 53], the fission barrier is modeled as a one-dimensional single- or double-humped inverted parabola. Even then, the fit of an actual fission barrier often requires additional approximations, such as assuming that the width of the barrier can be extracted from the Kramers formula for spontaneous lifetimes in the same fissioning nucleus. Based on this rapid presentation, it is clear that fitting the height of a barrier in a multi-dimensional EDF calculation to such a heavily processed pseudo-observable is questionable.

To conclude this section, one might attempt to establish a canonical set of observables that most practitioners could use for the calibration of EDF. Based on the existing literature, most fits are performed at the HF level, see e.g. [4, 8, 23, 28, 32, 54–57], or at the HFB level, see e.g. [9–11]. The case of modern covariant energy density functional (cEDF) is a little different from non-relativistic EDF. The functional DD-PC1, for instance, was fitted exclusively on the EoS of nuclear matter by mapping to DBHF calculations. Data in finite nuclei was used to validate the functional rather than determine it. In all these cases, the binding energies a few closed-shell nuclei and, sometimes, of some well-deformed nuclei are almost always included. Charge or proton radii are also almost always considered, since they can help control the incompressibility of nuclear matter. Most of the time, INM properties are either directly set, e.g. on values computed from *ab initio* models as in [56, 57], or are used to constrain the range of variation of the EDF parameters [9–11]. Information on the shell structure is often incorporated through the energy splitting of spin–orbit partners. This trick helps mitigate the pseudo-observable nature of single-particle energies and can control the spin–orbit strength. Few groups fit the particle–particle channel of the EDF simultaneously with the particle–hole channel [9–11, 55]. For those who do, the odd–even mass staggering as described by equation (9.13) is the most commonly employed indicator.

9.3 Uncertainties of EDF parameters

Like all approaches to nuclear structure, the parameters of EDF are not specified by the theory and need to be adjusted to experimental data. This process is often called simply a *fit* by nuclear theorists, an *optimization* process by mathematicians, and sometimes a *calibration* by statisticians. We will use these three terms interchangeably in the text. There are obviously infinitely many different ways to adjust the EDF parameters, depending on which EDF is actually used (Skyrme, Gogny, covariant EDF, others), what and how many experimental observables are used, at what level the calibration is performed (HF, HFB, QRPA, etc) and how calculations are actually performed. As a result, the EDF approach suffers from systematic, statistical and numerical uncertainties [58–61]. In this section, we will review briefly some of the statistical tools used to quantify these uncertainties. This is a relatively recent field in nuclear theory and for some time, only basic concepts of statistics such as covariance or correlation matrix were used. More recently, the availability of large-scale computing capabilities have made Bayesian methods increasingly popular.

Of the three types of uncertainties, the numerical ones are the easiest to understand. They simply come from performing calculation on a computer: when expanding HFB solutions on the eigenstates of the HO basis, for instance, the expansion is truncated. Systematic uncertainties are caused by the use of different models: different approximations (HF, HFB, QRPA, etc) or/and different functionals (Skyrme, Gogny, covariant, etc). Statistical uncertainties will be defined here as the consequence of picking up a given set of experimental data to calibrate the model: for the same model, two different data sets will give two different results on any prediction of the model, hence a source of uncertainties. As mentioned in the introduction to this chapter, in practice all three sources are mixed and it may not be easy to disentangle them.

9.3.1 Calibration of energy functionals

The simplest way to determine the parameters of a model such as an EDF is to build a χ^2 'objective' function and simply minimize it [59]. This simple, non-linear regression approach has recently been followed by nearly all practitioners of the EDF method both for Skyrme or Skyrme-like functionals [8–11, 54, 62], Gogny [63], covariant EDF [27, 64] and phenomenological EDF [56, 57, 65–67]. Note that the D1 [68], D1S [49], D1N [55] and D2 [69] parametrizations of the Gogny functional have been obtained with a hybrid (proprietary) procedure that also makes use of various analytical expressions; see the discussion in [69]. Further references for the strategies used for Skyrme EDF can be found in [70] and [71]. A standard form of the objective function is

$$\chi^2(\boldsymbol{\theta}) = \frac{1}{n_d - n_x} \sum_{k=1}^{n_d} \left(\frac{y_k - \eta_k(\boldsymbol{\theta})}{\lambda_k} \right)^2, \tag{9.14}$$

where $\boldsymbol{\theta}$ is the vector containing the n_x model parameters, y_k are the n_d experimental data points at which the model is being adjusted (or calibrated), and $\eta_k(\boldsymbol{\theta})$ is the model prediction corresponding to the data points y_k. In this formulation, λ_k are the weights of the data points. Both weights and data have the same units, making the whole χ^2 objective function dimensionless.

In principle, the weight λ_k should reflect an estimate of the full uncertainty for the data point y_k, including both the model and experimental uncertainties. For example, most EDFs currently on the market can typically reproduce nuclear binding energies at the level of about 1 MeV root-mean-square deviation. In heavy nuclei, this corresponds to an accuracy better than one part per thousand. This can be compared to experimental Penning-trap mass measurements, which can reach up to sub-keV level in accuracy [72]. In this kind of situation, the full uncertainty is entirely dominated by the model. An additional difficulty arises when different kinds of data are used, each with different units. In this case, the choice of the weight will also impact the relative importance assigned to each data type in the χ^2 function [60].

There are numerous algorithms to locate the minimum $\boldsymbol{\theta}_{\min}$ of the χ^2 function. For example, the UNEDF EDF models were adjusted by using the POUNDerS algorithm, which is a derivative-free quasi-Newton algorithm. In general, the choice of the algorithm depends on the nature of the optimization problem and on the availability of the derivatives of $\eta_k(\boldsymbol{\theta})$ with respect to the model parameters. Because of the size of the parameter space ($n_x \approx 10$–15) and of the limited numerical precision of EDF calculations caused, among other, by the self-consistency nature of the calculation, computing derivatives using, e.g,. finite-difference methods is not a robust option. A more promising route would involve automatic differentiation (AD), which provides derivatives at machine precision by using the chain rule to propagate a derivative at each step of a computational program. This approach has delivered remarkable results both in *ab initio* approaches to nuclear structure [73] and in electronic density functional theory (DFT) [74]. Implementing AD for the optimization of EDF requires substantial work on the existing EDF solvers.

The landscape around the minimum point of the χ^2 objective function contains valuable information about the model parameters. The purpose of uncertainty quantification is to use statistical methods to explore this landscape and quantify the uncertainties associated with each model parameter as well as the correlations among them. This is particularly important in view of propagating these uncertainties in model predictions, as we will see in the next section. The uncertainty of a model parameter can have two different causes (that can be combined):

- It may reflect the fact that the experimental data used in the calibration was not sufficient to constrain the parameter well enough. A typical example with EDF model calibration is the parameters associated with isovector properties. Data in very neutron-rich nuclei is rather scarce and does not extend up to the neutron drip line for medium-mass or heavy nuclei. As a result isovector parameters cannot be constrained as well as the corresponding isoscalar ones. As we will see in the next section, the impact of this lack of constraint shows up when uncertainties of model predictions are assessed for neutron-rich nuclei.
- It may also reflect an intrinsic limitation of the model. In the example of the UNEDF optimizations, all calculations were performed at the HFB level using mostly data on ground-state properties of finite nuclei. Fission isomer excitation energies, which were used for the UNEDF1 and UNEDF2 parametrizations, are not ground-state properties but are still obtained within a constrained HFB approach. It is quite well known that the incompressibility of nuclear matter K^{NM} cannot be really probed at that level and requires instead calculation of the giant dipole resonance within the QRPA framework [75–77]. As a result, the value of K^{NM} is ill-constrained because the model is not really sensitive to it.

The calculation of the full covariance matrix of the model parameters is usually a numerically challenging task and, therefore, various approximations are often employed to mitigate the cost. Recall that in elementary statistics, the covariance matrix of a random vector $\boldsymbol{X}$ (a vector made of n random variables X_i) is defined by

$$\Sigma_{ij} = E\Big[(X_i - \mu_i)(X_j - \mu_j)\Big],$$

where μ_i is the mean value of the random variable X_i and $E[]$ refers to mean value operation, i.e. $\mu_i = E[X_i]$. The covariance matrix thus generalizes the concept of variance (or standard deviation) to n-dimensional random variables. The covariance matrix of model parameters is often calculated in the linearized approximation, where the model prediction $\eta_k(\boldsymbol{\theta})$ is assumed to be linear with respect to $\boldsymbol{\theta}$ in the vicinity of the χ^2 minimum [59, 78]. In more detail, we start with the Taylor expansion of the χ^2 function around the minimum of the optimization,

$$\chi^2(\boldsymbol{\theta}) = \chi^2(\boldsymbol{\theta}_{\min}) + \frac{1}{2}\sum_{\alpha\beta}(\boldsymbol{\theta} - \boldsymbol{\theta}_{\min})_\alpha H_{\alpha\beta}(\boldsymbol{\theta}_{\min})(\boldsymbol{\theta} - \boldsymbol{\theta}_{\min})_\beta,$$

where the matrix $H_{\alpha\beta}$ is the Hessian matrix, i.e. the matrix defined by

$$H_{\alpha\beta}(\boldsymbol{\theta}) = \frac{\partial^2\chi^2}{\partial\theta_\alpha\partial\theta_\beta}(\boldsymbol{\theta}).$$

In the assumption of a linear regime, one assumes that the observables included in the χ^2 depend linearly on the parameters $\boldsymbol{\theta}$. A straightforward calculation then shows that the covariance matrix is given by

$$\mathrm{Cov}(\theta_\alpha, \theta_\beta) = [(J^T\Sigma J)^{-1}]_{\alpha\beta}, \tag{9.15}$$

where the $n_d \times n_x$ matrix $J_{k\alpha}$ is the Jacobian matrix that contains the derivatives of the model output $\eta_k(\boldsymbol{\theta})$ with respect to the parameter θ_α, and Σ contains the weights used in the χ^2 function,

$$J_{k\alpha} = \left.\frac{\partial\eta_k(\boldsymbol{\theta})}{\partial\theta_\alpha}\right|_{\boldsymbol{\theta}=\boldsymbol{\theta}_{\min}} \qquad \Sigma = \mathrm{diag}(\lambda_1^{-2}, \dots, \lambda_k^{-2}, \dots, \lambda_{n_d}^{-2}). \tag{9.16}$$

When the covariance matrix is known, the standard deviation of model predictions for any given observable can be calculated rigorously. This will be discussed in the next section.

The standard deviations of the model parameters and correlation matrix can be obtained from the covariance matrix. The correlation coefficient between model parameters α and β reads as

$$R_{\alpha\beta} = \frac{\mathrm{Cov}(\theta_\alpha, \theta_\beta)}{\sqrt{\mathrm{Var}(\theta_\alpha)\mathrm{Var}(\theta_\beta)}}, \tag{9.17}$$

where $\mathrm{Var}(\theta_\alpha) = \sigma_\alpha^2$ is the variance of the model parameter θ_α and σ_α is the corresponding standard deviation. The value of the correlation coefficient lies in the interval of $R_{\alpha\beta} \in [-1, +1]$. A value close to (−1) +1 means that two model parameters are strongly (anti-)correlated.

9.3.2 Bayesian inference techniques

The Bayesian approach to parameter estimation that we are going to briefly sketch in this section is centered on the empirical paradigm that the parameters of the model—sometimes, the model itself—should be treated as pure random variables. The philosophical, and at the same time practical, consequence of this statement is that one is ready to give up the very notion that there exist 'true' values of the parameters of our model. The most one can say about the particular value of a parameter is that it has a certain probability of being in some interval, given the existing body of experimental data and the particular model one is considering. In this section, we borrow from the presentations in [79–81].

One may identify three main reasons why the Bayesian approach is particularly appealing in the context of nuclear theory:

- There is no exact theory of nuclear forces, only approximations thereof. Even χEFT, which may be considered as the best and most promising approach for rigorously deriving nuclear forces, only provides an approximate expansion of nuclear potentials [1]. In practice, the convergence of this expansion is not entirely guaranteed since it depends on choices made about the breakup scale of the theory and the power counting of the theory [2]. In addition, at any given order, several expansion schemes are available depending on whether or not one includes, e.g., excitations of the nucleons, etc. As a result, the best one has in nuclear theory is a particular set of nuclear potentials at a given order in chiral perturbation theory, that is, a collection of operators $\hat{V}_{\rm NN}$, $\hat{V}_{\rm 3N}$, etc. Since this set is not uniquely defined, the concept of, say the 'true' value of the strength of a two-body spin–orbit potential is meaningless.
- Given nuclear potentials at a given order in perturbation theory, the truncation of the theory at that order implies that the potentials are not parameter-free. Since the theory does not provide the values of these parameters, one must invoke specific optimization protocols to fix them. This introduces systematic, statistical and possibly even numerical biases that further blurs the notion of true value [73, 82–85].
- Last but not least, even if the nuclear Hamiltonian were known exactly, solving the nuclear many-body problem with this Hamiltonian is, in and by itself, an impossible task—at least in a mathematically exact way. The analogy with the non-relativistic treatment of many-body electronic systems is particularly enlightening: for a system of N electrons in an atom treated at the non-relativistic, Born–Oppenheimer approximation, the Hamiltonian is exactly known. Yet, the eigenstates of this Hamiltonian can only be computed approximately using either various flavors of either direct *ab initio* methods or DFT methods. Nuclear systems are characterized by at least a similar level of many-body complexity. In fact, one may argue that the many-body physics of nuclei is a lot more complex than in atoms since specific N-body correlations can be induced by 2-, 3-, $\cdots$, $(N-1)$-, N-body terms in the Hamiltonian. In addition, atomic nuclei are self-bound and their components are not elementary particles like electrons, but composite many-body systems of bound quarks.

Bayes' theorem

The core of the Bayesian approach to parameter estimation is centered on the famed Bayes' theorem, which we recall here for a vector of continuously distributed random variables. Let us assume we have a model M characterized by a vector of parameters $\boldsymbol{\theta} \equiv (\theta_1, \ldots, \theta_{n_x})$. In the context of the EDF approach, we should define the 'model M' as the combination of (i) an approximation level such as HFB, QRPA, generator coordinate method (GCM), etc, (ii) a specific energy functional and (iii) in principle, a given numerical scheme and its characteristics, e.g. the HO basis, its oscillator lengths, number of shells and number of states. In practice, this latter aspect is rarely taken into account. We now assume that we have also gathered a vector $\boldsymbol{y} \equiv (y_1, \dots, y_{n_d})$ of experimental observations of quantities that can be computed by the model. The components of $\boldsymbol{y}$ need not be of the same type: for instance, in the context of nuclear physics, $\boldsymbol{y}$ may contain nuclear binding energies, proton radii, excitation energies of some excited states, transition strengths, etc. However, all these types of data must be calculable by the model M. The Bayes' theorem then states that the probability that the parameters of the model are between $\boldsymbol{\theta}$ and $\boldsymbol{\theta} + d\boldsymbol{\theta}$ given the data is given by

$$p(\boldsymbol{\theta}|\boldsymbol{y}, M)d\boldsymbol{\theta} = \frac{p(\boldsymbol{y}|\boldsymbol{\theta}, M)p(\boldsymbol{\theta}|M)d\boldsymbol{\theta}}{\int p(\boldsymbol{y}, \boldsymbol{\theta}|M)d\boldsymbol{\theta}}. \tag{9.18}$$

The probability $p(\boldsymbol{\theta}|\boldsymbol{y}, M)$ represents the conditional probability that the parameters take the values $\boldsymbol{\theta}$ given the experimental data $\boldsymbol{y}$ and the model M. It is called the posterior distribution. The goal of Bayesian inference is to compute this probability distribution function (PDF) given a model and a set of experimental observations. The figure 9.2 gives an example of such a posterior distribution for the particular case of the UNEDF1 parametrization of the Skyrme EDF.

In equation (9.18), the term $p(\boldsymbol{y}|\boldsymbol{\theta}, M)$ is the probability that the model M produces the data $\boldsymbol{y}$ given the parameters $\boldsymbol{\theta}$. This will be identified with the likelihood function. A typical choice is to set

$$p(\boldsymbol{y}|\boldsymbol{\theta}, M) \propto \exp\left(-\frac{1}{2}\chi^2\right), \tag{9.19}$$

where the χ^2 function has the general form

$$\chi^2 \propto \boldsymbol{Z}^T \Sigma \boldsymbol{Z},$$

with $\boldsymbol{Z}$ the n_d-dimensional vector collecting the deviation between the experimental value of the data and the theoretical prediction by the model, $\boldsymbol{Z} = (z_1, \dots, z_{n_d})$ with $z_k = y_k - \eta_k(\boldsymbol{\theta})$ with η_k the output of the model for that particular quantity. The $n_d \times n_d$ 'covariance matrix' Σ encapsulates the correlations between model errors. More precisely, this is the covariance matrix associated with the errors in predicting the observed values, sometimes also called misfits. We can also introduce another covariance matrix exclusively related to the parameters themselves. In the simplest form of parameter optimization, one may choose this matrix to be diagonal with the diagonal terms given by $1/\lambda_k^2$. In this case, λ_k^2 represents an estimate for the error in predicting the value of the data point y_k.

The term $P(\boldsymbol{\theta}|M)$ is the probability that the parameters take the value $\boldsymbol{\theta}$ irrespective of any observation. This probability distribution associated with this term is often referred to as the 'prior' distribution: it represents what we know about the parameters $\boldsymbol{\theta}$ before doing any calculation. One advantage of the Bayesian approach is that we can use as prior distribution the posterior distribution from a

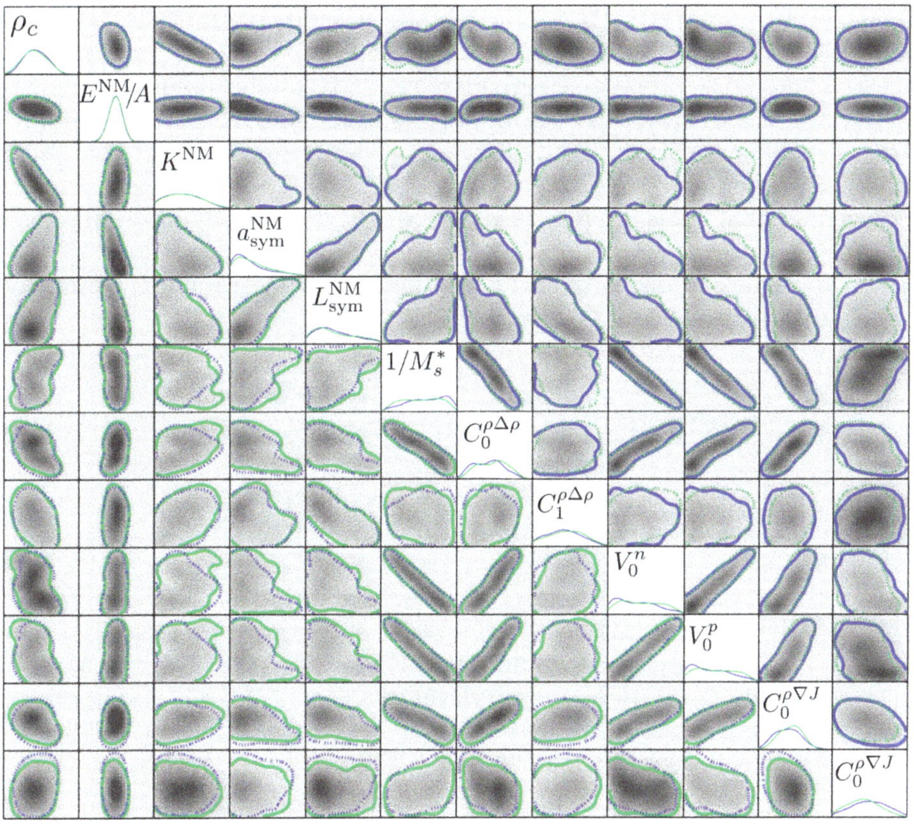

Figure 9.2. Bivariate representation of the PDF for the UNEDF1 EDF. The diagonal terms represent the PDF of each parameter, the off-diagonal terms' 2-by-2 correlation matrices. Details on the calculations can be found in [10, 86]. (Figure reproduced with permission from [87], courtesy of Higdon. Copyright 2015 The American Physical Society.)

previous analysis. Suppose that an analysis of the Skyrme EDF with a certain experimental data set $\boldsymbol{y}$ has resulted in a posterior distribution for the parameters of the functional. One may repeat the analysis with a new set $\boldsymbol{y}'$ and use as a prior distribution for the analysis the previous posterior. Although the obvious danger of this approach is introducing a bias—if, for instance, the old posterior was skewed towards a non-physical region of the parameter space—it can also significantly accelerate the analysis if it is properly handled.

Markov-chain Monte Carlo algorithms

In practice, the posterior distribution is obtained by sampling algorithms. For unknown multi-dimensional distributions, the most commonly used is the Metropolis–Hasting version of the Markov-chain Monte Carlo (MCMC) algorithm, which we will sketch briefly below; for more in-depth discussion of MCMC methods, the reader is referred to [81] and other statistics textbooks. The MCMC begins with knowledge of a 'jumping' or 'proposal' probability distribution q. This distribution has no other purpose than to allow generating samples of parameters. There is great flexibility in choosing it, as long as its analytical or numerical form is known. Choosing a proposal distribution that mimics the future posterior will improve the efficiency of the algorithm. Standard choices are the uniform or Gaussian random walks, where the current value is perturbed with a symmetric, zero-mean random variable.

Given an initial guess $\boldsymbol{\theta}^{(0)}$ for the parameters $\boldsymbol{\theta}$, the MCMC algorithm goes as follows. At iteration n,

1. Draw a new candidate value $\boldsymbol{\theta}^{*}$ from the proposal distribution q. This would typically involve writing something like

$$\boldsymbol{\theta}^{*} \propto q(\boldsymbol{\theta}|\boldsymbol{\theta}^{(n-1)}).$$

2. Compute the acceptance probability defined by

$$\alpha = \frac{p(\boldsymbol{\theta}^{*}|\boldsymbol{y}, \boldsymbol{\theta}^{(0)}, \dots, \boldsymbol{\theta}^{(n-1)})}{p(\boldsymbol{\theta}^{(n-1)}|\boldsymbol{y}, \boldsymbol{\theta}^{(0)}, \dots, \boldsymbol{\theta}^{(n-2)})} \frac{q(\boldsymbol{\theta}^{(n-1)}|\boldsymbol{\theta}^{*})}{q(\boldsymbol{\theta}^{*}|\boldsymbol{\theta}^{(n-1)})}.$$

   If the proposal distribution q is symmetric, then the second fraction is equal to 1. The first fraction involves the ratios of the target posterior distribution at the candidate point $\boldsymbol{\theta}^{*}$ and at the previous point $\boldsymbol{\theta}^{(n-1)}$. The key to computing this ratio is to recall Bayes' theorem (9.18), which states that the posterior distribution is proportional to the likelihood function and the prior.
3. Sample a uniform random variable $u \in [0, 1]$. If $u \leqslant \alpha$, then set $\boldsymbol{\theta}^{(n)} = \boldsymbol{\theta}^{*}$, otherwise set $\boldsymbol{\theta}^{(n)} = \boldsymbol{\theta}^{(n-1)}$.

In this way, the MCMC builds a Markov sequence of random vectors $\{\boldsymbol{\theta}^{(n)}\}$; see figure 9.3 for a simple illustration. One can show that this sequence converges to the target distribution—the posterior distribution $p(\boldsymbol{\theta}|\boldsymbol{y})$ of the parameters of the models given the data $\boldsymbol{y}$.

Gaussian processes

The MCMC algorithms are very powerful but may require many evaluations of the full physics model. This can sometimes be problematic for optimization problems when the number of parameters (i.e. the length of the vector $\boldsymbol{\theta}$) is large, when the parameter space itself is large, i.e. the range $[\theta_\alpha^{(\min)}, \theta_\alpha^{(\max)}]$ for each parameter θ_α is large, or when the χ^2 involves computationally expensive calculations. For example, in the case of the UNEDF energy functionals of [9–11], χ^2 requires of the order of 100 axially deformed HFB calculations and the number of parameters ranges from 12 for UNEDF0 and UNEDF1 to 14 for UNEDF2.

For these reasons, it is often advantageous to replace the evaluation of the true physics model in χ^2 by constructing an *emulator* of the EDF solver which is (much) faster to run, yet reproduces its values within a certain domain of the parameter space. One tool that is particularly ubiquitous in this context is Gaussian process (GP)S. There exist a rich literature on the GP and our goal here is not to enter into

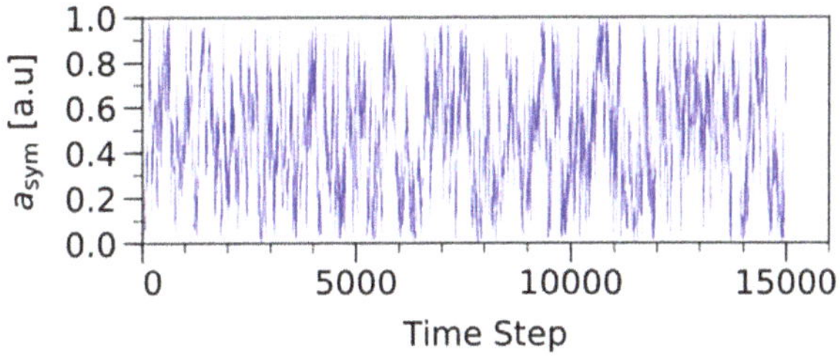

Figure 9.3. Schematic example of MCMC iterations. The figure represents the value of the $a_{\text{sym}}^{\text{NM}}$ parameter of the Skyrme EDF (normalized between 0 and 1) as a function of the MCMC time step. (Figure created by Earl Lawrence.)

too many details. Rather, we would like to highlight some of the main features of GP and illustrate why this technique could prove particularly useful for the problems involving nuclear EDF. For the sake of simplicity, let us first assume a simple one-dimensional problem characterized by an unknown scalar function f: $x \rightarrow f(x)$. We assume that we have observed m realizations of the function, $\boldsymbol{y} = (y_1, \ldots, y_m)$. Our goal is to build an emulator of the function f based on these observations. In other words, we want to determine analytically or numerically the actual function f such that $f(x_k) = y_k$. This is a standard problem of regression, and one may the usual methods of linear regression or spline interpolation. However, these methods are quickly limited for higher dimensional problems. The main idea behind the GP approach is that f can be thought of as the most probable realization of a family of functions $\{f_\alpha\}_{\alpha\in\mathbb{R}}$, and that Bayesian inference can be used to determine it based on the existing data.

In the following we will attempt to sketch how GP can be applied to the calibration of EDF. Our presentation is inspired by [86], where more technical details can be found. The function we want to emulate is the output η of an EDF solver. This output can be, for instance, a nuclear binding energy, a proton radius, etc. Generally, we will assume that for a given observable k, the output of the EDF solver is η_k and reads

$$y_k = \eta_k(\boldsymbol{\theta}) + \varepsilon_k, \tag{9.20}$$

with ε_k an error term that we will specify later. The model output function η is a scalar function $\mathbb{R}^{n_x} \mapsto \mathbb{R}$. The χ^2 objective function of equation (9.14), while itself a scalar function of n_x variables, χ^2: $\mathbb{R}^{n_x} \mapsto \mathbb{R}$, is composed of n_d terms each involving a given η_k. A GP approach will aim at emulating each of these n_d terms separately. To this end, we first run m actual calculations of each of the n_d terms in the χ^2. This corresponds to generating a set of m column vectors of length n_d. Stacking each of these columns side by side, we form the $n_d \times m$ matrix of simulation outputs Ξ. The column μ of this matrix contains the n_d model outputs $\boldsymbol{\eta}(\boldsymbol{\theta}_\mu) \equiv (\eta_1(\boldsymbol{\theta}_\mu), \ldots, \eta_{n_d}(\boldsymbol{\theta}_\mu))$ for the vector of model parameters $\boldsymbol{\theta}_\mu = (\theta_{1,\mu}, \ldots, \theta_{n_x,\mu})$.

Perhaps an easier way to think of the function η_k is to write it as $\eta(\boldsymbol{\theta}; x_k)$, where x_k would stand for a particular observable (energy, radius, etc) in a particular nucleus.

This notation would emphasize that there is a single EDF model, which can be run for an arbitrary nucleus and produce different types of outputs. From a statistics point of view, everything happens as if one had different models.

Let us now look more closely at the error term ε_k in equation (9.20). In order to handle it, we will assume that it is a random variable that follows a multivariate Gaussian distribution of mean 0 and variance Σ,

$$\eta_k \sim N(\mu, \Sigma).$$

Before going any further, we should introduce a prescription for the (statistical) correlation matrix Σ. This matrix represents the effect on the error of the model due to the correlation between any two parameter sets $\boldsymbol{\theta}_\mu$ and $\boldsymbol{\theta}_\nu$. We may choose that the correlation goes as

$$\Sigma_{\mu\nu} \sim \frac{1}{\lambda} \exp\left(-||\boldsymbol{\beta} \cdot (\boldsymbol{\theta}_\mu - \boldsymbol{\theta}_\nu)||\right),$$

where $|| \cdot ||$ refer to the norm of a n_x-dimensional vector. This would translate the fact that the correlation between the model output decreases as the 'distance' in parameter space increases. Let us illustrate this notion of exponential parameter correlation: there is no reason why the error on the binding energy of ^{208}Pb for some Skyrme EDF parametrization $\boldsymbol{\theta}_\mu$ should be particularly correlated with the error from another Skyrme parametrization $\boldsymbol{\theta}_\nu$—unless the two sets are not that different from one another. Other choices are of course possible and should in principle be explored. Note that the set of all ε_k for all m realizations of the parameters $\boldsymbol{\theta}$ form a matrix Z of deviations between model outputs and data points.

In the next step, we perform a change of variable $Z \to \mathcal{Z}$. One possible example of the transformation consists in doing a Cholesky decomposition of the matrix Σ (which should be positive-definite by construction),

$$Z = LL^T \Rightarrow \mathcal{Z} = L^T Z.$$

This transformation guarantees that the random variables in $\mathcal{Z}$ are independent and identically distributed. Therefore, $\mathcal{Z} \approx N(0, I)$. This is a key result since it ensures that the likelihood function of this matrix will take the standard form (9.19). Once this procedure has been established, one can set a prior distribution to all the parameters of the theory. These include now both the actual parameters $\boldsymbol{\theta}$ of the EDF and the parameters introduced to model the covariance matrix. As just mentioned, the likelihood is

$$\mathcal{L} \sim \exp\left(-\frac{1}{2}\mathcal{Z}^T \mathcal{Z}\right)$$

and we are thus fully equipped to run an MCMC algorithm.

A GP approach to parameter optimization has several key features. First, the likelihood function that is sampled in MCMC is parametrized both in terms of the parameters $\boldsymbol{\theta}$ of the EDF model and in terms of the parameters of the GP that aims at emulating it. In other words, the result of the MCMC produces a probability distribution function for the parameters that include uncertainties from the emulator itself. Second, it is possible to model explicitly the correlations between model outputs. This was done in the few studies of this type for EDF calibration [86, 87] and is certainly worth exploring. Third, the definition (9.20) could be extended to account for, e.g. model defects δ,

$$y_k = \eta_k(\boldsymbol{\theta}) + \delta_k + \epsilon_k,$$

that would account for a known bias or limitation of the model. For example, we know that a fit of the EDF at the HF approximation that would include open-shell nuclei (for instance the isotopic sequence of Tin isotopes) would risk introducing systematic bias, since these nuclei have large pairing correlations. However, based on our knowledge of how these correlations impact, say the binding energy, as a function of the number of particles (i.e. the index k), we could model them into the term δ_k. We could then follow the exact same recipe as presented above. The parameters resulting from this optimization would be valid at the HFB level—within some uncertainties that the GP would quantify through the posterior distribution of the relevant parameters. This approach would in fact be needed to guarantee that the error has a statistical mean of zero.

> The advantage of using model defects is to obtain a parametrization that effectively incorporates effects not captured by the model. In the example discussed in the text, one could do a fit at the HF level (where calculations are much faster) yet use the resulting parametrization with confidence at the HFB level.

9.4 Propagation of theoretical uncertainties

Given information on the statistical uncertainties of the model parameters as contained either in the covariance matrix or in the Bayesian posterior distribution, we are now in a position to compute how these uncertainties propagate to the calculation of other observables. To this end, we can first make a Taylor expansion of an observable O as a function of the model parameters. To first order,

$$O(\boldsymbol{\theta}) = O(\boldsymbol{\theta}_{\min}) + \sum_{\alpha} \frac{\partial O}{\partial \theta_\alpha}(\boldsymbol{\theta} - \boldsymbol{\theta}_{\min})_\alpha = O(\boldsymbol{\theta}_{\min}) + \boldsymbol{G}^O \cdot (\boldsymbol{\theta} - \boldsymbol{\theta}_{\min}),$$

where we denoted the vector of partial derivatives by $\boldsymbol{G}^O$. In this linear regime, the statistical variance of the observable O is given by

$$\sigma^2(O) = \sum_{\alpha\beta} G_\alpha^O \mathrm{Cov}(\theta_\alpha, \theta_\beta) G_\beta^O, \tag{9.21}$$

where $\mathrm{Cov}(\theta_\alpha, \theta_\beta)$ is the covariance matrix of equation (9.15) between model parameters θ_α and θ_β. In addition to the covariance matrix, one needs estimates for the partial derivatives of $O(\boldsymbol{\theta})$ with respect to the model parameters, at the minimum $\boldsymbol{\theta}_{\min}$ of $\chi^2(\boldsymbol{\theta})$. The validity of the linearized approximation depends on how linearly a given observable O behaves as a function of the model parameters in the vicinity of $\boldsymbol{\theta}_{\min}$ [59]. In the special case of Skyrme EDF, the nuclear binding energy is linear as a function of all coupling constants other than the power of the density dependence, and the covariance approach has been widely applied [88–90]. An example of this technique is given in figure 9.4, where the predicted statistical uncertainty of the nuclear binding energy for several neutron-rich isotopes is shown for the UNEDF0 functional. The fact that the uncertainty increases towards neutron-rich nuclei illustrates that for this particular EDF, the isovector coupling

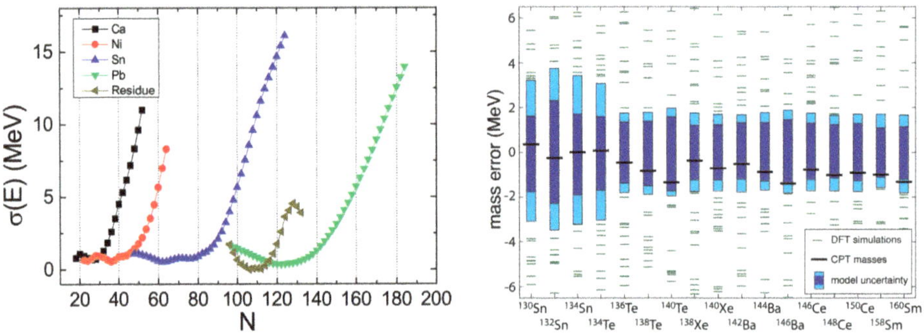

Figure 9.4. Estimates of statistical uncertainties of model predictions using either covariance methods (a) or Bayesian inference (b). (a) Predicted statistical uncertainty of the nuclear binding energy for the UNEDF0 EDF functional for the Ca, Ni, Sn and Pb isotopic chains, as a function of neutron number N. (Figure reproduced with permission from [88], courtesy of Gao. Copyright 2013 The American Physical Society.) (b) Deviation between experimental and theoretical masses for neutron-rich nuclei in the mass $A \sim 140$ region. The black lines represent the prediction from the UNEDF1 EDF. The dark blue line represents the standard deviation σ_k^2 for each data type as estimated by the GP. The light blue line accounts for the contribution of the emulator to the standard deviation. (Figure reproduced with permission from [87], courtesy of Higdon. Copyright 2015 The American Physical Society.)

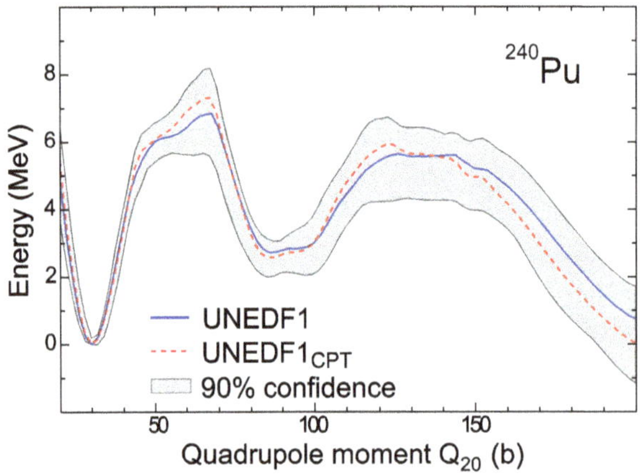

Figure 9.5. Fission barrier in ^{240}Pu obtained with the Skyrme UNEDF1 EDF (blue line) and refit of the EDF with the binding energy of a few more neutron-rich nuclei (dashed red line). The gray shaded area represents the spread of predictions coming from 50 different samples of the posterior distribution selected in the 90% confidence region. (Figure reproduced with permission from [87], courtesy of Schunck. Copyright 2015 by The American Physical Society.)

constants have large standard deviations, which propagates to observables when going further away from the valley of stability. In addition, it is interesting to note that the residues of Pb binding energies near the doubly magic ^{208}Pb are much larger than the computed statistical errors. This indicates that the model is lacking some important physics.

Practical calculations of the statistical variance require both the partial derivatives of $\partial E/\partial\theta_\alpha$ with respect to the model parameters, and the covariance matrix obtained from the sensitivity analysis. Partial derivatives can be obtained numerically, for example with a finite-difference method, but this implies that we have complete information about the χ_2 objective function. While this information has recently become more widespread, for instance for the Skyrme UNEDF [9–11, 88, 89] or SV families [4, 91, 92], and the FSUGOLD Lagrangian [64, 93–95], for many other interesting functionals, it is not available.

In the Bayesian setting, propagating uncertainties is pretty straightforward since we have, in principle, access to the full posterior distribution of the model parameters. Two kinds of uncertainty propagation can then be performed:

- For the small set of observables that were included in the calibration of the functional, the GP typically provides, in addition to the full posterior distribution, an estimate of the correlation matrix Σ. The diagonal term of this matrix represents a standard deviation σ_k for the data type k and is the analogue of equation (9.21). An example of such a calculation is in figure 9.4(b).
- For all other observables, it is still possible to estimate uncertainties by simply drawing samples of the posterior, say at the 90% confidence region. We can then perform calculation of any given observable with the original model. Figure 9.5 illustrates one such example for the posterior distribution of the UNEDF1 parametrization of the EDF obtained in [87]. The posterior distribution was generated following the approach outlined in the previous section. Fifty samples were then generated and the fission barrier for ^{240}Pu was computed.

References

[1] Epelbaum E *et al* 2012 Structure and rotations of the Hoyle state *Phys. Rev. Lett.* **109** 252501

[2] Machleidt R and Sammarruca F 2016 Chiral EFT based nuclear forces: achievements and challenges *Phys. Scr.* **91** 083007

[3] Bender M, Bertsch G F and Heenen P-H 2006 Global study of quadrupole correlation effects *Phys. Rev.* C **73** 034322

[4] Klüpfel P *et al* 2009 Variations on a theme by Skyrme: a systematic study of adjustments of model parameters *Phys. Rev.* C **79** 034310

[5] Lesinski T *et al* 2007 Tensor part of the Skyrme energy density functional: spherical nuclei *Phys. Rev.* C **76** 014312

[6] Zalewski M *et al* 2008 Spin–orbit and tensor mean-field effects on spin–orbit splitting including self-consistent core polarizations *Phys. Rev.* C **77** 024316

[7] Bender M, Duguet T and Lacroix D 2009 Particle-number restoration within the energy density functional formalism *Phys. Rev.* C **79** 044319

[8] Chabanat E *et al* 1997 A Skyrme parametrization from subnuclear to neutron star densities *Nucl. Phys.* A **627** 710

[9] Kortelainen M *et al* 2010 Nuclear energy density optimization *Phys. Rev.* C **82** 024313
[10] Kortelainen M *et al* 2012 Nuclear energy density optimization: large deformations *Phys. Rev.* C **85** 024304
[11] Kortelainen M *et al* 2014 Nuclear energy density optimization: shell structure *Phys. Rev.* C **89** 054314
[12] Dobaczewski J, Nazarewicz W and Werner T R 1995 Closed shells at drip-line nuclei *Phys. Scr.* **T56** 15
[13] Dobaczewski J and Dudek J 1996 Time-odd components in the rotating mean field and identical bands *Acta Phys. Pol.* B **27** 45
[14] Borisenko A I and Tarapov I E 1968 *Vector and Tensor Analysis with Applications* (New York: Dover)
[15] Riley K F, Hobson M P and Bence S J 2006 *Mathematical Methods for Physics and Engineering* (Cambridge: Cambridge University Press)
[16] Dobaczewski J *et al* 1996 Mean-field description of ground-state properties of drip-line nuclei: pairing and continuum effects *Phys. Rev.* C **53** 2809
[17] Perlińska E *et al* 2004 Local density approximation for proton–neutron pairing correlations: formalism *Phys. Rev.* C **69** 014316
[18] Landau L D and Lifshitz E M 1980 *Statistical Physics, Part I. Third* 3rd edn (Oxford: Butterworth-Heinemann)
[19] Parr R G and Yang W 1989 *Density Functional Theory of Atoms and MoleculesInternational Series of Monographs on Chemistry* (New York: Oxford University Press)
[20] Eschrig R 1996 *Fundamentals of Density Functional Theory* (Leipzig: Teubner)
[21] Sellahewa R and Rios. A 2014 Isovector properties of the Gogny interaction *Phys. Rev.* C **90** 054327
[22] Chappert F 2007 Nouvelles paramétrisations de l'interaction nucléaire effective de Gogny *PhD thesis* Université Paris Sud XI
[23] Lalazissis G A, König J and Ring P 1997 New parametrization for the Lagrangian density of relativistic mean field theory *Phys. Rev.* C **55** 540–3
[24] Lalazissis G A *et al* 2005 New relativistic mean-field interaction with density-dependent meson–nucleon couplings *Phys. Rev.* C **71** 024312
[25] Long W H *et al* 2008 Evolution of nuclear shell structure due to the pion exchange potential *Europhys. Lett.* **82** 12001
[26] Nikšić T, Vretenar D and Ring P 2008 Relativistic nuclear energy density functionals: adjusting parameters to binding energies *Phys. Rev.* C **78** 034318
[27] Nikšić T, Vretenar D and Ring P 2011 Relativistic nuclear energy density functionals: mean-field and beyond *Prog. Part. Nucl. Phys.* **66** 519
[28] Zhao P W *et al* 2010 New parametrization for the nuclear covariant energy density functional with a point-coupling interaction *Phys. Rev.* C **82** 054319
[29] Wang M *et al* 2017 The AME2016 atomic mass evaluation (II). Tables, graphs and references *Chin. Phys.* C **41** 030003
[30] Angeli I and Marinova K P 2013 Table of experimental nuclear ground state charge radii: an update *Atom. Data Nuc. Data Tab* **99** 69
[31] Huang W J *et al* 2017 The AME2016 atomic mass evaluation (I). Evaluation of input data; and adjustment procedures *Chin. Phys.* C **41** 030002
[32] Brown A B 1998 New Skyrme interaction for normal and exotic nuclei *Phys. Rev.* C **58** 220

[33] Satchler G R 1978 Outline of the development of the theory of direct nuclear reactions *Rev. Mod. Phys.* **50** 1

[34] Kramer G J, Blok H P and Lapikás L 2001 A consistent analysis of (e,e'p) and (d,^{3}He) experiments *Nucl. Phys.* A **679** 267

[35] Duguet T and Hagen G 2012 *Ab initio* approach to effective single-particle energies in doubly closed shell nuclei *Phys. Rev.* C **85** 034330

[36] Koopmans T 1934 Über die Zuordnung von Wellenfunktionen und Eigenwerten zu den Einzelnen Elektronen Eines Atoms *Physica* **1** 104–13

[37] Vautherin D and Brink D M 1972 Hartree–Fock calculations with Skyrme's interaction. I. Spherical nuclei *Phys. Rev.* C **5** 626–47

[38] Dobaczewski J, Flocard H and Treiner J 1984 Hartree–Fock–Bogolyubov description of nuclei near the neutron-drip line *Nucl. Phys.* A **422** 103–39

[39] Satuła W, Dobaczewski J and Nazarewicz W 1998 Odd–even staggering of nuclear masses: pairing or shape effect? *Phys. Rev. Lett.* **81** 3599

[40] Rutz K *et al* 1998 Odd nuclei and single-particle spectra in the relativistic mean-field model *Nucl. Phys.* A **634** 67

[41] Bender M *et al* 2000 Pairing gaps from nuclear mean-field models *Eur. Phys. J.* A **8** 59

[42] Dobaczewski J *et al* 2001 Odd–even staggering of binding energies as a consequence of pairing and mean-field effects *Phys. Rev.* C **63** 024308

[43] Duguet T *et al* 2001 Pairing correlations. II. Microscopic analysis of odd–even mass staggering in nuclei *Phys. Rev.* C **65** 014311

[44] Duguet T and Lesinski T 2008 Non-empirical pairing functional *Eur. Phys. J.* A **156** 207

[45] Duguet T *et al* 2001 Pairing correlations. I. Description of odd nuclei in mean-field theories *Phys. Rev.* C **65** 014310

[46] Schunck N *et al* 2010 One-quasiparticle states in the nuclear energy density functional theory *Phys. Rev.* C **81** 024316

[47] Perez-Martin S and Robledo L 2008 Microscopic justification of the equal filling approximation *Phys. Rev.* C **78** 014304

[48] Bartel J *et al* 1982 Towards a better parametrisation of Skyrme-like effective forces: a critical study of the SkM force *Nucl. Phys.* A **386** 79–100

[49] Berger J F, Girod M and Gogny D 1991 Time-dependent quantum collective dynamics applied to nuclear fission *Comput. Phys. Commun.* **63** 365–74

[50] Nikolov N *et al* 2011 Surface symmetry energy of nuclear energy density functionals *Phys. Rev.* C **83** 034305

[51] Bjørnholm S and Lynn J E 1980 The double-humped fission barrier *Rev. Mod. Phys.* **52** 725

[52] Herman M *et al* 2007 EMPIRE: nuclear reaction model code system for data evaluation *Nucl. Data Sheets* **108** 2655

[53] Koning A J, Hilaire S and Duijvestijn M C 2008 TALYS-1.0, Proceedings of the International Conference on Nuclear Data for Science and Technology (Nice: EDP Sciences) p 211

[54] Tondeur F *et al* 2000 Towards a Hartree–Fock mass formula *Phys. Rev.* C **62** 024308

[55] Chappert F, Girod M and Hilaire S 2008 Towards a new Gogny force parameterization: impact of the neutron matter equation of state *Phys. Lett.* B **668** 420

[56] Fayans S A *et al* 2000 Nuclear isotope shifts within the local energy-density functional approach *Nucl. Phys.* A **676** 49

[57] Baldo M, Schuck P and Viñas X 2008 Kohn–Sham density functional inspired approach to nuclear binding *Phys. Lett.* B **663** 390–4
[58] Schunck N *et al* 2015 Quantification of uncertainties in nuclear density functional theory *Nucl. Data Sheets* **123** 115
[59] Dobaczewski J, Nazarewicz W and Reinhard P-G 2014 Error estimates of theoretical models: a guide *J. Phys. G: Nucl. Part. Phys.* **41** 074001
[60] Schunck N *et al* 2015 Error analysis in nuclear density functional theory *J. Phys. G: Nucl. Part. Phys.* **42** 034024
[61] Schunck N *et al* 2015 Uncertainty quantification and propagation in nuclear density functional theory *Eur. Phys. J.* A **51** 1
[62] Chabanat E *et al* 1998 A Skyrme parametrization from subnuclear to neutron star densities Part II. Nuclei far from stabilities *Nucl. Phys.* A **635** 231–56
[63] Goriely S *et al* 2009 First Gogny–Hartree–Fock–Bogoliubov nuclear mass model *Phys. Rev. Lett.* **102** 242501
[64] Chen W-C and Piekarewicz J 2014 Building relativistic mean field models for finite nuclei and neutron stars *Phys. Rev.* C **90** 044305
[65] Fayans S A *et al* 1994 Isotope shifts within the energy-density functional approach with density dependent pairing *Phys. Lett.* B **338** 1
[66] Baldo M *et al* 2013 New Kohn–Sham density functional based on microscopic nuclear and neutron matter equations of state *Phys. Rev.* C **87** 064305
[67] Bulgac A *et al* 2018 Minimal nuclear energy density functional *Phys. Rev.* C **97** 044313
[68] Dechargé J and Gogny D 1980 Hartree–Fock–Bogolyubov calculations with the D1 effective interaction on spherical nuclei *Phys. Rev.* C **21** 1568–93
[69] Chappert F *et al* 2015 Gogny force with a finite-range density dependence *Phys. Rev.* C **91** 034312
[70] Stone J R and Reinhard P-G 2007 The Skyrme interaction in finite nuclei and nuclear matter *Prog. Part. Nucl. Phys.* **58** 587
[71] Bender M, Heenen P-H and Reinhard P-G 2003 Self-consistent mean-field models for nuclear structure *Rev. Mod. Phys.* **75** 121–80
[72] Kankainen A, Äystö J and Jokinen A 2012 High-accuracy mass spectrometry of fission products with Penning traps *J. Phys. G.: Nucl. Part. Phys.* **39** 093101
[73] Carlsson B D *et al* 2016 Uncertainty analysis and order-by-order optimization of chiral nuclear interactions *Phys. Rev.* X **6** 011019
[74] Ekström U *et al* 2010 Arbitrary-order density functional response theory from automatic differentiation *J. Chem. Theory Comput.* **6** 1971
[75] Blaizot J P 1980 Nuclear compressibilities *Phys. Rep.* **64** 171
[76] Colò G *et al* 2004 Microscopic determination of the nuclear incompressibility within the nonrelativistic framework *Phys. Rev.* C **70** 024307
[77] Piekarewicz J 2010 Do we understand the incompressibility of neutron-rich matter? *J. Phys. G: Nucl. Part. Phys.* **37** 064038
[78] Brandt S 2014 *Data Analysis—Statistical and Computational Methods for Scientists and Engineers* (Berlin: Springer)
[79] Gregory P 2005 *Bayesian Logical Data Analysis for the Physical Sciences* (Cambridge: Cambridge University Press)
[80] Gamerman D and Lopes H F 2006 *Markov Chain Monte Carlo: Stochastic Simulation for Bayesian Inference* (London/Boca Raton, FL: Chapman and Hall/CRC)

[81] Gelman A *et al* 2014 *Bayesian Data Analysis* (Boca Raton, FL: CRC Press)
[82] Ekström A *et al* 2015 Statistical uncertainties of a chiral interaction at next-to-next-to leading order *J. Phys. G: Nucl. Part. Phys.* **42** 034003
[83] Navarro Pérez R, Amaro J E and Ruiz Arriola E 2015 Error analysis of nuclear forces and effective interactions *J. Phys. G: Nucl. Part. Phys.* **42** 034013
[84] Furnstahl R J, Phillips D R and Wesolowski S 2015 A recipe for EFT uncertainty quantification in nuclear physics *J. Phys. G: Nucl. Part. Phys.* **42** 034028
[85] Melendez J A, Wesolowski S and Furnstahl R J 2017 Bayesian truncation errors in chiral effective field theory: nucleon–nucleon observables *Phys. Rev.* C **96** 024003
[86] Higdon D *et al* 2015 A Bayesian approach for parameter estimation and prediction using a computationally intensive model *J. Phys. G: Nucl. Part. Phys.* **42** 034009
[87] McDonnell J D *et al* 2015 Uncertainty quantification for nuclear density functional theory and information content of new measurements *Phys. Rev. Lett.* **114** 122501
[88] Gao Y *et al* 2013 Propagation of uncertainties in the Skyrme energy-density-functional model *Phys. Rev.* C **87** 034324
[89] Kortelainen M *et al* 2013 Neutron-skin uncertainties of Skyrme energy density functionals *Phys. Rev.* C **88** 031305
[90] Reinhard A *et al* 2013 Self-trapping in an array of coupled lD Bose gases *Phys. Rev. Lett.* **110** 033001
[91] Erler J, Klüpfel P and Reinhard P-G 2010 Misfits in Skyrme–Hartree–Fock *J. Phys. G: Nucl. Part. Phys.* **37** 064001
[92] Erler J *et al* 2012 The limits of the nuclear landscape *Nature* **486** 509–12
[93] Todd-Rutel B G and Piekarewicz J 2005 Neutron-rich nuclei and neutron stars: a new accurately calibrated interaction for the study of neutron-rich matter *Phys. Rev. Lett.* **95** 122501
[94] Fattoyev F J and Piekarewicz J 2011 Accurate calibration of relativistic mean-field models: correlating observables and providing meaningful theoretical uncertainties *Phys. Rev.* C **84** 064302
[95] Fattoyev F J and Piekarewicz J 2012 Neutron skins and neutron stars *Phys. Rev.* C **86** 015802

www.ingramcontent.com/pod-product-compliance
Ingram Content Group UK Ltd.
Pitfield, Milton Keynes, MK11 3LW, UK
UKHW051909100926
8556IPUK00005B/815